MOSQU

C. MARTIN SHARP
and
MICHAEL J. F. BOWYER

FABER AND FABER LIMITED
3 Queen Square London WC1

First published in 1967
by Faber and Faber Limited

Printed in Great Britain by
W. & J. Mackay & Co. Ltd, Chatham

All rights reserved

First published in this edition 1971

ISBN 0 571 04750 5
(Hard Bound Edition)

ISBN 0 571 09531 3
(Faber Paper Covered Edition)

CONDITIONS OF SALE

This book is sold subject to the condition that it shall not, by way of trade or otherwise, be lent, re-sold, hired out or otherwise circulated without the publisher's prior consent in any form of binding or cover other than that in which it is published and without a similar condition including this condition being imposed on the subsequent purchaser

© 1967, 1971 by C. Martin Sharp
and Michael J. F. Bowyer

Contents

Foreword by Sir Geoffrey de Havilland	page 17
Preface	19
Acknowledgements	21
Introduction	23

Book One

1. PROJECT	29

Book Two

DEVELOPMENT

2. W4050 and the Bombers	43
3. W4051 and the Photo-reconnaissance Mosquitoes	51
4. W4052 and the Fighters	54
5. Against the High Raider	62
6. The Sabre Mosquito Project	68
7. Highball	74

Book Three

PRODUCTION

8. Britain	81
9. Canada	96
10. Australia	109

Book Four

OPERATIONS

11. Photographic Reconnaissance	117
12. Fighters	148
13. Day Bombers	186
14. Meteorological Reconnaissance	214
15. Middle East	218
16. Fighter-Bombers	235

Contents

17.	Far East	*page* 260
18.	Coastal Strike	270
19.	Bomber Support	287
20.	Light Night Striking Force	301
21.	Mosquito Airliner	328
22.	Intruders and Rangers	338
23.	Pathfinders and Target Markers	354
	Appendices (for list see page 15)	375
	Index	469

Illustrations

See also Appendix 1, pages 375–90, for full photograph captions.

Between pages 32 and 33

1. Prototype wing and details, 16 June, 1940
2. The mock-up, 16 June, 1940
3. Prototype assembly, October, 1940
4. EO234 awaiting engine runs, November, 1940
5. Hereward de Havilland and Alan S. Butler, 19 November, 1940
6. A. S. Butler and F. T. Hearle, 19 November, 1940
7. The prototype brought to Hatfield, 3 November, 1940
8. After the second flight, 29 November, 1940
9. Slipstream rectifiers, 10 January, 1941
10. Half-extended engine nacelle, 12 February, 1941
11. The prototype on 21 November, 1940 in flight test shed
12. After the first flight, 25 November, 1940
13. Turret and cannon mock-up on W4050
14. The design team, May, 1943
15. His Majesty at Hatfield, 15 August, 1940
16. Fighter prototype demonstrated, 5 September, 1941
17. The Prime Minister at Hatfield, 19 April, 1943
18. W4050 fitted with Merlin 61 engines
19. W4051, the first P.R. Mk. 1
20. First Mosquito fighter, W4052
21. W4052 after forced landing at Panshanger, April, 1942
22. W4065, PRU/Bomber Conversion during A.F.D.U. trials
23. PRU/Bomber Conversions awaiting delivery

Between pages 48 and 49

24. The first high-altitude Mosquito fighter, MP469, on 16 September, 1942
25. MP469 after conversion to Mk. XV standard
26. A standard production NF. XV
27. Timber to season—for a long war
28. Making the double-skin fuselage shell
29. Lifting the shell complete with bulkheads
30. Simple wing of robust wooden scantlings
31. Assembly of wing fittings

Illustrations

32. Women on drop-hammer work
33. The compression-rubber undercarriage leg
34. Hatfield fighter assembly line, August, 1943
35. Mosquitoes pouring out at Hatfield, 2, February, 1944
36. Production of Mk. II and III at Leavesden, June, 1943
37. Production at Standard Motors
38. Salvage hangar, Mosquito Repair Organisation, Hatfield, 18 May, 1944
39. A job for the D.H. Mosqu to repair organisation
40. Cockpit layout in the fighter prototype
41. Cockpit layout in the PRU/Bomber Conversion
42. Cockpit layout in the Sea Mosquito Mk. 37, November, 1946

Between pages 112 and 113

43. Undercarriage compression leg assembly
44. Engine nacelle assembly
45. From well-bombed London
46. Canopies by Perfecta of Birmingham
47. Tailplanes and wings by Hoopers, coachbuilders
48. Mrs. Hale's group; the smallest subcontractor
49. First Australian Mosquito flies, 23 July, 1943
50. Pat Fillingham: Flight-test method in Australia
51. Australian FB. Mk. 40 formates on British-built trainer
52. The Sydney assembly line
53. The last Australian-built PR. Mk. 41
54. First Canadian Mosquito flies, 24 September, 1942
55. An F-8 being towed into a hangar at Toronto
56. Russ Bannock and a flight test engineer (left) by a Mosquito destined for China
57. The crew of LR503, veteran of 213 sorties, being greeted in Canada
58. KA102, a Mosquito Mk. 21
59. KA970 after its accident on 17 April, 1945
60. DK296 waits at Errol for delivery to the U.S.S.R.
61. The sort of women they were
62. The sort of men they were
63. Victor Ricketts at Hatfield, March, 1942
64. Low oblique of the Renault works, Billancourt, March, 1942
65. Merifield and his navigator
66. LR432 a Mk. IX of 544 Squadron

Between pages 144 and 145

67. High-level P.R. photograph of Gdynia, 2 June, 1942
68. Gnôme Rhône Limoges factory, 10 March, 1944
69. Enemy night movement in 1944, photographed using photo-flash
70. RDF station at Bergen-Am-Zee taken using forward facing camera
71. Geoffrey de Havilland checks his flight test log after flying a Mk.VI

Illustrations

72. La Pallice photographed during the first Mosquito operational sortie
73. The Philips works, Eindhoven, after 2 Group's raid in December, 1942
74. Two Mk. IV bombers of 105 Squadron, DZ353 and DZ367
75. DD712 of 23 Squadron, a standard Mk. II
76. DD673 at Manston after crashing into a steam roller
77. DZ238 back from service in Malta
78. A standard Mk. II night-fighter, DD750
79. The turret fitted to the first turret Mosquito
80. HP850 a Mk. VI supplied to the Ranger Flight of 157 Squadron
81. NF. XIII HK428 'K' of 29 Squadron with thimble nose radome
82. MM466 of 409 Squadron, possibly the top-scoring Mosquito
83. A Mk. XIII of 264 Squadron after flying through burning debris
84. 105 Squadron line-up at Marham, December, 1942
85. Sqn. Ldr. Ralston and his navigator, December, 1942
86. Wg. Cdr. Wooldridge and a group of 105 Squadron crews, 28 June, 1943
87. Crews back from Berlin on 31 January, 1943
88. Railway workshops and power station under attack Trier, 1 April, 1943

Between pages 208 and 209

89. Trier on 1 April, 1943—a few moments later
90. A line-up of 139 Squadron at Marham
91. Wg. Cdr. Peter Shand briefing 139 Squadron crews
92. One of 60 Squadron's PR. IIs, DD744
93. Lt. Archie Lockhart with the Mosquito engaged by an Me 262 in August, 1944
94. 'Lovely lady', a PR. Mk. IX of 60 Squadron, S.A.A.F.
95. An early production Mk. VI fighter-bomber
96. Amiens prison under attack
97. The prison photographed later from DZ414
98. LR366 of 613 Squadron undergoing operational turn-round
99. Low-level day attack by 613 Squadron on the Hague
100. Mosquitoes bombing Aarhus, October, 1944
101. A PR. Mk. XVI of 684 Squadron in aluminium finish
102. VL619, a PR. Mk. 34 with 200-gallon wing tanks
103. The summit of Mount Everest photographed by a camera in a Mosquito
104. Mosquito VI PZ446 of 143 Squadron being re-armed
105. A ship under attack at Sandshavn, 23 March, 1945
106. Rockets, long-range tank, paddle-bladed airscrews and nose camera on a Mk. VI RS625
107. 235 Squadron attacking shipping on 19 September, 1944
108. A shipping strike at Nordgulen, 5 December, 1944
109. Banff Wing attack on U-boats, 9 April, 1945
110. The first Mk. XVIII 'Tsetse'
111. A Mk. IV during *Highball* trials at Loch Striven
112. Mk. 30 NT585 of the type used in later stages of bomber support

Illustrations

Between pages 304 and 305

113. The trial-installation Mk. IV with bulged bomb bay, in November, 1943
114. Berlin being marked by P.F.F. Mosquitoes
115. Loading a 4,000 lb. bomb on PF432 for the Berlin raid of 21 March, 1945
116. Dicing tonight
117. Unarmed merchantmen of B.O.A.C.
118. B.O.A.C. Mosquito G-AGGD comes in at Leuchars
119. Mk. IX ML897 of 1409 Flight lands at Wyton after its 153rd sortie
120. Oboe-equipped Mk. IX LR504 after its 190th sortie
121. Mk. IX ML914 veteran of many operations
122. A line-up of 109 Squadron Oboe Mosquitoes at Marham, 1943
123. DZ319:H of 109 Squadron after its 101st sortie
124. A Mk. VI of 21 Squadron photographed from a 109 Squadron Mosquito during a *Noball* operation
125. Loading target markers into a B. Mk. XX, KB162:J of 139 Squadron
126. Searchlight tracks and markers around Berlin on 11/12 August, 1944
127. ML963 'K-King' of 571 Squadron in September, 1944
128. MM156 Mk. XVI, bought with contributions from the Oporto British
129. 464 Squadron Mk. VIs at Hatfield, 2 June, 1944
130. A P.R.34 with swollen 'bomb bay' and 200-gallon wing tanks
131. NF. Mk. XIX, MM652
132. HK382, a Mk. XIII *Night-Ranger* aircraft of 29 Squadron
133. W4087, a Mk. II with Turbinlite airborne searchlight
134. A Mk. XVI converted to carry *H2X* for the U.S. 8th A.A.F.
135. A Mk. VI supplied, after refurbishing, to the Turkish Air Force
136. The second pre-production T.R.33 for the Royal Navy
137. A Mosquito T.T.39, PF606, before delivery to the Royal Navy
138. TW240 after conversion into the prototype T.R. Mk. 37

SOURCES OF PHOTOGRAPHS

Listed here are the sources of photographs, by photograph number.

Aircraft Production: 28, 33
Allan W. Gifford, Toronto: 54
British Aircraft Corporation: 111
Brigdens, Ltd., Toronto: 55
Cpl. Petty (via de Havilland records): 80
Crown Copyright: 20, 22, 25, 26, 64, 66, 67, 68, 69, 70, 71, 72, 73, 76, 88, 89, 96, 97, 99 to 102, 103, 105, 107, 108, 109, 112, 115, 119, 120, 126, 131, 132, 137, 138
de Havilland: 1 to 18, 21, 23, 24, 32, 34/35, 36, 38 to 48, 60, 63, 65, 75, 77, 78, 79, 81, 83, 86, 92, 95, 106, 110, 113, 116, 117, 121, 125, 127 to 130, 134 to 136
de Havilland (Canada): 56 to 59
Flight International: 85
Flt. Lt. A. Howard: 124
Fox Photos: 114
General Electric, Ltd.: 133
Illustrated: 90, 91, 104
Keystone Press: 62, 98
Milton Kent, Sydney: 49 to 53
P. F. Woolland: 122
Photopress: 87
Royal Canadian Air Force: 82
South African Air Force: 93, 94
Standard Motors, Ltd.: 37
Sport & General: 61
The Aeroplane: 19, 31, 74, 84, 118

Maps and Diagrams

Fuselage construction	*page* 36
Main plane construction	37
High-altitude Mosquito fighters	64–65
Highball installation as adapted for the Mosquito	76
Mosquito Bomber/P.R./Trainer deliveries 1941–December 1945	90
Mosquito Fighter deliveries 1942–December 1945	92
Mosquito deliveries 1946–50	94
Acceptance of Mosquito aircraft from D.H. Canada (Toronto) Works by R.A.F. and R.C.A.F.	108
Early photographic reconnaissance Mosquitoes	122–123
Two-stage Merlin reconnaissance Mosquitoes	140–141
Mosquito II Fighters	158–159
Positioning and strength of Mosquito Night Fighter Squadrons at 2000 hrs. on 6 June, 1943	164
Positioning of Mosquito Fighters at 2200 hrs. on 21 January, 1944	169
Two-stage Merlin Mosquito Fighters	180–181
Mosquito Mk. IV Bombers	202–203
Mosquito Fighter-Bombers	246–247
Survey of Mosquito operations and sorties despatched between 2200 hrs. on 5 June, 1944 and 2200 hrs. on 6 June, 1944	*facing page* 248
Mosquito Unit dispositions and strengths 0001 hrs. 6 June, 1944	*facing page* 248
Mosquitoes for the Royal Navy	282–283
Mosquito Night-Fighters	296–297
Two-stage Merlin Mosquito Bombers	326–327

Appendices

1. Full photograph captions *page* 375
2. Abbreviations 391
3. Mosquito Genealogy 392
4. Summary of Mosquito Variants 393
5. Mosquito Operational Performance and Loads 401
6. General arrangement drawing of Mk. VI and aerodynamic data, Mosquito Mk. II cut-away drawing 404
7. Mosquito performance graphs:
    i. Prototypes W4050, W4052 406
    ii. B. Mk. IV DK290 speed/altitude 407
    iii. Comparison of maximum speeds of Mosquito B. IV and FW190 A-3 407
    iv. Mosquito FB. VI speed and climb 408
    v. Mosquito PR. XVI prototype performance 408
8. General statement of Weights, Armament, Bomb Load and performance for late production FB. VI, Mk. 33 and Mk. 35 409
9. Rolls-Royce Merlin Progressive Improvement 412
10. Notable Mosquitoes 413
11. Versatile Mosquito 417
12. Salient Data Relating to German Aircraft Engaged by Mosquitoes 418
13. Disposition, losses and disposals of Fighter and Reconnaissance Mosquitoes 419
14. Increase in Mosquito Distribution 1942–43 420
15. Mosquito Build-up 1942–44 421
16. Confirmed sinkings of U-boats by Mosquitoes 422
17. Survey of Mosquito Production and Serial Numbers 423
18. Dispersal of Work People from Hatfield, 1941; comparison of labour force dispersed and at Hatfield 432
19. Makers of the Mosquito 433

Appendices

20. Mosquito Squadrons, Units, Bases *page* 440
21. Brief Notes on the Fighter Interception Unit and its Equipment 452
22. Enemy Aircraft Destroyed in Air-to-Air Combat by Mosquito Fighters during Operations over Britain and the Continent 454

Foreword

by Sir Geoffrey de Havilland
O.M., C.B.E., A.F.C., Hon.F.R.Ae.S., Hon.F.A.I.A.A.

It is a good thing that the history of the Mosquito is to be published. I still think that it ought to have appeared when first projected just after the war, as an immediate tribute to many brave people whom one can never thank, also as a timely lesson in the importance of encouraging personal enterprise.

One was told that the Mosquito would be seen in better perspective after some years had passed; and perhaps today the lesson is more necessary than ever.

It was not only a war-time lesson. The Mosquito was part of a broader picture embracing the attitude to airliner development before the war. Encouragement to commercial air transport in the years following the world trade depression, and consistently, would have given us an export market then, also more advanced aircraft in 1939, and in 1946, and to-day.

Knowing that the facts about the Mosquito will be set out plainly and fairly one hopes that young people who read the book will determine to be self-dependent, and will see how vital it is to foster competitiveness in industry.

Highlands House, Hatfield.
11th May, 1964.

Preface

Our aim has been to set down the facts about the Mosquito, as authenticated in the records of the industry, the Royal Air Force and other authorities. The research has thus been broadly shared between us, and we have endeavoured to set a standard of thoroughness, so far as the prescribed size of the book permitted.

The work is offered in tribute to the clever and the courageous, those who bore responsibility in the designing, building, flying and servicing of an outstanding aeroplane, which for two-and-a-half years was faster than any other in the fray, friend or foe.

Our hope is that this plain narrative will revive memories in the minds of such persons, and will move them to take up their pens and bring the story to life.

London, 30 May, 1967   C.M.S.   M.J.F.B.

Note

Mosquito won the C. P. Robertson Memorial Trophy for 1967. This award annually commemorates the first press officer of the Air Ministry, and is made for the best interpretation of the Royal Air Force to the public during the year, be this by written or spoken word, by flying exploit or display performance, by stage presentation, television or motion picture or other means.

Acknowledgements

The authors appreciate the help which many have given in seeking out records of more than a quarter-century ago, to supplement the fund of information which each of us has for so long collected.

The Air Historical Branch of the Ministry of Defence could not have been more generous or enthusiastic in providing facilities to trace information in their records. We are especially grateful to Mr. L. A. Jackets, Head of the Branch, to Mr. W. J. Taunton, Mr. E. H. Turner, Mr. S. L. H. Bostock, Mr. G. Gateley and Mrs. G. A. Fowles.

Over the years considerable useful material has been furnished by the Ministry of Aviation from its Technical Library. Information has been given by the Royal Navy, also by British Overseas Airways Corporation, who made available their full documentation on Mosquito operations. The Royal Australian, Canadian, New Zealand, South African and United States Air Forces and the air forces of other Commonwealth and foreign nations have assisted liberally.

Daily working association with de Havilland colleagues in all departments, throughout the evolution and career of the Mosquito, has afforded the background for the industrial side of the history. All the de Havilland archives, and the records of Rolls-Royce and other companies, have always been freely available. Of the many who were responsible for the aircraft it is difficult to single out individuals to thank for their generous help and patience; however, especial thanks are due to Sir Geoffrey and young Geoffrey de Havilland, to the Chief Engineer C. C. Walker, and Assistant Chief Engineer R. M. Clarkson, D. R. Newman (aerodynamics), to R. E. Bishop (Chief Designer), C. T. Wilkins and W. A. Tamblin (design), H. Povey, C. G. Long, L. C. L. Murray, S. R. Rudge (production), W. P. I. Fillingham, G. H. Pike (test flying), A. J. Brant (servicing), P. C. Garratt, W. D. Hunter, A. Murray Jones, M. M. Waghorn and J. Byrne (overseas manufacture) and Hereward de Havilland whose records of his contacts with the Royal Air Force units using Mosquitoes have proved invaluable. Some of these men, alas, are no longer with us. M. Herrod-Hempsall kindly steered the authors through the extensive technical records.

Special thanks are due to Alfred M. Alderson and George Burn

Acknowledgements

who, between them, produced the charts, drawings and maps which illustrate this book.

Air Marshal Sir Peter Wykeham, K.C.B., D.S.O., O.B.E., D.F.C., A.F.C., Order of Dannebrog, United States Air Medal, A.F.R.Ae.S., and presently Deputy Chief of Air Staff, contributed to the record of operations in the Middle East a graphic contemporary account.

To Group Captain John Cunningham, D.S.O., O.B.E., D.F.C., Deputy Lieutenant, County of Middlesex, we extend special thanks for assistance with the chapter on fighter operations.

Captain C. N. Pelly, O.B.E., of B.O.A.C., co-operated ably on the Mosquito airliner story.

To Bruce Robertson we are indebted for his readiness to help with some of the most tedious aspects of compilation.

The use of de Havilland photographs and other illustrations and data, and the co-operation of John Scott, Oliver Tapper and Nigel Bacon are much appreciated.

Wg. Cdr. R. W. Bray, D.F.C., Keith Braybrooke, Peter M. Corbell, Flt. Lt. P. F. Dillon, D.F.C., Wg. Cdr. R. P. Elliott, D.S.O., D.F.C., Wg. Cdr. J. H. Ford, D.F.C., Roger A. Freeman, A. D. Grant, Flt. Lt. F. E. Hay, D.F.C., Flt. Lt. A. Howard, Flt. Lt. T. Mason, Flt. Lt. W. T. Mason, D.F.C., Flt. Lt. L. J. Porter, John D. R. Rawlings, Wg. Cdr. H. E. Tappin, D.F.C., Wg. Cdr. J. E. Tipton, D.F.C., E. A. Tyrell, H. Widdop, P. F. Wolland, Alan J. Wright and Gerrit J. Zwanenburg have all provided useful material for this work.

The authors gratefully acknowledge permission given to reproduce Crown Copyright photographs and material, and photographs from many sources. Thanks are especially due to John F. Golding of the Imperial War Museum for his kind assistance.

Decorations and ranks listed are those held at the time of the events recorded. Every attempt has been made to ensure accuracy. Claims referring to enemy aircraft destroyed have wherever possible been checked against German and British records, or accepted in good faith for the last year of hostilities concerning which relevant Luftwaffe records no longer exist.

Introduction

The Mosquito was the outcome of a revolutionary concept which held that, on a given engine installation, a bomber would be so much faster if relieved of defensive guns and gunners and the structure and fuel load needed to carry them, that it would be safer, and would perform more efficiently. In the 1930s this was contrary to accepted policy.

This narrative of industry and war shows how difficult it was to argue the little Mosquito into existence. Similar concepts, applied on a larger scale, would have been entirely rejected. Had the Mosquito design been ordered in 1938, when it was mooted, and proved itself by 1941, then a four-Merlin unarmed bomber to follow by 1943 would have been so superior that it would have added tremendous impetus to the bomber offensive.

Such a machine would have carried a heavier bomb load for the same range, or flown further with the same load, faster than did the armed four-Merlin bombers. At a time when forces of a thousand aircraft were being marshalled for raids, and traffic congestion and collision danger were serious, fast bombers flying at a great variety of heights would have lessened the risks and afforded far greater tactical flexibility.

Even the little twin-engined unarmed Mosquito carrying a 4,000 lb. bomb to Berlin, 600 miles away, at around 300 m.p.h. at 27,000 feet compared well with four-Merlin bombers carrying 8,800 lb. at 240 m.p.h. at 21,000 feet. With a flight time of 3·8 hours against 4·7 and a crew of two instead of seven, it delivered twice the bomb load per crew man-hour, 12 per cent more per engine hour, 18 per cent more per pound tare weight (that is to say, for the same expenditure on production and maintenance) and 27 per cent more on a gallon of fuel.

Through 1944 the Mosquito carried as large a bomb load to Berlin as did some four-engined heavy bombers with crews of up to eleven men, and closely escorted by fighters, such as the Flying Fortress. The Mosquito was a much smaller investment, exposed to much less risk and for much shorter times. Because casualties were fewer, crews did more sorties, fifty per operational tour, and the all-round saving in crew training was great. Some Mosquitoes flew over two hundred sorties, achievement unmatched by other Allied bombers. Mosquitoes were chosen to carry *Oboe* the most important pathfinding equipment, and

Introduction

marked for many of Bomber Command's heaviest and most important attacks, by day and night. They led the tremendous assault on enemy rail communications prior to D-Day, made possible precision attacks on V-weapons sites and depots and, in the hands of some of the R.A.F.'s most skilled crews, marked the smallest of targets almost from ground-level to six miles high.

Being as small and manoeuvrable as a fighter, and as fast, the Mosquito was from the start a nimble performer, which rapidly developed into an all-purpose aircraft, a fact which allowed increased economy in factories in Britain, Australia and Canada where it was built.

The Mosquito could sneak in beneath the radar screen and place bombs on to pin-point targets from roof-top height, and even lob 4,000-pounders into railway tunnels. It was equally able to cross the Continent at great altitudes in almost complete immunity. It intruded throughout Europe as a fighter and fighter-bomber, by night and day. Ranging freely against every kind of enemy activity, it afforded army or bomber support, and destroyed fighters and bombers at their bases.

As a night-fighter it took over the defence of Britain in 1943 and, fitted with remarkable British-invented radar, virtually prevented night intrusion over Britain. Mosquitoes had claimed 659 enemy aircraft by the end of November, 1944, and brought down over five hundred flying-bombs in the first and worst sixty-one nights of this fresh scourge. Over Normandy night-fighter Mosquitoes protected the build-up of the Allied armies, and, using their long range, escorted Lancasters on daylight raids.

With the accurate six-pounder gun, or using rocket projectiles corresponding to a salvo from a six-inch cruiser, Mosquitoes bombarded ships, U-boats, harbours and various installations. Off Norway they used the six-pounder to break up formations of enemy fighters. They laid mines in enemy waters, and blocked the Kiel Canal. The Mosquito was able to carry *Highball*, a naval weapon based upon the famous mine used against the German dams by 617 Squadron, which, in the Mosquito, found an aeroplane with unequalled qualities to mark targets for later highly specialized attacks. Towards the end of the war the Mosquito was given folding wings and hooked, to become a twin-engined deck-landing aircraft able to carry a 15-inch torpedo.

The Mosquito was the Allies' only really long-range photo-reconnaissance and weather-reporting aeroplane. It photographed and surveyed the whole of Europe, almost to the borders of Russia. With photo-flashes night pictures were secured and a watch over German movement eastward from Normandy in 1944 was achieved, in darkness as well as in daylight. The enemy, on the other hand, never flew effective reconnaissance missions, not even over our concentrations before D-Day.

Mosquitoes discovered the existence of the V-weapons, their launching sites and associated depots. They even flew into quarry workings to secure ground-level pictures. A Mosquito found the *Tirpitz* in a Nor-

Introduction

wegian fiord; and photographs were brought back to England from sorties over Konigsburg, Belgrade, Genoa and Toulon; from Sumatra to Ceylon; from the East Indies to Australia. The weather was reported ahead of every major R.A.F. and U.S.A.A.F. attack, and when the Americans formed their own meteorological unit they chose Mosquitoes: they also used them to contact enemy agents in the occupied countries. Loads were parachuted to the underground movements by R.A.F. Mosquitoes. B.O.A.C. used a fleet of Mosquito airliners for two years on the service flown nightly between Scotland and Sweden through the enemy zone. Winston Churchill depended upon Mosquitoes for courier mail services during some of his overseas conferences.

Now the principle of the Mosquito is generally accepted, for the four-jet Vulcans and Victors flying fast and low or high have no defensive armament.

'Mosquito' tells the story of a British endeavour by a team that was dedicated to civil aviation and is now sadly being eliminated as a commercial name at a time when the spirit which created the Mosquito, and the courage of those who flew it, are what Britain needs more than anything else.

Book One

PROJECT

Chapter 1

Project

On 8 November, 1934, within three weeks of the D.H.88 Comet winning the MacRobertson Air Race—11,000 miles from Mildenhall to Melbourne in 71 hours—a fast airliner with the Comet's clean form was proposed by de Havilland, but the Government attitude to airliner development held back its realization, until 1937. The Americans were building fast, modern monoplanes, and getting export orders for them. The D.H.91 Albatross emerged in 1937, first as a North Atlantic mail carrier with 1,000 lb. payload for a range of 2,500 miles against a 40 m.p.h. headwind. Cruising at 210 m.p.h. it yielded 36 gross ton-miles per gallon of fuel. In 22 years since the 1915 D.H.1 fighter three times its speed was achieved by an airliner, and on about a quarter of the fuel per gross ton-mile.

The mail-carrier version of the Albatross at 32,000 lb. had a payload of 6,000 lb. with fuel enough to fly to Berlin and back at 11,000 feet. This capability was mentioned in Parliament, and was greeted with cries of 'Shame', and 'Why Berlin?' But de Havilland had been giving thought to the most effective contribution they could make to re-armament. The company were among those invited to tender to military specifications devised by the Air Ministry, but generally, because of experiences in the 1920s and wariness of government specifications, preferred to build civil aircraft for the open market.

However, Specification P.13/36, issued 24 August, 1936, was given attention when the threat from Hitler began to appear grave. The Albatross airliner and the D.H.94 Moth Minor light aeroplane (90 h.p. and £575) were the company's newest projects at that time. These aircraft flew on 20 May and 22 June, 1937, respectively.

P.13/36 called for 'a twin-engined medium bomber for world-wide use', and continued: 'It should be an aircraft that can exploit the alternatives between long range and very heavy bomb load which are made possible by catapult launching in heavily loaded condition. During all operations it is necessary to reduce time spent over enemy territory to a minimum. Therefore the highest possible cruising speed is necessary.' The specification called for nose and tail gun turrets, horizontal bomb stowage in tiers if needed, suitability for outdoor maintenance at home or abroad. 'It appears', it said, 'there is a possibility of combining

Project

medium-bomber, general-reconnaissance and general-purpose classes in one basic design,' and it suggested that two 18-inch torpedoes might be carried. A top speed of 275 m.p.h. was required at 15,000 ft., and a range of 3,000 miles with a 4,000 lb. bomb load. Consideration could be given to remotely controlled guns.

Speed through aerodynamic cleanness and minimum skin area was equally sound for a bomber, so that a military Albatross was one time-saving possibility. In-line engines showed promise of advance in output, especially at altitude. A twin-Merlin Albatross estimate appeared in April, 1938, with Hercules HE 1M and Sabre comparisons. Other formulae were calculated. On 7 July a letter was sent to Sir Wilfrid Freeman, Air Council Member for Research and Development, discussing the Specification and arguing for wood construction, in case war should break out; except in torsion its strength for weight was as great as that of duralumin or steel. De Havilland felt, however, that the P.13/36 requirements would produce a mediocre aeroplane. They suggested a different approach.

In a second letter, sent 27 July, they concluded that the specification could not be met on two Merlins. If speed was to be paramount, then only half the specified load could be carried. If load was paramount then a larger, slower aircraft would result; on two Merlins it would have a range of 1,500 miles, with a 4,000 lb. bomb load, max. speed 260 m.p.h. at 19,000 ft., cruise speed 230 m.p.h. at 18,000 ft.

'A good bomber *could* be produced using Merlins,' (here is the first twinkle of the idea that became the Mosquito) 'but to meet this particular specification double the power would be needed.' A compromise bomber might be considered if 'some of the Appendix A stuff (it weighs a ton) be dispensed with, the speed asked for reduced, height of operation increased.' A two-Merlin compromise bomber was arrived at on 11 August, with a top speed of 260 m.p.h. and a range of 1,500 miles. But Hatfield disliked the compromise.

During the Munich crisis in September, 1938, reflections on the Summer schemes became suddenly serious. Trenches were being dug around Hatfield factory, sandbag walls were going up by the power house and key buildings, and typists assembled tens of thousands of gas masks. Hatfield now firmly felt that the most useful aircraft the company could contribute was a bomber in which armament was altogether sacrificed for speed. It would not be an Albatross adaptation. Capt. de Havilland considered it should be small, have two Merlins and a crew of two.

He and C. C. Walker went to the Air Ministry early in October to make this proposal, remarking afterwards that an air of urgency and apprehension was everywhere detectable in the London streets. Again they advocated using wood construction as developed in the Comet and Albatross, reckoning thereby to save a year in the prototype stage, and to speed up every subsequent variant. De Havilland had always argued

Project

that, in war, all the metal industries would be overtaxed whilst wood workers would not be fully employed.

Their proposal was quite out of the general scheme upon which the R.A.F. was being expanded, out of keeping with officially expressed views, and it came from a firm with little experience of working with the Ministry. The Blenheim, Whitley, Wellington and Hampden were by now in production, heavily armed metal bombers, and the trend was towards larger four-engined machines. The Ministry was preoccupied with approved designs, the Shadow Factory scheme, eight-gun fighter production. Would de Havilland take on building wings for one of the other bombers? The two men were taken aback.

De Havilland's suggestion aroused little or no interest. There were further conversations as to the likelihood of getting 'fighter speed' with a twin-engined bomber-reconnaissance formula. Then the project was set aside.

The twin-Merlin Albatross, meeting P.13/36, would be much slower. The papers of 4 October, with Merlin Xs, showed a top speed of 300 m.p.h. (exhaust propulsion would account for 10 m.p.h. of this). Cruise would be 268 m.p.h. at 22,500 ft. Needing a crew of three it would have six or eight forward-firing guns, and one or two manually operated guns. A tail turret was provided. Equipment would be simple. Weight was 19,000 lb., wing span 61·5 ft. Wood construction was retained.

Naturally, adaptations of the later and smaller de Havilland airliner, the D.H.95 Flamingo, were looked into. This all-metal 12-to-20 passenger aircraft flew on 28 December, 1938, and the first calculations for a possible bomber, on 28 October, showed that with a smaller, 47 ft. fuselage, two front guns, an under-belly gun, and a tail turret, it would have an all-up weight of 17,794 lb. with provision for four 500 lb. bombs, three crew, fuel for 1,500 miles. It featured a then-novel nosewheel landing gear. On 11 November, the length had become 51·75 ft., the span 65 ft. and of course the speed was uninteresting.

On 16 December, a new Specification, B.18/38, was looked at but was not of interest; from it eventually stemmed the Albemarle.

De Havilland continued through 1939 trying to evolve a worthwhile formula that the Ministry would look at.

In the conventional bomber defensive guns, with gunners, and the extra structure and fuel needed to carry them, could amount to one-sixth of the total weight. And guns were defensive only against fighters; for this purpose escort fighters—offensive defence—might be needed anyway. Guns were no defence against anti-aircraft artillery. Speed and manoeuvrability, with height, evaded both fighters and ground guns.

The larger structure took more man-hours to build, to service and to fly. The fast bomber would be a smaller investment, with smaller crew, exposed to smaller risk and for shorter time, would make more trips and deliver more bombs per month.

So familiar have we become with this logic since it was demonstrated

Project

twenty-five years ago that we must carefully record why it was not accepted in official circles then. One reason was that they suspected estimates of the speed of the clean aircraft based on the performance of the Comet and the Albatross. Another was that they feared that the Germans might produce even faster fighters than expected. Where would the unarmed bombers be then? Being committed to several armed bomber designs made it easier to harbour these doubts.

The unarmed design made no headway right up to the outbreak of war; and the Hatfield assembly lines then comprised about forty little Moth monoplanes, the seventh Albatross, most of the sixteen Flamingoes, hundreds of Tiger Moths, Queen Bees and Oxfords—but no front-line war machine. It felt unrealistic indeed to be delivering a Moth Minor to Brussels on 4 July for the Exposition de l'Aeronautique—at which Belgian and German workmen came to blows.

There was no future for the Albatross and Flamingo once war was declared on Sunday, 3 September. Design staff were thus freed. The following Wednesday de Havilland approached the Ministry again. The attitude was a little less unreceptive but still sceptical. Would fighters really be outpaced? Would not a crew of two be overworked and fatigued? The navigator would have to operate the radio, aim the bombs, keep a look-out for enemy fighters and his own vapour trails. The pilot would have no relief at the controls. True, his total time on a sortie would be less. Getting through gunfire and fighter zones quickly would cause less strain at the time, and for less time. But—Berlin only two hours away? (At any rate nobody *now* asked 'Why Berlin?')

Fresh estimates during September included a small unarmed twin-Merlin aircraft to carry 1,000 lb. bombs for 1,500 miles; and on 4 October a slightly larger aircraft to be 10 m.p.h. slower was estimated (the Hatfield formula of October, 1938), also a twin-Griffon adaptation, thus:—

	Twin-Merlin No. 1	Twin-Merlin No. 2 (larger)	Twin-Griffon
Total weight lb.	15,075	15,600	17,000
Wing area sq. ft.	360	410	410
Span ft.	48	51·25	51·25
Take-off power b.h.p.	2,600	2,600	3,200
Cruise at 15,000 ft. m.p.h.	332	325	334
At b.h.p./engine	715	715	800
At percentage max. b.h.p.	55%	55%	50%
Max. speed m.p.h.	419	409	433
At b.h.p./engine	1,145	1,145	1,445
At altitude ft.	20,000	20,000	18,500
Take-off run yds.	363	349	374

Studies based on two 'large' Daggers, and on one Sabre, were among

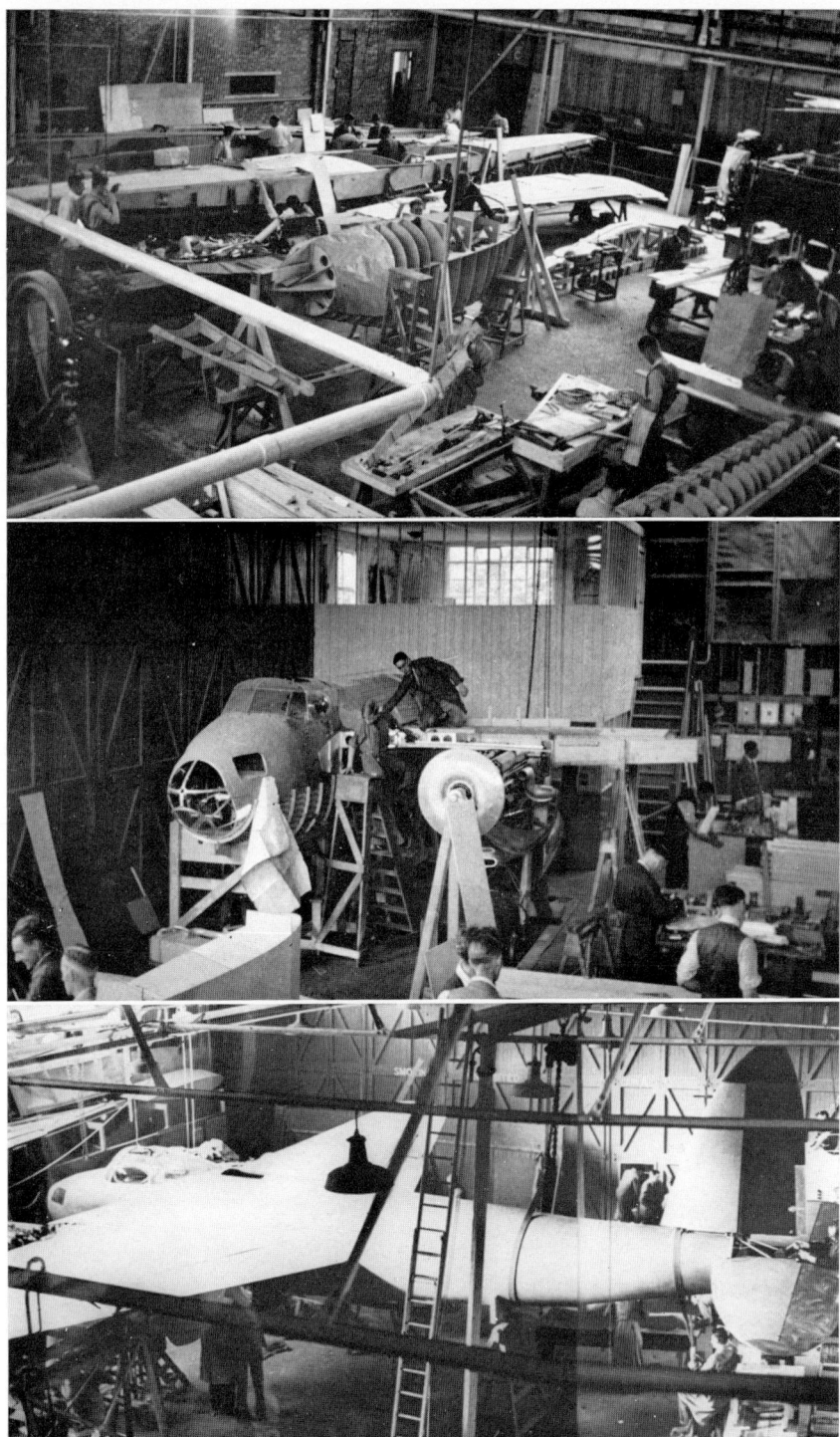

1. Prototype wing and details, 16 June, 1940
2. The mock-up, 16 June, 1940
3. Prototype assembly, October, 1940

4. EO234 awaiting engine runs, November, 1940
5. Hereward de Havilland and Alan S. Butler, 19 November, 1940

6. A. S. Butler and F. T. Hearle, 19 November, 1940

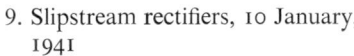

7. The prototype brought to Hatfield, 3 November, 1940
8. After the second flight, 29 November, 1940
9. Slipstream rectifiers, 10 January, 1941
10. Half-extended engine nacelle, 12 February, 1941

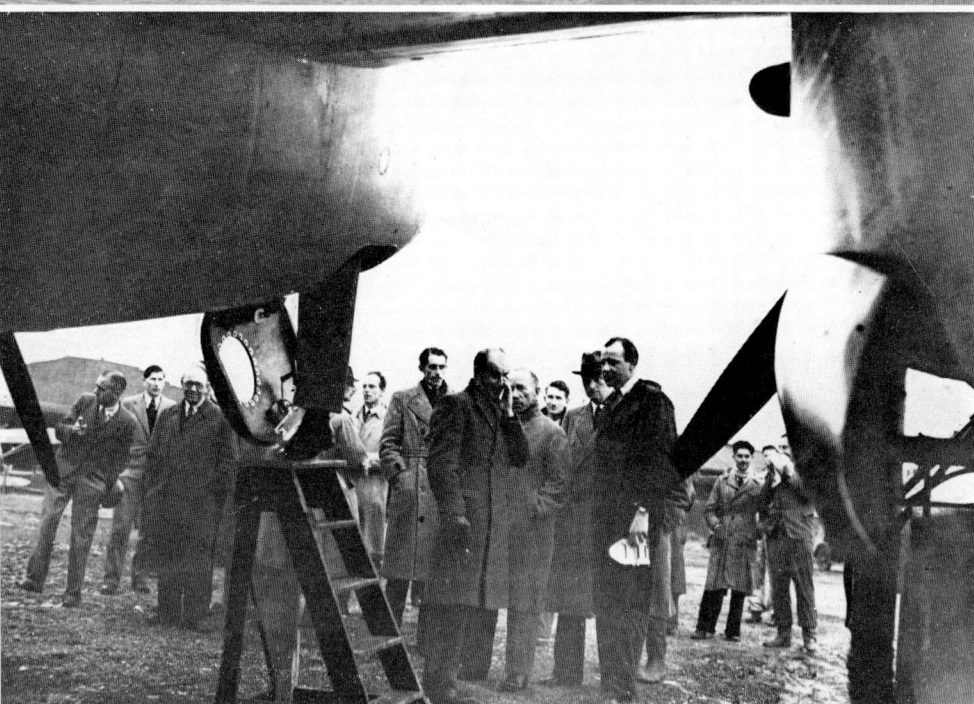

11. The prototype on 21 November, 1940 in flight test shed
12. After the first flight, 25 November, 1940
13. Turret and cannon mock-up on W4050

14. The design team, May, 1943

15. His Majesty at Hatfield, 15 August, 1940
16. Fighter prototype demonstrated, 5 September, 1941
17. The Prime Minister at Hatfield, 19 April, 1943

18. W4050 fitted with Merlin 61 engines
19. W4051, the first P.R. Mk. 1
20. First Mosquito fighter, W4052

21. W4052 after forced landing at Panshanger, April, 1942
22. W4065, PRU/Bomber Conversion during A.F.D.U. trials
23. PRU/Bomber Conversions awaiting delivery

Project

others that were compared. Of course the Merlin would be quickly available and the prospects of early development were good. The speed effect of wing-thickness variations was considered 9 October. Drag was re-examined, also exhaust propulsion, and the effect of adding armament. A 2-gun turret would cost 500 lb. and 20 m.p.h., at least. The weight, etc., for an overload range (2,000 miles) at 55 per cent and 40 per cent power was calculated.

On 16 October, the cost of a 2-gun powered turret with third crew member, and having one traversing gun beneath, was worked out. It would be about 915 lb. for guns and gunner, fuel 570 lb., oil 36 lb., armour 200 lb., structure 206 lb., and details, putting the take-off weight up from 15,600 lb. (larger aircraft) to 18,000 lb. Take-off run became 412 yds. A run of 350 yds. could be obtained at 18,000 lb. with a span of 55·25 ft., wing area 475 sq. ft., take-off power 2,600 b.h.p.

Drag and weight estimates for this larger armed aircraft were made more closely on 17 October. The performance became poor; cruise at 15,000 ft. 279 m.p.h., max. speed at 19,000 ft. 355 m.p.h. (as against 325 m.p.h. and 409 m.p.h. for the unarmed aircraft allowing for exhaust propulsion in each case).

Calculations were made late in October on twin-Dagger E108 and twin-Napier E112 formulae, of lower powers and weights, slightly better specific consumptions, interestingly close in most aspects.

Scope for a long-range and escort fighter development was reviewed mid-November and that month reconnaissance, bomber and fighter variants all looked good—at Hatfield.

A conference with the Air Member for Development, Sir W. Freeman, was called for 22 November to consider the company's findings. De Havilland and C. C. Walker attended, and strongly pressed the argument against the turret. Clarkson asserted that fighter speed *would* be obtained without it and most interceptions could be avoided; the turret, he said, spoilt the aircraft. How would scare guns affect the speed?— not much, if in nacelles, but the penalty in range or load seemed undesirable.

Not only at Hatfield but among pilots in the squadrons the idea of compromise was coming to be abhorred, and at this conference Sir Wilfrid Freeman in particular was steadfastly for the unarmed formula. He virtually put an end to all rear-turret talk. He and de Havilland stressed reconnaissance as well as bombing roles. But there was still some pressure for the third crew member, and still scepticism about de Havilland estimates.

Measurements of the drag of turrets on the Don and Oxford were now passed to the R.A.E. for comparison with other data, and models of the D.H.98 were sent for wind-tunnel tests.

The de Havilland case was that the two-crew unarmed wooden bomber with two Merlins carrying 1,000 lb. for 1,500 miles, having twice the wetted surface area of the Spitfire and twice the power, would

Project

nevertheless be 20 m.p.h. faster, by reason of less back-pressure power loss, higher altitude, better Reynolds number, faired propeller blade roots, and ducted radiator contributing more thrust than drag.

R. E. Bishop, the Chief Designer, with an eye on basic versatility, all the time made sure that there would be space beneath the cockpit floor for four 20 mm. cannon.

At this stage two cases had to be refigured for the Ministry:
(a) Three-man bomber with third crew member aft of the wing, having windows to look out rearward and downward, and a load of 2×500 lb. or 4×250 lb. bombs, which could be replaced, in a fighter version, by four 20 mm. guns, ammunition drums to be changed by the wireless operator.
(b) Two-man fighter-reconnaissance aircraft with crew in tandem, four 20 mm. guns or three F24 cameras; to ensure a good view the pilot would sit over the wing.

The project nearly came to grief again on 12 December at a further conference attended by the Assistant Chief of Air Staff, the Director-General of Research and Development, the A.O.C.-in-C. Bomber Command and others. The compromise armed bomber was still out of favour, but expressions like 'no use for the unarmed bomber' were heard. However, the C.-in-C. agreed about unarmed reconnaissance.

As ever, Sir Wilfrid championed the cause, and a decision was minuted to order a prototype. Whilst reconnaissance was probably the role that kept the project alive, bomber and fighter variants were not to be excluded. There were more talks before Christmas, a mock-up conference on 29 December, and Specification B.1/40 was drawn up on this basis:

Two RM3SM engines with ducted radiators, 1,280 b.h.p. each at 12,250 ft., 1,215 b.h.p. at 20,500 ft., structure weight 4,319 lb., tare weight 12,674 lb., P.R. normal loaded weight 17,150 lb., bomber overload weight 18,845 lb., max. level speed 397 m.p.h. at 23,700 ft. at 1,217 b.h.p., cruising 327 m.p.h. at 26,600 ft., range on full tanks, 555 gal., 1,480 miles at 343 m.p.h. at 24,900 ft., or 1,500 miles at full power, service ceiling 32,100 ft., landing run light 637 yds.

The company, not hampered by detailed specification requirements, hoped to keep clear of these, but the fear hung over them. Rolls-Royce co-operated closely through their A. T. Henry and R. F. Messervy. Cyril Lovesey was Chief Development Engineer for the Merlin, working with John E. Walker, D.H. engine installation designer. In January–March 1940 not more than fourteen senior designers, aerodynamics and stress men were on Mosquito work, rising to thirty in the second three months, and fifty-five by November when the prototype flew. In the prototype shop twelve men were at work in January, thirty in February, fifty-five in March, 180 in November.

For secrecy and safety from air attack the nucleus Mosquito design team moved on Thursday, 5 October, 1939, to an historic old country

Project

house, Salisbury Hall, near to London Colney and about five miles west of Hatfield. They comprised: R. E. Bishop, R. M. Clarkson, C. T. Wilkins, W. A. Tamblin, D. R. Adams, M. Herrod-Hempsall, R. M. Hare, J. K. Crowe, R. Hutchinson and Mrs. D. Ledeboer. They were followed by the main design and drawing office staff, and a hangar was erected there, disguised as a barn, where the prototype was built with F. Plumb in charge.

Contract 69990 was issued 1 March, 1940, for fifty bomber-reconnaissance aircraft, to include the prototype, to Spec. B.1/40.

The greater bulk of wood, for a given strength, had advantages. A stressed skin was thick, and therefore stiff without the need for much internal reinforcement, leaving clear spaces for tanks, bombs, guns and equipment. For lightness the wing was made in one piece from tip to tip, stressed to carry 82 tons. It accommodated ten tanks, 539 gallons, close to the centre of gravity, their weight spread economically along the span.

Shell holes and bullet holes would represent a smaller percentage of the mass of a bulky wooden member; a shell fragment that might sever a strong metal member would scarcely weaken a thick, continuous wood shell or a stout wood spar. Active service soon verified this. Buoyancy of wood was an advantage. No higher fire risk was expected.

To develop compression in birch plywood double skins were employed, separated by spruce stringers in the case of the upper surface of the wing, by a thick interlayer of feather-weight balsa wood in the fuselage. Thick wood structures had been developed in the Comet wing and the Albatross wing and fuselage. Surface smoothness possible with wood was exploited.

Adaptability of wooden units was proved as variants were built in 1940. Wood also lent itself to the making of the fuselage in halves, as a lobster is served, convenient for plumbing and installing equipment before boxing—especially as the fuselage was small. Wood made for ease of repair, in field or workshop. And as expected, woodworkers were readily available. Weathering qualities were known from experience with the Albatross. A notable improvement was the use of formaldehyde cement instead of casein for surface jointing; this synthetic resin was unaffected by moisture or micro-organisms, and it was a pity that it could not be available from the start.

As the design was being worked out and the mock-up was taking shape thoughts again turned to a long-range fighter, the need for which was brought home during the battle of France. After the Dunkirk withdrawal home defence became the immediate concern. Three times Lord Beaverbrook told Freeman that he had better stop the D.H.98 work, but he omitted to put out a firm instruction. He said that Beaverbrook was not interested in anything that would not be in use against the enemy by early 1941. 'To keep the project alive', said L. C. L. Murray, General Manager, 'I had to promise fifty Mosquitoes by July 1941!'

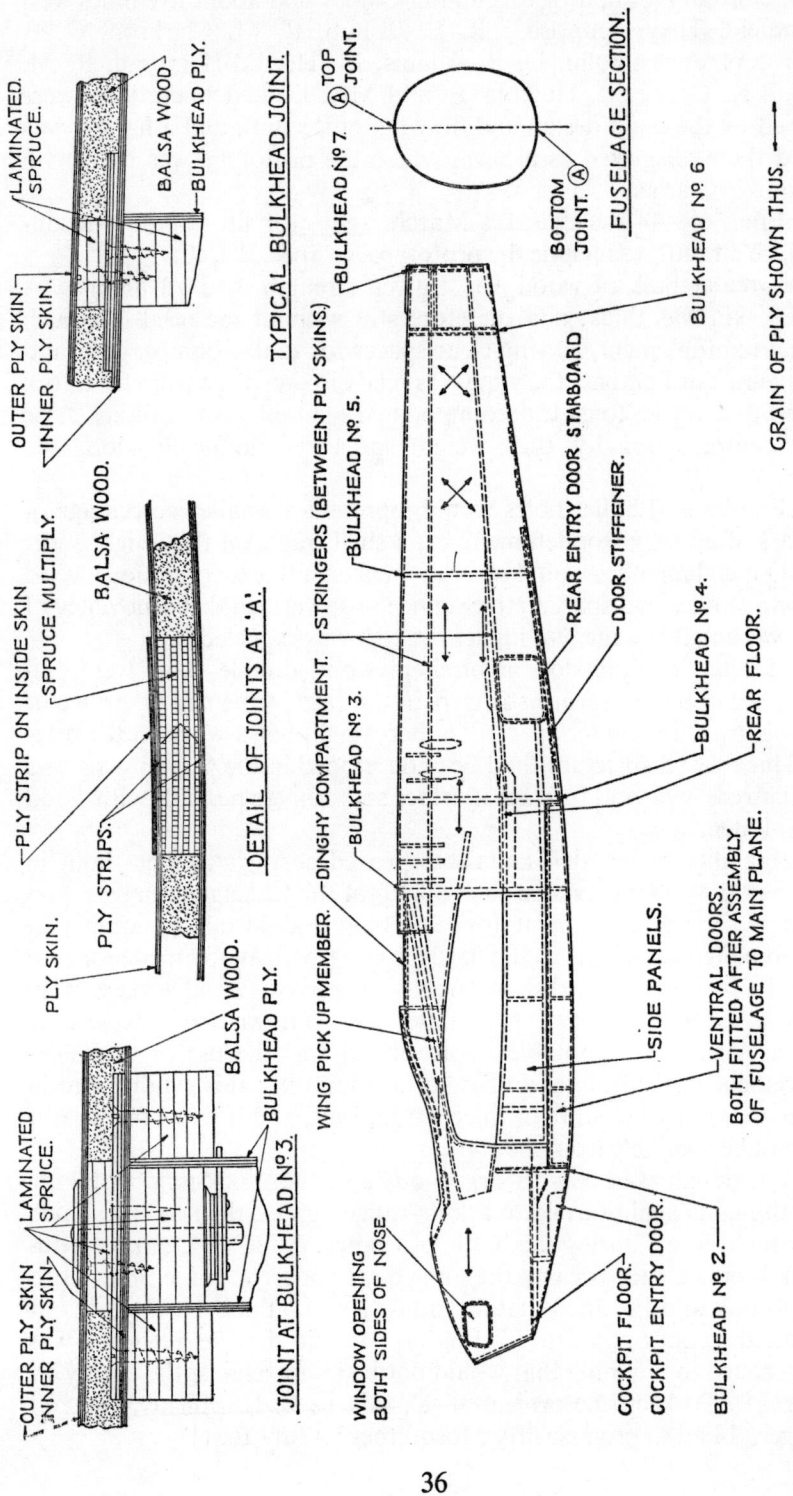

Project

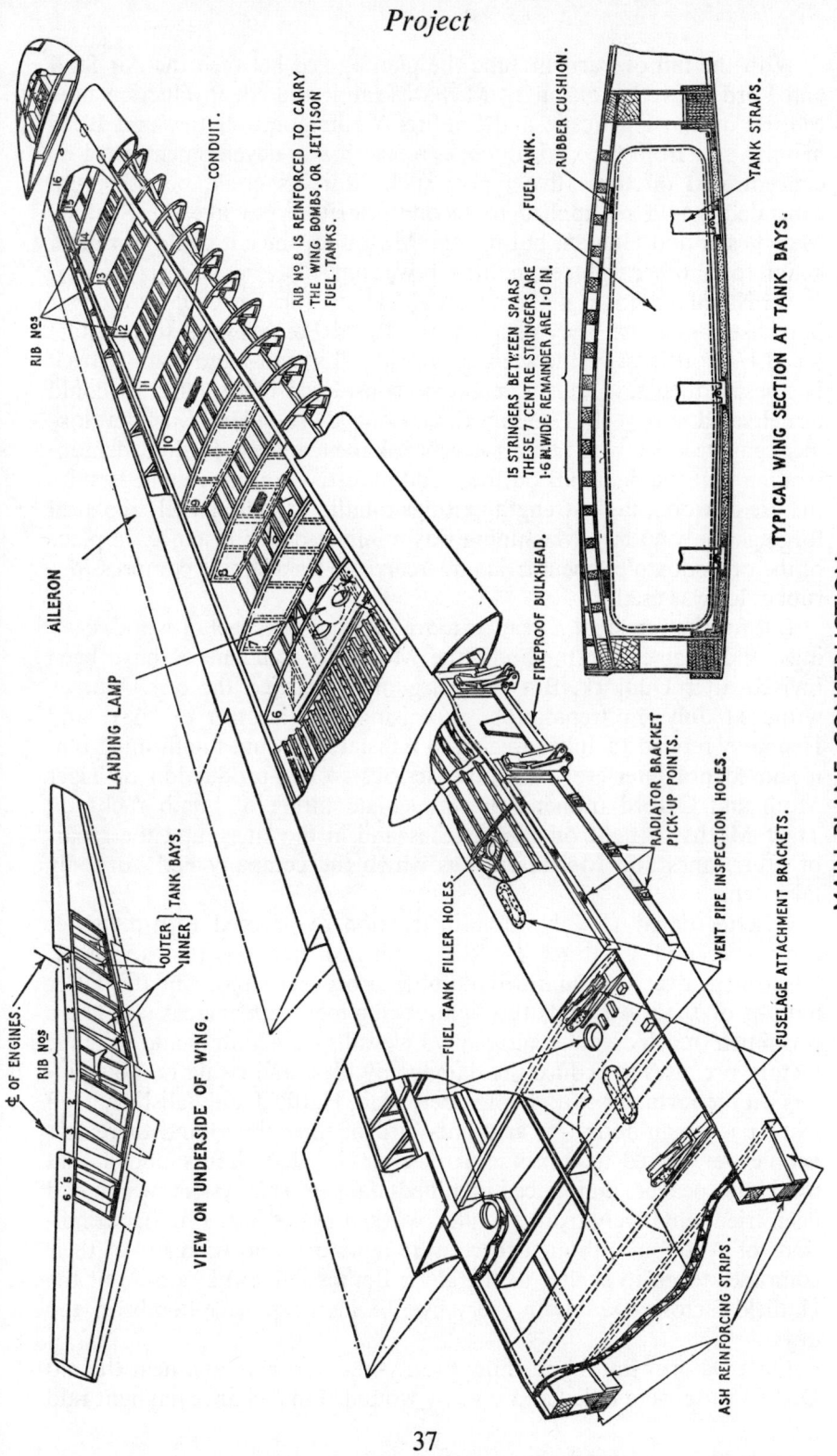

Project

Project

With the fall of Paris in June the plan agreed between the Air Staff and Lord Beaverbrook on 15 May was implemented. Production now centred on the Hurricane and Spitfire, Wellington, Whitley and Blenheim. Apart from theoretical work, research and development must be concentrated on these five types; such effort as could be otherwise expended would be applied to second-priority types like the Stirling, Manchester and Halifax, but not the Mosquito, since it could not be in action for another eighteen months. It was out.

De Havilland could make no progress now, for they could no longer purchase materials. They pleaded with Lord Beaverbrook to reinstate the B.1/40, at least in the second priority. His right-hand man, Patrick Hennesey, then said that it might be considered if fifty aircraft could be delivered in 1941. In London, C. G. Long, de Havilland chief development engineer, showed Lord Beaverbrook the material schedule, demonstrating that the demands on the metal industry would be light. Besides the use of wood, high-strength castings totalling 250 lb. largely replaced forgings (only 30 lb.). Machining was minimized; for instance, in place of the orthodox oleopneumatic undercarriage leg a simple compression-rubber leg was used.

But for the foresight a year or more earlier, in respect of wood, castings, and engineering methods, the Mosquito could never have been revived after Dunkirk. But these arguments carried the day. Murray wrote 11 July, in trepidation, promising fifty aircraft in 1941, and Hennesey replied 12 July agreeing to reinstatement but re-affirming that it should not interfere with more serious work—production of Tiger Moth and Oxford trainers, the immediate fitting of bomb racks on Tiger Moths to fight on the beaches and in the lanes, and the repair of Hurricanes and Merlin engines which the company had hurriedly taken on.

A letter dated 18 July, gave instruction to proceed with a fighter prototype, and Contract 135522 confirmed this on 16 November. Uneasiness over the unarmed bomber arose once more in the latter months of 1940. Air raids throughout Summer were increasing. There was much trench digging, and many low walls were built in and near the factory premises to reduce the damage by blast and rising bomb splinters. In the fortnight to 13 September, 1940, Hatfield and Salisbury Hall lost 84,309 man-hours—nearly one-third of their time—and thereafter people were asked to continue work after the alert siren sounded, and until loudspeakers announced imminent danger. This system, which had been tried tentatively from 26 June, worked better in the harder conditions of Autumn, although there were agitators who tried to get their comrades to go to shelter at the alert. Bombs fell within a mile of the Hatfield factory one day in every five, 68 high-explosive bombs in 100 days.

On 3 October in mid-morning twenty-one people in and near the old D.H.94 shop were killed and seventy wounded in a chance daylight raid

Project

by a single Ju 88 at 100 ft., which was shot down nearby.[1] About 80 per cent of the raw materials, and much work in progress for the Mosquito, was lost. A lot of steel plate was then used for strengthening above-ground shelters, and steel helmets were issued that month to all employees. Saturday night, 7 December, was recorded as being the first night since 7 August, when there was no alert throughout Britain.

Taxiing trials with W4050 were begun during the afternoon of 24 November, and next day Geoffrey de Havilland, chief test pilot since 1 October, 1937, made a short hop at 14,150 lb. Then at 3.45 p.m. on an overcast afternoon, eleven months after the start of design work, he took off on a 30-minute flight. John Walker was in the right-hand seat. They flew at 220 m.p.h. indicated. The undercarriage was retracted. Handley Page slots had been fitted but were locked for this flight, and as there was no wing-drop tendency near the stall they were judged unnecessary. 'Freeman's Folly' gave every ground for optimism, and was a joy to behold.

[1] At 10.30 a.m. on 3 October, 1940, Ju 88A 3Z+BB of Stab. 1/KG 77 set off from Laon, France, to bomb Reading. Target visibility being poor, the pilot sought another and by chance found Hatfield. Running in low from the East he was engaged by small arms fire which set his starboard engine ablaze. On his second run he dropped four large H.E. bombs shortly before 11.30 from about 60 feet, and machine-gunned workers running for shelter. The bomber crashed at Hertfordbury and was burnt out. Its crew were oblivious to the important nature of the havoc they had wrought.

Book Two

DEVELOPMENT

Chapter 2

Development: W4050 and the Bombers

Trials of W4050 proceeded through Winter weather. It was to be the principal experimental Mosquito. Second and third flights were made on 29 November, 1940, carrying respectively, J. E. Walker and R. E. Bishop, the chief designer, as observers. There were six flights the following week. Captain de Havilland flew with his son. Clarkson, Wilkins and Tamblin flew. In gusty weather on Friday, 6 December, manoeuvrability promised well. Controllability was good. On the ninth flight on 6 December, Bishop and Gardiner crouched in the aft fuselage to retract the tailwheel and observe wool-tuft movement on inner sides of the nacelles. Tufts on outboard sides of the nacelles, observed from a Hurricane, to check tailplane airflow, showed them well streamlined. However, for airflow control, a Flamingo-style 'slot' had been fitted around the inner section of the rear of the engine nacelle. On 11 December the first cowling trouble was encountered, one of the few snags that was to afflict Mosquito fighters in the months ahead. The starboard cowling blistered and the aircraft flew home on one engine. W4050 had now made 30 flights.

The D.H.98 made a good impression at Langley Sunday, 29 December, at a show put on for Lord Beaverbrook and other Ministers. At de Havilland invitation, through correct channels, the Hon. C. D. Howe, Canadian Minister of Munitions and Supply, was among them, and it was felt that the D.H.98 might be built also in Canada. Sir Wilfrid Freeman was now Vice-Chief of the Air Staff, and so Sir Henry Tizard, who had taken over his R and D duties, came to the display. He was an old friend.

The other aircraft demonstrated were the Hawker Tornado (Vulture), Gloster F.9/37 fighter (2 handed Peregrines), Philips and Powis fighter (Merlin XX), Short Stirling bomber (4 Hercules) and Handley Page Halifax (4 Merlins). One recalls a heavy incendiary-bomb raid on London City that evening, until midnight.

The next day Grp. Capt. H. E. Forrow, M.A.P. Overseer at Hatfield, told de Havilland that 150 more Mosquitoes were to be ordered and that additional floor space, materials and subcontracting capacity were to be sought, with a view to 'further extensive orders'.

On 11 January, 1941, de Havilland were told that a reconnaissance

Development: W4050 and the Bombers

prototype was to be built and that the remaining 47 aircraft of the contract 69990 were to be finished as 19 reconnaissance and 28 fighters—no bombers. But bombers were likely to figure in the next 150 to be ordered, so development flying continued with all three roles in mind, for a fighter prototype had been ordered in July, 1940.

W4050 reached 22,000 ft. on 17 January, at which height it was expected to perform best, but its engines had peaked at 19,000 ft., a matter of adjustment. Slipstream rectifiers were now fitted also beneath stub wings, but failed to cure tailplane flutter. Two weeks of bad weather then grounded the machine, but taxiing tests were done to try out the castoring tailwheel and engine cooling. Hot exhaust manifolds demanded forward intakes for cooling air. Early in February, the aircraft flew with engine nacelles lengthened and intakes modified, and on 10 February with still longer nacelles. Successful, the modification was applied to all but the first few Mosquitoes. It called for divided flaps, additional torque tubes and levers.

Before despatch on 19 February to the Aeroplane and Armament Experimental Establishment, Boscombe Down, for official trials, the yellow prototype had its upper surfaces sprayed dark green and brown, and was reweighed. It had flown 35 hours and characteristics were summarized:—

1. Stability (a) Lateral and directional stability satisfactory.
    (b) Unstable at C.Gs. aft of 14 inches behind datum—mods. in hand.
2. Controls satisfactory.
3. Slight instrument vibration, cured by cutting three inches off airscrew radius.
4. Tail dither caused by disturbed flow off trailing edge of nacelles above 240 I.A.S., cured by nacelle extension (precluding use of flaps for the time being).
5. Tailwheel castoring unsatisfactory.
6. Speed tests conducted at 16,000 and 17,000 lb., and up to estimate (386 m.p.h.).

At Boscombe Down the lack of urgency surprised Clarkson, eager that W4050 should prove the de Havilland case, and his estimates. When after irritating delay the aircraft began flight trials there was frank surprise at its capabilities, so that its speed was double-checked. When the results were confirmed visitors arrived hastily from London, for the Mosquito was about 20 m.p.h. faster than the Spitfire for the same engines, with double the wetted area. Among the visitors were officials who had shown no interest or confidence in the unarmed aircraft idea. Then, disaster. As the prototype taxied across the bumpy, muddy field on the evening of 24 February for the 57th flight its tailwheel jammed due to castoring trouble, and the fuselage fractured.

De Havilland, Walker and Clarkson were to meet W. S. Farren

Development: W4050 and the Bombers

(D.T.D., becoming Director of the R.A.E. that year) at Boscombe and felt that the fracture and the tailwheel trouble would kill the Mosquito. But Farren made light of it all. 'You'll soon get that right,' he said, 'What interests me is this remarkable performance.' That was a great moment.

On 7 March, Plumb despatched to Boscombe Down the fuselage of the P.R. prototype, replaced in W4051 by the first production fuselage. It took only a few days for the small working party to fit the new fuselage. W4050 returned to Hatfield 15 March.

It went back to Boscombe Down 19 March, after adjustments, only to return five days later with engine starter trouble. Meanwhile on the production line incorrect gluing of bulkheads and principal fuselage members had called for fuselages to be re-jigged. More important was the Initial Handling Report 767 dated 3 March, of the Boscombe Down pilots. 'The aeroplane is pleasant to fly', it read, 'Aileron control light and effective. Take-offs and landings are straightforward. The aircraft stalls at 105 m.p.h. I.A.S. with flaps up, 90 with flaps down, and was flown at up to 320 I.A.S.' The best rate of climb in M.S. blower was 2,880 f.p.m. at 11,400 ft., and 2,240 f.p.m. in F.S. gear at 18,100 ft.

Top speed in F.S. gear was found to be 388 m.p.h. at 22,000 ft. Estimated service ceiling was 33,900 ft. the greatest height reached being 29,700 ft. Tests were conducted at 16,767 lb.

Commenting generally the report stated 'If attention had not been drawn to buffeting by the firm's pilot, it is likely that adverse criticism would not have been made'. It concluded that if the machine was to become a fighter it would require a different door, and that the opening of the existing door from the outside only, by use of a key, was unacceptable. The heating system was found to be good, seating a little cramped. Of the peep hole of armoured glass in the armour plate behind the observer the report commented 'this is an excellent feature'. The undercarriage was pleasant to travel on, the brakes efficient and the aeroplane handled well when taxied. 'It accelerated rapidly as the undercarriage was retracted and its rate of climb was high. The aeroplane handles well at all speeds.' Commenting on stability it pointed out that as no auto-pilot was fitted the aircraft would need an improvement in stability for long endurance flights. From this time the official view was positively for the Mosquito. At Hatfield the report was considered 'a most favourable and heartening document', and on 14 April W4050 made her 100th flight.

A display of eighteen new aircraft was held at Hatfield Sunday, 20 April, 1941, primarily to show Americans that the reason why their aircraft had not been figuring in operational news was the need to modify them extensively for combat duty. Mr. Wynant, American Ambassador, was there leading American Army Air Corps experts, also Sir Henry Tizard, Air Chief Marshal Sir Charles Portal, and prominent R.A.F. and Ministry officials. The eighteen aircraft were the Spitfire,

Development: W4050 and the Bombers

Hurricane, Beaufighter, DB-7 night-fighter, Mosquito, Typhoon, Tornado, Battle, Buffalo, DB-7A, Hudson, Martlet, Tomahawk, Mohawk, Maryland, Stirling, Manchester, Halifax. The performance of the day came from Geoffrey de Havilland and W4050, which he demonstrated superbly, showing its climb, speed, manoeuvrability and small turning circuit.

In a letter to the Chief Superintendent, Royal Aircraft Establishment, on 14 May, 1941, de Havilland wrote 'You may be interested in the speed which we measured on the prototype before it went to Boscombe Down; the maximum level speed at F.T. height in F.S. gear was 392 m.p.h. at 22,000 ft., and all-up weight was 16,000 lb.' This later was confirmed by Boscombe Down and was above estimate. W4050 had gone to Boscombe Down on 4 May, for tests of tropical suitability, cooling on the climb, etc. A report on 18 May said that the tailwheel shock absorption was inadequate and the rear fuselage had been damaged on overload tests. W4050 returned to Hatfield 23 May, and A.A.E.E. tests were completed on other Mosquitoes.

Aerodynamic test under Clarkson in June included stalls, trials with the (discarded) slots, effects of flying with bomb doors open, and airflow investigation, particularly over the tailplane, when the projected dorsal turret was fitted for the fighter. Three more pilots were now flying Mosquitoes—Pat Fillingham, George Gibbins and John de Havilland.

W4050 emerged from the shops on 23 July after modifications to tailwheel leg, engine and radiator. It was flown with bomb doors open at speeds up to 360 m.p.h., at which there was some fairly bad bomb-door flutter, observed by lying in the rear fuselage. On 24 July W4050 flew with a mock-up turret with four guns, facing aft, then sideways, then forwards and upwards, with bomb doors closed, also open. Handling was satisfactory, but the turret stole about 20 m.p.h.[1] Next it was fitted with ejector-type exhausts.

In the Experimental Department the bomber prototype, W4057, originally laid down as a reconnaissance aircraft, was taking shape; it was flight tested during the third week of September and delivered to Boscombe Down 27 September.

In accordance with the modified programme the first of the production bombers, W4064, was delivered there on 18 October, and with the issue of three that week a dozen Mosquitoes were in Service hands,

[1] Performance with turret:

		IAS	TAS
Guns aft	11,960 ft.	278	340
	14,050 ft.	277	352
Guns side	12,100 ft.	265	326
	14,050 ft.	263	334
Guns forward	12,000 ft.	269½	331
	14,000 ft.	266	337

Tested at 14,700 lb.

Development: W4050 and the Bombers

leaving fourteen in the shops, W4066 on engine runs and two on flight tests, 4063 and 4065.

Back again in the experimental hangar at Hatfield the bomber prototype W4057 was the object of experimentation, but mainly on the ground, as Tamblin's team studied the possibilities of its carrying 2 × 1,000 lb. and 2 × 500 lb. bombs. On 12 November, future development was visualized as leading to an unarmed B.Mk.VB bomber carrying 6 × 250 lb. or 4 × 500 lb. bombs. In October C. T. Wilkins had the idea of telescopic fins, to accommodate 500 lb. bombs. W4064 was retrieved and fitted to carry them. The fins automatically extended when the bomb was released, and Wilkins watched successful drops at Hatfield on 31 October. Simpler still, a bomb with short vanes was devised and test drops were made from W4057. Wilkins' suggestion that ballistics were not impaired was not popular at A.A.E.E., but after many weeks of tests they adopted 'the short bomb' as standard.

W4050 was taken into the factory late in October for fitting two-stage Merlin 61 engines. It was not practicable to land a Mosquito at Salisbury Hall, so the wing was sent there by road on 31 October for the installation. There was time to spare; the first 61 engine came to Salisbury Hall early February. On 25 March the wing was reunited with its fuselage, but air coolers were yet to arrive. Four-bladed propellers also were on hand for the machine, which began engine runs with Merlin 61s 17 June, 1942. Flame dampers specially designed to cover the six stubs of the new machine became too hot and were removed before the first flight was made on 20 June. On the second the same day Geoffrey de Havilland with John Walker attained 40,000 feet, the highest a Mosquito had reached. Tail buffeting was encountered, for 4050 still had short nacelles and a small No. 1 tailplane; at the end of June long nacelles and a No. 2 tailplane were fitted.

An investigation into stability at fairly high weight (20,700 lb.) was made on DK290 carrying 4 × 500 lb. bombs and on a Mk. II DD643. The extra weight should make the bomber less stable and so the design office commenced work early on a 10° dihedral tailplane to be above the slipstream. In the closing days of June, 1942, DK290 flew with its dihedral tailplane, but it was better with the standard tailplane, and in fact Boscombe Down cleared the aeroplane thus for full bomb loads.

Contract SB.24916 of 3 July covered development of the Merlin-61 Mosquito with a pressurized cabin. Tests with W4050 on 21 July were made with 4-blade Hydromatic propellers. At about 15,000 feet surging occurred, at first put down to a new constant-speed unit, so three-bladed airscrews were refitted. Spitfires with Merlin 61s suffered similarly. Twice radiators cracked and were repaired, and three and four-bladed airscrews were compared, along with tropical intakes.

During July the pressure-cabin bomber prototype MP469 was completed, but all the available Merlin 61s were now going into Spitfires, so

Development: W4050 and the Bombers

pressurization ground tests were made before engines or radiators became available. On 8 August it flew, and after clearing initial snags it went to Farnborough. In the fine weather of 27 August, weighing 22,350 lb., it flew over-load and stability tests from the long runway at Hartford Bridge, and airborne pressure-cabin trials were now being made. On 4 September, when flying at 38,500 feet Geoffrey de Havilland lost consciousness, and Brian Cross, observing, took over the controls for half a minute. He turned on the pressure system which quickly brought Geoffrey round. A few days later, 7 to 14 September, MP469 underwent urgent unexpected modification to a high-altitude fighter, as detailed in Chapter 5.

The 250th Mosquito was delivered in September 1942, and the 346th in October, at which time one of the biggest decisions on production was being taken—when to introduce the Merlin 61 aircraft on to the lines. Trials with W4050 in Autumn 1942 still pointed to the need for satisfactory inter-cooler radiators for these aircraft. Following fuel-consumption tests it was planned in October to fly W4050 with greater wing span than any other Mosquito, 65 feet, but wing strength would not allow beyond 59 feet 2 inches, the half-way stage. The extensions were intended for the NF. Mk. XV. Meanwhile W4057 was being treated to fuselage bending experiments, in the course of which the tailplane incidence was found to be likely to change $\frac{1}{4}$ degree in flight, and this partly explained the Mosquito's sensitivity to tailplane incidence. Window freezing problems called for investigation and a heated cavity windscreen was tested on W4072. At the same time drag was compared on three exhaust types—the shroud, saxophone and P.R. type.

With W4050 on 18 July, 1942, a true air speed of 413 m.p.h. at 17,000 feet had been attained and on 23 July, after all possible corrections, 432 m.p.h. High Merlin 61 speeds continued. In October Merlin 77s were installed, and flying resumed on 8 October when W4050 flew with wings of 59 feet 2 inches. Second tests took place on 12 October, by John de Havilland, flying at full-throttle level speed. Absolute ceiling reached was 37,800 feet with 700 f.p.m. climb at 31,000 feet. Next month the highest speed ever attained by a Mosquito in level flight, 437 m.p.h., was reached.

Comparison of three and four-bladed airscrews followed, there being little to choose between them. Development of the high-altitude fighters with pressure cabins, and the PR. Mk. VIII, soon relieved pressure on W4050, which spent much of 1943 testing modifications to exhaust stacks.

On 14 April, 1943, R. M. Clarkson wrote for the record which becomes this book 'We measured 437 m.p.h. at 29,000 feet on the prototype Merlin 61 Mosquito on 18,000 lb. but we have never been able to repeat it on this or any other aircraft. A maximum speed of 421 m.p.h. was reached on several Mk. VIIIs at 29,000 feet on 18,000 lb. We measured 424 m.p.h. at 26,200 feet on 18,000 lb. on the first Mk. IX

24. The first high-altitude Mosquito fighter, MP469, on 16 September, 1942
25. MP469 after conversion to Mk. XV standard
26. A standard production NF. XV

27. Timber to season—for a long war

28. Making the double-skin fuselage shell

29. Lifting the shell complete with bulkheads

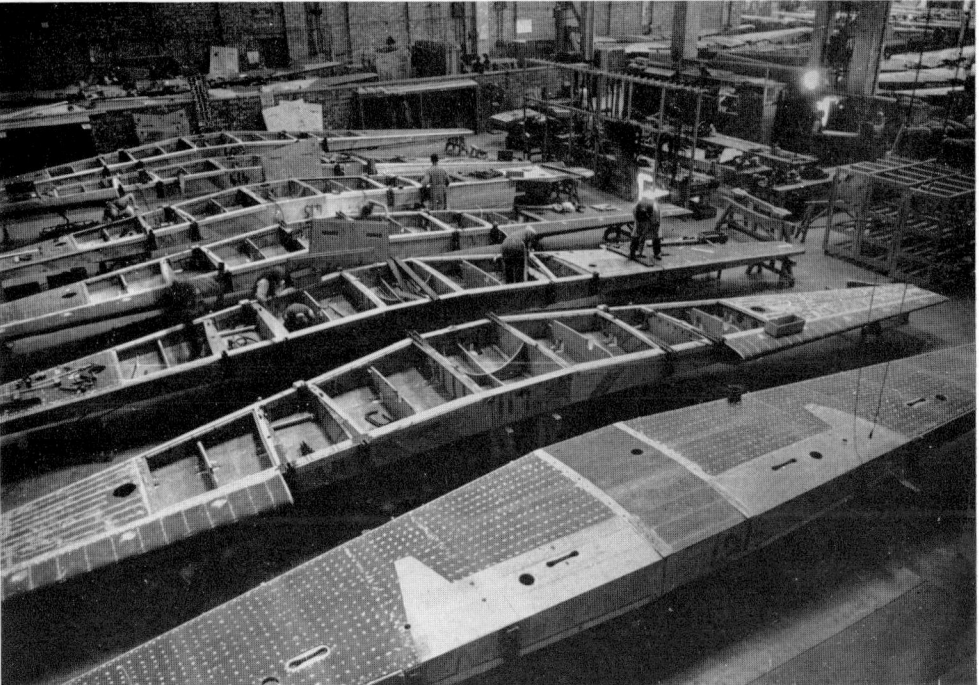

30. Simple wing of robust wooden scantlings
31. Assembly of wing fittings

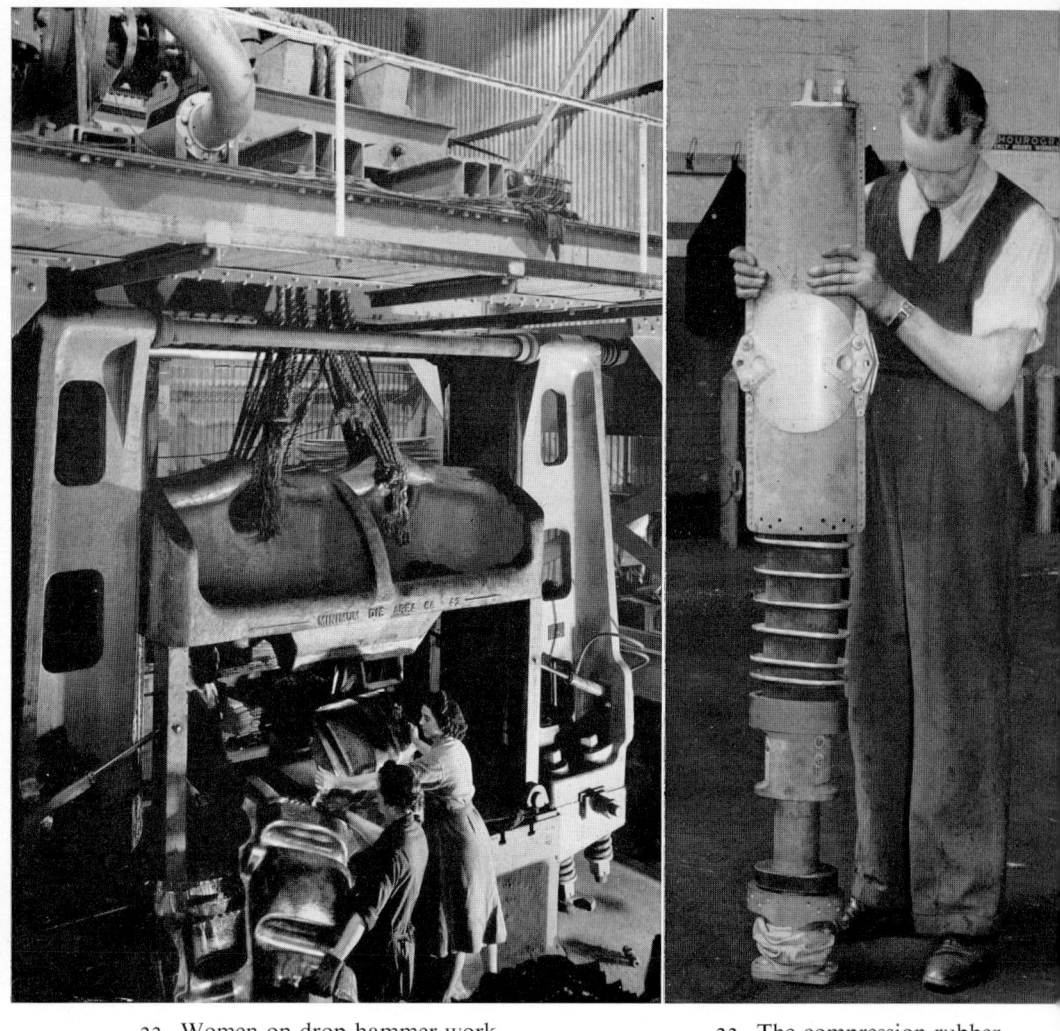

32. Women on drop-hammer work

33. The compression-rubber undercarriage leg

34. Hatfield fighter assembly line, August, 1943
35. Mosquitoes pouring out at Hatfield, 2 February, 1944

36. Production of Mk. II and II at Leavesden, June, 1943

37. Production at Standard Motors

38. Salvage hangar Mosquito Repair Organization, Hatfield, 18 May, 1944

39.
A job for the D. H. Mosquito repair organization

40.
Cockpit layout in the fighter prototype

41.
Cockpit layout in the PRI/Bomber Conversion

42.
Cockpit layout in the Sea Mosquito Mk. 37, November, 1946

Development: W4050 and the Bombers

(Merlin 72).' At this time A.A.E.E. trials with a Mustang produced 430 m.p.h. at 22,000 feet with a Merlin 66. A P-47 Thunderbolt at A.F.D.U. attained 416 m.p.h. at 25,000 feet. Clarkson added: 'I think you would be quite safe to say (the Mosquito is) the fastest operational aeroplane in the world'.

Spiral slipstream led exhaust gases into the starboard radiator, causing heating problems. There was a need for flame damping on exhaust stubs which, on Merlin 61/72 aircraft, were further forward than on earlier versions. It was many months before these difficulties were overcome, and the decision to convert W4050 into a pressure-cabin machine had to be set back in January, 1943. Frank Halford of the Engine Division commented 'Next time it would be better to design an aeroplane around the exhaust system'.

Three production two-stage-Merlin Mk. IXs, LR495-497, were introduced on to the final assembly lines at the beginning of March, 1943. On 24 March, LR495 made its first flight and in April LR405 and '406 (the first PR. IXs) emerged. After initial tests LR495 flew to Farnborough where on 4 April it took off at 23,000 lb., the highest weight at which a Mosquito had yet flown. It was carrying 4×500 lb. bombs and a 500 lb. bomb on each wing rack, 540 gallons of fuel and 200 lb. of desert equipment.

These trials were important, for it was realized late in April, 1943, that the Mosquito was now to carry a 4,000 lb. bomb, as related in Chapter 20. Additional special equipment would increase the weight. Before April was out 109 Squadron had received two Mk. IXs, LR496 and 497. Thereafter the Mk. IX—faster, and able to fly higher than the Mk. IV and with increased load—was introduced first to 109 and then to 105 and 139 Squadrons.

Mr. Winston Churchill saw the Mosquito demonstrated—also the Hawker Tempest and the Gloster E28/40 (Whittle jet)—at a Hatfield line-up of some forty types on Monday morning 19 April. The Prime Minister had asked for this to get a better idea of the work involved in aircraft development. He walked through the Hatfield factory.

Final major stage in bomber development came when a pressure cabin was wedded to the Mk. IX, design work upon this project taking place from April to July, 1943. Late in July, a converted Mk. IV DZ540 made its first flight as the pressure-cabin production prototype, strictly speaking a PR. Mk. XVI. Soon it was reaching 40,000 feet and a watch was being maintained for pressure leakages. Customary trials at A.A.E.E. followed and the Mk. XVI was accepted for service as bomber and reconnaissance aircraft. The first P.R. production machine flew in October, at the same time as the first B. Mk. XVI ML926 emerged. Production deliveries commenced in December to 109 Squadron. These machines were Merlin 72/73-powered, and on 2 March, 1944, orders were received to fit the first 80 B. Mk. XVIs with these engines and the remainder with Merlin 76/77s. The first 12 Mk. XVIs had provisioning for 500 lb.

49

Development: W4050 and the Bombers

bombs, but thereafter all had a bulged bomb bay to take the 4,000 lb. cookie.[1,2]

Similar modifications were made to some of the B. Mk. IXs[3] in service in 1944. Mks. IX and XVI carried out many of the Mosquito operations in 1944–45, their payloads, apart from bombs, being *Oboe, H2S* and a host of special installations required by 8 Group.

The final bomber variant,[4] the B.35, first flew on 12 March, 1945. It resembled the B. Mk. XVI apart from its Merlin 113/114 engines, and it had a top speed of 422 m.p.h. at 30,000 feet. Nine machines were brought to assembly state in March, 1945: by 31 May, 57 had been delivered from Hatfield. But they were too late to see operational service. In Bomber Command B.35s equipped the pathfinding element of two squadrons until 1953.

[1] ML937, first flown by Geoffrey de Havilland 1.1.44, delivered to A.A.E.E. 5.2.44, first with 4,000 lb. bomb provisioning.

[2] Tare weights 4,000 lb. aircraft: Mk. IX 14,644 lb. Mk. XVI 14,901 lb.

[3] ML915 was the T.I. conversion aircraft, also used for tests with the special Avro 6-store carrier.

[4] At a conference held at Hatfield 20 March, 1946, General Aircraft Ltd., discussed an idea of using the Mosquito Mk. XVI as the basis for a naval target tower, the T.T. Mk. 39 with a long nose affording an excellent view, and only six inches propeller clearance. Of the many converted or earmarked for conversion at Hanworth a few saw squadron service. The T.T. Mk. 35 was introduced in the early 1950s. Two types were developed, with and without external winches. Sqn. Ldr. L. V. Worsdell of Marshalls Flying School made the initial flight trials off the Norfolk coast with VR793 in July, 1952.

Chapter 3

Development: W4051 and the Photo-reconnaissance Mosquitoes

January 11, 1941, was the date when the fighter and reconnaissance versions and still no bombers figured in the revised programme for the contracts covering fifty D.H.98s. A prototype W4051 and nineteen production reconnaissance aircraft were wanted.

When W4050 was damaged at Boscombe Down 24 February, the fuselage of the reconnaissance prototype W4051 was sent there, and was replaced by a production fuselage—a fact which enabled W4051 later to fly on operations.

Production of the first nineteen P.R. aircraft was almost on schedule at the end of February, although many drawings and modifications were still being fed to the shops; bomber-type hydraulic pipes were all changed to reconnaissance type, and the adoption of a metal-skin aileron meant scrapping seven completed pairs. Prototype and production aircraft emerged together.

Assembly of W4051 was completed on 24 May, 1941, on the same day as the first production P.R. Mk. 1 W4054 entered the paint shop, and nine days after the fighter prototype W4052 had flown. Next week W4051 was weighed and on 10 June it flew for a stability check prior to Boscombe Down trials. On 12 June—such was the eagerness of the P.R.U. at Benson—Wg. Cdr. G. Tuttle came and flew it. On 25 June it reached Boscombe. There on 28th the tailwheel failed to lock down on landing, and the tail cone was damaged. W4054 had meanwhile flown and John Walker was investigating engine cut-outs around 30,000 feet. Blazing hot weather obscured the diagnosis and this trouble plagued him and Lovesey of Rolls-Royce for months.

In the Summer of 1941, using both aircraft, camera installations were checked, climb trials, consumptions and radiator tests were finished. By 13 September Benson had five P.R.1s and the sixth had flown.

On 8 October, 1941, top speed in high supercharger gear at 22,000 feet on the Merlin XXI was figured at 384 m.p.h. for the P.R. Mk. 1 loaded weight, 19,360 lb., 378 m.p.h. for the F.21/40 fighter at 18,437 lb., and 362 m.p.h. for the turret fighter weighing 18,812 lb.

On 20 September, 1941, W4060, the first long-range P.R.1 was

Development: W4051 and the Photo-reconnaissance Mosquitoes

flying. It carried 700 gallons of fuel, the extra tanks feeding first. On 30th it went to Boscombe Down for full-load tests. The second, W4061, was delivered 15 October.

Two tropicalized long rangers, W4062 and '63, with different radiator flap settings, short nacelles and No. 1 tailplanes, left for Malta and Egypt in January, 1942. Stability tests had cleared the long-range version in November.

Benson started operating Mosquitoes 17 September with W4055 as Chapter 11 relates, and the suitability of the design was at once evident. So was the scope for the bomber version. A struggle began. Ten P.R. Mosquitoes had flown 640 hours by 24 January, 1942, and very successfully. The second ten of the original twenty P.R.1s were to be finished as bombers. Benson argued that Beaufighters could look after the night-fighter role for the time being, leaving only P.R. and light bombing duties to be filled by Mosquitoes. But fighter production was well under way and deliveries were beginning. Forty-seven Mosquitoes had been delivered by the end of February, 17 that month. Benson had then flown 88 P.R. sorties and lost one Mosquito.

P.R.U. managed to obtain two of the 28 Mk. IIs which were in Maintenance Units being fitted with A.I. These, DD615 and DD620, were fitted with cameras on 12 and 17 April. Then, following a visit to Hatfield on 13 April by Air Cdre. Breakey and Grp. Capt. Laws, two of the B. Mk. IVs W4067 and DK284 were converted also.

The P.R. Mk. 1, with little development except tankage, served well, but longer penetrations into Europe in 1942, and the threat of Fw 190s, demanded better performance. Merlin 61 inter-cooled engines with two-speed two-stage superchargers were flown in W4050 from 20 June, 1942, and by 20 October a P.R. Mk. IV DK324 likewise powered was airborne as prototype for the P.R. Mk. VIII. On 2 October orders were confirmed for ten P.R. VIIIs (later reduced to five) having 2×50-gallon drop tanks, besides 26 converted P.R. Mk. IVs with Merlin XXIs and drop tanks. The P.R. pilots felt they had a future.

Contract SB 24916 of 3 July covered not only the development for Merlin 61, but a pressure cabin to follow. Jettison tests with DK324 showed up flow-sequence troubles, which were cleared in November. This aircraft went to Boscombe Down 28 November for all-round acceptance while a second Mk. VIII, DZ342, had flown at Hatfield two weeks earlier and was tried out by P.R.U. The pilots were pleased with the advance, for theirs were the first high-altitude Mosquitoes.

There was general re-allocation of Mosquito production by March, 1943, by which time it had been decided to build a large number with two-stage Merlins as Mk. IXs—25 bombers and 375 P.R. IXs. Similar to the Mk. VIIIs, except for their Merlin 72/73 engines, the Mk. IX had 18 lb. maximum boost, with 12 lb. for take-off, and extra fuselage tanks which, with 2×50-gallon drop tanks, gave them a range of over 1,500 miles.

Development: W4051 and the Photo-reconnaissance Mosquitoes

Before completion of these the pressure cabin became available, further contract adjustments were made, and 90 P.R. IXs were eventually completed, delivery from Hatfield ranging from May to November, 1943. The first P.R. IX flew at the end of April, 1943.

On 3 April, 1943, priority was given to the pressure-cabin version of the Mk. IX. Again the reconnaissance variant took priority over the bomber. Its pressure cabin was based upon that of the high-flying Mk. XV fighters, and on the first pressure-cabin Mosquito MP469, which flew 8 August, 1942. DZ540, the prototype reconnaissance aircraft, flew in July, 1943, reaching 40,000 feet; production deliveries began in November. A persistent trouble with this variant was the misting of the sandwich windscreen. Not until February, 1944, did Mk. XVIs start operations, then they served in a big way.

The next stage in development was to lighten the aircraft and fit extended wing-tips for operations around 40,000 feet in an attempt to evade jet and rocket fighters. Only five such Mosquitoes were built as P.R.32s, and were used in the closing stages of the war in Europe. Interest was then centring on the Mk. 34 ultra-long-range version of the P.R. XVI, with a bulged bomb bay similar to that of 4,000 lb. bomb carriers, but full of fuel. Wing drop tank capacity increased from 50 to 100 and 200 gallons. With a total of 1,269 gallons the range rose to 3,600 miles on a cruise at 25,000 feet and with T.A.S. 300 m.p.h. All-up weight had risen from the 19,413 lb. of the P.R.1 to 25,500 lb. on the P.R.34, which at this weight would cruise at 326 m.p.h. T.A.S. at 31,500 feet. Calculated on 25,500 lb. it produced nearly 33 gross ton-miles per gallon of fuel—comparable with the Albatross airliner.

Chapter 4

Development: W4052 and the Fighters

Contract 69990, placed 1 March, 1940, had been for one bomber prototype and 49 production bombers. A letter on 18 July instructed de Havilland to finish one as a fighter prototype and the company replied 22 July allocating W4052, which was to have four fixed cannon and four Browning machine-guns, in the nose. A separate Contract 135522 for W4052 and other fighters was issued 16 November, 1940, so the bomber order remained at 1 plus 48 only for a short while.

Meanwhile studies of turret variants continued with urgency. The 8-gun Spitfire and Hurricane were satisfactory day fighters, but improvisations like the Blenheim fighter needed to be followed by a modern night-fighter design, for which the Mosquito had promise—whether with fixed guns or dorsal turret.

Much thought was given also to the development of long-range fighters to combat the Focke-Wulf Condors which were harassing our shipping west of Ireland. Lord Woolton, Food Minister, came to the Langley display of Sunday, 29 December, 1940, and pressed for help in this direction. Estimates for long-range convoy-escort fighters and home-defence night-fighters had been summarized on 23 August, 1940, and Specification F.18/40 for a defending and intruding night-fighter was issued in October. For these needs the D.H.98, with its exceptional speed and range, using the best available power unit, was a unique candidate.

Although at this time it was likely that D.H.98 fighters, beyond the prototype, would be needed, and de Havilland were finalizing design work based on a higher wing-strength factor, no decisions affecting the first 48 production airframes were made until after the display of 29 December. The following week it was said that about half of them were to be converted to fighters.

The convoy-escort fighter version was estimated on 30 December, 1940, as having a take-off weight of 20,124 lb. with 600 gallons, 2 crew, 4×200 rounds of cannon ammunition and 4×700 rounds of Browning ammunition.

A corresponding calculation on 20 January, 1941, for the fixed-gun fighter with 550 gallons and 4×150 plus 4×500 rounds gave 19,606 lb.; and for the Bristol turret fighter with guns in the turret instead of the

Development: W4052 and the Fighters

nose, with 500 gallons, crew of three, and 4×60 rounds for the cannon and 4×500 rounds for the machine-guns, it gave 19,380 lb.

The turret version was all right for weight; its maximum speed on Merlin XXIs each giving 1,300 b.h.p. for take-off was reckoned to be 12 m.p.h. less than the fixed gun version, i.e. 382 m.p.h. and 370 m.p.h. in the configuration stated. At a maximum economical cruise at 20,000 feet, at a mean weight of 17,500 lb. for both, the turret version would have 4·6 hours and 1,380 miles (304 m.p.h.) against 5 hours and 1,570 miles (314 m.p.h.) endurance and still-air range. The speed difference mattered, and if the turret were not faired it would be 10 m.p.h. worse. Two-cannon and four-cannon turrets were considered, also six fixed upward-firing guns, as later tried in a Havoc 1 at the Fighter Interception Unit.

The company were instructed in the second week of January, 1941, to convert 28 bomber airframes to fixed-gun fighter in addition to the fighter prototype, and the remaining 19 to photo-reconnaissance. After further discussions the company were told mid-April, 1941, to finish two of the fighters with four machine gun Bristol turrets. Some fighters were to have dual control, the quantity to be settled later.

The fourth Mosquito W4053 and the 24th W4073 (the first production fighter) were allocated for turret fittings. Modified wing skins were in hand by 19 April.

Before dawn, Tuesday, 13 May, when the new fighter was within two days of flying, Karel Richard Richter, a Sudeten German spy, born at Kraslice, 29 January, 1912, parachuted into a field near Salisbury Hall. He spent Tuesday and Wednesday in a wood, burying his parachute, some food, a pistol and ammunition, a radio transmitter and a trowel. Within twenty minutes of emerging on to the road Wednesday evening he was in trouble. A lorry driver asked him the way, then asked P.C. Alec Scott, who happened to be nearby. The driver made a remark about the unhelpful stranger whom the constable then questioned. He took Richter to a police station where he was found to be carrying a considerable sum of British and United States currency, map and compass. A search of the countryside revealed the buried articles. Richter was executed at Wandsworth on 10 December, 1941.

On the evening of Thursday, 15 May, W4052 was taken from its hangar at Salisbury Hall and flown from the adjoining field by Geoffrey de Havilland, thereby saving nearly a month's dismantling and re-erection at Hatfield. To make the first flight with the machine from a 450-yard sloping field never previously used for flying showed immense confidence, not to mention courage. Lightly loaded the aircraft was in fact rigged one wing low. Its place was taken in the jigs by the first turret fighter W4053. Fred Plumb was passenger on the take-off. Having bet Geoffrey £1 that he wouldn't fly it out he could hardly refuse to go with him!

Cannon were fired on the Hatfield butts 20 May; Bishop was pleased

Development: W4052 and the Fighters

with the structural and cartridge feed aspects. There had been talk of reducing the fighter range, which would move the centre of gravity forward, but when squadron pilots flew it at Boscombe, on its arrival there Monday, 23 June, they pronounced the general stability on 18,460 lb. and 303 C.G. aft limit satisfactory for night flying. By 5 July, however, Geoffrey felt the longitudinal stability—comparable with the Beaufighter—was not good enough. Boscombe agreed—because the R.A.F. would be sure to use eventually the provision for 150 gallons aft of the cannon, as well as the standard 410 gallons of fuel in the eight wing tanks (118 outboard, 292 inboard), in all 560 gallons. By way of comparison the P.R. prototype W4051 had 410 gallons plus two 70-gallon fuselage tanks high between the spars, a total of 550 gallons.

Clarkson's brief pre-Boscombe speed checks with W4052 showed it about 6 m.p.h. slower than W4050, at the same weight—18,500 lb. Estimate had been 10·5 m.p.h. slower, so the fighter speed in F.S. gear at 22,000 feet should be about 378 plus or minus the effect of flame damping. Mosquito speed was sensitive to boost control.

Small snags were few. In the course of loops and rolls on Friday, 6 June, the emergency hatch flew off—it needed stronger fixing. Geoffrey blacked out that evening on a Hatfield airfield-defence Hurricane in comparing its elevator control with that of the stick-type grip, new to the Mosquito, which, he felt, was geared to be too stiff at $3\frac{1}{2}$ G.

By 29 July W4052, tests finished, was back at Hatfield, to have A.I. fitted. The saxophone flame shroud exhaust was not fitted now, for it was burning out on the Hucknall test beds. A.I. flight tests with a Blenheim as target were done by 5 September, and that evening Geoffrey de Havilland demonstrated a low half-roll followed by a long inverted climb.

A week later Hearle was informed that 20 fighters were to have dual control, and in the meantime large fighter orders were confirmed and big Mosquito production was planned in England and Canada, justifying the development programme.

The first turret fighter W4053 was flown out of the Salisbury Hall field on Sunday, 14 September, and shed part of its turret on the way to Hatfield. George Gibbins flew the second, W4073, out of the same field 5 December, with a mock-up turret. But soon this was removed and faired over at Boscombe Down. The turret tests were over. Having dual control W4073 was taken to 157 Squadron, Castle Camps, by Wg. Cdr. Slade on 17 January, 1942, to become a trainer. The turret fighter was out, primarily because there could be no doubt of the effectiveness of fixed guns—given deceleration.

The first hopes for deceleration, when coming up behind an enemy aircraft, were pinned to the Youngman frill brake. This had been tried on Spitfires. It encircled the rear fuselage like a cake frill and was operated by a bellows and venturi. Between 1 January and 1 August, 1942, the brake was tried in many forms, without gaps, with gaps, with

Development: W4052 and the Fighters

chords of 10, 13 and 16 inches, but compared with lowering the undercarriage (despite door problems at high speed) it was not a worthwhile complication.

All through 1941–42 W4052, in conjunction with other Mosquitoes, was used for innumerable developments and tests, large and small, for instance:

> exhaust stacks and flame shrouds, cowlings, front-glow shrouds, intake guards, under-wing bomb racks and tanks, larger rudder tab (for one-engine case below 140 m.p.h.), dropped ailerons for flap effect on landing, convex trailing edge ailerons, tailplane sizes, nose searchlight, balloon cutters, lamp-black finish, undercarriage doors, propeller governors, Mk. XIV bomb sight, braking propellers, radar for NF. XIII, 1143 radio, R3090 radio, 40 mm. cannon, three downward identity lights, U.V. lighting, fuel coolers, negative-G carburettor, F. Mk. II long-range tanks, Hamilton propellers and sandwich windscreen.

The tailplane issue was complex, affecting all Mosquitoes, with their different loadings, and the short nacelles of the first P.R. batch. The exhaust and shroud problems were persistent. The lamp-black finish, tested with W4082 and W4078, cost 26 m.p.h., an increase in drag of 20 per cent. The shrouded exhaust system, compared on bomber W4072, was only 2 m.p.h. slower than the standard P.R. and bomber ducted type, and removing the shrouds lost a further 7 m.p.h., a reason enough for standardizing. Wing bomb carriers cost 5 m.p.h.—or 9 m.p.h. with the bombs and 1 m.p.h. for weight of bombs. Doping over the cannon slots gained 4 m.p.h. Any production inaccuracy in intake would lose full-throttle altitude; 1,000 feet loss from 22,000 feet on full throttle meant a loss of 8 m.p.h. in speed above that height—a point carefully watched in the factories. The pitot static flush vent hole, standardized in case the ordinary head should ice up, was highly sensitive to local surface, varying as much as 10 m.p.h. This sometimes led to false claims about 'fast' individual aircraft.

As performance was measured on more and more aircraft it appeared that production B. Mk. IVs at the end of 1941 were doing 382 m.p.h. on 18,000 lb. at 22,000 feet, 5 m.p.h. less than had been measured on the prototype with faired nacelles. Fighters were doing 378 m.p.h. (352 m.p.h. with matt finish), the greater drag of their flat windscreen, plus A.I. and larger tailplane, being offset by their better nacelle fairings at that time—until the final nacelles became standard. Variations of this kind explain slight discrepancies in performance quotations through the present book.

To pile up hours on the new cowling to carry the weight of the night-fighter shroud, consumption tests were done with W4052 in April, 1942. They were interrupted on Sunday, 19, when Geoffrey, with Brian Cross, at 20,000 feet over Bedford, had to feather for a vibrating port engine.

Development: W4052 and the Fighters

Baulked by a Proctor taking-off from Hatfield, and unable to open up his starboard engine enough in a right-hand turn, he crabbed across Welwyn Garden City rather low, with flaps and undercarriage rising painfully slowly, and belly-landed at the de Havilland flying school—'Hogsnorton', now Panshanger. Jacked up on Monday, the undercarriage responded to the handpump, and W4052 was flown out on 5 May.

It served as a rehearsal for Friday, 29 May, when Geoffrey wrestled half an hour to lower the undercarriage of a B. Mk. IV DK291 (jammed due to jack glands that made even the handpump inoperative) and then sat its belly neatly down on the Hatfield grass.

The word 'Intruder' came into vogue in 1941 when the squadrons, the designers and the Air Staff were enthused with the possibilities of offensive sweeps by Mosquito fighters without night-fighter equipment carrying 2 × 250 lb. bombs behind the cannon, and the same under the wings. Discussions were going on in October, when Wilkins' bright proposal to clip the bomb vanes was disturbing ballistic traditions. It meant that there was room for 2 × 500 lb. bombs behind those cannon.

The airframe picked for fighter-bomber conversion, DZ434, was also fitted with gas sprayers—two under-wing cylinders, which the standard bomb fairings would not fit. DZ434 flew 1 June, 1942, and went to Boscombe Down 13 June, renumbered as HJ662. The report, 1 October, was satisfactory, and especially the finding that the Mk. VI was exceptionally manoeuvrable.

The mid-1942 plan envisaged the following useful range of variants at this stage of development, based on Merlin XXI 2 × 1,300 b.h.p. for take-off, plus 12 lb., 3,000 r.p.m. and 2 × 870 b.h.p. at 19,000 feet F.S. gear, plus 4 lb., 2,650 r.p.m., at the tankages shown:—

	PR.Mk.V	B.Mk.V	NF.Mk.VI	FB.Mk.VI	LRF.Mk.VI
Normal all-up weight lb.	19,670	20,540	18,630	19,820	19,620
Fuel gal.	668	539	403	403	520
Bomb load, lb.	nil	2,000 (4 × 500 fuselage)	nil	1,000 (2 × 250 fuselage) (2 × 250 wing)	nil
Cannon 20 mm.	nil	nil	4	4	4
Guns 0·303 in.	nil	nil	4	4	4
Cameras	2 × F24 1 × K8	1 × F24	nil	nil	nil
Max. still-air range, 215 ASI 22,000 ft., miles	2,200	1,720	1,280	1,190	1,640

Max. speed Mk. V standard bomber/reconnaissance at 18,000 lb., 382 m.p.h. at 22,200 feet F.S. gear.

Development: W4052 and the Fighters

To increase fighter range two tanks, fitted behind the cannon in W4096, gave 151 gallons, a smaller 50-gallon tank being possible, with 2 × 250 lb. bombs, in the intruder.

At this stage a 'basic wing' was evolved, strengthened for the heavier loads that were being visualized, and applicable with simple adaptation to all kinds of Mosquitoes.

Merlin development, which so improved the PR. and bomber, was available for improving the fighter and fighter-bomber as and when needed. How the Merlin 61, on test in a pressure-cabin bomber MP469, enabled a 45,000-ft. fighter to be evolved in a week, 7–14 September, 1942, is told in the next chapter. This, becoming the NF. Mk. XV, was waiting for high-flying bombers, but in vain, and no large quantity was ever needed.

Replacing the 4 Browning guns by nose radar provided the NF. Mk. XII, a highly potent intruder-fighter system as radar developed, which was not ineffective at low altitude. An F. Mk. II DD715 had been taken off the line as early as 22 July, 1942, for this 'thimble' radome installation taking initially A.I. Mk. VIII. (It was the nose of DD715, with 4 Brownings, that was grafted on to MP469 in a hurry to produce that 45,000 ft. high-altitude fighter in September.)

After trials with DD715 at Defford in September it was decided to modify similarly 97 F. Mk. IIs, flown from Leavesden to Marshall's Flying School at Cambridge, where the A.I. VIII 'thimbles' were fitted. First arrivals, on 2 January, 1943, were HJ945 and 946, the latter proceeding to Defford on 13 February. The first production conversion was delivered to 85 Squadron 28 February, and the Mk. XII was an aircraft which squadrons operated almost to the end of the war.

A similar four-cannon radome fighter developed from the FB. Mk. VI with Merlin 21 or 23 (Roman figures were becoming as unmanageable as the very variety of marks!), and with the 'basic wing', became the NF. Mk. XIII, and offered greater tankage, 716 gallons instead of 547 gallons.

Merlin 25s and a new Universal Nose to take either British or American radar, turned this into the NF. Mk. XIX, which came to be known as the 'bull-nosed' Mosquito. The conversion began late in January, 1943, when a white Wellington from Defford unloaded at Hatfield A.I./SCR 720 radar and a working party, with the object of testing this equipment in flight as much as of seeing whether it would fit into a Mosquito (DZ659). At the final Ministry conference on it, 27 March, the Americans claimed that production was further advanced than with British A.I., and that 2,000 sets could be available almost at once. The British authorities argued that the vehicle far and away best fitted to make use of it, in the interests of the Allies, was the Mosquito.

It was an unlikely suggestion that one British fighter could be singled out in that manner, but the record of operations shows what part the

Development: W4052 and the Fighters

Mosquito fighter variants did play in the whole offensive against Germany.

The conversion aircraft DZ659 was flown away from Hatfield 1 April, 8 weeks from the start of work. Tests were done at Ford, and DD715 with A.I. Mk. VIII was there as well. Sir Wilfrid Freeman inspected the two aircraft, back at Hatfield, on 6 May.

The Merlin 25 development, giving 18 lb. boost for improved low-level and medium-level performance, was being installed at the same period. The first intruder to have this, an FB. Mk. VI, HJ679, flew on 5 April at Hatfield, and was through its trials, including cooling and speed checks, by 17 April, ready for Boscombe Down.

Mk. IIs fitted with A.I./SCR 720 (known to the British as A.I. Mk. X) became NF. Mk. XVIIs, and because fighters had been pouring out of the factory a batch of 98 of these started arriving at Marshall's for radomes—almost as soon as the firm had converted Mk. IIs into Mk. XIIs.

The FB. Mk. VI converted to take the six-pounder 57 mm. gun was named FB. Mk. XVIII. The big accurate gun replaced the 4 cannon, the 4 Browning guns being retained. The gun was brought to Hatfield 15 April, 1943. 'Where are the horses?' asked Rex King as he had it limbered towards HJ732. Installation was finished at the end of May and on the butts, Sunday, 6 June the firing, feed, blast and recoil appeared satisfactory. It was flown 8 June and delivered 12 June. The R.A.F. did air firing at Boscombe Down and Exeter—800 rounds were fired by 7 August and Hatfield were fitting the third barrel. Three production aircraft by then had armour plate fitted around cockpit and engines, and soon the four were in service; the Mk. XVIII was in action from 24 October; 50-gallon or 100-gallon drop tanks, or 500 lb. wing bombs, or eight 60 lb. rocket projectiles could be carried. The gun fired 25 shells in 20 seconds. Merlin 25s gave the necessary performance low down. Maximum tankage was 668 gallons.

Because of reluctance to interrupt the now fast and rhythmic flow of effective aeroplanes, for the introduction of a change for a special role, it was decided not to marry the two-stage Merlin to the Mk. XIX airframe until 1944. The first such, MM686, appeared in April, 1944; the first production aircraft, NF. Mk. 30 (not XXX), had Merlin 72s: and later on Merlin 76s and 113s edged the performance higher. Exhaust shrouds were a problem of the Mk. 30, doing much to keep them out of the war until late in 1944.

Thereafter they proved the most efficient Allied night-fighters, serving with Fighter Command, 100 Group and the U.S. Air Force in North-West Africa and all through that winter. They remained in R.A.F. service until 1949, and on into the 1950s in the Belgian Air Force. The number Mk. 31 was reserved for Packard Merlin-engined Mk. 30s, which however were never produced.

Final variant of the Mosquito fighter to see squadron service was the

Development: W4052 and the Fighters

NF. Mk. 36, which differed in having higher-powered Merlin 113s and various forms of improved A.I. radar.

The NF. Mk. 38 with Merlin 113/114 and later A.I., although given trials at C.F.E., West Raynham, never entered service due to stability troubles. Of the 101 built 60 were sold to Jugoslavia.

Chapter 5

Development: Against the High Raider

Early in August, 1942, the strength of 4/(F) Ob. d.L. mixed long-range reconnaissance unit based in France consisted of He 111s, Ju 88s, Do 215s and Bf 109s. By 10 August three experimental Junkers Ju 86 high-altitude aircraft had been added to unit strength for operations against Britain. These were old aircraft re-engined, fitted with pressure cabins and had increased span with pointed wing tips. Since 1941, and the appearance of similar machines in the Middle East, they had been expected over England, and they commenced lone operations here in mid-August, dropping single 250 kg. bombs in the Bristol, Cheltenham and Newport areas. Elaborate plans to catch the raiders centred on the pressure-cabin Spitfire VI with low-altitude engine and extended wings. The Spitfire VII with two-stage engine was not yet ready for service, but a few unpressurized Mk. IXs with Merlin 61s were available. Thus, defence rested mainly with the Spitfire V equipping many squadrons.

On 24 August the contrail of an enemy aircraft was seen over Berkshire at an estimated height of 38,000 feet. A bomb, believed to be of 50 kg., fell at Camberley from what was almost certainly a Ju 86 P high-altitude raider. A second aircraft, originally considered to be a Dornier 217, which climbed to over 37,000 feet above Southern England, was later identified as a Ju 86 P. Next day an enemy aircraft climbed to 34,000 feet, but flew over Britain mainly between 15,000 and 20,000 feet. It dropped one bomb, estimated to be of 250 kg., at Stanstead St. Margarets, Hertfordshire, not so very far from Hatfield.

An attempt to intercept the raiders occurred when two Spitfire Vbs of 310 (Czech) Squadron saw an unidentified aircraft at 32,000 feet observed by another pair of Spitfires of the squadron, and itself almost certainly a Ju 86 P. Two Spitfire Vbs of 313 (Czech) Squadron saw the machine flying at 33,000 feet, but due to its vapour trails positive identification was impossible. Sighting of another bomber at 30,000 feet was made by two Spitfires of 308 (U.S.A.) Squadron, but it is believed this was not a Ju 86 P.

Fighter Command was determined to halt the high-level raids and when a Ju 86 P came in on 29 August, two Spitfire IXs of 401 Squadron scrambled, only to record that the bomber was at 41,000 feet, where it appeared to be labouring at its ceiling. Two Spitfire VIs of 124 Squadron

Development: Against the High Raider

were despatched from Debden, but the raider, then at 39,000 feet, escaped them and dropped a bomb near Chivers factory at Histon, Cambridgeshire. Contrailing, the Ju 86 turned for home. Two Spitfires of 121 Squadron were sent after it, and located it at 40,000 feet by its vapour trails. As they climbed to intercept they saw it diving towards Belgium. A patrolling Spitfire of 332 Squadron watched the attempt, its pilot reporting that the bomber dived to 500 ft. before pulling out over the sea. So far attempts at interception had proved fruitless. It was clear that no British fighters could catch the raiders, whose operations intermittently continued.

September 5, 1942, was a bright, sunny day, conditions over Britain being ideal for Ju 86 attacks. One of the bombers wended its way to Luton delivering a 250 kg. bomb in Midland Road, damaging the L.M.S.R. depot and houses. As it flew home its course took it across the de Havilland works at Hatfield where, below, Clarkson amongst others watched its progress. He discussed with his colleagues then and there modifications that would probably produce a gunned Mosquito of rivalling altitude performance. It was 11.30, and ten minutes earlier a 124 Squadron Spitfire VI AB516 flown by Plt. Off. W. Hibbert specially positioned scrambled from West Malling to 25,000 feet to await the raider. He was vectored on to it and chased it towards Dover, climbing as he went. At 500 yards range over Boulogne and at 36,500 feet he fired at what he identified as an He 177 on account of its wing span, but which was really a Ju 86 P. As he fired the Spitfire stalled—it seemed impossible to destroy the new foe.

This was almost the last attempt the enemy made to perfect high-altitude day bombing against Britain. On 9 September two Spitfires of 616 Squadron managed to reach 38,000 feet after a Ju 86 flying over Ramsgate at 42,000 feet. After twelve operations, however, the weather broke and raids by the five Ju 86 Ps now available were halted. For the enemy this was fortuitous.

On the afternoon of 7 September, N. E. Rowe, D.T.D., had arrived at Hatfield to discuss the D.H.102. More urgently he told de Havilland that the pressure-cabin Mosquito was now immediately needed by the R.A.F. as it was the only aircraft that could fly and fight high enough to destroy the Ju 86s. At once the design staff considered modifications needed to convert MP469 into a fighter. In the experimental shop the staff worked flat out that week fitting the nose and four Browning guns that had been cut from DD715 when that machine was adapted to carry centimetric A.I. radar. A fighter stick replaced the bomber's wheel and a duralumin bulkhead replaced the front armour plate; a plywood panel took the place of the pilot's back armour, but the observer's was retained as part of the structure. After flight tests four-bladed propellers were snatched from W4050 on 13 September.

By then MP469 had been lightened as much as possible. All bullet proofing had been taken off fuel and oil tanks; outer and fuselage tanks

High-altitude Mosquito Fighters

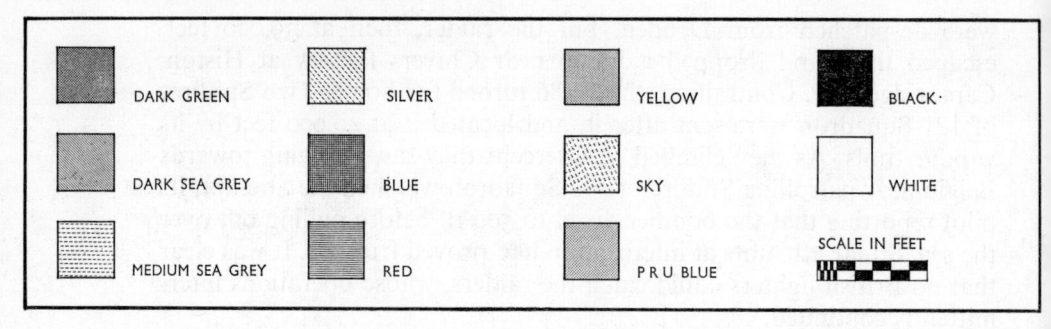

DARK GREEN	SILVER	YELLOW	BLACK
DARK SEA GREY	BLUE	SKY	WHITE
MEDIUM SEA GREY	RED	P R U BLUE	SCALE IN FEET

Mosquito NF. XV (Merlin 61) DZ366 as in February, 1943. The original high-altitude fighter, MP469, had dark green and dark sea grey camouflage on its upper surfaces and sides. Under surfaces were PR. Blue. Roundels on wings and fuselage were as applied to the Mosquito bombers. Initially the NF. Mk. XVs were painted PR. Blue overall, a grey-blue tint chosen after much experimentation. A further change at the end of 1942 was to Deep Sky Blue, a shade in which blue predominated in place of grey. This was applied to the Mk. XVs. In March, 1943, an instruction was issued that for night operations the aircraft were to be black overall, but it was countermanded in August, 1943, when the NF. XVs reverted to Deep Sky.

Mosquito NF. XV (Merlin 61) DZ366 as in February, 1943.

Mosquito Pressure-cabin Bomber Prototype MP469 (Merlin 61) after conversion to fighter, as in November, 1942.

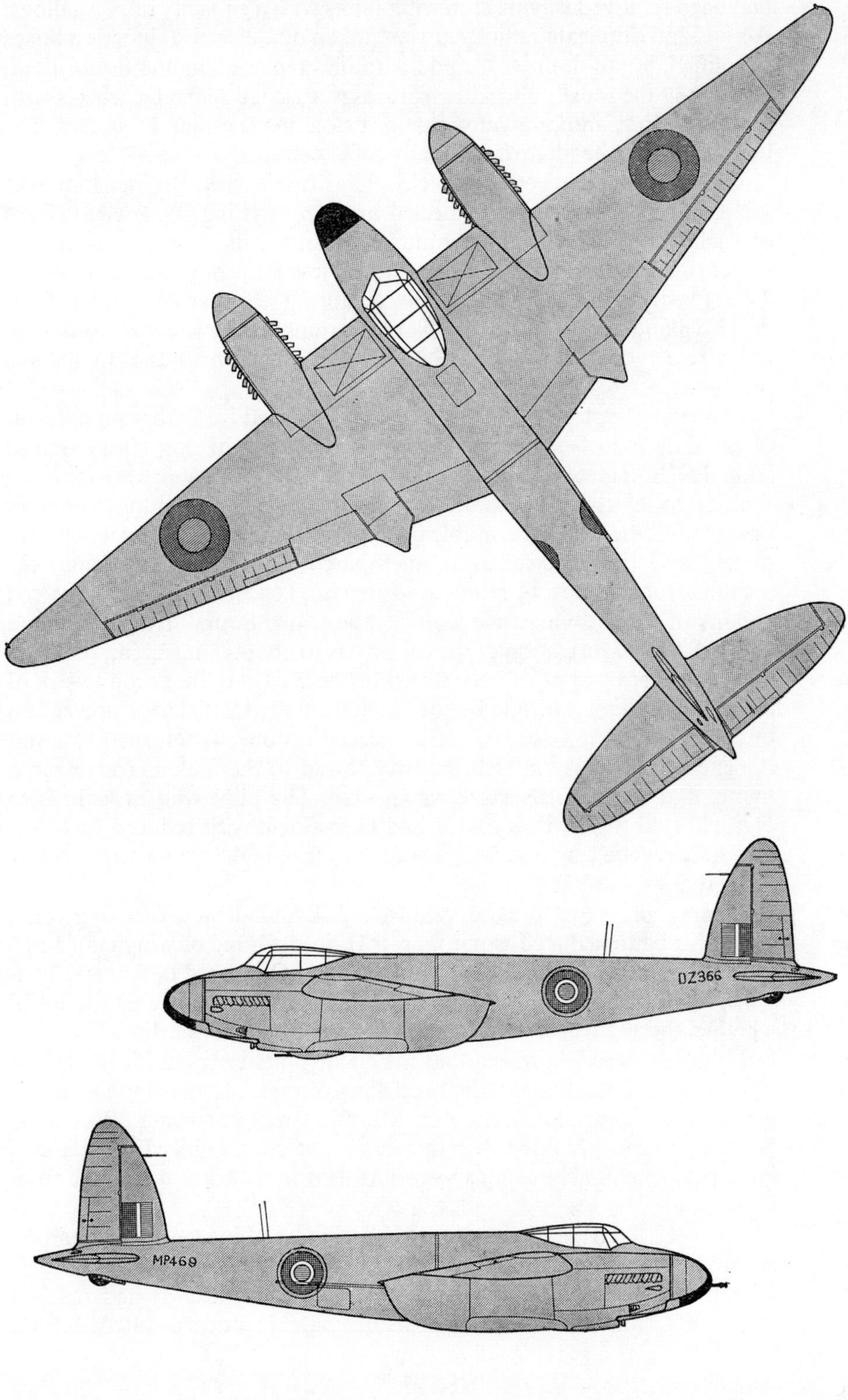

Development: Against the High Raider

had been removed leaving stub wing tanks with a capacity of 287 gallons. Aerials and duplicate radio gear were taken out, electrical junction boxes simplified, bomb-door jacks and hydraulics replaced, lighter doors fitted. Externally the most noticeable features were small diameter wheels with treadless tyres, and pointed wing tips extending the span to 59 feet. The hope was that the aircraft would fly to 43,000 feet.

On 14 September, only a week after Rowe's visit, the machine was miraculously complete and wheeled out for weighing. With 287 gallons of fuel and 18 of oil, 2,000 rounds of ammunition, and the pilot, it tipped the scales at 16,200 lb.—2,300 lb. less than the standard Mk. II. Two twenty-minute flights followed during the evening, with John de Havilland at the controls. Next morning he reached 43,500 feet, the aircraft behaving extremely well. The answer to the Ju 86 was produced.

During 15 September Flg. Off. Sparrow, picked by Embry on account of his skill and of having 150 Mosquito hours in his log book, arrived from 151 Squadron. In the evening he flew MP469 and next morning took it to Northolt to await the enemy. Waiting, waiting: Summer passed, and the weather suitable for high-altitude attacks dissolved completely into the mists of Autumn. Sparrow on test flights found the machine took about 35 minutes to reach 42,000 feet, using about 105 gallons of fuel, leaving sufficient for about two hours' flying at around 360 T.A.S. Pressurization of the cabin was to about 2 lb./sq. in., reducing altitude effect by about 12,000 feet. For three days in the second week of service with High Altitude Flight MP469 was at Hatfield for prevention of oil leaks and replacement of a cracked canopy. It returned to await the enemy, but early in October was passed to the makers for research flying, and Flg. Off. Sparrow was posted. The pilot who brought it to Hatfield told the makers that it had been flown with reduced fuel load and had reached 45,000 feet, about 300 feet higher than the recently delivered Spitfire VII.

Meanwhile, as a 'private venture', 264 Squadron under Wg. Cdr. Kerr also lightened a Mosquito, an F.II, in the hope of winging a Ju 86 should one cross to the West Country again. Outboard tanks, machine-guns, radio equipment—all were removed, but it seems doubtful whether success would have come without two-stage Merlins.

New fears were expressed that the enemy might exploit his tactics for special night operations to distort defensive radar. Some contended this was already being done, but A.I. Mk. VIII was not being affected, so MP469 was quickly fitted with this in a radar nose. Under Fred Plumb's guidance, four machine-guns were installed in a blister under the fuselage.

The revised MP469 was wheeled out for weighing on 15 November. All-up it weighed 17,400 lb. now that two 24 gallon wing tanks had been added, and a second crew member. A week later, after it had reached 42,000 feet, climb rate was still 500 feet/minute. It proceeded to A.A.E.E.,

Development: Against the High Raider

then to radar trials at Defford. On 28 November it was decided to build four similar aircraft now designated NF. Mk. XV.[1]

DZ366, the first, was flying by Christmas, but it was held up when the radome cracked on 24 December. The second machine, DZ385, also the remainder, was fitted with Merlin 77s, similar to the Merlin 61 but with a Bendix-Stromberg carburettor with metering fuel injection into the supercharger eye. The engine change made little difference, nor did it matter whether three or four-bladed airscrews were used. After brief tests the fighters in deep blue finish were delivered by mid-March, 1943, to the Fighter Interception Unit at Ford.

On 21 March two Mk. XVs arrived at Hunsdon and were quickly followed by the others, which immediately entered 85 Squadron, then under the command of Wg. Cdr. John Cunningham, and were placed in a special 'C' Flight which at once began operational trials. Initially, handling flights to over 40,000 feet were undertaken, followed by A.I. practice interceptions; 43,800 feet was reached by Flt. Lt. E. N. Bunting on 30 March and he took MP469 to 44,500 feet on 10 April; 44,600 feet was the highest reached, again by Bunting in MP469. Between 30 April and mid-August, 1943, 183 hours of day trials were flown and two hours of night operational flying, without any contact with the enemy. For most of the time only two machine-guns were installed, and the climb improved to 30,000 feet in ten minutes. The absence of high flying raiders brought squadron use to an end. The Royal Aircraft Establishment found the aircraft useful for pressure-cabin research, and DZ417 with MP469 spent the late summer months at Turnhouse awaiting possible raids on Scotland.

The story of the high-altitude Mosquito fighter shows the versatility and competitive merit of the design. Born as a bomber MP469 within a week became the most outstanding fighter the Royal Air Force possessed.

[1] Histories of these aircraft are:—
DZ366 initially to F.I.U.; to Hunsdon 2.3.43, 85 Sqn. 23.3.43 until 8.43. Used later at S. of T.T. Cranwell.
DZ385 first flew 3.1.43; to A.A.E.E., F.I.U., Hunsdon 21.3.43, 85 Sqn. 23.3.43 until 8.43. Later to S. of T.T. Cranwell.
DZ409 Defford 24.2.43, F.I.U. 7.3.43, 85 Sqn. 23.3.43, R.A.E. 8.43, to No. 8 S. of T.T. 14.10.44. SOC at Halton as 4888M on 29.5.47.
DZ417 Defford 18.2.43, F.I.U. 19.2.43, Hunsdon 22.3.43, 85 Sqn. 22.3.43, Turnhouse 22.8.43, R.A.E. 3.11.43. Written off there 25.1.44.
MP469 Chief of Res. & Dev. D.H. 14.9.42, High Alt Flt. Ftr Cmd. (Northolt) 16.9.42, D.H. 10.42, A.A.E.E. 20.11.42, S.I.U. Defford 21.11.42, A.A.E.E. 12.42, F.I.U. 4.2.43, 85 Sqn. 23.3.43, Turnhouse 22.8.43, D.H. for repairs 9.3.44. Became 4882M.

Chapter 6

Development: The Sabre Mosquito Project

Faced with developing the Mosquito for various roles, and later on to take high-altitude engines and pressure cabin, de Havilland, early in 1941, also felt that in a long war there would be need for a scaled-up Mosquito night bomber, perhaps with twin high-altitude Sabres.

The de Havilland Halford team had also undertaken to design Britain's first production jet engine, the H 1, and were intent upon designing a single-engine fighter around it.

It was thought, at that early date, that the associated Airspeed design staff might, under Bishop's guidance, take on part of the programme, for instance the jet fighter, which would give them metal experience, also high interest.

Talks with D.T.D. were fairly frequent, with Linnell and Rowe, occasionally with Freeman. All were understanding men. On 16 October, 1941, de Havilland and Bishop proposed the twin-Sabre unarmed bomber, and the idea was well received in principle, though it was suggested that it might be approached more conservatively with existing engines rather than with a high-altitude engine yet to be developed. That, however, would not have interested the company. On 13 November Walker wrote this letter to Air Marshal F. J. Linnell, who had been appointed C.R.D. in June:—

CCW/ELC.   13th November, 1941.

Air Marshal F. J. Linnell, C.B., O.B.E.
Controller of Research and Development,
Ministry of Aircraft Production,
Millbank,
London S.W.1.

Dear Air Marshal Linnell,

On looking more closely into what can be done with the design of a high-speed undefended bomber, in the shape of a twin-Sabre aeroplane developed from the Mosquito, one cannot help being struck by the fact that not only can it (under most operational conditions) surpass the heavy bomber as such but it possesses a versatility which fits it for any course which the air war may take, having twice the operational height, 150 m.p.h. greater speed and a superior bombing achievement for the same man-hours in manufacturing, maintenance and flying-crew effort—

Development: The Sabre Mosquito Project

that is, reckoning two twin-Sabre D.H.99's as equivalent to one heavy bomber like the Stirling.

Two exceptional circumstances make it possible for these advantages to become available by the summer of 1943, namely (a) following closely an existing design—the Mosquito, (b) using wood construction, which permits the production stage to be reached more quickly than metal. Wood construction has two other advantages in that it taps labour and material markets which are much less exploited; there is a surplus of wood-working capacity. It must be remembered that the design and production of a four-Sabre metal bomber would take a very long time and could not be considered for 1943.

It is understood that heavy bombers are needed not because they are big and heavy, but because they can carry big bomb loads in relation to the manufacturing and maintenance effort and the effort and risk of skilled flying crews—and because they can carry big single bombs.

Designing specifically to meet these requirements, we find that the fast undefended bomber is more efficient because it is relieved of the load comprising armament and air gunners, also the airframe structure and fuel to carry them. This relief gives it better all-round performance—higher speed at any height, higher altitude, better manoeuvrability and speed range for evasive action, etc. Besides producing greater bombing achievement for the same 'cost' these qualities (and the fact that it presents a smaller target) make it safer from ground attack. As the enemy's methods of detection and interception improve this may become an increasing asset.

Our assumption that two D.H.99's can be manufactured for the same cost in man-hours and material as one Stirling can be estimated by comparing the tare weights. The tare weight of two D.H.99's will be about 5 per cent greater than that of one Stirling, but owing to (1) the much greater simplicity of the undefended bomber and (2) the use of wood construction, the production man-hours will be smaller. Nevertheless, we base the following figures on the conservative assumption that two D.H.99's are obtained for the same 'cost' as one Stirling.

The logic of the figures works out as follows:—

The total bomb capacity of two D.H.99's is a little less than one Stirling, but this is only felt in short-range bombing. In long-range bombing their bomb load is much greater.

The crew of two D.H.99's is less, in the ratio of 6:7, as they carry more bomb load for less crew.

The two D.H.99's fly fewer hours for the same bombing operation, in the ratio of $4:6\frac{1}{2}$.

Thus they carry more bomb load to the same target in less time, for less crew hours, for less journey hours and less engine hours, and with much higher performance and manoeuvrability, though without guns. Put another way, on all but the short ranges the bomb load delivered by two D.H.99's may be expressed thus:—

Development: The Sabre Mosquito Project

Per number of crew	—nearly double.
Per flying hour	—nearly three times as great.
Per man/hour of crew's flying time	—nearly three times as great.
Per engine hour	—more than double.
Per hundred gallons of fuel	—30 per cent greater.
Per ton of tare weight (or per man/hour of manufacturing and maintenance effort)	—30 per cent greater.

The figures in full are:—

Aircraft	*Two* D.H.99's	*One* Stirling
Engines	Two Sabre N.S. 8SM.	4 Hercules XI
Maximum speed	392 m.p.h.	250 m.p.h.
Cruising speed	300 m.p.h.	186 m.p.h.
Max. Operating Altitude (full load)	27,000 ft.	15,000 ft.
Ditto (less bombs and 1/3rd fuel)	33,500 ft.	?
Crew	6	7
Largest individual bomb	4,000 lb.	2,000 lb.
Max. bomb capacity (short range)	16,000 lb.	19,000 lb.
Bomb load delivered, with 2,200 miles range—full tanks—(600 miles target radius)	16,000 lb.	10,000 lb.
Ditto per member of crew	2,670 lb.	1,430 lb.
Ditto per journey flying hour	4,000 lb.	1,550 lb.
Ditto per crew-hour	640 lb.	230 lb.
Ditto per engine-hour	960 lb.	400 lb.
Ditto per 100 gallons of fuel burnt	1,250 lb.	950 lb.
Ditto per 1,000 lb. of tare weight (i.e. per unit of constructional man/hours)	330 lb.	250 lb.

Note. With an ultimate range of only 1,500 miles, equivalent to a target radius of 360 miles, the figures for the D.H.99's are still superior.

Future Engine Developments.

The fitting of Napier Sabre N.S. 19 SIM, two-stage, three-speed (32,000 ft.) engine with 15-ft. counter-rotating airscrew, when this becomes available, will give the D.H.99 improved performance as follows:—

Operating altitude on full load	32,000 ft.
Ditto when bombs and one-third of fuel are disposed of	38,500 ft.
Maximum speed	430 m.p.h.
Cruising speed	328 m.p.h.

I have written to Mr. Rowe in identical terms.

Yours sincerely,
C. C. Walker.

Development: The Sabre Mosquito Project

Provisionally called the D.H.99 the twin-Sabre study was renamed the D.H.101 because the figures 99 were a code prefix to do with spare parts, and this number would therefore better be applied to some study that was not going to materialize—in fact to a light civil twin by J. P. Smith's D.H.94 team which the war had thwarted. That twin had been resoundingly called the D.H.100. But the jet fighter which was being investigated would suit that number well since it would be opening a new era in propulsion. The twin-Sabre was therefore (in December, 1941) called the D.H.101, and that put them all in chronological sequence again.

At a Hatfield meeting of 3 February, 1942, de Havilland were warned not to take on too much, and not to let any jet fighter explorations (D.H.100) interfere with the twin-Sabre D.H.101. It might be better to leave the jet aircraft now, and later to do a twin-jet bomber. This, it seemed to de Havilland, would hand the necessary single-jet experience to a competitor. A twin-jet, with different C.G. and other features, would not follow a twin-Sabre straightforwardly. On 7 February, 1942, recording this discussion, Sharp wrote: 'One imagines the historic importance in these conversations on 3 February, for example when the above sentences are read say three years hence. These changes of plan in the early stages of a project or projects jar and upset a concern like ours. Time has been lost since April, 1941, when we were first talking and thinking about the E.6 jet fighter, and a fair amount of exploratory work has been done since then.'

Gently pointing this out, also that any other firm taking on the jet fighter then would be six months behind in the investigatory work which de Havilland had done, C. C. Walker obtained an instruction 18802 of 14 April, 1942, to proceed with the D.H.100 design, with a high tail on booms, and to build two aircraft.

Meanwhile on 4 April the sad news came that Sabres could not be seen as becoming available for the D.H.101, and that Griffons were to be considered. De Havilland argued that these would not yield the same fast aircraft; the D.H.101 was dropped before the month was out. Merlin engine development was going apace, and the company were asked to work on a twin-Merlin 61 'Mosquito Series II', larger, to take 4,000 lb. internally and probably two 500 lb. wing bombs. Specification B.4/42 was issued.

This, too, lacked appeal, because it would be slower than the Mosquito, not faster. In July it was called the D.H.102, and the 'Mosquito replacement', not the Series II, for fear of confusing it with the Merlin 61 high-altitude Mosquito that was programmed.

By August design work was moving ahead, without enthusiasm, the fuselage was being lofted and the wing was taking shape, also a full-scale centre section for laboratory tests. Fuselage halves were being equipped by October.

There were talks at this time about a twin-Merlin single-seat long-

Development: The Sabre Mosquito Project

range fighter with intake arranged between the cylinder banks as in the pre-war Schneider Trophy engine. This could give virtually the ultimate in piston-engine speed, and de Havilland were asked for studies, which were called D.H.103.

In September, 1942, Hatfield had to put such urgency upon the high-flying Mosquito fighter based upon MP469 that work on the D.H.102 suffered, and appeals were made to the M.A.P. for more woodworkers. By 10 October the D.H.102 was almost halted through shortage of labour.

At the Ministry Hearle sought the help of a production order, but there was no sign of urgency there, which puzzled him.

Discussions early in November brought out that a suggestion from Walker on 22 October for a swollen Mosquito to take a 4,000 lb. bomb had not found favour, although 'it was much appreciated', for it did not appear to open the way to a design for the twin-Merlin special fighter in which the Ministry were interested. The company were to proceed with the bigger D.H.102 and the jet fighter D.H.100. There was still no production order for the 102, however, and it did not seem to be badly needed.

Captain de Havilland wrote 6 November that a modified, swollen Mosquito for 4,000 lb. would save a lot of time, as well as freeing people to press ahead quickly with the jet fighter ('the only way to settle the numerous queries connected with the jet'), and before they would be required to design and produce the special twin-Merlin fighter.

De Havilland saw Freeman on 14 November and went over the situation. On 21 November he wrote 'He is the man for us.' In December de Havilland was told that the D.H.102 would likely be dropped, but that there was no discussion yet about the twin-Merlin fighter. A letter 32786 of 30 December, 1942, confirmed the decision not to proceed further with the B.4/42 (i.e. D.H.102).

At Hatfield the 103 had reached the mock-up stage by 23 January, 1943, and Rolls-Royce were keenly interested in it, also in applying the slender engine to the Mosquito.

But still lacking any decision on it de Havilland wrote on 1 March that all work had stopped on the D.H.103 and that the company did not wish to do any further work unless the twin-fighter was so important that a production order for a reasonable number of machines was decided on.

Late in April a swollen Mosquito to take 4,000 lb. was ordered, and on 1 June, de Havilland and Hearle were told that the M.A.P. wished them to proceed with the twin-Merlin fighter.

The swollen B. IV Mosquito to take a 4,000 pounder DZ594 was out and flying by July, 1943. The D.H.100 jet fighter (Vampire) with de Havilland H1 Goblin jet engine was out and flying by 20 September, 1943. The D.H.103 Hornet single-seat twin-Merlin fighter was out and flying by 28 July, 1944.

Development: The Sabre Mosquito Project

This is not the place to detail the weights and performances of the other aircraft, except to say that the Hornet showed a higher performance than any other piston-engined aircraft, climbing at over 4,500 ft./min., flying level at over 470 m.p.h., having an operational ceiling around 35,000 feet, and a range with drop tanks exceeding 2,500 miles.

To this day Charles Walker, still attending his office in 1967 (in his ninetieth year), regrets that he did not get the chance to prove his point with the large clean unarmed bomber. Apart from anything else, aircrew casualties might have been much less in 1943–44.

Chapter 7

Development: Highball

At Skitten on 1 April, 1943, No. 618 Squadron formed under Sqn. Ldr. C. F. Rose, D.F.C., D.F.M. Its aim, to sink the Tirpitz in Alten Fiord using a special weapon, code-named *Highball* and designed to be dropped from the Mosquito. It resembled *Upkeep*, the weapon devised by Dr. Barnes Wallis and used to breach the Mohne Dam, and was initially developed alongside it. *Highball* was an anti-shipping store commonly referred to as 'the Naval Weapon', the programme for which envisaged its use almost concurrently with the dams raid, although the Tirpitz attack was to be in daylight when the delivery method—highly secret—would be too evident to permit a repeat.

Two spherical mines were to be carried in tandem, back spun to nearly 1,000 revs. by a belt drive connected to an air turbine as the aircraft descended to 25 feet, then released almost a mile from the target. The mines would bounce along the water behind one another, leaping nets and booms around the ship. On striking its hull the weapon would rebound and sink. At selected depth beneath the hull its 600 lb. Torpex charge would be detonated by hydrostatic pistol. In November, 1942, the Beaufort had been selected as carrier, but the Mosquito's performance made it the ideal choice.

On 24 March, 1943, DK290 flew to Heston for the trial installation. Target date for the operation from Sumburgh was set at 15 May. Time was short. As 618 Squadron formed intensive weapon development was underway. On 9 April, a week after 618 Squadron borrowed Mosquitoes from the squadrons at Marham for flying training, a conference decided Sqn. Ldr. Longbottom of Vickers-Armstrong and Sqn. Ldr. Rose should fly trials from Manston, making test drops at Reculver. Between 13 and 29 April twenty-three wooden-skinned weapons were dropped with varying success as techniques were tried and whilst the squadron made 130 low-level sorties against the target ship 'Bonaventure' in Loch Cairnbawn. On 19 April releases were made at Ashley Walk against an armoured wall with the store in a new case packed with aerated resin. Stronger, it stood up well on impact. After further tests at Angle, South Wales, weapon shell and plates were thickened.

618 Squadron now sent a detachment to Turnberry, there to maintain more test aircraft. An M.A.P. London conference on 5 May decided to begin trials of steel-cased weapons immediately.

Development: Highball

Attacks delivered between 50–60 feet at 360 m.p.h., called for skilful yet dangerous flying. Stores rotated at 700 r.p.m. were dropped from 1,200 yards range and set to explode at 27 or 40 feet. Sqn. Ldr. Rose released 10 stores on 10 May in Loch Striven, but half were lost due to mechanical failure of the release gear. It was now obvious the target date could not be met, yet agreed tests should continue since the enemy would be unlikely to learn the weapon's secrets from the dams raid. Indeed, German conclusions were that rocket bombs were employed. 618 Squadron received its first pair of modified Mosquitoes on 14 May and had fifteen by the end of the month. Twenty-three stores released at Reculver between 1 and 23 May had wooden cases which broke upon water impact. Detail features of *Highball* all needed testing. Throughout June tests were undertaken at Reculver. On 21st the axis of the rotation was changed under elbowing motion of the store, and various driving wheels and balance weight positions on the weapons were tried a week later. For these tests DZ530, 533 and the special prototype DZ471 were used.

A meeting at Burhill on 7 July was told of excellent recent releases, including double drops. Steel covering the 32-inch-diameter store was now $\frac{1}{8}$ inch thick. All trials were again to be at Reculver. New cases were proven, and attention was next devoted to sighting.

On 17 August an entirely unexpected decision was announced to 618 Squadron. To give it a chance to operate, the stores were to be used as depth charges released from Mosquitoes! Thirty-five trials were hurriedly made between 24 and 26 August in Sinclair Bay, and nine Mosquitoes loaded for anti-submarine operations with live weapons. Almost as suddenly as it came, the order was withdrawn. The end for 618 seemed near, and on 13 September it was ordered to despatch its Mosquitoes for storage, mainly at Shawbury.

Ironically, tests had reached a new fever of excitement when, on 7 and 8 September, ten stores were released against the 'Corpet' in Loch Striven, some behaving exactly as intended. It was at this time that another use for *Highball*, foreshadowing the operation of 8 Group in 1945, envisaged their being rolled into a tunnel, experimentally near Haverford West. Sqn. Ldr. Longbottom was the demonstrator and 4/12th of his stores ran the length of the tunnel. To show their value two were blown up in a reinforced tunnel in North Wales.

Consideration of the entire programme was now undertaken and because of the perfection of the weapon it was decided to reassemble 618 Squadron at Wick to re-train. An order for 250 steel-covered weapons was placed with Vickers on 12 November, 1943, and fifty old ones were allotted for six squadron aircraft to train with from Turnberry. Attention was devoted to sighting, a ring aperture being designed to fit on to the pilot's helmet. Hand sights and a fixed one for the navigator also were developed but, ultimately, good height judgement was the paramount requirement. Absence of operational flying was affecting squadron

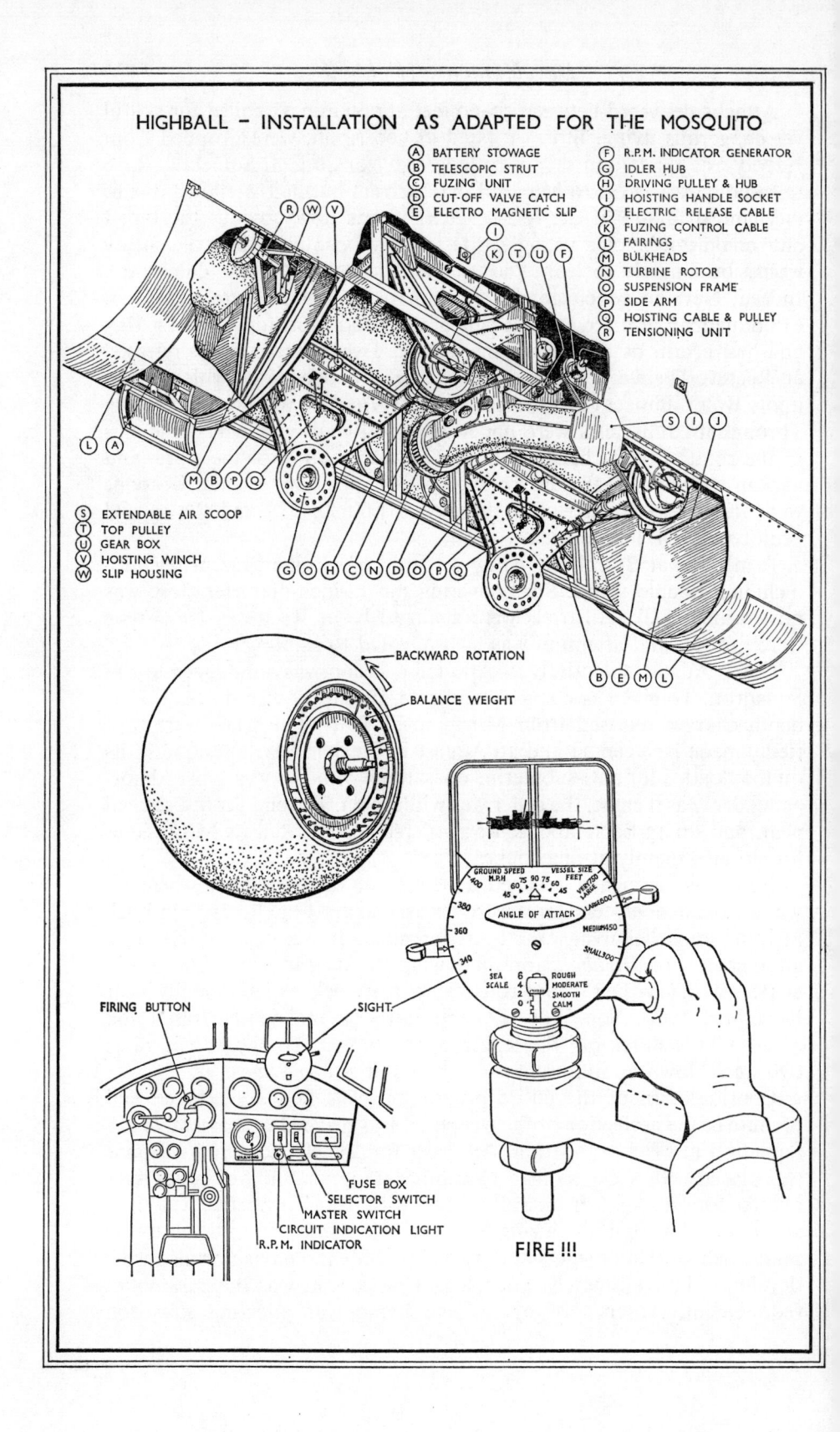

HIGHBALL – INSTALLATION AS ADAPTED FOR THE MOSQUITO

- (A) BATTERY STOWAGE
- (B) TELESCOPIC STRUT
- (C) FUZING UNIT
- (D) CUT-OFF VALVE CATCH
- (E) ELECTRO MAGNETIC SLIP
- (F) R.P.M. INDICATOR GENERATOR
- (G) IDLER HUB
- (H) DRIVING PULLEY & HUB
- (I) HOISTING HANDLE SOCKET
- (J) ELECTRIC RELEASE CABLE
- (K) FUZING CONTROL CABLE
- (L) FAIRINGS
- (M) BULKHEADS
- (N) TURBINE ROTOR
- (O) SUSPENSION FRAME
- (P) SIDE ARM
- (Q) HOISTING CABLE & PULLEY
- (R) TENSIONING UNIT
- (S) EXTENDABLE AIR SCOOP
- (T) TOP PULLEY
- (U) GEAR BOX
- (V) HOISTING WINCH
- (W) SLIP HOUSING

BACKWARD ROTATION
BALANCE WEIGHT

GROUND SPEED M.P.H.
VESSEL SIZE FEET
ANGLE OF ATTACK
SEA SCALE — ROUGH / MODERATE / SMOOTH / CALM

FIRING BUTTON
SIGHT
FUSE BOX
SELECTOR SWITCH
MASTER SWITCH
CIRCUIT INDICATION LIGHT
R.P.M. INDICATOR

FIRE !!!

Development: Highball

morale, so detachments were made to 248 Squadron where 618's crews flew the early sorties of the Tsetse Mosquitoes, and Sqn. Ldr. Rose was killed.

Manufacture of *Highball* ran into difficulties early in 1944 as firms became overloaded with priority needs. May saw the development of air-turbine gear, and release of 54 weapons during sight trials. On one occasion Sqn. Ldr. Longbottom managed to hole the target ship as Royal Navy officers watched; another time a weapon bounced into a side opening and spun around inside, causing utter confusion. Merlin 25s were an obvious requirement for the remaining eleven modified Mosquito IVs. Once more the whole situation was entering the melting pot. By this time the American TBF Avenger and B-25 Mitchell had been adapted to carry the weapon, and plans existed for B-26 Marauders to carry four, portraying American interest in the weapon.

On 9 July, 1944, an exciting decision was made, to reconstruct 618 Squadron at Wick for anti-shipping operations in the Pacific area under the code name *Oxtail*. Airspeed and Marshalls of Cambridge would now fit Merlin 25s, armour plating, air turbines, motive power for stores spinning, new wind screens and arrester hooks allowing carrier-borne operations. Ten Mosquito VIs arrived for crew training. Deck landings were practised using Barracudas at Crail before landings were made on H.M.S. *Rajah* off Rothesay. Six Barracudas were damaged and one fell into the sea off Troon.

To Beccles the squadron moved in August, whilst its Mk. IVs were being modified. DZ537 flew to Renfrew for loading trials on H.M.S. *Implacable*. On 7th Wg. Cdr. Hutchinson attacked H.M.S. *Malaya* in Loch Striven; on 15th the first of 68 drops were made at Wells-next-the-Sea, where at low water the stores could easily be recovered. Beccles was marked out for deck landing training, and a ship was painted on the runway. As a Mosquito made a dummy attack it dropped a sashlite bulb. Its run and height were then checked. After finally being fitted out at Weybridge the 24 Mosquito IVs[1] returned to the squadron in September. Three PR. XVIs NS729, 732, 735 were added to squadron strength to search for Japanese shipping. PZ276, a Mk. VI, was the subject of sealing experiments for embalming the aircraft for the journey East. On 2 October the squadron was ordered to Turnberry for live weapon training, completed by the 13th. It had been decided to deck land the Mosquitoes on the carrier at Scapa, but the accident risk was responsible for its cancellation. Nevertheless Flt. Lt. T. M. Clutterbuck, D.F.C., Flt. Lt. D. G. Rochford and Flg. Off. E. G. Bull made very successful landings and take-offs on the carrier in PR. XVIs on 10 October.

At Beccles a full-scale practice attack on Scapa was meanwhile mooted, but prudence prevailed! The carriers *Fencer* and *Striker* received 24 Mosquito IVs at George V Dock, Glasgow, in October and the 3 P.R. machines. During the morning of 31 October, 1944, the

[1] See Appendix 17.

Development: Highball

carriers sailed, reaching Melbourne on 23 December. The squadron made for Narromine and further training, whilst the ships took the stores to Sydney for safe keeping at the Royal Australian Navy Arms Depot, Auburn. After servicing at Fishermen's Bend the Mosquitoes stood by for action. Waiting . . . waiting . . . through March, April . . . whilst no clear arrangements for integrating the force into the campaign at sea could be arrived at among the Allies. In April the P.R. Mosquitoes flew some long trips, but training constituted the main employment, whilst tempting targets floated by. Then Admiral Portal of the British Pacific Fleet told the squadron on 29 June what it feared, that it was to disband. To all ranks the event was a decided shock, a bitter disappointment. What this meant to those who perfected *Highball* can but be imagined. By an order of 10 July, 125 *Highballs* were statically destroyed for security reasons, rattling the windows of Sydney and causing consternation amongst the residents. To 618, close sister of the famed 617 Squadron, the Vice-Chief of Air Staff sent a message of condolence.

This was not quite the end of the *Highball* story, for, unused, it remained a valuable and effective anti-shipping weapon. Its Mk. II form, designed to fit into an easily removable crate for the Sea Mosquito 33 and the Sea Hornet was the subject of experiments in 1945 initially on a Mosquito VI, PZ281, equipped also to carry bombs and rockets making it an extremely potent weapon. *Highball* was curiously ill-fated, for it was never used mainly because of its potential if ever the idea were copied by the enemy.

Book Three

PRODUCTION

Chapter 8
Production: Britain

F. T. Hearle headed the de Havilland administration right through to August, 1944, then handed over to W. E. Nixon. Both men bore great responsibilities.

6,710 Mosquitoes were built during the war. Of these 3,054 came from de Havilland's Hatfield factory, at an average of two Mosquitoes a day for nearly four years. De Havilland's Leavesden shadow factory—built and run by the company with government money—produced 1,390, Standard Motors 916, Percival Aircraft 198, Airspeed 12, de Havilland Canada 1,032, de Havilland Australia 108.

The yearly totals were:—

	1941	1942	1943	1944	1945	Total
de Havilland, Hatfield	20	389	806	1,202	637	3,054
de Havilland, Leavesden	—	54	379	586	371	1,390
Standard Motors, Coventry	—	—	42	470	404	916
Percival Aircraft, Luton	—	—	—	49	149	198
Airspeed, Portsmouth	—	—	—	—	12	12
de Havilland, Toronto	—	—	90	419	523	1,032
de Havilland, Sydney	—	—	1	27	80	108
YEARLY TOTALS	20	443	1,318	2,753	2,176	6,710

Mosquito production in U.S.A. was seriously contemplated, but the four-way division of effort—early enough—would have been unwise. Also, the Mosquito was not American.

When large numbers of Mosquitoes were in combat the repaired output became significant. De Havilland alone, at Hatfield and in war theatres, repaired 1,252, a quantity equivalent to one-sixth of the entire wartime manufacture at home and overseas. One in every four Mosquitoes going into operations 1941 to 1945 came from a repair depot; de Havilland shops did major modifications to 1,131 Mosquitoes. Every Mosquito unit in the R.A.F., Commonwealth or United States forces, anywhere from Canada through Europe to the Pacific, had its de Havilland and Rolls-Royce technical representative, attached or travelling, the D.H. men directed by A. J. Brant and supervised by Hereward de Havilland, and they modified 3,847 Mosquitoes at operating bases. Martin Hearn repaired 258 aircraft, and 43 Group, R.A.F., repaired 564.

Production: Britain

De Havilland factories, under John Brodie and Eric S. Moult, also overhauled 9,022 Merlin engines; this little contribution alone sufficient to enable 4,500 Mosquitoes to take-off, with confidence.

De Havilland manufactured more than 12,000 Mosquito propellers, indeed the major share (102,245) of all the propellers of all types of aircraft used by the R.A.F. throughout the war, and made 2,006 of these in Sydney. John J. Parkes bore the main responsibility for this production, and sent Ian Spittle to generate it in Australia.

De Havilland overhauled and repaired 40,708 propellers, nearly half as many as were built, such was the rate of battle damage. Charles Burgess organized this operation. The ductile duralumin blade straightened well, time and again, and often went back into the fray with a couple of polished-up bullet holes (if holes can be polished).

In 1941, with Northern Aluminium Company and the Ministry, the company had to set up the de Havilland Forge, so as to have enough propeller blade capacity; it forged 67,178 blades, fifty per cent more than all the Mosquitoes ever needed—but in other shapes and sizes also—a wartime output of 2,987 tons at the pre-machined weight. Tom Wallace, in charge here, turned out in one month 5,531 blade forgings.

The growth in the de Havilland payroll was nothing like might be expected, because an enormous proportion of Mosquito work was subcontracted. The aggregate of home and overseas establishments, including the shadow factories at Lostock, near Bolton, and Leavesden, near Watford, rose from 10,515 in 1939 to 38,311 in 1944. It was 37,105 in 1945 and quickly dropped to about 10,000.

The Mosquito, the propellers and Merlin repairs accounted for most of the activity, but it must be remembered that de Havilland also built 3,744 Tiger Moths and 110 radio-controlled pilotless Queen Bees, 200 Dominies, 87 Dragons (in Sydney), 375 Ansons (in Toronto) and 1,440 Oxfords. Stag Lane, Edgware, built 10,212 Gipsy engines, and a further 1,300 Gipsy Majors were made by General Motors—Holdens in Australia. The repair organization at Hatfield and Witney turned out 905 overhauled D.H. aircraft as well as the 1,252 Mosquitoes, not to mention 470 Hurricanes and 335 Spitfires, a total of 2,962 aircraft repaired.

Considering all this even the cash turnover was moderate. For the home companies not including the shadow factories it rose from £5,000,000 in 1939 to nearly £25,000,000 in 1944.

Subcontracting on a huge scale, like the dispersal of departments into requisitioned premises away from Hatfield, Witney, Stag Lane and Bolton, reduced the bombing risk. It was necessary also for providing the sheer capacity needed when the Mosquito became really wanted. From hard-earned experience at the end of the war in 1918–19, also, the de Havilland team knew the folly of building up their own company and having to discharge thousands of people when peace was declared. W. E. Nixon planned instead to release subcontractors when they would be eager to return to peace-time trade. Thus F. E. N. St. Barbe, when

Production: Britain

relieved of Mosquito and other war business, would be free to revive world sales.

But in 1940 de Havilland were warned not to tool elaborately, because the Mosquito was no more than an interesting development, and even if it went on many changes would be needed and several variants. How right.

Despite needing to minimize original tooling Hatfield found themselves in July, 1940, having to promise the impossible to keep the project alive. Fifty Mosquitoes by December, 1941! Why, the prototype couldn't be airborne much before Christmas, 1940!

In 1941 there was a tendency internally to criticise the management for having set such a target, since non-fulfilment reflected on the company's reputation. One Friday, 28 November, Larrard from the Directorate of Aircraft Production had been pointing out to Murray that he was behind programme. The same evening, after a thirteen-hour day, Murray went over to the old London Aeroplane Club, which had become the E.F.T.S. pupil pilots' mess, where the supervision staff were having a frugal party. He decided to tell these chaps why it was he had had no alternative but to make the rash promise, and then to maintain the pretence through the subsequent seventeen months. Even so, he said, had the first 50 aircraft been kept as bombers—instead of altered to 30 fighters and 20 reconnaissance—they would have come close to delivering them all by the end of 1941.

That day he had given Larrard a private programme for passing on to the A.O.C.s of the Commands to help them in forming their squadrons in 1942.

De Havilland actually completed twenty Mosquitoes in 1941, and fifty by mid-March, 1942—$2\frac{1}{2}$ months late on a wild target set just before the Battle of Britain. The first twenty comprised the bomber and PR. prototypes W4050 and 4051, nine PR. Mk. 1s (short nacelles) W4054–56 and 58–63, and eight B. Mk. IVs W4057, 64–68, 70–71, also the first production F. Mk. II (single control) W4076. This was the third fighter, for the first and second W4053 and 73 were finished off with turrets. W4054 was signed for on 22 July, W4050 and 51 on 31 July, W4055 (first operational Mosquito) on 7 August. The first production bomber W4057 went on 23 September. Leavesden's first aircraft, W4075, first dual-control fighter or T. Mk. III, hopped across to Hatfield for test in the last week of January, 1942.

After taking their fiftieth aircraft in mid-March, 1942, the R.A.F. had fifty more in the next two months, the 101st by 30 May. They had the 408th by 28 November (Hatfield 366, Leavesden 42), and the 1,000th in June, 1943 (Hatfield 775, Leavesden 198 by 21 June). Thus the production rate was one a week in late 1941, three a week in early 1942, eight a week by midsummer, twelve to fourteen a week by the end of 1942, and double that a year later from all sources.

The first hope of quantity production and more sophisticated tooling came at the display of Sunday, 29 December, 1940, and next morning Hatfield were told that 150 more would be ordered at once, and capacity

Production: Britain

was to be sought with a view to 'further extensive orders'. The headache was that the division between fighters and bombers was being discussed that Monday, even for the first fifty, and by the time this was settled (three prototypes, 19 PR. Mk. I, 28 F. Mk. II) a week later, the bomber wings were too far advanced to convert to fighter type, and there were 45 completed bomber fuselage shells of which 28 noses had to be replaced by fighter noses.

Straightaway the facility of altering wood structures (argued in Munich week, October, 1938) was paying off.

On 17 January Murray was phoning and writing to J. S. Buchanan, Deputy Director General of Production, about a target to make the first fifty by 19 July. They still pressed for five a week by then. The same day Murray was begging for the new contract, to be able to order up materials for 150 more, and a great struggle was being made to tell him at least the contract number, so that he could quote that to suppliers as some sort of authority for ordering. Otherwise Murray foresaw a gap in production after the first fifty. There was at that date no indication how these 150 were to be divided between bomber/PR. and fighter. But Murray was striving to make feasible a target to produce them all by 6 December, rising to 8 a week by then. By 1 February it was said that the contract for 150 would be No. 555. Hennessey was a realist. He was driving de Havilland no more than he was the other manufacturers. He arranged for the Director of Sub-contracts to help find suitable capacity in various industries, to assist the Hatfield staff, which C. G. Long was quickly expanding.

This was a time when a lot of fairly young men who wanted to fight were held back for recruiting and teaching new employees. Those in the production engineering and control sections had indispensable experience. Less technical men—in the business departments—were valuable too, simply because they knew the ropes and could go right to it and get things done. There was a de Havilland code, or cameraderie, in those days, which could accomplish anything, and the new recruits were just swept up and carried along while they caught the idea.

The Director of Contracts, Harrogate, on 11 March confirmed that they wanted an output of 35 Mosquitoes a month by September, and all 197 delivered by December, 1941! This was in addition to 70 per month Oxford Mks. I and II, and 10 per month Dominies. The production of 120 per month Tiger Moths had fortunately passed to Morris Motors.

On 21 April Contracts Department wrote to say they had decided to order a further 50 fighters, beyond the 28 against Contract 69990 and the 150 (all now to be fighters) against Contract 555. There were still the 19 PR.s against Contract 69990 and the three main prototypes—altogether orders for 250, but not yet any bombers beyond the prototype. The 50 more fighters were to be added on Contract 555, and the paper followed dated 19 May.

This straight run on fighters and the sanguine expectation of large

Production: Britain

quantity orders made it easier to plan subcontracting, and the dispersal of Hatfield's departments against the bombing risk. It also helped settle the form of the Second Aircraft Group, a completely duplicated shadow organization, which had been in preparation since mid-1940—before it could even be guessed whether it would be needed to make Mosquitoes.

In fact de Havilland had been approached to take on Albemarles, and that had been switched to Wellingtons, to rise to 300 bombers a month! The company had to find an airfield, design and build a factory, get dispersal premises requisitioned, and to staff an entirely separate production organization—as had been done for propellers at Lostock in 1937. In taking on this task in 1940 there was the hope that Wellingtons would never be built, but Mosquitoes instead, and so it came to pass.

The airfield selected had been Leavesden, and by 16 August, 1940, the building there was taking shape. Also two dispersals had been opened up, the old Alliance factory (of the 1914–18 war) on Western Avenue, Acton, where 171 people were then at work, and the Aldenham bus depot, where the first ten men had moved in. At that date there were 4,738 people at Hatfield and Salisbury Hall and 86 on Hurricane and other repairs at Witney; Stag Lane and two dispersals had 3,482 on engines and propellers, and Bolton 2,224 on propellers. With 182 people at the other E.F.T.S., at White Waltham the de Havilland pay-roll with shadow factories had 10,893 names.

In May, 1941, when the Second Aircraft Group were doing Oxford components, Mosquito wings at Western Avenue and fuselages at Aldenham, it seemed clear that this could be built up as a large separate Mosquito organization. It could be devoted to one type, probably fighters.

Quite apart from that, the basic Hatfield organization had gone some way with departmental dispersal, responding to M.A.P. word at a meeting 23 December, 1940, that 60 per cent of production would have to leave Hatfield. By 10 May, 1941, fourteen outside premises had been taken over, all within a few miles, yielding 154,180 sq. ft., and 1,600 people had moved out to them. That left 3,548 at Hatfield (day and night shifts) plus No. 1 E.F.T.S.

By May, 1941, subcontracting was in full swing. Among the first firms, introduced in 1940, were the furniture makers of High Wycombe, E. Gomme Ltd., and Dancer and Hearne Ltd., making spars, wings, jigs, etc. Van den Plas, Hendon, were on wing coverings. H. J. Mulliner, of Chiswick were on details, Bell Punch were doing control columns, Pollards took on wing noses, Lostock handled machining.

The old 94 shop, bombed on 3 October, 1940, had been partly rebuilt, and was quickly turned into a supplementary Mosquito assembly line—until it should become needed for the Mosquito Repair Organization.

On 10 May, eight PR. wings from Western Avenue were at Hatfield, six joined to fuselages. Six fuselage shells had been converted to fighter.

Production: Britain

The third fighter wing was leaving the jig. A trouble with defective plywood, a glueing problem and a new metal aileron had been settled, and only the sheet-metal cowlings were slow. Harry Povey, chief production engineer, was fixing with A.I.D. the interchangeability standards, applicable from the seventh PR. aircraft.

12 May was the day when the company brought in an engineer accustomed to automobile quantities. Ernest Grinham, a senior director of Standard Motors, was generously made available to take control of D.H. production. Povey was able to remain in England as chief production engineer for a few more weeks before needed in Canada. Jock Allardyce as works superintendent upheld the discipline and spirit of the factory. G. K. Carlson, as chief inspector, continued responsible to Murray.

Lord Beaverbrook handed over as Minister of Aircraft Production at the beginning of May to Colonel Moore-Brabazon. He, eight days later on 10 May, lunched at Hatfield and sized up the task. With him came Sir Henry Tizard, T.C. Westbrook and others. At lunch were four of the A.T.A. women pilots. These ladies had made Hatfield their base and lent an air of charm to the place, as they modestly flew everything from fighters to the biggest bombers. The four at lunch were Miss Pauline Gower, Mrs. Winifred Crossley, Mrs. Alan Butler and Mrs. Farrer. Miss Gower by then had about 40 lady pilots.

F. T. Hearle took stock of the planning with Murray and Grinham, difficult as it was with no real idea of the quantities to be needed a year ahead, and no order whatever for the bomber, though surely it was going to be needed. Now he was told that the 'ultimate standard fighter' was to have dual control. Ten single-control fighters were to be converted (September–October delivery) and 40 single-control fighters were to be completed as such. With the utmost optimism as regards freedom from bomb damage and further design changes, Hearle stated on 16 May that they might accomplish 140 aircraft by December, 1941, and an output then of five weekly from Hatfield, two weekly (all dual-control fighters) from Leavesden. But, candidly, these figures seemed unrealistic, all but impossible. Perhaps their main virtue, and object, was to urge everyone concerned.

Sir Charles Craven became Controller-General at the Ministry of Aircraft Production on 21 June, also a member of the Aircraft Supply Council, and Air Marshal F. J. Linnell took over as Controller of Research and Development.

Within a week Hearle was told by telephone that orders would be increased to about 500 Mosquitoes, and with an output probably 50 a month, but no news of bombers.

On 10 July it was put in writing that an output of 150 Mosquitoes a month was needed, 80 from Hatfield, 30 from Second Aircraft Group, 40 from Canada (in addition to any Canadian Government orders for the R.C.A.F.). Any easement on the total requirement should be taken

Production: Britain

mainly off the Hatfield output, and anyway production of Oxfords was to be reduced, from August.

Meanwhile on 21 June the contracts had been tidied up thus:—
Contract 69990 of 1.3.40 to cover bomber prototype W4050, PR. prototype W4051, and 19 PR.s W4054–72 (Total 21).
Contract 135522 of 16.11.40 to cover fighter prototypes W4052, W4053, W4073, and 26 Fs, W4074–99 (two of these to have dual control and Mk. IV A.I.) (Total 29).
Contract 555 of 9.2.41. to cover 45 Fs, DD600–644; 33 Fs, DD659–691; 48 Fs, DD712–759, and 24 Fs DD777–800 (total 150 Fs), also 50 aircraft of variant yet to be decided DK284–303, 308–333, 336–339. (Total 200).

A cable to P. C. Garratt in charge of the Canadian company, at Downsview, Ontario, on 5 July brought an eager reply, plans were afoot, and Murray flew to Canada 30 July.

Drawings were sent to U.S.A. for possible manufacture there, which did not come about.

Bombers figured eventually in the week ended 19 July, 1941, when the last 10 of the PR. aircraft to Contract 69990 had to be converted to unarmed bombers, and the last 50 of the 200 on Contract 555 likewise—but on these it was 'hoped' to carry some fixed forward armament.

The whole plan had to be revised again, and any prospect of delivering even 50 aircraft in the remaining 23 weeks to end of year disappeared. On 2 August, C. G. Long was begging for instructions to permit him to relieve subcontractors of 400 Oxfords and an Oxford spares commitment, and for contract cover for bigger quantities of Mosquitoes, as material suppliers could not act without this. Contracts Department were able on 5 August to provide a letter confirming the need for 150 a month (including any that Canada could contribute) and promising a further contract for 400, making 650.

That morning Sharp was at the Ministry pressing for contracts for these 400, and 1,000 more to prevent a gap in production twelve months later, since the programme would use up 650 aircraft in that time and neither material suppliers nor subcontractors (pressed with other top-priority commitments) could favour de Havilland without authority.

On 8 August the Ministry required the programme to be changed from 150 per month to 160, and possibly 200. Other manufacturers would need to be brought onto Mosquitoes. Bomber output, beyond the 10 conversions, was required to start January, 1942.

Murray arrived back from Toronto 7 September, and accompanied St. Barbe to a meeting called by Sir Charles Craven on 11 September. Murray presented the Canadian plan. He said that Canadian output could start at 2 in September, 1942, but the first 25 (perhaps more) would embody British-supplied equipment, and the conversion to parts and materials made in North America was difficult to estimate in months. Time in transit—and losses—on data, jigs, samples, equipment could only be guessed. New machine tools had to be obtained. Anson

Production: Britain

and Tiger Moth production was to be eased out. Would there be setbacks through changes in mark of aircraft, engines etc., as had happened at Hatfield? A programme to work to had to be set, and the aim was 50 aircraft monthly by May, 1943.

Sir Charles then asked whether de Havilland could increase production to reach, with or without Canadian contribution, 200 a month—and with emphasis on the bomber. It had been made clear that 20 per cent extra production, equivalent to making a total of 240 per month, would be needed for spares. Murray recalled that bombers had only appeared in the programme on 19 July, and went away to tackle a bigger task than he had ever faced.

The organization was stressed to the limit in the subsequent months. Engineers travelling to and from subcontractors and dispersals on programme changes had a rough time, especially at night. Hatfield workers were averaging 52 to 53 hours a week and were asked to make 55 hours. Grinham had driven for 60 hours. Executives worked a twelve-hour day, often longer, taking a half-day off only when really necessary. Housing and transport were becoming a bottleneck. But with the situation changing on the war fronts, as it was on the test beds, no stability of programme could be hoped for.

A new Mosquito programme on 17 September allowed the English fighter output to level off at 90 a month late in 1942 while bombers must build up to 110 a month, including planned Canadian production—plus spares, 15 per cent.

Tiger Moth production having gone to Morris Motors late in 1940, Dominies occupied only a corner at Hatfield. Oxfords were being cut from 80 a month in August, 1941, to finish with 30 in December—by a helpful Ministry instruction on 12 August.

A letter on 28 January, 1942, increased the orders and confirmed them thus:—

Contracts B69990, B135522 & 555 Hatfield	Contract 1576 S.A.G. Leavesden
3 Prototypes	96 F. Mk. II
10 PR. Mk. I	20 T. Mk. III
60 B. Mk. IV	334 FB. Mk. VI
420 Mk. V B or PR.	
110 F. Mk. II	
325 FB. Mk. VI	
928	450

Additionally there were the 400 aircraft, mark undefined, on de Havilland Canada, against Contract 1576. Contracts totalled 1,778.

The disruption caused by programme changes was explained to Mr. Winston Churchill, who, on returning from U.S.A., had expressed concern at the shortage of Mosquito fighters.

The second assembly line at Hatfield came into operation at the end of February, for bombers, and there were three 'fuse to wing' by 7 March.

In February and March, 1942, arrangements were agreed for de Havilland Australia to take on Mosquito production. This involved another task of liaison, and of shipping quantities of parts, all through 1942 and onwards. Fortunately Australians, like Canadians, were resourceful.

On Sunday 12 April, Hearle at a London meeting was asked how much he could shift effort from fighters to bombers without an overall loss of output, and Grinham did what he dared in the way of moving labour and putting a priority on materials for bombers.

An addition of 202 aircraft on Hatfield (total now 1,130) was notified by letter on 4 May, 1942, making in all 1,980 on order. Then 59 were added to S.A.G. contract, making it 509 and the total (16 May) 2,039. Canada were advised to expect a contract for a further 1,100 to make 3,139. On 28 May this quantity was augmented by 350 Mk. VI on Hatfield and 360 on S.A.G., total 3,849. And Grinham was still struggling to achieve double figures per week off the Hatfield line! He made 33 in June; Leavesden and Hatfield together had built 145.

On 11 July a new M.A.P. programme dropped the F. Mk. II at November, and changed over from the B. Mk. IV to the FB. Mk. VI between December and April, 1943, then adding the B. Mk. IX and PR. Mk. VIII. There should be enough Merlin 61s by April to accomplish this, and enough by the end of 1943 to change over the F. and FB. lines to NF. Mk. X and FB. Mk. XI. The introduction of Merlin 61s inevitably would reduce the output by 30 monthly, to recover in 6 months, and all through September D.A.P. tried to assess whether the Commands could accept the set-back.

Meanwhile at Leavesden, with the new A.I. Mk. VIII radar coming in, an effort would be made to introduce it at the same time as the Merlin 61.

In this situation, puzzling even to read about, Hearle and colleagues sought a more specialized organizer, and were told that T. C. L. Westbrook, a demon at production engineering and trouble shooting, might be spared for a while from the Ministry. It says something for him that he was willing to come. He did come, on 4 August, co-ordinating Grinham's efforts at Hatfield and Roy Walker's at Leavesden. Anyone could have refused to keep switching the programme: the result would have been many more aircraft, a lot of them less effective. Also it must be made plain that not every Ministry change was purely to improve the aircraft or even to arm the neediest squadrons; some were essentially to *help* production.

As with the progress of the war, the opening of 1943 saw an improving situation in Mosquito production, and Westbrook had to marshal the burgeoning outflow from previous preparations. The struggle that Long's men had had to find subcontractors, and to inspire them in the face of repeated design changes, was beginning to pay off in the need for more

MOSQUITO BOMBER / P.R. / TRAINER DELIVERIES 1941 - DEC. 1945

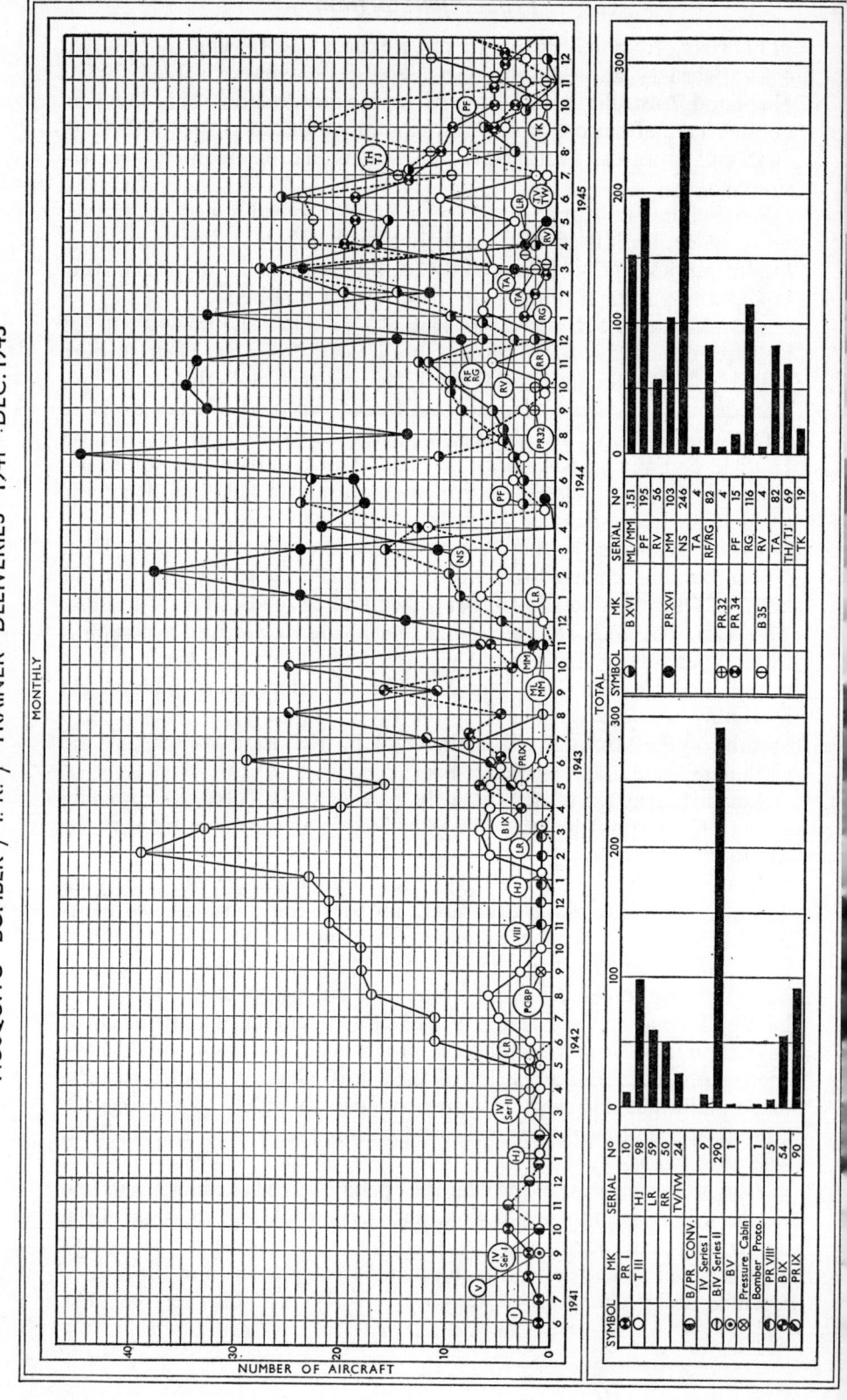

Production: Britain

and more space to house their output near the assembly lines and in the R.A.F. spares depots.

The Mosquito, it can be said, was unique in being manufactured, in the main, not by established concerns of the traditional aircraft industry, but by hundreds of companies, large and small, most of which had no previous experience of aircraft. They ranged from ecclesiastical ironwork craftsmen through the bicycle trade and the coachbuilders and furniture makers to garden-shed groups of patriotic housewives. The Mosquito was designed to be simple enough to utilize such ordinary resources for most of the work other than the engines and specialized components. It meant, however, that these concerns had each to be educated, quickly yet painstakingly, to unaccustomed techniques, disciplined to precision tolerances and strict inspection; and the whole heterogeneous organization, spread widely over the country, drew its supplies and returned its products through de Havilland, who were responsible to assemble them into the fastest, most versatile aircraft operating in the war.

Soon there were four hundred firms working on Mosquito components, and (as we had to say, in commentator's language, when creating the Mosquito film) 'Nowhere could the enemy bomb out a vital flow'. There were still only 4,698 people in the Hatfield factory when Westbrook came, and 1,984 were dispersed into 23 other places. Leavesden and three dispersals had 2,933 people.

The programme of 2 October, 1942, provided a more gentle introduction of the Merlin 61 from March, 1943, which was thought the best all-round solution, bearing in mind there were now ten marks plus the PR. conversions. The order confirmation of 10 October, shows a new situation, with altered quantities, and a total increased by 35 (at Leavesden).

HATFIELD: 1,480 aeroplanes: Prototypes 3: PR. Mk. I 9: F. Mk. II 360; T. Mk. III 4; B. Mk. IV 300 (includes 26 to be converted to PR.); FB. Mk. VI 360; PR. Mk. VIII 10; B. Mk. IX 150; PR. Mk. IX 250; FB. Mk. X 34.

LEAVESDEN: 904 aeroplanes: F. Mk. II 230; T. Mk. III 150; NF. Mk. XIII 370; NF. Mk. XIV 154.

Hatfield and Leavesden now had 2,384 on order. With Toronto's 1,500 it made 3,884, Australian Government orders not being included.

The programme for 500 from Standard Motors, after changes, was laid down in December; FB. Mk. VI 200; FB. Mk. X 300; 1 in April, 1943, rising to 35 in December.

St. Barbe had talks with D.G.A.P. about bringing in Airspeed as Mosquito builders, and it now seemed that they would be needed later; Hatfield had a big job to bring in Standards.

On 23 November Colonel Llewellin left the M.A.P. and his place as Minister was taken by Sir Stafford Cripps.

MOSQUITO FIGHTER DELIVERIES 1942 – DEC. 1945

Production: Britain

In February, 1943, Merlin 72 supplies ahead looked better so the FB. Mk. X and NF. Mk. XIV were to be brought in a little sooner. Then on 6 March a further switch on Mk. IXs was made in favour of reconnaissance, now to be B. Mk. IX 25, PR. Mk. IX 375. On 17 March fresh orders were placed on Hatfield: PR. Mk. IX 250; FB. Mk. X 250; and on S.A.G: NF. Mk. XIV 300, adding 800. Some had the feeling that it could be the last order of the war, but Percival were to be brought in, and later Airspeed; and the pressure-cabin versions were yet to be put into big production. Percival's order for 250 B. Mk. IXs, Contract 3047, was placed 4 May, 1943. The first off Standard's line, HP848, was flown that day. They came along with 2 in June, 3 in July, 3 in August, liaison excellent.

With production now buzzing E. G. Grinham asked to return to the motor industry, and all were sorry to see him depart on 15 May, knowing how much, in two years, he and his track assembly system had contributed. Hatfield produced 71 that month, Leavesden 36. Their totals topped a thousand in June.

On 10 May came instruction to finish 30 FB. Mk. VIs with the 6-pounder anti-tank gun, as FB. Mk. XVIII; on 13 July this was cut to 3 aircraft and 27 sets of parts. On 22 July all the 304 FB. Mk. Xs of the Hatfield contracts were re-allocated, to be FB. Mk. VIs making 641 plus the 3 fitted with the 6-pounder. Standard, instead of Hatfield were taking on 300 FB. Mk. Xs as well as 200 FB. Mk. VIs, the huge numbers of which were needed for the build-up of varied and profitable low-level operations.

On Monday afternoon, 23 August, 1943, two Mosquitoes HX849–850 test-flying in cloudy conditions collided between Hatfield and Salisbury Hall. All four men were killed. The pilots were John de Havilland (the third son) and George V. Gibbins, and their observers were G. J. Carter, flight-shed superintendent, and J. H. F. Scrope, an aerodynamicist.

The orders for pressure-cabin PR. Mk. XVIs came in August and on 20 September—amounting to 368 plus 200, from Hatfield. Indeed, the PR. aircraft, IXs and XVIs, and the versatile death-dealing FB. Mk. VIs, were the marks in largest ordering when the war was at its height. The NF. Mk. XIVs were all re-allocated on 22 November to be NF. Mk. 30s, with some NF. Mk. XIXs, and a further order for 350 aircraft was placed on the Leavesden organization. The order book was at its most interesting that day:

Hatfield, 2,548 aircraft; Contracts 69990, 135522, 555: Prototypes 3; PR. Mk. I 9; F. Mk. II 360; T. Mk. III 4; B. Mk. IV 300; FB. Mk. VI 641; FB. Mk. XVIII 3; PR. Mk. VIII 10; B. Mk. IX 25; PR. Mk. IX 625; PR. Mk. XVI 568.

Leavesden S.A.G., 1,554 aircraft; Contract 1576: F. Mk. II 230; T. Mk. III 200; NF. Mk. XIII 150; NF. Mk. XIX 170; NF. Mk. 30 804.

Standard Motors, 500 aircraft: FB. Mk. VI 200; FB. Mk. X 300.

Percival Aircraft, 250 aircraft, all B. Mk. IX.

R.A.F. MOSQUITO DELIVERIES 1946-1950

Production: Britain

The FB. Mk. VI with Merlin 25 went steadily ahead through to early 1946, with an output away above 2,000 aircraft, thus dominating the armed-Mosquito production after the F. Mk. II phase. The other Merlin 25 aeroplane that did well in the period 1944-45 was the NF. Mk. XIX; useful with the Universal Nose, 280 were built. The T.III trainer continued through to 1948.

Not having the pressure cabin, the NF. Mk. 30, of which more than 500 were built in the twelve months from before D-Day to VE-Day, was not difficult as a follow-on to the NF. Mk. XIX, for its high-altitude Merlin 76 installation was accomplished fairly quickly, and 280 NF. Mk. XIXs with Merlin 25 were built concurrently. The higher-powered NF. Mk. 36 had Merlin 113 and up to 27 January, 1947, Leavesden had turned out 162. The NF. Mk. 38, which went into production a year later, was the last Mosquito variant on the shop floor at Chester, at the end of 1950.

The change to pressure cabin, major concern of the Bomber/PR. production engineers in the last phase of the war, was taken in the factories' now vigorous stride, and 640 Mark XVIs had come off the lines by 21 April, 1945, when production ceased. With the extra performance of the Merlin 113/114, offered from the previous Winter, PR. Mk. 34s appeared in January, 1945, and B. Mk. 35s a couple of months later. In those last few months of the war, and the remainder of 1945, 116 of the PR. Mk. 34s were turned out, their speed and range making them invaluable for photographing the Pacific campaign, which might well have gone on for another year. Likewise with the portly bomber (but handsome is as handsome does) 182 B. Mk. 35s were manufactured in 1945, and these were regarded as much needed equipment in the unsettled post-war years. A halt to production was called before the new Chester factory, brought in by de Havilland during 1948 to cope with world-wide rearmament orders, could get on to it. By then well over 7,000 Mosquitoes had been built.

Chapter 9

Production: Canada

Before war was declared Hatfield were thinking that it would be wise to build operational aircraft in Canada, and that the Mosquito was eminently suitable. After Dunkirk it was visualized that the main de Havilland establishment might have to move to Canada.

The Canadian Minister of Munitions and Supply, the Hon. C. D. Howe, was in England in December, 1940, discussing what Canadian industry could best contribute. At Langley on Sunday, 29 December, he saw the Mosquito fly.

Production in U.S.A. also was considered. Packard had arranged in 1940 to manufacture the Merlin, and had two running by 2 August, 1941. When the American Ambassador, Mr. Wynant, with U.S. Service chiefs, watched the Mosquito and eight American types fly at Hatfield Sunday, 20 April, 1941, General Arnold took Mosquito data back with him. In June Curtiss-Wright men spent some time at Hatfield. The estimate of 14,000 man-hours per airframe was given them for building 1,000 Mosquitoes at 140 a month. They returned to the States 27 June. Ten days later the talks on building the aircraft in Canada came to a positive decision; the American plan did not. Americans would not want to build a British aircraft in this war. Also, it would not have been easy to spare Hatfield engineers for U.S.A. as well as for Canada.

The Downsview factory, outside Toronto, were finishing a contract for 404 Tiger Moths, and were taking on more (they built 1,520), also 375 Ansons, but these plans could be adjusted.

The British Government would order Canadian Mosquitoes, and perhaps the R.C.A.F. would also. P. C. Garratt (who headed de Havilland Canada from 1936 to 1965) had cabled Hatfield 9 July that Downsview with subcontractors could do 40 Mosquitoes a month—this came usefully at a time when Hatfield were being pressed to work up to 150 a month, with or without Canadian contribution; the stage was set.

Lee Murray flew to Toronto 30 July and spent a month working out plans with Garratt. Back at Sir Charles Craven's London meeting 11 September he outlined the proposals, and they were approved. A target of two aircraft by September, 1942, and 50 monthly by 1943 was set—with full awareness that it depended upon factors which nobody could foretell or control. An order for 400 was to be placed by M.A.P.

Production: Canada

on de Havilland Hatfield, Contract 555, and subcontracted to de Havilland Canada. Twenty-five sets of embodiment-loan and bought-out items were to go from Britain.

Five days after that meeting W. D. Hunter, senior Hatfield designer who had been working with W. A. Tamblin, left for Toronto to take charge of the design there, adapting it to local material specifications and equipment, yet striving to minimize non-interchangeability. With him went Harry Povey, chief production engineer at Hatfield, to organize production at Downsview.

First batches of drawings, etc., had gone off ahead, 3,000 were on the way by the end of October. Micro-negative drawings that Povey took were processed in Toronto in seven days. Fuselage and wing jigs were waiting for shipping space by the end of September, also sample parts. There were bad shipping delays. A long search traced batches of drawings in December to vaults in Montreal, others to Washington. Twenty further cases of parts, castings, forgings, double-curvature members sent in January, 1942, were received in Downsview in March. By the time fuselage jigs arrived Downsview had made their own. The British jigs were forwarded to Australia, but were sunk by the Japanese on the way.

Canada were to build the bomber version, to which emphasis had been directed since July by M.A.P., the purchasers, but the Canadians then asked for fighters. Hatfield asked M.A.P. to switch the last 100 of the 400 to Mk. VI fighter, and sent Mk. VI drawings. This arrangement was cancelled in May, 1942—a Canadian contract was to be issued for fighter-bombers.

April found Hatfield in difficulty to meet all Povey's requests. He needed 25 sets of further parts, hydraulics for example, which he could not get locally so readily as he had hoped; Kelsey Wheels of Windsor had a difficult job. His request for certain ballraces from England was declined by M.A.P., which meant redesigning to take American races.

Povey was grateful to Charles Misfelt for locating a drophammer. He also acquired a 300-ton stretcher press and developed techniques for making plaster moulds and dies, for fettling, lofting and jig making. A large extension of the main assembly shop was quickly erected, also a building for the drophammer, foundry etc. Northern Aluminium gave stalwart help. Tanks were a development task. Povey praised the receptive ability of Canadians. Infra-red fuselage drying, and concrete fuselage jigs (Povey's idea) were developed, which later benefited English production.

Tamblin visited Canada in March, taking twelve days via Brazil, and worked with Hunter from 16 March to 15 April on design adaptations. He took out information on modifications, and on new variants.

Ralph P. Bell, Canadian Director of Aircraft Production, visited the plant at the end of March.

Garratt flew to Prestwick early May, for nine days in London and Hatfield, with news that he was to receive a contract for 1,100, versions

Production: Canada

to be decided, to be financed by U.S.A. under Lend-Lease agreement. This was confirmed in June, making 1,500 on order at Downsview.

Photographs reaching Hatfield 1 August showed the first Canadian bomber with fuselage on wing. Based on the B. Mk. V powered by Packard Merlin 31, it was provisionally called B. Mk. VII, but later B. Mk. XX. Marks XX to XXIX were reserved for the Canadian variants. North American suppliers had been found for all but three or four of the embodiment-loan and proprietary equipment items and would 'come in' later on the line. General Motors, Oshawa, were making fuselages, as were de Havilland at a Dupont Street dispersal. Massey Harris were making wings. Canadian Power Boat made flaps, Boeing Vancouver made tailplanes.

Hunter and Povey worked 21 hours a day trying to get the first aircraft flying within the year of their arrival on 21 September.

The B. Mk. IV DK287 which Hatfield were shipping to Toronto was ready to leave Liverpool 13 September. Burrell, prototype flight-shed engineer at Hatfield, left by ship 7 September to be in charge of it. Having this aircraft in Canada would have been more useful but for delays in shipping it, and its damaged state when unloaded at Halifax.

Burrell accompanied Ralph Spradbrow, chief test pilot at Downsview, on the first flight of the Canadian B. VII KB300 on 24 September. It was a fine flight. Ralph Bell was there and cabled England praising the work of Garratt, Hunter and Povey.

The previous day, after a maddening weather delay, Geoffrey de Havilland left Poole for Baltimore; he arrived in Toronto by bus from Buffalo at 2.30 a.m. Sunday, 27th. He was disappointed to have missed the party, and his suitcase had gone to West Africa. Tuesday with Burrell he flew the Canadian aircraft and cabled that he was very pleased with it. He demonstrated it to senior people two days later, including a roll with one feathered. Thereafter he gave a series of shows up and down the continent, including Wright Field on 29 October, Ottawa (before the Earl of Athlone) on 21 November.

The question of building in U.S.A. arose again, but by now Australia was taking it on, and a further spread of effort might have yielded less output.

Murray, visiting Canada, was held up by the same weather but saw Geoffrey's first flight. Alan S. Butler, de Havilland Chairman, went across 24 October. Murray reported, that week, every reason for confidence in future production, considering those early unavoidable difficulties, but there must be a gap after the 25th aircraft, with the introduction of local supplies. De Havilland had no power to dictate priority for any of these supplies.

With Hatfield's agreement Garratt had now appointed Hunter and Povey to his own company in complete control of the D.O. and the production respectively. Tyler went over from Hatfield to look after Inspection under Don Long.

Production: Canada

Early in October Povey had started asking Hatfield for supplies for the fighter-bomber version and was concentrating much effort upon this, also on a trial installation with the two-stage Merlin. Furthermore, Canada were being asked to produce their own dual-control trainer, and to pass the benefit of experience to de Havilland Sydney.

There was another delay over drawings. Five batches were lost at sea and had to be repeated; then in November batches 174 and 178 went the same way, through enemy action.

Spradbrow, arriving in England 1 December, told how one of his pilots, Fisher, doing the first flight with Canada's second bomber 21 November had two hydraulic pump failures and belly-landed at Malton, (repaired 20 January), and it had nearly happened on the first flight of the first aircraft.

Murray joined de Havilland Canada in November as assistant managing director, eager to work on production control systems. Geoffrey returned to Hatfield by Liberator via Prestwick 4–5 January, 1943. Spradbrow was taken ill in England and Pat Fillingham went across to Canada 4 February, taking Wooll, a P.R.U. Mosquito pilot, who had been attached to Hatfield after being exchanged for Me 109 pilots following a forced landing in Switzerland. George Errington, Airspeed chief test pilot, helped out at Hatfield.

Ralph Bell in January, 1943, pressed for armed-version drawings and sample parts; 98 per cent of drawings had gone, 85 per cent had been received in Toronto, and a fair start had been made in the despatch of parts. Bell, however, wanted things to move still faster.

After long declining, M.A.P. agreed about 20 January to send a sample T. Mk. III aircraft, drawings and some parts to Canada, to build their own dual-control trainers. Fillingham took the control conversion drawings with him. The sample was to be used for training Ferry Command crews for bringing Mosquitoes to Europe, also for converting R.C.A.F. pilots. Late February M.A.P. released a further five T.IIIs to hasten the training. Another five were notified in April. This, it was hoped, would relieve Downsview of the dual-control conversion, for they faced the change-over to the fighter-bomber, F.B. Mk. XXI, also the change to two-stage Merlin. The first 50 bombers were to be retained in Canada for O.T.U. work in the R.C.A.F.

On 22 February Garratt wrote to C.D.M.S. explaining hold-up problems, not that the department could help very much to solve them. On 9 March it appeared that Ralph Bell would like to have the help of T. C. L. Westbrook for a few months on these production complexities, but this was not immediately practicable, and Westbrook was helping from the English end. The problems building up had all been foreseen and were as inevitable as war casualties. Typical was the difficulty in getting supplies of jacks in North America when it coincided with a shortage in England and a need in Australia. Three further lists of parts wanted at Downsview—some to replace losses and damage in transit,

Production: Canada

for the first 25 aircraft—arrived at Hatfield in March, the most urgent of which (29 items) went off on 2 April.

Change of variant, however, was the most disconcerting thing. It happened again by 12 April, when it appeared that the O.T.U. was to take the first 70 aircraft and then about 100 were to be converted for use by U.S.A.A.F. reconnaissance units.

Murray and Hunter flew to England mid-April for discussions, returning early May. At mid-April Downsview had built only 17 Mosquitoes. A Lancaster sample, for helping Lancaster production in Canada, was also to carry Mosquito and Lancaster parts westbound, but hopes that it would save awful delays were not fulfilled, and later the official service became quite good.

In June P. C. Garratt was made a director of the parent company in England, responsible for all de Havilland interests in Canada and U.S.A.; and the Canadian authorities appointed Grant Glassco as Controller of Mosquito Production at Downsview. This might have given a little more weight in matters of industrial priority, but the rise in Mosquito output a few months later was (as it had been in England a year earlier) the outcome of the hard work which had gone before. There was no easy way around engineering problems. An example was the change to formaldehyde cement. No magic could avoid some set-back on this score. An engineer, W. Houston, was sent to Hatfield in July to study the technique. Tego-film plywood was another exacting development introduced from England.

Garratt was at Hatfield in August, discussing the major issues, and by that time the supply position of special parts from England was 'generally satisfactory'. Glassco came across late in September to go into numerous details, and it was then clear that troubles with North American subcontractors were the greatest bottleneck.

The slow start had been expected; it was typical of every aircraft project, in every country. Downsview's performance compared well with Hatfield's.

By the end of 1942 four Canadian-built Mosquitoes were flying. The output month by month through 1943 was 2, 3, 4, 6, 9, 5, 11, 10, 13, 5, 6, 14, making 67 twelve months after the first flight, and 92 by the end of the year. These were (except for two fighter-bombers) all B. Mk. XXs, with Merlin 33 of 14 lb. boost, and the flow improved through the first six months of 1944 to 10, 21, 33, 41, 45, 44; the growing total was 286 at the end of June. Packard started delivering the Merlin 225 of 18 lb. boost, and the 271st aircraft, KB370, that month, started the series B. Mk. XXV. It was accepted 7 July.

Two pre-production FB. Mk. XXI (Canadian equivalent of the FB. Mk. VI) with Merlin 33 were out in September–October, 1943, but production concentrated on bombers, and when the next two fighter-bombers emerged, in October and November, 1944, they, like bomber 271 three months earlier, had Merlin 225 and so were called FB. Mk. XXVI.

Production: Canada

In mid-1943 Downsview were after all required to produce dual-control trainers, which they accomplished with sets of control conversion parts from Hatfield. They engineered it as a variant of their FB. Mk. XXVI, calling it the T. Mk. XXII with Merlin 31 or 33, and the T. Mk. XXVII with Merlin 225. The trainer became urgent because there was a shortage of cannon and radio in Canada, and unarmed fighter-bombers could be finished as trainers rather than stand uncompleted. By March, 1944, 75 sets of most of the trainer (dual-control) parts had been shipped from Hatfield.

After initiating the Canadian pilots in production testing of the Mosquito Fillingham left Toronto in October for Sydney, to repeat the good work there.

Work on the two-stage-supercharger version at Downsview had been aimed at deliveries from mid-1944, but Hereward de Havilland realized that, in face of the demands for low and high flying the Packard Merlin 69 medium-altitude engine looked like being unsuitable. Hatfield raised this to C.R.D., and M.A.P. held a meeting of Sir Wilfrid Freeman, Air Marshal Sorley, Hives of Rolls-Royce and the M.A.P. engine-supply men.

A plan emerged to continue Packard Merlin 225 engines in fighter-bombers until the new 14 SM two-stage engine could be started at Packard. This had a fairly high altitude rating but was slightly more efficient low down than the single-stage 18 lb. engine, although less so than the latest 25 lb. single-stager being developed by Rolls-Royce. Packard were then asked to build an equivalent of the Merlin 76 (by a change of supercharger gear ratio) to give Canadian bombers a satisfactory high-level performance.

The Canadian bomber with two-stage engine would be the B. Mk. XXIII, and the fighter-bomber the FB. Mk. XXIV.

Hereward de Havilland went on 26 February, 1944, to Canada and U.S.A. to discuss the two-stage engine, also the use of B. Mk. XXs by the U.S.A.A.F. for reconnaissance. Harry Povey, his main task accomplished, returned with his family to England on 11 March. Lee Murray visited England 5-12 May, and in March, 1945, moved to the Australian company. Pat Fillingham returned to England from Australia after seventeen months away, delivering a Mosquito bomber KB221 from Canada on his way, by 26 June, 1944. (Arriving over Greenland at midnight, weather forced him back to Goose Bay.)

B. Mk. XXVs were the main production through July-December, 1944, the month-by-month output being 17 (hydraulics shortages), 53, 31, 43, 24 (cowlings troubles in November), 62; the growing total was 516 at the end of the year.

Beyond the 516 accepted there were 87 aircraft parked at Downsview awaiting cowlings, which soon turned up.

T. C. L. Westbrook left England 21 March to advise de Havilland Canada in the large-quantity phase of their project.

Production: Canada

The 1945 programme for Downsview was:

For the R.A.F.	Jan.	Feb.	Mar.	April	May	June 1945
B. Mk. XXV	42	69	28	10	7	0
FB. Mk. XXI	2	2	1	20	28	55
T. Mk. XXVII	0	0	8	10	10	8
For the R.C.A.F.						
B. Mk. XXV	0	0	0	0	0	0
FB. Mk. XXI	1	7	40	34	40	8
T. Mk. XXVII	4	2	4	6	5	4
	49	80	81	80	90	75

The month-by-month deliveries achieved in January–May were, 64, 68, 80, 76, 83, total at 31 May, 887.

This change-over to fighter-bombers, also the two-stage supercharger development, etc., were discussed at Hatfield 24 January, 1945, with Fred Smye of Federal Aircraft, the Agency set up to supervise Canadian production in place of C.D.M.S. He visited squadrons near Huntingdon which were using the Canadian bombers. The boys were pleased with them.

The first two had crossed to Prestwick early in August, 1943, KB162, *New Glasgow*, flown by American civilians, and KB328, *Acton*, flown by R.C.A.F. Transport Command, Flg. Off. J. G. Uren and Flg. Off. R. C. Bevington. They came into Hatfield 12 August, having staged in Greenland and Iceland to check airfields there. Hatfield, Boscombe Down and the Air Ministry approval conference all eyed them critically, and were well satisfied with the Canadian product.

Five B. Mk. XXs were in Britain by November, 1943, being brought up to operational standard. After careful consideration 13 M.U., R.A.F. Henlow, had been selected as the reception base; it was handy for Hatfield, for instance when KB329 needed its electrics rewired. Interchangeability and non-common spares provisioning as between the British and Canadian marks had been gone into that Autumn.

A snag arose: the British would not accept Mosquito XXs with Canadian solid windscreens. Hastily Triplex made up sandwich windscreens for the first Mk. XXs in Britain.

KB161 reached 139 Squadron at Wyton on 11 November, 1943, KB162 on 23rd. Flt. Lt. Salter was the first pilot to fly a Canadian Mk. XX on an operation when on 2 December in KB161: XD-H he bombed Berlin, despite engine trouble and frozen controls. A broken windscreen grounded KB162 until January.[1] KB329 briefly served 627 Squad-

---

[1] KB162 rejoined 139 Squadron 29.5.44 and was damaged in action 20.7.44 raiding Hamburg, was repaired and damaged again a week later on a sortie to Stuttgart. From August, 1944, again it was busily engaged until severe damage on 14 October brought its end. KB161 was written off on 11.5.44 after raiding Hanover and suffering damage.

Production: Canada

ron, then joined the Mk. XXs on 139 Squadron. KB161 had flown 14 sorties by 29 February, 1944, seven of these to Berlin, KB162 ten (four to Berlin), and KB329 two (one to Berlin).

The British Government in October, 1943, had agreed to supply a large number of Mosquitoes (120) to the U.S.A.A.F. for reconnaissance. But Britain needed Canadian bombers, so a readjusted allocation of 90 B. XXs from Canada to the U.S.A., as F-8s, had followed; the balance were to be 30 fighter-bombers from Britain for the 8th Air Force.

Three squadrons in 8 Group—139, 608, 627—and 1655 M.T.U., used Canadian Mosquitoes, and 11 F-8s were now with No. 375 Servicing Squadron, U.S.A.A.F., Watton. Early machines, these, all had been maintained at Rome/Romulus and Hunter Field, and they were all that came from 25 scheduled. Cracked side cowlings, damaged leading edges and control surfaces, oil leakages, all were apparent. They were passed to the R.A.F. for renovation in exchange for PR. XVIs.[1]

R.A.F. personnel expressed satisfaction with Canadian aircraft, but 139 and 627 Squadrons' pathfinder equipment forced the C.G. position aft. For some time 8 Group agitated for their Canadian aircraft to have the 4,000 lb. bomb bay, so that they could carry new-type target indicators, and perhaps 4,000-pounders. Because of the C.G. problem the only way safely to make the modification would have been by fitting two-stage engines. 800 Packard Merlin 68s, originally intended for Spitfires, were located in an M.U. These, it was decided, could be modified into Merlin 69s if they had reversed cooling. They could then fit into Mk. XXV bombers. After October deliberations it was agreed that Marshalls of Cambridge could do conversions, although difficulty with supplies of cowling, radiators, heaters etc. was foreseen. Spares problems with ordinary Canadian aircraft being difficult, modifications could only make things worse.

Henlow meanwhile continued their vital work and by 31 December, 1944, had completed 212 operationally fit bombers. Earl Ferguson, sent over by Burrell, arrived at Hatfield 26 June, 1944, as liaison officer on maintenance and spares problems between the squadrons, the de Havilland Mosquito Repair Organization and Downsview. R. B. Nelson arrived on 26 August to work with White on Canadian modifications.

When all was going well there were losses on Atlantic crossings, via both Gander–Prestwick and Bermuda–Natal–Dakar–St. Mawgan. Four aircraft were missing by January, 1945, of which two probably succumbed to tropical storms. Seventeen Mosquitoes had come across to Prestwick in December, then they were all grounded while investigations were made. By then 230 had come across, and the crossings built up heavily again by February.

Flight trials of Marshall's special conversion began in January, 1945.

[1] Overhauled, these old machines gave good service. KB146 and 189 operated for two months in 1945 with 608 Squadron. KB148 and 156 likewise worked for 139 Squadron. KB185 failed to return from Berlin on 2–3.4.45.

Production: Canada

Cowlings were cut and extra panels inserted. Existing single-stage radiators, cabin heaters and oil coolers were retained, and coolant temperature was reckoned satisfactory. Approval was given on 16 January, but the changing war situation caused the programme to be questioned. It would be many months before these aircraft could be available in quantity. Consideration was given to converting the bombers before leaving Canada. Again, there would be delays. Eventually it was decided to modify a few to have 4,000 lb. bomb doors, accommodating $2 \times 1,000$ lb. target indicators or $2 \times$ Type 7 mines, for 627 Squadron. Engine changes were not proceeded with.

Production in Canada was now mounting, 80 Mosquitoes being built in March, 1945, making the total 728. 103 bombers arrived in Britain that month, also the first FB. Mk. XXVIs came, practically all using the Iceland route. Mk. XXVs were being prepared at Henlow for 614 Squadron in the Middle East. In April 66 B. Mk. XXVs and 9 FB. Mk. XXVIs came across, and in May only 9 bombers and 48 fighter-bombers, making altogether 430 Canadian bombers and 59 fighter-bombers in Europe at 31 May, out of 887 produced at Downsview.

In the squadrons there were now many Canadian Mosquitoes credited with 50 or more sorties. They played a large part in the thirty-one consecutive nights of attacks on Berlin, with 139, 162, 163 and 608 Squadrons. Serviceability reached a high level. Admiralty received some XXVs for special duties, and later used the machines in the Middle East.

As a by-product of Atlantic crossings there were numerous record flights. In May, 1944, Wg. Cdr. John Wooldridge, D.S.O., D.F.C., D.F.M., flew from Labrador to Prestwick coast-to-coast in 5 hours $39\frac{1}{2}$ minutes in a bomber, 6 hours 46 minutes for the runway-to-runway crossing. In October Maurice Gill beat this by one minute. A month later he flew from Goose Bay to Prestwick in 6 hours 8 minutes. By April, 1945, this had been reduced to $5\frac{1}{2}$ hours, at a ground speed of 390 m.p.h. Fast flights over North America included Jim Follett's when, as chief test pilot at Toronto, he flew from there to La Guardia, 377 miles, in 55 minutes averaging 411 m.p.h. He flew from Fort William, Ontario, to Toronto in 83 minutes, a 570-mile journey averaging 412 m.p.h. at 20,000 feet.

KA970 was the subject of a hair-raising accident on 17 April, 1945, when an air bottle in the starboard side of the fuselage exploded. After crossing the Atlantic at about 15,000 feet in $-30°$C. a regulator apparently froze up following a climb to 19,000 feet when avoiding a front. Seventy-five miles from Prestwick, on descending, the pilot heard a loud explosion. His feet were knocked from the rudder pedals, and as the starboard engine ran roughly he feathered the propeller. Despite twenty minutes hand pumping he failed to lower the undercarriage. A belly landing at Prestwick followed, a small fire soon being put out. On examination Nos. 3 and 4 bulkheads were severed for about twelve inches of their circumference, No. 5 was torn in half, the hydraulic oil

Production: Canada

tank smashed, No. 6 bulkhead and rudder lever distorted. A hole about seven by two feet had been blown in the rear starboard fuselage skin, yet the crew escaped unhurt.

357 Mosquitoes had reached the R.A.F. from Canadian sources by the end of May, 1945. Work started at Henlow on preparing fighter-bombers, increasing numbers of which were arriving. But, with the war in Europe ended, there was little need for further deliveries. By VJ-Day there were many Mosquitoes in the flight shed awaiting delivery from Downsview with many more at London, Ontario, where the despatching pool was organized. Production ceased in October at 1,034 aircraft of which 1,032 were completed by 15 August, 1945, VJ-Day.

New employment for the Canadian fighter-bombers arose in 1947. Some 200 taken from storage were shipped to China. At Shanghai an assembly factory was set up. Eddy Jacks was shop superintendent. By 12 November, 1948, when the last fuselage was on the line 179 Mosquitoes had been erected. Communist forces were closing in, and KA440, tested on 18 November, was the last Mosquito flown before the factory shut down. As many as possible were passed to General Chiang Kai Shek's forces in Taipeh, serving until American aircraft replaced them. Fred Plumb, who had been with the Canadian Company, was the Chinese Government's acceptance engineer in charge, and returned then from Taipeh to Toronto.

The Luftwaffe never made a determined attack on the British aircraft industry after 1940. Had it done so Canadian Mosquito output would have been vital. One seventh of Mosquito production took place in Toronto; Canadian Mosquitoes played a major part in the light bomber night offensive. Three squadrons for over a year were frequently the spearhead for such attacks.

Mindful of the immense production of their American neighbours the Canadian staff worked out a figure of merit, military effectiveness per pound of aeroplane produced. This was arrived at by multiplying the weight of bombs by the striking range and figuring per man of crew per tare pound. The ratio was Fortress 24, Liberator 26, Halifax 62, Lancaster 68 and Mosquito 80. The Canadians felt they had backed a winner.

Notes on Canadian-built Mosquitoes

B. Mk. VII   KB300-324 (Merlin 31) delivered Canada only 30.12.42–6.10.43. To U.S.A.A.F.:— KB306 (43-34931), 312 (43-34924), 313 (43-34925), 315 (43-34926), 316 (43-34927), 317 (43-34928).

B. Mk. XX   KB100–179 (Merlin 31), KB180–299 and KB325–369 (Merlin 33). 135 to U.K. 12.8.43–7.4.45. KB113, 119, 196, 216, 220, 230, 296, 340 all crashed on test or delivery. To U.S.A.A.F. as F-8:— KB130, 131, 132* (became 43-34932

Production: Canada

to '934); KB138*, 139, 140*, 141 (43-34936 to '939); KB145 (43-34935), KB146* to 152 (43-34946); KB154 to 159 (43-34948 to '953); KB171; KB180–189* (43-34954 to '963); KB190 (43-34947); KB328.
* returned to R.A.F. in U.K. 9.44 to 2.45, also KB148, 156, 158, 182, 185, 188.

FB. 21   KA100–102 (Merlin 31) delivered 31.10.43, 24.2.44, 29.11.44.

T. 22   KA873–876, KA896–897 (Merlin 31/33) delivered 7.9.44 to 6.6.45.

FB. 24   KA928–929 both cancelled 2.44, also the Mk. 23.

B. XXV   KA930–999, KB370–699 (Merlin 225) delivered 6.7.44 to 1.7.45. 343 to U.K. 70 subsequently handed to Royal Navy. 19 written off in testing and delivery. 4,000 lb. bomb bay conversions: KB409, 416, 490, 561, 625.

T. 27   KA877–895, KA898–927 (Merlin 225).

F.B. 26/T. 29   KA103–773 ordered; KA431–32, 434, 438, 442–43, 445–49, 451–773 all cancelled. Delivered as FB. 26 (Merlin 225) except for these T. 29s:— KA117, 120–22, 137–39, 141, 149–50, 158, 166–67, 172–74, 202–3, 206–7, 221, 232–4, 242–3, 280–1, 290, 297–301, 312–14. KA153, 197, 237, 259, 260, 316, 317 crashed, etc., prior to delivery.
FB. 26s delivered to China, and the Chinese Air Force numbers included:— KA203 (038); 232, 233, 469, 540, 470, 471, 516, 394, 515, 408, 488, 536, 409, 456, 498, 474, 499, 423, 526, 499, 462, 435, 252, 466, 463, 421, 467, 437, 403, 457, 436, 447, 481, 401 (044 to 077 inclusive); 453, 441, 450, 451, 468, 512, 477, 155, 504, 231, 458, 595, 538, 522, 420, 235, 311, 428, 207, 222, 332, 388, 427, 460, 439, 205, 297, 534 (078–105 inclusive); 422, 444, 372, 411, 429, 414, 518, 425, 454, 517, 537, 415, 313, 381 (109–122 inclusive); 384, 443, 539, 400, 421, 528, 483, 432, 486, 371, 434, 523, 242, 521, 501, 376, 472, 505, 386, 352, 485 (124–144 inclusive); 497, 461, 382, 446, 399, 550, 507, 433, 536, 495, 530, 391, 440, 445, 478, 424, 464, 448, 430, 506, 528, 410, 390, 419, 340, 347, 336, 385, 397, 379 (146 to 175 inclusive); 288 (178).

Delivery of Mosquitoes to the Royal Canadian Air Force:—
T. Mk. III   Delivered 23.4.43–21.3.44:—
HJ857, 866, 869, 871, 872, 874, 876, 879, 880, 882, 883, 958, 959, 965, 966, 974, 988, 989, 995, 996, 998, 999, LR533, 536.

Production: Canada

Total deliveries of Mosquitoes to the R.C.A.F.:—

T. Mk. III	24	FB. Mk. 26	191
B. Mk. VII	16	T. Mk. 27	19
B. Mk. 20	94	T. Mk. 29	34
FB. Mk. 21	1	TOTAL	420 aircraft.
T. Mk. 22	3		
B. Mk. 25	38		

Delivery of FB. Mk. 26 to the Royal Air Force, Middle East:—
KA134–136, 140, 151, 152, 154, 156, 160, 161 (249 Sqn.), 162, 164–165, 169, 171, 175, 178, 190, 191, 198, 200, 214, 215, 217, 227, 248 (249 Sqn.), 256, 258, 259, 262, 263, 264, 266, 273, 277, 282, 284, 286, 294 (249 Sqn.), 304, 308 (249 Sqn.), 309, 310, 322 (249 Sqn.), 324, 325, 329, 333, 337, 338, 341, 342, 343, 349, 351, 353, 354, 356 (249 Sqn.), 362, 364, 368, 370, 373, 378 (249 Sqn.), 389, 406, 407, 412, 413, 416, 417 (249 Sqn.).

The only T. Mk. 29s used by squadrons were:—
KA158 (151 Sqn. 13.6.46 to 11.4.47), KA280 (25 Sqn. 21.3.46 to 26.9.46 as ZK:Z), KA290 (264 Sqn. 11.4.46 to 9.7.46), KA117 (85 Sqn. 27.2.46 to 24.10.46), KA119 (29 Sqn. 28.3.46 to 26.9.46 as RO-Z), KA120 (219 Sqn. 16.5.46, to 64 Sqn. 27.6.46 until 26.11.46), KA138 (151 Sqn. 18.4.46 to 16.5.46) and KA139 (65 Sqn. 16.7.46 until 24.4.47).

ACCEPTANCE OF MOSQUITO FROM D.H. CANADA (TORONTO) WORKS BY R.A.F. & R.C.A.F.

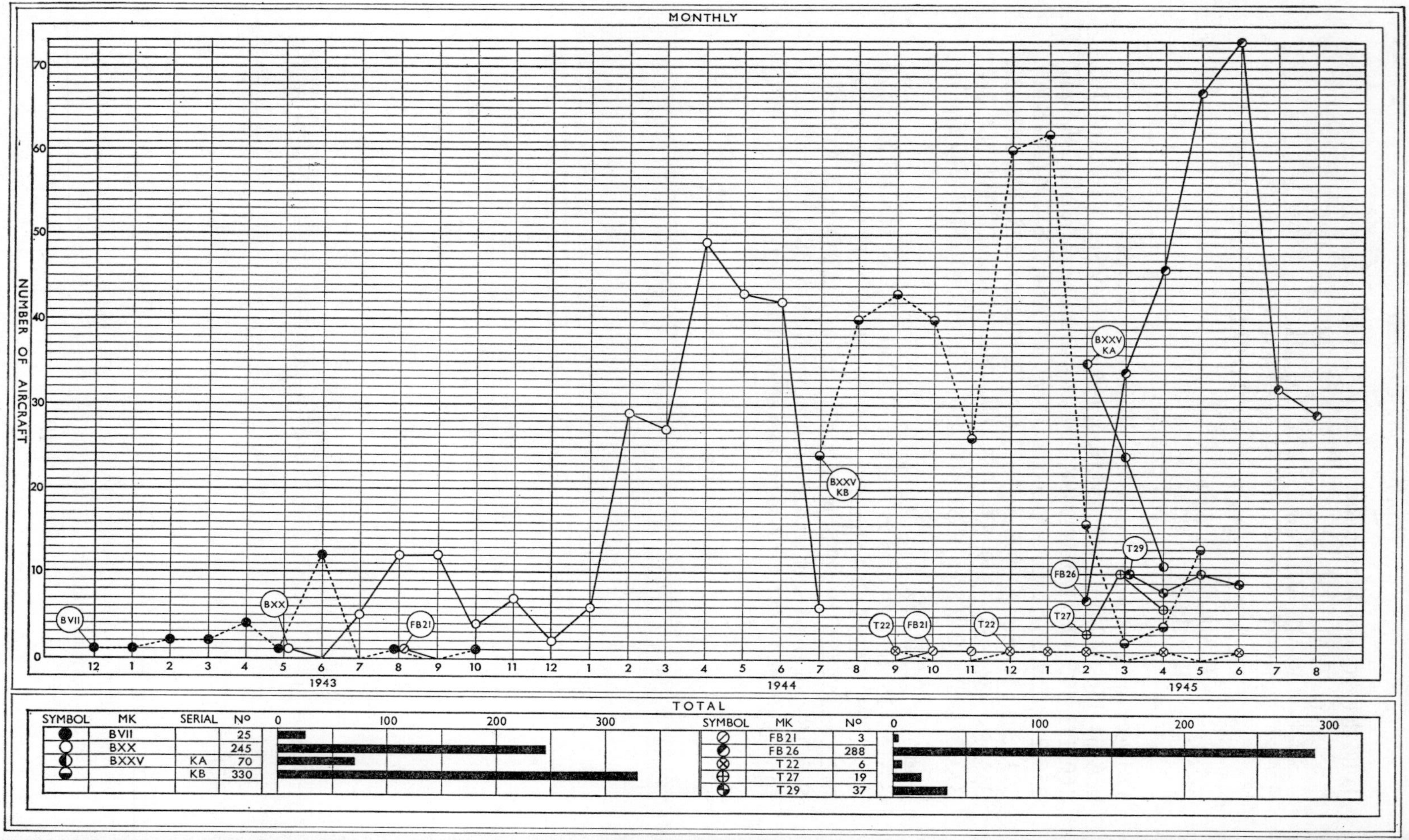

Chapter 10

Production: Australia

Aircraft production in Australia was first a problem of importing across the oceans not only engines but the bulk of proprietary components and, initially, materials. Allan Murray Jones, who from the early 'thirties headed the oldest overseas de Havilland company (it had been formed by Hereward in 1927) was not to be easily put off. In November, 1940, he submitted a report to the Australian authorities weighing the possibilities of building operational aircraft; a year later, when the Packard Merlin project gave hope of engine supplies from America, the Mosquito was seriously considered. Hatfield, hard pressed, were not hopeful at first, and advised 'M.J.' to sound de Havilland Canada for supplies, even for drawings.

In January, 1942, John Byrne and John Mills, supplies director and a senior engineer of the Australian company, went to U.S.A. and Canada. Soon they were cabling Hatfield for 50 sets of embodiment-loan and bought-out components, 3 at once and 47 over several months. Hatfield could supply, M.A.P. assented. Back came a cable from Byrne 26 February: 'Engine supply now satisfactorily arranged and project practically certain'.

Early in March the Australian Government instructed de Havilland Sydney to produce Mosquito fighter-bombers for the R.A.A.F. England were asked for one sample aircraft; an F. Mk. II, DD664, was packed at Sealand in June, and shipped. Ian Spittle, in Britain on propeller matters, pleaded that Hatfield, rather than Downsview, provide the drawings, which was now possible. In fact, from early March, 1942, Hatfield was able to co-operate eagerly. Since de Havilland Australia could not finance, over long periods, the purchase of supplies from Hatfield, a contract by M.A.P. on Hatfield, with reimbursement from the Australian Government, was suggested; Contract 2419 materialized in September.

At an M.A.P. meeting in London on 31 March Lee Murray estimated that the first Australian Mosquito, embodying much imported equipment, might fly by August, 1943, but obviously it depended on uncontrollable factors, the same sort of imponderables as were then facing Hatfield and Downsview.

M. M. Waghorn, a young Hatfield engineer, was seconded to Australia and set about studying and gathering data, which he took out in June.

Production: Australia

Byrne went back from Toronto, but Mills came through England in April, to study the task and chase up the supplies, working with men with whom he had been trained there before the war. Canada had promised certain jigs, samples and materials; for a further considerable schedule of parts he pressed Hatfield for 80 sets. They could not then promise more than 34 sets. Mills left Britain early in May. He sent ahead 3,000 Mk. VI drawings, 6,000 sheets of schedules, 250 photographs, a mass of notes, and 30 lb. of 35 mm. film, all by the ATFERO Atlantic Liberator service. It was July when Sydney had their hands on the drawings. Little detail engineering could be tackled until this data was received, and the subcontracting problem was more difficult than in Canada, for there were not the same engineering resources.

Not until 17 December, 1942, at Bankstown, did the sample DD664, now A52-1001, fly in Australia, and Bruce Rose was the pilot. De Havilland then installed a pair of Packard Merlin 31s, flew it in that form on 23 March, and delivered it at Mascot for assessment. At that time drawings were requested also for the dual-control conversion, although, as in the case of Canada, it was hoped that trainers might be supplied from England, and soon a promise of eight T. Mk. IIIs was extracted. The conversion drawings were sent through R.A.A.F. channels in February, 1943.

A month earlier Sydney had cabled for larger supplies, up to 180 sets, as local manufacture was proving difficult; fortunately the items sought were for the most part fairly easy ones, which Hatfield were able to ship. In February 95 per cent of all the British supplies were going on time and 3 per cent were in fair prospect, while 2 per cent were in very bad shape. These were such items as chassis jacks, radiators, oil coolers and cabin heaters, and their delay looked like affecting Australia's programme. Just then, and through March, Sydney had to ask for 180 sets of a much longer list, now almost 200 items. There were also 1,800 smaller items of which Sydney needed 40 to 60 sets.

The Australian Air Board had cabled their representative in London, A. E. Hyland, in January, 1943, that they hoped to introduce the two-stage Merlin 69 at the end of the year, and the next month Hatfield agreed to send Mk. X drawings as these became available. Rolls-Royce arranged to send out Frank Wormald to assist, and he left England 23 August, travelling via Toronto.

By the end of April a new assembly hangar and flight shed of 115,000 sq. ft. at Bankstown was ready for partial occupation. There was delay in obtaining machine tools, but by May this and the subcontracting troubles began to clear—indeed, everything cleared except the rainy Autumn weather. The Packard Merlin 31s were removed from the British sample Mk. II and put into the first Australian-built FB. Mk. 40, A52-1, which was flown 23 July, 1943, by D.H. test pilot Wg. Cdr. Gibson Lee. (This aircraft came to grief later at R.A.A.F. Laverton when an air bottle burst while being refilled on 14 June, 1944.)

Production: Australia

In August Hatfield advised that John de Havilland should go to Australia to explain the Hatfield test-flying routines; but because of his death on the 23rd Pat Fillingham went instead, from Toronto, arriving in Sydney 19 October.

In Britain by 11 August 156 sets of most items had been despatched and 80 per cent of required components were going on time; the 400th case of supplies was ready to ship. But only 11 sets of chassis jacks had reached Sydney, and these were short on the English assembly lines. Losses through enemy action, fortunately, were less than expected, and shipments of most parts just managed to keep pace with Bankstown's production flow through this critical period.

The first dual-control trainers (without engines) arrived in Sydney in the last months of 1943, and seven were being assembled by November. To install the Packard Merlin 31s was rather more work than had been forecast. As Britain was reluctant to spare more trainers the R.A.A.F. requested sets of dual-control conversion parts, but eventually the R.A.A.F. received in all 14 R.A.F. T. Mk. IIIs, numbered A52-1002 to 1015.

It had appeared that the Merlin 69 change-over would take place with the 256th airframe, so on 10 September, 1943, the Bankstown project engineer, S. Newbigin, left for Canada to study the installation, working with the Rolls-Royce man. He went on to arrive at Hatfield 6 December.

Four versions of the basic FB. Mk. VI were now planned for the R.A.A.F., the fighter-bomber, long-range fighter, reconnaissance and night-fighter. Interest was shown also in the Mk. XVIII variant with 57 mm. gun, and in rocket-projectile and 40 mm. gun versions.

By the end of October, 1943, five fuselages had been boxed, but shortages still delayed the second machine, the most troublesome now being plywood, fuel pipes, oil and coolant pipes, fairings and cowlings. The radio required alterations. The change to formaldehyde cement, produced locally by Elliott Brothers, was then about to be made, and was considered vital before much production took place; this was going to call for expert liaison. An overall problem was the shortage of labour, especially skilled men, and this long persisted. Nevertheless, plans that had been made, without any ground for confidence, a long time previously were at this stage almost being realized.

The first delivery to the R.A.A.F. took place 4 March, 1944, the second later that month; one went in April, two in May, one in June. A detail cementing problem arose, associated with the careful fitting of a 'blind' joint between the top of the spar and a wooden member that formed the edge of the top skin in that vicinity. A gap could occur instead of a cemented joint. After the first few wings had been built by de Havilland, and wing assembly was being done at the Pagewood plant of General Motors Holden, they experienced difficulty, and production was halted while non-destructive testing was done. The Director

Production: Australia

of Aeronautical Inspection called for modifications to the 49 wings built and the 50th, then in the jig. Since 22 Mosquitoes were almost complete this had disastrous effects upon deliveries.

The cementing problem became confused by reason of a fatal failure of the wing of aircraft A52-12 on pre-acceptance flight, witnessed by Mills, Waghorn and others, which however was found to be due to wing flutter under conditions of high speed and high G, not to cementing or jointing. A modification was introduced concerning the fixing of the detachable wing-tip, and later the cementing of it.

No aircraft were delivered in July. One, incorporating the mandatory wing modification, was delivered in August and two went in September.

Conversion to the PR. role had already been made, with A52-2, in May, 1944. Type trials followed, and six PR. Mk. 40s were delivered by October (A52-2, 4, 6, 7, 9, 26). The trials report, studied at Hatfield in December, showed a range of about 3,000 miles. Internal fuel capacity had been increased from 513 to 636 gallons, and 100-gallon drop tanks were fitted. Unarmed, these aircraft had the vertical nose camera and two more cameras aft, supplemented by two oblique cameras. These PR. Mk. 40s saw much useful service reconnoitring and photographing north and west of Australia.

Late in 1944 Murray Jones felt that Lee Murray's Mosquito production experience in England and Canada would be a considerable help in Australia, and arranged for him to come as general manager of the Aircraft Division at Bankstown; thus Murray with his family moved in March, 1945, back to the land of his birth.

Another accident, from no clear-cut cause, but under severe manoeuvring loads, occurred with A52-18 on 8 November, 1944, at Bankstown. Investigation was reopened into aspects of high-speed flutter, torsional stiffness, aileron balance, plastic wing-tip attachment, and again of cementing technique. But meticulous examination of the aircraft found the cementing to be satisfactory. Once more production was set back. Hereward left for Australia 20 January, 1945, with information and specimens of the latest British cementing practice. A few weeks later Mills arrived in England for comparative studies. Other experts went to Australia, and by the end of May all problems were solved. The German problem itself was solved by that time, but of course the Japanese might well hold out for a prolonged period, so that Mosquitoes were still needed.

The month-by-month deliveries from March, 1944, to May, 1945, were: 2, 1, 2, 1, 0, 1, 2, 6, 5, 7, 9, 10, 11, 6, 12—total 75 at 31 May. At the sudden surrender of Japan on 15 August it was 108, and production continued up to 212 aircraft. By this time the basic aircraft was almost entirely of Australian construction; all the castings and almost all the forgings were being made in the country, all sheet-metal and fabricated parts, cowlings and control-system parts. Radiators and oil coolers were eventually made there.

43. Undercarriage compression leg assembly
44. Engine nacelle assembly
45. From well-bombed London

46. Canopies by Perfecta of Birmingham

47. Tailplanes and wings by Hoopers, coachbuilders

48. Mrs. Hale's group; the smallest subcontractor

49. First Australian Mosquito flies, 23 July, 1943
50. Pat Fillingham: Flight-test method in Australia
51. Australian FB. Mk. 40 formates on British-built trainer

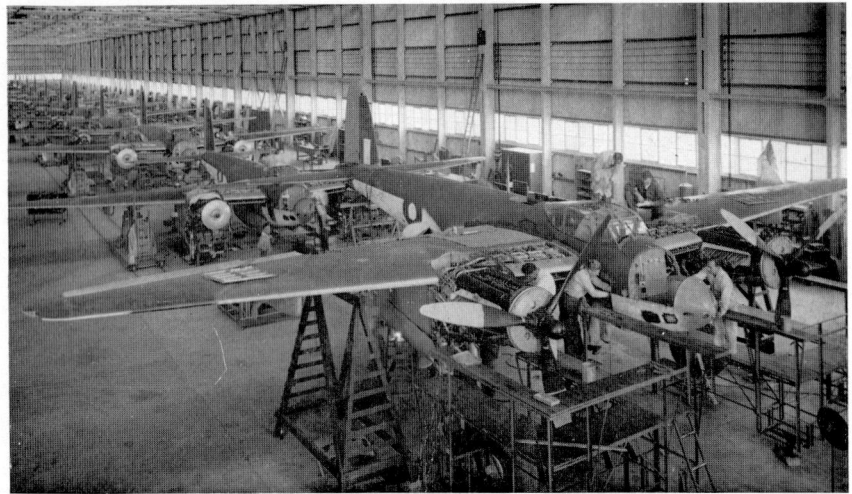

52. The Sydney assembly line
53. The last Australian-built PR. Mk. 41
54. First Canadian Mosquito flies, 24 September, 1942

55.
An F-8 being towed into a hangar at Toronto

56.
Russ Bannock and a flight test engineer (left) by a Mosquito destined for China

57.
The crew of LR503, veteran of 213 sorties, being greeted in Canada

58. KA102, a Mosquito Mk. 21
59. KA970 after its accident on 17 April, 1945
60. DK296 waits at Errol for delivery to the U.S.S.R.

61. The sort of women they were
62. The sort of men they were

63. Victor Ricketts at Hatfield, March, 1942

64. Low oblique of the Renault works, Billancourt, March, 1942

65. Merifield and his navigator

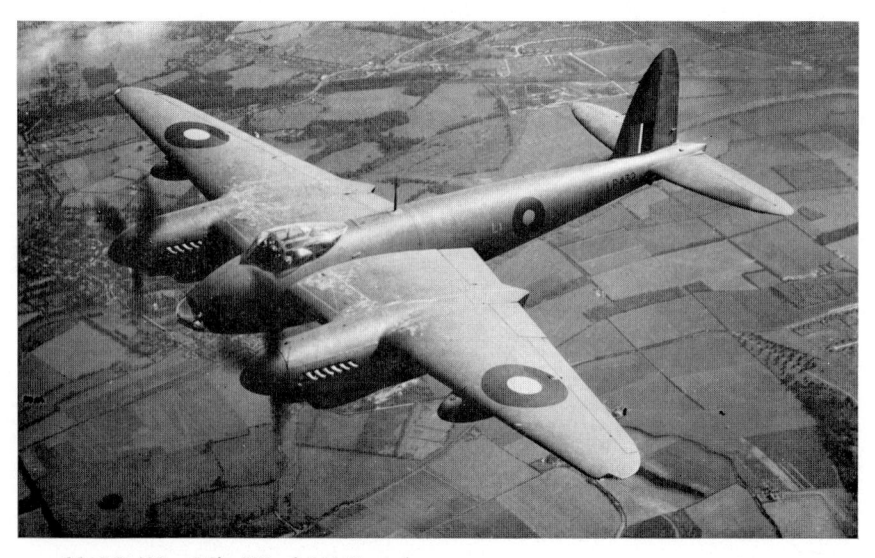

66. LR432 a Mk. IX of 544 Squadron

Production: Australia

Of the 212 aircraft A52-12, 18 and 24 crashed before being delivered. A52-24 was destroyed in a landing accident. The R.A.A.F. also used 76 British-built Mosquitoes, namely the sample F. Mk. II DD664, the 14 T. Mk. IIIs, 38 FB. Mk. VIs (A52-500 to 537) and 23 PR. Mk. XVIs (A52-600 to 622).

The FB. Mk. 42 with Merlin 69s was not proceeded with. Instead, the PR. Mk. 41 conversion with 69s, A52-90, from the last batch of Mk. 40s, was produced; it had extra radio gear, long-range oil tanks, additional oxygen and enlarged radiators. A PR. Mk. 41, A52-62, renumbered A52-324, (but with Merlin 77s) was entered for the England-to-New Zealand Air Race as VH-KLG, but Sqn. Ldr. Oates was obliged to ditch it on 3 October, 1953, in the Indian Ocean off the Burma coast, prior to the race. Another A52-210, renumbered A52-319, was allotted to Capt. Woods for the same race, but did not compete; it was registered VH-WAD. Most of the 28 PR. Mk. 41s built were used for a large-scale post war air survey of Australia. They were A52-90, 192 to 211, 41, 45, 49, 62, 64, 83, 36, and they were renumbered A52-300 to 327.

A further development of the FB. Mk. 40 was the T. Mk. 43 with Packard Merlin 33s, which was the dual-control trainer with dual elevator trim-tab controls. The 22 so converted (A52-3, 16, 17, 19, 20, 10, 8, 11, 21, 22, 25, 27, 28, 30 to 33, 37 to 39, 42 and 44) were renumbered A52-1050 to 1071.

1 Squadron R.A.A.F. used Mosquito 40s based on the Halmaheras and later in Borneo, shooting up and bombing Japanese positions. 87 and 94 Squadrons, 78 Wing, 1 A.P.U., A.R.D.U., C.F.S., C.C.U., 5 O.T.U. and 1 P.R.U. all used Mosquitoes. The survey work continued to 1953, and then a few Mosquitoes went to the R.N.Z.A.F. 1 Squadron were proud that they were chosen to escort General Yamanura when he flew in to surrender to the Allied forces.

Meanwhile, Murray Jones had arrived in England on 7 May, 1945, to discuss the future of the de Havilland world enterprise, to see the D.H. 104 Dove, and to hear more about the proposed jet airliner.

Survey of Mosquito production and delivery in Australia

Mosquito FB. 40
A52-1 to A52-212. First 100 had Merlin 31, with tapered-blade propellers, remainder Mk. 33. with paddle-blade propellers. Delivery 1944–45. Aircraft of the first 100 that were converted to T. Mk. 43 had engines changed to Merlin 33, with paddle-blade propellers.

Mosquito PR. 40
A52-2, 4, 6, 7, 9, 26 delivered May–October, 1944. Merlin 31.

Mosquito PR. 41
A52-300 to A52-327. Merlin 69. Delivered 29.5.47–22.7.48.

Mosquito FB. 42
A52-300 only. Merlin 69.

Production: Australia

Mosquito T. 43
A52-1051-1071. Merlin 33. Delivered 27.6.44–2.5.47.

Mosquito T. Mk. 3
A52-1002 to A52-1015. Merlin 31 or 33, 8 assembled by D.H., 3 by R.A.A.F. Delivered November, 1943–August, 1944. Aircraft in order of Australian registration were:—HJ891, HJ893, HJ963, HJ960, HJ968, HJ975, HJ984, HJ987, LR568, LR569, LR572, LR573, LR577, LR578.
1002–1009 incl. used by No. 5 O.T.U., 1010 by 87 Sqn., 1014 by 94 Sqn.
To R.N.Z.A.F.:–1003 (became NZ2304), 1005 (NZ2303), 1006 (NZ2302), 1015 (NZ2301).

Mosquito PR. 16
A52-600 to A52-622. Imported complete from Britain for R.A.A.F. Aircraft, in order of Australian registration were:—NS631, NS659, NS660, NS679, NS680, NS681, NS694, NS695, NS697, NS726, NS727, NS728, RG122, NS813, RF969, RF975, RG120, RG123, RG130, RG151, RG152, RG153. With the exception of A52-607, 612, 615, 617, 620, 621 and 622 these aircraft were used by No. 87 Sqn., R.A.A.F.

Mosquito Mk. VI
A52-500 to A52-537. Imported complete from Britain for R.A.A.F. Aircraft, in order of Australian registration and all prefixed 'HR' were:—HR302, 304, 312, 333, 335, 336, 281, 306, 307, 310, 412, 450, 463, 499, 370, 408, 410, 440, 460, 461, 467, 488, 501, 502, 504, 505, 506, 510, 503, 508, 485, 445, 587, 509, 517, 511, 513, 577. All of these aircraft were used by No. 1 Sqn., R.A.A.F.

Transfers to R.N.Z.A.F.
As *T. Mk. 43*:—A52-16 (became NZ2307), A52-17 (NZ2306), A52-19 (NZ2305), A52-20 (NZ2308).
As *FB. Mk. 40*:—A52-101 (NZ2320).

Delivery of FB. Mk. 40 to R.A.A.F. units
To No. 87 Sqn., R.A.A.F.:—A52-6, 8, 9, 11, 26, 28, 36, 64, 83, 192–199, 201, 203, 204.
To No. 94 Sqn., R.A.A.F.:—A52-52, 84, 85, 94–100, 102–106, 108–110, 112–122, 124, 125.
To No. 1 P.R.U.:—A52-2, 4 (written off after action 15.8.44).
To No. 5 O.T.U.:—A52-29—35, 39, 40, 46–52, 63, 65–82, 87, 88, 192–199.
To Survey Sqn.:—A52-101, 192, 193, 194, 195, 196, 197, 198, 199, 600, 603, 607, 608, 610, 615, 620, 1002, 1009.
To CFS:—A52-37, 135, 183, 1015.
To C.C.U.:—A52-67, 68, 71, 72, 73, 74, 76, 77–82, 87, 1006, 1007.

Book Four

OPERATIONS

Chapter 11

Operations: Photographic Reconnaissance

Neither Britain nor Germany entered the war with effective reconnaissance aircraft. Britain urgently developed this role, especially with the Spitfire, and later overwhelmingly with the Mosquito, so that throughout Europe and the Mediterranean, and later in the East, enemy activity was always thoroughly photographed. Germany failed in reconnaissance, even before D-Day. Afterwards Germans told us 'Our filing and indexing were good, but our photographic operation was bad'.

The first operational formation to receive Mosquitoes was No. 1 Photographic Reconnaissance Unit, based at Benson, Oxfordshire. Its two leading pilots were Flt. Lt. Alastair Taylor, D.F.C. and two Bars, and Sqn. Ldr. Rupert Clerke. The Unit had made long photographic reconnaissance flights with Spitfires, but needed the extra range, twin-engined security, navigational radio and an operator, which the Mosquito would afford. Also, P.R.U. was much interested in the two extra cameras which would be carried. The navigator's excellent view would clearly improve the accuracy of operational missions.

By refuelling in the Shetlands their Spitfires could reach the Norwegian Coast as far North as Trondheim; Namsos if conditions were right. There was little fuel margin, however, and a wind from a southerly quarter could easily carry a pilot returning from Norway past the northern tip of Scotland, forcing him to alight in the sea. To the south, targets photographed included Genoa and Spezia.

Spitfires were preferable in some respects to the Mosquito over hotly defended zones, for a single-engined machine excited the defenders less than a twin. The Spitfire was thus useful for near regions, such as the Ruhr and Rhineland, remaining so throughout the war. But there was much long-range work to do. The Mosquito, as Sir Wilfrid Freeman had foretold, was in great demand from the first, and ever after.

On the sunny afternoon of 13 July, 1941, only eight months after the prototype's maiden flight, and nineteen months from the start of design, the first Mosquito, W4051, arrived at Benson from Boscombe Down. W4054 arrived on 22 July and W4055 on 7 August. Flying personnel took to them well; Hereward de Havilland, responsible for liaison with the Service, reported in light vein, 'in my experience this is the only aircraft which, initially, has not been branded by the pilots as

Operations: Photographic Reconnaissance

a death trap in one way or another'. On the other hand its reception by engineering personnel was not warmly favourable. Ground crews of the younger generation were inclined to regard wooden construction as retrogressive. They became less averse after a few months' experience, and prejudice passed in less than a year. The extra bulk of the wood scantilings made combat damage less weakening, and repairs were especially easy. Metalworkers were in short supply, but there were plenty of wood workers.

Much time was now spent testing camera installations, changing steel mountings for wooden ones to damp out vibration, and fitting a control to the camera heating. Consequently the earliest Mosquitoes plied to and from the works, hopping across the Chilterns in a few minutes. On 13 September Sir Archibald Sinclair visited Benson, being flown for fifteen minutes in a Mosquito by Sqn. Ldr. Clerke—'and was impressed'. Good photographic results were obtained on practice flights, but oil tanks were mysteriously swelling sometimes fouling the undercarriage gear, this trouble emerging in mid-September when operations were imminent. At de Havilland the tank shop worked round the clock to complete two modified tanks, flown to Benson on the evening of 14 September, and fitted to W4055 during the night. All next day she was thrown about the sky and thumped hard on landing. The modified tanks seemed satisfactory, so the ground crews busily prepared the aircraft for the big event.

At 11.30 hours on 17 September, Sqn. Ldr. Clerke set out in W4055 to photograph Brest and the Spanish-French frontier. He flew over Bordeaux and La Pallice at an average height of 24,000 feet. W4055 functioned correctly in all respects, arriving home at 17.45. During the previous day's tests however a generator had overheated because the blank had not been removed from the cooling air duct, and electrical trouble arising prevented the cameras working. The second operation by W4055 came on 20 September, when, with Flt. Lt. Taylor, D.F.C. as pilot and Sgt. Horsfall as navigator, a four-hour photographic reconnaissance mission to the Sylt-Heligoland area was successfully flown.

During October sixteen successful P.R. sorties were flown by three aircraft detached to Wick. All were to Norway, where Stavanger, Bergen, Trondheim, Kristiansund and Oslo were covered. Sqn. Ldr. Clerke impressively demonstrated a Mosquito before the King and Queen at Watton, having a mock battle with a Spitfire flown by Wg. Cdr. Tuttle, O.B.E., D.F.C. On 15th, Sqn. Ldr. Clerke made a record-breaking flight from Wick to Benson in 1 hr. 32 minutes.

Early November four Mosquitoes with Flt. Lt. A. L. Taylor in charge, were transferred to Wick in Scotland, whence flights were made over Norway. An exception came on 18th, when W4055 flew to Kiel, only to find the target obscured by cloud. She flew one more successful mission to Norway, then on 4 December with Sqn. Ldr. A. L. Taylor, D.F.C., and Sgt. Horsfall aboard took-off from Wick at 10.15 for a sortie

Operations: Photographic Reconnaissance

to the Trondheim-Bergen area, from which she failed to return. 'Benedictine', as she was colloquially known, was presumed shot down by gunfire over Bergen on her fifteenth sortie.

Bad weather interfered with operations in December and the next two months; then followed twenty-six successful sorties in March, three months after the Wick detachment had moved to Leuchars. Almost all flights were to Denmark or Norway, but on 15 January came the first attempt, by Flt. Lt. Merifield in a long-ranged aircraft, W4061, to reach Gdynia in Poland and Danzig, both of which were found to be cloud clad. Kiel, too, evaded the cameras twice in January. A very-long-distance mission by W4051 on 20 February took her to the French-Spanish border zone, and marshalling yards and airfields in the Toulouse region. Two days later she was flown by Flt. Lt. Victor Ricketts and Sgt. Boris Lukhmanoff to Cuxhaven, then to Kiel, where magnificent photographs of the *Gneisenau* in dry dock were secured on the 24-inch camera.

A Spitfire followed up the success, finding the *Scharnhorst* at Wilhelmshaven, likewise undergoing repairs following the 'Channel Dash'. Further pictures of the *Scharnhorst* were obtained on 2 March by Flt. Lt. Ricketts flying W4060, the first long-range aircraft, whilst W4059 surveyed the other vessel. W4060 and W4051 photographed the French coast prior to the Commando raid on St. Nazaire, and on 3 March Flt. Lt. Merifield at last photographed the Danzig-Gdynia region. His account follows:—

'I took off from Leuchars at 10.25 on 3 March with the intention of photographing Copenhagen, Danzig, Gdynia and possibly Koenigsburg. We set course for Copenhagen, climbing to 20,000 feet over the North Sea and cruising at 280 m.p.h. We sighted the Danish coast at Esbjerg, visible through patches of medium cloud, and photographed the town and harbour. We then climbed to 23,000 feet and flew towards Copenhagen. Medium and low cloud increased to 10/10th, and we were soon flying through a layer of cirro-stratus. Copenhagen was not seen, so we altered course for Gdynia on E.T.A. Soon afterwards all low and medium cloud disappeared, though high cloud persisted, but we could see the Baltic Sea underneath us, covered with large patches of ice. We were making intermittent condensation trails in the cloud, so climbed to 24,000 feet where the trails ceased. After two hours fifty minutes flying time we crossed the German coast at Leba, and about five miles west of Gdynia the cirro-stratus thinned, and over the target disappeared altogether. We ran over Gdynia with cameras on, but while doing so noticed that a persistent trail was forming behind us, so descended to 23,000 feet and did another run. We then proceeded to Danzig and photographed the town and harbour with Neufahrwasser from the same height. Koenigsburg was the next objective, and it was photographed, with adjacent aerodromes, also from 23,000 feet. From here we could see over a hundred miles to the east, well into Lithuania.

Operations: Photographic Reconnaissance

'After 3½ hours flying we set course for Leuchars, encountering such a strong head wind that it was twenty-five minutes before we passed Gdynia. Accordingly we descended to 18,000 feet and cruised at 2,000 r.m.p. in M.S. gear to economize in fuel. Fortunately as we flew west the wind decreased, and we left the Danish coast at 4.30 p.m., photographing Esbjerg again on our way out. After an uneventful trip across the sea we landed at Leuchars at ten past six.'

Writing of his early Mosquito sorties Flt. Lt. Ricketts had this to say:— 'I usually climb on weak mixture at about + 3 boost, and maintain this as a rule for about twenty-five minutes, at 180 indicated or less, levelling out at about 24,000 feet. We photograph usually from 22,000 feet, but have done so from 17,000, 15,000, 14,000, 12,000, 2,500 and 400 feet depending on the weather. If one has flown 600 or so miles to a target one doesn't feel like coming home without a picture merely because of cloud—it depends on how brave you feel at the time. Soon after leaving the enemy coast we like to come down to 15,000 feet or less, the idea being to take our oxygen masks off and eat raisins and chocolate. The only evasive actions so far have been light changes of course and height to avoid A.A. fire. On such changes I increase throttle, for example from zero boost to − 2 and then suddenly to + 2 with a turn of 20 degrees and a steep climb or descent of 2,000 feet. By the way, the St. Nazaire Commandos were much obliged to old W4051 for preliminary information received.'

Two of the earliest encounters with fighters were described by Flt. Lt. Merifield after a flight from Scotland:— 'We were flying in W4061 on 30 March, 1942, over Trondheim at 18,000 feet in F.S. gear at 2,400 revs. at the time. I noticed an Me 109 in my mirror about half a mile behind and 500 feet above. It was making a trail of black smoke, presumably because it was at full throttle. I increased revs. to 3,000, switched over to M.S., pulled the cut-out and dived gently. My observer reported another 109 on our starboard quarter about the same distance behind. We levelled off at 14,000 feet but did not seem to draw ahead. Observation of the enemy aircraft was difficult because they were dead astern and we were making a lot of black smoke ourselves. After a quarter of an hour they were no longer to be seen, so boost was reduced to 6 lb. and revs. to 2,700. Shortly afterwards my observer reported one aircraft crossing our tail 400 yards astern but no fire was opened. Thereupon I opened up again and flew out to sea towards cloud which was entered ten minutes later. I could not say what speed was reached during the dive but at 14,000 feet the I.A.S. was 320, which was afterwards computed to be 395. Engine temperature remained below 100°C. and oil temperature below 80°C. the whole time. The port engine had to be changed because of an internal glycol leak.

'The second encounter took place off Statlandet in W4060, when a single-engined aircraft was observed a mile astern, also making black

Operations: Photographic Reconnaissance

smoke. The same tactics were employed and the aircraft easily outdistanced. About the trailing aerial—all our observers say that it is unnecessary and is never used, as our range is sufficient with the fixed aerial.'

On the night of 3–4 March, 1942, Bomber Command despatched 235 bombers in good weather to the Renault works at Billancourt, near Paris, on a most important raid. Evidence of its success was vitally necessary and on 4 March Ricketts took W4060 to gather the material. He talked to de Havilland workers at Hatfield on 11 March describing his low-level mission of seven days earlier, for which he and Lukhmanoff were decorated:—

'I would rather fly to Paris again than go through this! I thought when I came along that you might like to hear the story of the Paris flight, which was unusual in a way because it was the first time a survey had been made of R.A.F. blitzes, and also the first time it had been made at such a low level, as some of you will have seen from photographs on the board. You will remember last week, one moonlight night, Bomber Command went across and, in their own words, "pranged" the Renault works. Command said this was very likely, but we must have pictures to prove this.

'The job came to our unit, and it was just about weather like to-day, and it was considered for a time that we could not cope. However, the Mosquito as we use it is equipped with wireless, whereas most P.R.U. aeroplanes are not. After talking things over with my observer, I decided that we could probably get home safely using wireless, so we offered to have a stab at this job. Well, we left our base (and most of you know the name of it). At the time of leaving we could not see across the aerodrome. I thought "Are you a sucker to take this on?" but it was too late to turn back then, as we had lost sight of the ground, and we did not see ground again until my observer, Boris, who is a Russian and a very keen type, said "Well, I think we are somewhere near Rouen, not very far from Paris, and it is time we came down." We still could not see anything and it was raining like Hell and there were thick clouds.

'We went down to 1,000 feet. At 900 we caught sight of the ground and he was reasonably happy. We found the River Seine and, as it was quite obvious that we could not navigate to Paris by that weather, we decided to fly along the twisting river. The only way we could do it was for Boris to lie on his nose in the nose of the aeroplane, saying "Turn right, now left etc.", as the river twisted. Very soon we found ourselves over the roofs of houses and saw people in the streets running for cover, thinking the bombers had come back again. We could not find the works now, so we went on and came out of clouds at 600 feet, to see people running like mad. Finally Boris said "There it is" and just caught a glimpse of roofs full of holes 500 feet below.

'We had not time to take a picture, so Boris got his cameras ready and

Early Photographic Reconnaissance Mosquitoes

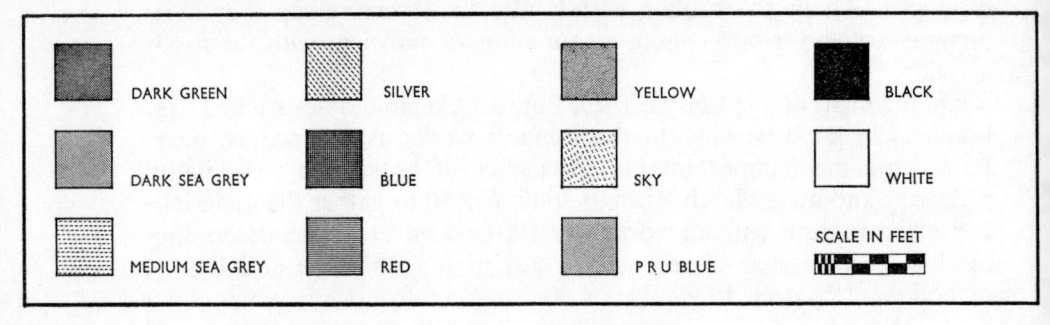

DARK GREEN	SILVER	YELLOW	BLACK
DARK SEA GREY	BLUE	SKY	WHITE
MEDIUM SEA GREY	RED	P R U BLUE	SCALE IN FEET

Mosquito PR. 1 wearing P.R.U. Blue finish. The tone of blue-grey applied was arrived at after much research, particularly on Spitfires which experimentally were painted in shades of white, grey, pink and blue. From the start the Mosquito PR. 1s wore P.R.U. Blue finish.

Mosquito IV (Merlin 21) DK310 converted to PR. IV reached Benson 19.7.42 and joined 1 P.R.U. in August. On 13.8.42 it flew to Gibraltar via St. Eval. Next day it made its first operational sortie in connection with the forthcoming North African landings. It flew two more sorties before returning to Benson on 18.8.42. As recounted on page 126 it force-landed in Switzerland on 24.8.42. Most PR. IVs were painted P.R.U. Blue overall, unlike DK310.

Mosquito PR. 1 (Merlin 21) W4059 reached 1 P.R.U.13. 9.41, crash landed 23.9.41. Repaired, passed to 540 Squadron 30.9.42, to 8 OTU 28.7.43, damaged 8.44 and written-off 20.9.44. Unit codes depicted were pale blue-grey. Mosquito PR. 1 front view is taken as typically representing the head-on appearance of the Mosquito.

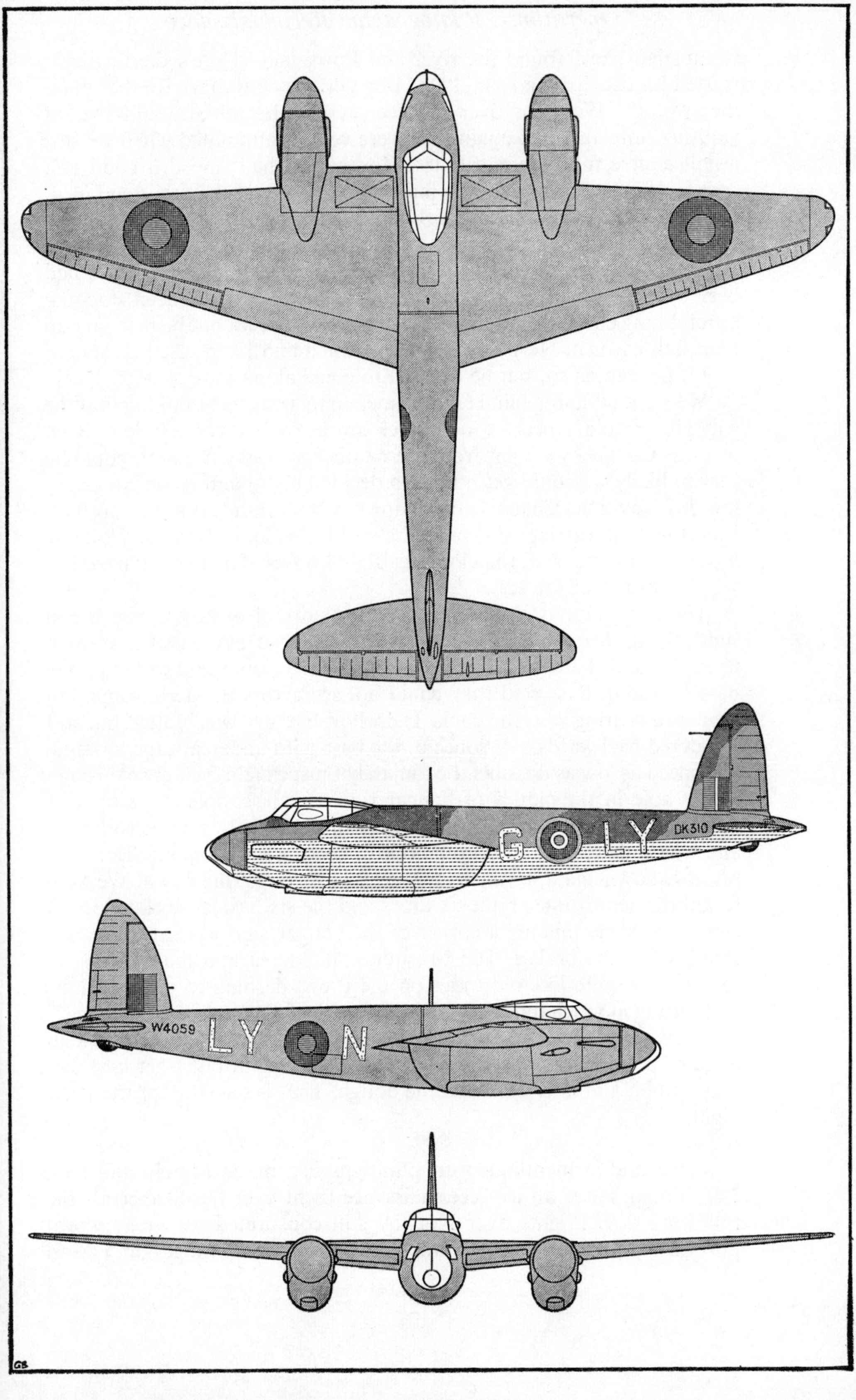
G
LY
DK310
W4059
LY
N
GB

Operations: Photographic Reconnaissance

we hurried back, found the river and Boris said "Here's the factory", started his cameras and said "Oh! Boy Oh! Boy did they give that place the works!" We went over it once, across the middle, and then it vanished into the mist again. We were very disappointed when we saw a golf course full of bomb holes. However, it had ruined a good golf course. We flew back over the factory, trying to get the other end of it. Boris said "I don't want you to run into the Eiffel Tower, which is about 1,000 feet high." We flew back up the river and had to come down to 400 feet. We tried to do it a fourth time but lost it. Boris said we had been over for thirty-four minutes now. I thought we had been stooging around long enough, and that the Huns would just about be bringing up their flak by then. He was very disappointed and I suggested he should stay if he wanted to, but he decided to come along too.

'We went off flying blind all the time, up to 5,000 feet thinking that the wily Hun would expect us to go back low as we had come in low. Back over the Channel we knew from the weather we had left that it would be very unlikely we could get home, so decided to try and come back very low down over the Channel and sit down on one of the coast aerodromes. I put the undercarriage down in case we hit the South Downs. We came lower and lower, until the clock registered 0 feet. Then we managed to see the surface of the sea.

'The Mosquito is rather fast for that sort of weather. The beach suddenly flashed underneath the wheels and we gave up the idea of trying to land there. We shot up again into the clouds and called up the base by radio. Base said they could not see across the aerodrome, but gave us a bearing to come back. The whole country was blotted out so I wandered backwards, got south of the base with undercarriage lowered, and speed as low as possible. Fortunately I just caught a glimpse through a tiny hole in the clouds of hangar roofs and a couple of aeroplanes parked. I thought where there are aeroplanes there's an aerodrome, and where there's an aerodrome that's where we are landing. Boris got his nose down again, trying to see a clear way of getting down. We went round the aerodrome about six times and the sixth time we came out of cloud, narrowly missing a corner of the hangar, and just sat down and stamped on the brakes. The Mosquito shuddered and finally came to rest. It was quite like old times on the Press, dashing to the telephone and getting in touch quickly with my place. I said "We found the target." My boss said "At 3,000 feet?" and I said "No, 400 feet." He said that none of the pictures would come out at that height, but I bet him that they would, and later collected the dough. That is the story of the Paris flight.'

Stettin and Swinemünde were photographed on 22 March, and Sqn. Ldr. Young made an ice reconnaissance flight over the Skagerrak the following day. Flights over Norway still constituted the greater proportion of operations. W4056 was shot down over Stavanger on 3 April

Operations: Photographic Reconnaissance

and the crew taken prisoner. Ricketts made the first flight to Bavaria on 21 April, but the weather precluded photographs so he settled for the St. Hubert area only. On 24th he again took W4059 to Augsburg, getting some fine photographs showing the results of the Lancaster raid of a week previously, and turned his cameras on Mulhouse and Stuttgart as he headed for Benson.

P.R.U's urgent need for more Mosquitoes was partly met by diverting two Mk. IV bombers and two Mk. II[1] fighters to Benson, where unit personnel installed cameras. DK284 made the first operational flight by a PR. Mk. IV,[2] in the hands of Ricketts on 29 April. He took off at 08.05, photographed Augsburg, Stuttgart and Saarbrucken and landed back at 13.35. It was this machine that on 7 May made the deepest penetration into enemy territory so far when, on a six-hour flight, it reached Dresden, Pilsen and Regensburg—only to find the weather consistently bad. On 14 May DD615, the 'PR. Mk. II', flew its first operation during which Wg. Cdr. Ring photographed Alderney. Eleven days later Ricketts took her to Billancourt, Poissy and Le Bourget, and two days later Flt. Lt. Wooll successfully photographed Saarbrucken and Amiens, using this machine. Lack of long-range tanks unfortunately limited the range of the Mk. IIs.

The new 36-inch F. 52 camera developed by R.A.E. and manufactured by Williamson was now entering service. With a normal operation of 1/300 second at f5·6 it gave about 1/8000 scale at 24,000 feet, with sharpness that made clearly visible even the German markings on an aircraft carried on the *Prinz Eugen*. Plt. Off. Sinclair and Plt. Off. Nelson in W4060 obtained perfect cover of the whole port of Gdynia with this camera on 2 June, 1942, revealing there the aircraft carrier *Graf Zeppelin* and the *Scharnhorst* with her front turret dismantled—most useful information.

A 7¾-hour sortie was flown on 10 June, the longest yet, to Spezia, Lyons, Marseilles and back to Benson. Wg. Cdr. Ring took the other Mk. II DD620, on a 2½-hour tour of Holland on 29 June bringing back a wide assortment of photographs. 8 July witnessed the first of many flights to Russia and back in a day, made in this instance by Flg. Off. Bayley and Flt. Sgt. Little. They were to find and photograph the *Tirpitz*, known to be hiding somewhere north of Narvik. Arrangements

---

[1] In all, four F. II were converted into P.R. IIs, details:— DD615 to PRU 16.4.42, 540 Sqn. 30.9.42, suffered engine fire and forced landing 18.11.42, then overhauled. Reached 141 Sqn. 21.3.44. FTR 30.4.44. DD620 PRU 17.4.42, crashed 19.7.42. Repaired, reached 456 Sqn. 1.5.43, 13 MU 27.2.44, at 51 OTU 4.1.45–6.7.45. DD659 PRU 7.6.42, 540 Sqn. 19.10.42, 8 OTU 7.5.43, became 5084M 24.3.45. W4089 PRU 1.6.42, FTR 13.7.42.

[2] Two 'PR. IVs' preceded the conversion of a batch of B. IVs into PR. aircraft. DZ411 and '419 were delivered first, in 12.42. Others delivered:— 4 in 1.43, 5 in 2.43, 9 in 3.43. 7 in 4.43, 1 in 5.43, 1 in 6.43. 16 were in all used by 540 Sqn. Later aircraft went to 8 OTU, from July, 1943. DZ592 delivered to 540 Sqn. 27.5.43 was used until 1945, and FTR 10.4.45.

Operations: Photographic Reconnaissance

were made for them to land at Murmansk and operate from there until the job was completed. Having found the *Tirpitz* on their way out, and photographed her thoroughly, they went on to Murmansk, refuelled and flew straight back to Leuchars, covering 3,000 miles in a day.

11 July was a distressing day for P.R.U., for Victor Ricketts and Boris Lukhmanoff failed to return from a photographic sortie in the armed Mk. II W4089 to Strasbourg and Ingolstadt.

Flt. Lt. Wooll was approaching Venice from England, hoping to obtain pictures of gondolas and much more on 24 August, when his starboard engine developed a glycol leak. Being near his objective he carried on, but failed to cover Trieste, Fiume or Pola, which were his other targets, because he had to shut down the starboard engine. Shortly after, he noticed the port engine's temperature begin to soar, possibly due to an oil leak. He turned DK310 about, and it soon became necessary to force land. He chose Belp airfield close to Berne, Switzerland. As they came to rest Sgt. Fielden immediately endeavoured to destroy the Mosquito with the incendiary equipment provided but was unable to do so—and not for the want of effort or dexterity—so the aircraft was interned in a hangar, alongside two Bf 109E's. The crew were treated with hospitality and correctness. After negotiation with the German Government, they were exchanged for two Messerschmitt pilots and returned to England in December. A few weeks before their internment the crew had used DK310 for a number of vital flights from Gibraltar, taking pictures for use in planning the North African landings. Flt. Lt. Wooll became a test pilot for de Havilland (Canada) in 1943, after repatriation.

DK310, incidentally, remained at Berne until 1 August, 1943, when it was agreed by Britain and Switzerland that the aircraft could fly in Swiss markings. It was used to train Swiss pilots, then handed to Swissair on 13 October, 1944, for night mail flights under the registration HB-IMO. Six pilots trained upon it, but the scheme was abandoned in February. DK310, registered B-4, was again put in Air Force hands. It was flown for various purposes, then modified as a test bed for the Swiss N-20 aircraft. Records held by the Swiss Authorities show that between 6 September, 1943, and 8 March, 1953, it made 185 flights and logged 76·20 hours.

On 13 August, 1942, the 'P.R.U. Detachment in North Russia' sailed from the Clyde; arriving at Vaenga in North East Russia on 23 August. Its mission was to obtain photographic coverage of the anchorage in which the *Tirpitz* lay. Three Spitfires were initially used, but on 23 September, Sqn. Ldr. Young brought W4061 to the field with instructions for reconnaissance sorties for Operation *Jupiter*. Flights were made by the Spitfires, then the unit disbanded on 18 October, the cameras and equipment being handed over to the Russians. Five days later the personnel sailed for Britain and the Mosquito flew home.

At home a similar event took place. Such was the success of the

Operations: Photographic Reconnaissance

Photographic Reconnaissance Unit's operations that it was decided to re-establish it as four squadrons, and on 18 October place its dozen Mosquitoes into 540 Squadron. Before this happened DK320 had secured the required pictures of Fiume, Pola, Trieste and Venice by making a two-day round trip on 4–5 October. The following day Plt. Off. McKay took DK315 to Prague, whilst DK314 roamed the sky above Lake Constance. On the final day of operations, 17 October, W4058 flew to Oslo, but failed to return. P.R.U. losses proved to be one per 470 hours of operational flying.

540 Squadron formed on 19 October from 'H' and 'L' Flights of No. 1 P.R.U. at Leuchars, under Sqn. Ldr. M. J. B. Young, D.F.C., remaining under control of Coastal Command's 16 Group. The Squadron's first operation was a highly successful sortie from Sumburgh, during which the *Tirpitz* and the *von Scheer* were photographed in Ofot Fiord. 'B' Flight moved shortly after to Benson, from where Flt. Lt. Acott and Sgt. Leach flew to Milan, Genoa, Savona and Turin to assess the results of recent attacks, photographing Le Creusot on return. Several detachments to Malta were made at this time, and to Gibraltar in connection with the North African landings. Now the Mosquitoes faced irritating attacks by Vichy French fighters.

No. 8 P.R. Operational Training Unit, based at Fraserburgh, Aberdeenshire, received its first Mosquito in December, 1942. Task of the unit was the training of crews for P.R. squadrons. Wg. Cdr. Malcolm Douglas-Hamilton, unit commander, wrote to de Havilland telling them that he and his pilots thought the Mosquito to be the finest aircraft yet produced. He also related the difficulties at Fraserburgh, where there was no proper hangar accommodation and the climate was rigorous. In spite of this the only trouble encountered in the winter was madapolam peeling off the wing.

Before the end of 1942 540 Squadron had despatched its aircraft on numerous missions across occupied Europe to Malta, keeping watch thereby on the Italian naval bases of Leghorn, Pola, Spezia and Fiume. Two Mosquito IVs fitted with multiple ejector exhausts in place of the saxophone type reached the squadron in November. It was decided that the extra 10 m.p.h. or so obtained from propulsion effect was worth more than flame suppression, and future aircraft were similarly equipped.

The prime need was for more ceiling, and in January—before any Mk. IXs came off the assembly line—two of the five Mk. VIIIs built were delivered to 540 Squadron.[1] These were IVs fitted with two-stage-supercharged Merlin 61s, and fitted with two 50-gallon drop tanks. The VIIIs filled a gap just when high-altitude German fighters were becoming

[1] DK324, the first VIII, came on charge 28.11.42 for tests. DZ342 on 15.12.42 was the first to reach Benson and was used by 540 Sqn. 14.1.43–24.12.43. DZ364 reached 540 Sqn. on 22.1.43. DZ404 was used by 540 Sqn. 4.2.43–24.4.44. DZ424 served on 540 Sqn. 28.3.43–19.12.43.

Operations: Photographic Reconnaissance

a problem. A serious position arose early in February due to a large number of P.R. aircraft being unserviceable—about 50 per cent—as a result of water soakage, badly fitted No. 7 bulkheads, and delay in modifying elevator noses.

Sqn. Ldr. G. E. Hughes and Sgt. H. W. Evans were the first to fly an operational sortie in a Mk. VIII, taking DZ342 to La Rochelle and St. Nazaire on 19 February, 1943. Unfortunately the mud flap over the rear camera lens did not open, rendering the mission fruitless. Another Mk. VIII, DZ364, made one of the squadron's first damage-assessment flights over Germany, when Flt. Lt. K. H. Bayley, D.F.C., flew to Frankfurt whilst DZ342 visited Emden and Bremen, both on 27 February. On her second sortie, on 2 March, DZ364 was despatched to Nuremburg, and good shots of Mannheim were also obtained. Her mission the following day was to Berlin, when she seemed certain to win the distinction of being the first P.R. Mosquito to photograph the city. W4059 had tried on 19 January, but was intercepted at 17,500 feet North of the Capital and had to be satisfied with pictures of Odense, Stralsund, Sylt and Peenemünde, the value of which was not then apparent. Yet luck was again out for, although Sqn. Ldr. Hughes and Flt. Sgt. Chubb made seven clear weather runs, and only encountered flak on the final, their F.52 camera lens had an oil coating rendering all pictures useless.

Next day Wg. Cdr. Young flew from Leuchars to Knaben to assess the result of the previous day's Mosquito raid, only to find a layer of strato-cumulus cloud preventing photography. A second fruitless attempt, again using DZ342, came on 7 March. Young took DZ404 on its first operation on 8 March and headed for Berlin. Beyond Ijmuiden he found Europe completely covered by a layer of alto-cumulus until he was sixty miles from the city. He obtained photographs, in the face of heavy flak, a Bf 109G, and an Fw 190 which climbed ahead of him to intercept. Cruising at 26,000 feet, he opened up to 410 m.p.h. T.A.S., leaving them both comfortably behind.

Plt. Off. McLeod and Plt. Off. Leach aboard DZ342 were unsuccessfully intercepted by four single-engined fighters whilst returning from an abortive trip to Berlin on 23 March; the Merlins were run at +16 lb. boost, 3,000 r.p.m. for twenty-three minutes, the pilot's report of the operation being as follows:—'I was detailed to fly to Wunsdort, Berlin. After passing Hanover the condensation level fell below 24,000 feet and it was decided to return to base. At 15.41 hours at 24,000 feet over the Deventer area, east of the Zuyder Zee, T.A.S. 360 m.p.h., G.S. 330 m.p.h., true track 270°, the navigator observed two S/E aircraft about 1½ miles astern following us at approximately the same height. I immediately opened the throttles fully, and C.S. controls to fully forward, keeping same height and course. Three minutes later I saw an Fw 190 diving from the north and from above. I turned towards him and he pulled up in front of us into a half roll, dived out endeavouring

Operations: Photographic Reconnaissance

to get onto our tail, but failed to do so. This left him slightly below and approximately 1,000 yards behind.

'Seven minutes after first sighting enemy aircraft a fourth S/E aircraft approached from the North and joined the chase with the others, but stayed at about two miles range.

'We crossed the Dutch coast ten minutes after the beginning of the first interception, with the nearest aircraft 1,000 yards astern and slightly below. Two minutes later the last aircraft to intercept broke off the chase. At 16.04 hours the remaining three aircraft turned away, and I went back to normal cruising revs. and boost.'

Reconnaissance Mosquitoes were now penetrating regularly deep into Europe. DZ473 went to Darmstadt, Würzburg, Pilsen and Prague on a successful tour on 12 March. The following day Flg. Off. P. Hugo set out for Brux, but contrails at 20,500 feet caused him to turn for home and, as he flew over Calais he was fired upon and forced out of control into a dive to 4,000 feet. With his starboard engine out of action he made a forced landing at West Malling. Five days later Flt. Sgt. M. Custance took off for Brux but he and DZ364 failed to return, this being the first loss of a Mk. VIII.

Clouds persistently covered Berlin, as on 29 March when Plt. Off. R. A. Hosking found layers between 18,500 and 28,000 feet. Vienna and Wiener Neustadt bared themselves to the cameras of a Mosquito on 13 April, and Pilsen five days later, when DZ523 was flown through cumulus clouds en route towering to 32,000 feet. Still the sorties over the Norwegian coast and Denmark were maintained, and sometimes operations of a less usual nature, as on 4 May when Sqn. Ldr. M. D. S. Hood made a low-level trip to Bordeaux. His route to the target was above 10/10th cloud. Over the town cloud base was at 15,000 feet; so runs were made immediately below, operating a fourteen-inch camera. This was the first occasion on which a Mosquito IV with a small blister added to the roof to improve the view to the rear was operated. A damage-assessment flight by DZ424 to Pilsen came on 15 May. Flying at 30,000 feet the Mosquito was intercepted in the target area by two Fw 190s, which chased it for about fifty miles before admitting defeat. Over Nürnberg a Bf 109 resumed, unsuccessfully, the chase.

Douglas-Hamilton at this time pointed out that the long-range missions which his squadron were being called upon to carry out were, in some cases, really beyond his unit's capability. Some were of great importance and therefore there was a pressing need for extra fuel tanks —drop tanks—of the largest possible capacity. Talking to Hereward de Havilland he cited as examples a flight to Gleiwitz in Poland and two to 'secret destinations' 800 miles away. He also related the trouble encountered in the northern sector of operations from contrails which on some days extended down to 16,000 feet.

In the winter and early months of 1943 two Mosquitoes were used by Flt. Lt. Merifield and his navigator Flg. Off. W. N. Whalley, with Flt.

Operations: Photographic Reconnaissance

Lt. R. L. C. Blyth, D.S.O., D.F.C., and his navigator Flt. Lt. Leach, D.F.C., to develop the art of night photography. Use was made of the American M.46 photoflash of 600,000 candlepower—three times as bright as the British equivalent—which the Mosquito could conveniently carry. Problems inherent in the use of long-focal-length cameras used on this work were solved with remarkable ingenuity, and by Spring, 1943, good night photographs were being obtained. Mk. IVs, increasingly vulnerable on daylight missions, were switched to night tasks. Operational trials were conducted by Sqn. Ldr. W. R. Acott and Flt. Lt. E. G. C. Leatham of 544 Squadron in DZ538 on 26 March, but bad weather interfered. Next night all went well over a target in Northern France. During April DZ538 was used four more times. By 8 September, when the Mk. IVs operated by night for the last time, 29 sorties had been flown for the loss of two aircraft. One was DZ600 which fell to British night fighters six miles from Benson on one of the distressing occasions when wrong identification occurred. After the summer of 1943, when the Mk. IX Mosquitoes were available in useful numbers, night reconnaissance was increasingly used for tactical intelligence purposes.

A special day-reconnaissance Norwegian squadron, 333, was formed at Leuchars in May under Commander Lanbrechts, and supplied with Mk. IIs with long-range tanks. Mosquitoes equipped 'B' Flight, the remainder of the squadron using Catalinas based at Woodhaven. Special training was given at 8 O.T.U. Then, the Norwegians specialized in day reconnaissance flights along the coastline of their homeland, observing enemy shipping movements and engaging such enemy aircraft as ventured near enough. The first sortie was flown on 27 May, 1943, along the coast between Stavanger and Lister. As often, it was a misty day and there was no enemy activity to report. A Dornier 24 was shot down by G-George west of Karnoy on 13 June, the first enemy to fall to the squadron. 333 Squadron carried out similar duties under Coastal Command almost to the end of the war, and used Mk. VIs from September, 1943, onwards.

LR405, the first P.R. Mk. IX, with Merlin 72 two-stage engines, but no pressure cabin, was delivered to 540 Squadron on 29 May, 1943, and was joined by LR406 the same day. A standard feature of the Mk. IX was the small blister on the cabin roof. 540 Squadron, now with a third Mk. IX on charge, LR408, introduced the new version to operations on 20 June when Flg. Off. Clutterbuck set out for Zeitz and Jena in LR406. Smoke poured into the cabin from behind the transmitter as the aircraft crossed the Dutch coast, so the afternoon's sortie had to be abandoned. Flg. Off. R. A. Hosking soon after took off for a damage-assessment flight in LR405, successfully photographing Augsburg and the Ober-Faffenhofen airfields. Not until 26 October was a PR. IX, LR420, lost on operations.

544, the second Mosquito P.R. Squadron commenced operations on

Operations: Photographic Reconnaissance

13 September sending LR431 flown by Flt. Lt. R. L. C. Blyth to Vannes.[1] This was a night mission but use of the Mk. IX made daylight operations possible. The first of these, to the Caen area, came on 15 September. Six days later LR431 became the first Mk. IX to operate over Norway, flying from Leuchars.

544 Squadron ranged far in October, airfields around Berlin and Rechlin being photographed by Merifield, on 4 October. A notable flight by Merifield on 12 October took him to Trier, Regensburg, Linz, Vienna, Budapest and back to Vienna, Sarbono, Bucharest, Foggia and to Catania, where LR417's engines stopped as he taxied into a dispersal bay. His entire 760-gallon fuel load had been consumed. Much of the flight was made at 30,000 feet at 2,300 r.p.m., with $+2$ boost for two hours, at a true speed of 340 m.p.h., the remainder being at 2,100 r.p.m. at zero boost. 1,900 miles were covered in six and a half hours at an average speed of 292 m.p.h. and a fuel consumption of 2·5 miles per gallon. No enemy fighters were seen. Five days later Merifield reconnoitred Poznan.

During the rest of the month 544 Squadron's Mosquitoes reached Toulouse, Schaffhausen, Munich, Trento, Lyons and the Pyrenees. 'B' Flight of the squadron was absorbed by 541 Squadron at Gibraltar on 17 October and a new 'B' Flight was formed in Britain, two days after 544 had ended operations with Spitfires, subsequently using only Mosquitoes.

Following the attack on the Antheor viaduct in the extreme south of France by 617 Squadron, Wg. Cdr. D. C. B. Walker flew to photograph the area in LR478 on 11 November, but failed to return. Evidence of the effectiveness of this, and another attack made on Modane, were especially required to assess the value of using the stabilized automatic bomb-sight. Several other missions were accordingly flown until 29 November, when MM247 returned with photographs of the Antheor viaduct, and of a bridge near Cannes bombed in error.

For their night operations late in 1943, 544 Squadron stipulated runs of two minutes at 23,000 feet to suit the cameras and flashes. The south of France and Northern Italy were now the areas against which the night sorties were directed.

One of the most valuable services that the Mosquito reconnaissance squadrons performed during 1942 and 1943 was the photographing of the V-weapon development centres and launching sites. Vague reports from agents led the R.A.F. to send Mosquitoes from 540 Squadron to areas near the Baltic Coast, where the enemy was believed to be developing a novel long-range bombardment weapon. Thus it was that the importance of the research station at Peenemünde was discovered. Plt. Off. W. J. White, sent to Stettin and Politz on 22 April in DZ473, returned with some photographs. Others were taken of Peenemünde on

[1] 544 Sqn. received its first PR. IX LR478 on 22.10.43. Both 540 and 544 Sqns. often used Benson pool aircraft.

Operations: Photographic Reconnaissance

2 June from DZ419, and on 12 and 23 June from DZ473. Prints were closely scrutinized, and revealed two objects resembling large rockets. Consequently Bomber Command despatched 597 aircraft to Peenemunde on 17–18 August to destroy the establishment and its staff. Flg. Off. Hosking in LR413 left Leuchars at 07.00 on the 20th for Wismar and returned home at lunchtime, with photographs of the results of the Peenemünde raid, and of a strip in the Gieswald area.

Varied reports filtered through during the summer suggesting that, in addition to the long-range rockets, the Germans were developing a small pilotless aeroplane. Trials were, it was said, being undertaken from Zempin, near Peenemünde, which was frequently photographed by Mosquitoes. On 3 October Flt. Off. Babington-Smith, working at Medmenham, observed a small aircraft on the edge of Peenemünde airfield, and when earlier photographs were re-examined others were found. In the third week of October reports came to hand of a concrete construction in the Bois Carré, ten miles north-east of Abbeville, photographed on 28 October by the redoubtable Hosking in LR424. Analysis showed an object resembling the shape of a ski, really a long concrete platform, and three huts.

Reconnaissance Mosquitoes and Spitfires were immediately switched in an attempt to photograph the whole of northern France. Discovery of a large number of similar 'ski sites' followed. These scouring flights, part of Operation *Crossbow*, began on 30 October when Flg. Off. S. I. Baird in DZ424 covered the Pas de Calais and Dieppe regions. It was at this time that 140 Squadron, which concentrated on tactical photographic reconnaissance, began using Mosquitoes. Its first sortie was flown by Wg. Cdr. Bowen, who took MM248 to Port en Bassin on a very-low-level reconnaissance on 11 November. By the end of the month ten sorties had been flown by 140 Squadron all to the Low Countries and Northern France.

Bomber Command was meanwhile engaged in the so-called Battle of Berlin. Damage-assessment photographs were urgently needed, but persistent cloud over Berlin prevented their being taken. Twice on 24 November 540 Squadron tried, and again on 28th, but to no avail. Merifield flying LR428 on the latter occasion decided to take a look at Peenemünde which he found to be clear, likewise Zempin. His photographs were hurried to the C.P.I., where upon examination the pictures revealed a 'ski site' like those seen in Northern France. Furthermore, one of the small aircraft could be seen on the ramp.

Peenemünde consequently came under frequent observation. Merifield flew there in LR424 on 8 December after again finding Berlin cloud clad. This time he reported seeing vapour trails in the vicinity of the research station, conceivably left by a rocket. He flew towards Wismar, Rostock, Warnemünde then Esbjerg, whence he reported seeing two trails. The Germans tried too late to conceal their work, but on 20th succeeded in placing a smoke screen over the area of interest. So, LR424,

Operations: Photographic Reconnaissance

on that date, contented its masters by obtaining proof that the *von Scheer* and *Leipzig* had both weighed anchor in their Baltic lair.

Apart from *Crossbow* reconnaissance operations Mosquitoes had been making other important flights. Innsbruck was visited from Benson on 15 July and again on 3 August, when LR415 overflew the Brenner Pass and Verona to land at La Marsa and return to Benson the following day—via Gibraltar. DZ473 covered the Narvik and Bodo regions on 3 September and Dresden on the 6th. Plt. Off. P. J. Hugo set out for Belgrade on 18 September, but had engine trouble and aborted. Next day he made the flight, refuelled at Bo Rizzo, and returned via Gibraltar.

The final PR. IV sortie came on 10 September, when Sqn. Ldr. Hughes made a shipping reconnaissance flight in DZ473 off Kristiansund and Aalesund. Air Cdre. J. N. Boothman, A.F.C., with Flg. Off. J. F. Fielden as his navigator, was last to take a Mk. VIII on operations, when he flew to the Roanne area and continued to North Africa from where he returned after a short stay. Whereas a year before the Mosquitoes were making a few sorties per week, on this day, 4 October, 1943, seven sorties were despatched by 540 Squadron alone, to such varied places as Munich assessing bomb damage, Breslau, the Zuyder Zee, Toulouse and Foix-Grenoble. The immunity and success of the Mosquito were abundantly clear now, to friend and foe.

As winter developed the stratus base fell to around 23,000 feet, and high-altitude Mosquitoes could not avoid leaving trails. 1409 Meteorological Flight was now operating at 30,000 feet irrespective of contrails, for these had the advantage that enemy fighters could be seen and action could thus be taken if necessary. P.R. aircraft then adopted the practice of flying at about 33,000 feet to ensure that any enemy fighter climbing after them would be as conspicuous as they. Mosquito contrails also attracted increasingly accurate flak, but of course there was no way of preventing them.

Engine failures during December further aggravated the operational outlook, and failure of fuel to reach the carburettor became so serious that the Mk. IXs were taken off operations until modifications, including the blanking of the fuel cooler to prevent freezing up, had been incorporated. Of five Mk. IXs flown over Scotland at 30,000 feet in December at a temperature of −50°C. four had engine failures and from one the crew baled out. These annoying troubles were cured by the end of January, 1944. As early as 29 November Sqn. Ldr. B. G. Aston had flown LR425 to 34,000 feet with the modifications mentioned, and encountered no trouble.

On 1 December, 1943, 544 Squadron tested the use of *Gee*,[1] but two days later the Mk. IXs were grounded for the linking of fuel supplies

---

[1] *Gee* was a system by which the navigator could calculate the position of his aircraft by observing the time taken to receive pulse signals from 3 different ground stations. The navigator had a chart showing curves of distances from the stations.

Operations: Photographic Reconnaissance

from main and long-range tanks, and the blanking. Hoar frost on the main filter, and pressure differences were found responsible for the troubles, which fortunately came at a time when the weather over Europe would have made effective sorties less likely. Interception avoidance techniques were meanwhile worked out over East Anglia with the help of Spitfire HF VIs of 124 Squadron. In an attempt to discover the effectiveness of *Wurzburg* radar five crews were trained under 106 Group instructions to use special gear named *Boozer* in an effort to discover whether the German equipment was locking on to Mosquitoes.

In the struggle for extra ceiling a Mk. IX was experimentally fitted with Merlin 76/77 engines and paddle-blade airscrews at Benson in January, 1944. By the end of February four such conversions had appeared, affording a noticeable improvement in ceiling and speed at high altitudes. Four more conversions soon followed.

Persistent trouble was being encountered by canopies icing up, as the clamour held for higher Mosquito ceiling than the 35,000 or so feet obtainable with the Mk. IX. Unfortunately the Mk. XVI with pressure cabin, which entered production in November, 1943, offered no advantage in its present form. Examples first released in December, 1943, to 140 and 400 Squadrons were without cockpit heating due to non-delivery of heat exchangers, and crews complained of the cold after one hour at 30,000 feet. Operational flights were usually planned for at least three hours' duration. Squadrons reckoned the ceiling of the Mk. XVI to be about 36,000 feet, but before these aircraft could operate they needed to be fitted with top and side rear view cabin blisters. Without them the squadron commanders were adamant about their unsuitability for operations. *Boozer* and *Rebecca H* were fitted at this time to the XVIs of the Tactical Air Force squadrons.

First deliveries of PR. XVIs to the United States 8th Air Force in Britain, mainly for weather reconnaissance, occurred in February, 1944, MM310 reaching Langford Lodge on 2nd, MM338/G Burtonwood on 9th and MM308 Alconbury on 22 February. Soon a handful of the twenty XVIs received by May were at Alconbury and used for experimental radar work by the 482nd Bomb Group, which later tested *H2X*, the American form of *H2S*, in the noses of six specially modified Mosquitoes.

Meanwhile the Mk. IXs were getting back into their stride and on 4 January, 1944, W. Off. W. Kennedy in MM242 was surprised northwest of Beauvais by nine Fw 190s. He flew off at full power, miraculously escaped and then photographed Paris, Chartres and Châteaudun. Flg. Off. A. P. Morgan, a navigator, on 544 Squadron, had an even more frightening experience on 23 February when the oxygen tube broke away from his mask. Throttles were quickly closed and the pilot, W. Off. Kennedy, replaced the tube and quickly brought LR434 into Manston.

In March and April of 1944 serviceability was seriously affected by

Operations: Photographic Reconnaissance

internal misting and the formation of ice crystals on the pilot's and observer's windscreens, and the icing-up of the non-sandwich windows. Urgent work was put in hand to cure the troubles, at a time when pressure of work on the squadrons was almost at its greatest. Douglas-Hamilton tested LR434 on 5 March, as soon as it had been fitted with paddle-blade propellers, and found that whilst these slowed the aircraft at lower levels, at 35,000 feet there was an increase in speed. Next day this aircraft was despatched on operations and obtained photographs of Swinemünde, Peenemünde and Copenhagen.

Ten days later MM240 with Merlin 76/77s and paddle-blades was found to have a ceiling 3,000 feet greater than earlier aircraft, but a second test four days later showed a need for a C.S.U. to cure surge trouble. Views on the efficiency of the XVIs at this time were mixed. 544 Squadron considered them inferior, for still there were neither side nor top blisters, and frosted windows made operations impossible. The value of a pressure cabin for heights below 38,000 feet was doubted. MM352 now was set aside for cockpit de-misting and icing problems, and not until 18 May were these confirmed as cured.

During the early weeks of 1944 Mosquitoes were busy photographing the assortment of workings to be found in Northern France which 544 Squadron began to review on 28 October. 140 Squadron worked out a technique to deal with them, whereby its Mosquitoes arrived in the target area at about 20,000 feet and dived at 410–430 I.A.S. to 9–10,000 feet, took photographs, and continued in a controlled descent back to base at Hartford Bridge. 140 Squadron despatched thirty such sorties during December, 1943, and January, 1944. One of the targets in a quarry at Wizernes called for low-level oblique photography, better achieved with a forward-facing camera. This approach had two advantages over that of the Spitfire's use of oblique cameras; it was easy to sight by flying straight at the target, and there was hardly any movement because the target came towards the camera instead of across it. On 4 February 140 Squadron became the first to operate a Mk. XVI, when MM279 flew to the Cabourg-St. Aubin area. Regrettably she abandoned her mission due to presence of American Fortresses, fighters and clouds.

An urgent need of photographs of Berlin in late 1943 has already been referred to. For several weeks the weather was persistently bad and none could be obtained. Fifty-eight sorties, of which fifteen were unsuccessfully intercepted, were made to the City before any satisfactory pictures were obtained. During this period as many as six German fighters were once sighted in formation near Berlin at 42,000 feet. Spitfires were now being sent there because they could operate at around 40,000 feet, although on their fuel load they could only just reach the City. The extreme altitude had lately caused some of the pilots physiological troubles—several could not remember to where they had flown. The first pictures of Berlin from a 540 Squadron Mosquito were obtained by an Australian, Flg. Off. Holland, whose navigator was Flg.

Operations: Photographic Reconnaissance

Off. Bloomfield. LR424 returned with the results of four clear runs over Berlin on 19 February. She had left on full power, pursued by flak bursts. LR428, flown to Berlin by Flt. Lt. A. C. Graham the same day, was intercepted by two enemy aircraft at 26,000 feet when only ten minutes from the City. Forced to turn back, drop tanks were jettisoned. As the starboard one fell away it damaged the tailplane and nacelle, but by opening up Graham left the enemy aircraft behind.

The three Mosquito reconnaissance squadrons assigned to the 2nd Tactical Air Force were busily engaged in the early months of 1944. 4 Squadron was equipped with Mk. XVIs and concentrated on targets in Northern France. Conversion training was undertaken at Aston Down, and on 18 February it was planned to move the unit ('A' Flight of which flew Spitfires only) to North Weald. The intended home, Sawbridgeworth, was described as 'mud and marshland'. However, it was into this that 4 Squadron moved on 3 March, four Spitfires leading the six Mosquitoes which were followed by eight Spitfires and four Mustangs, making a stylish fly-by on arrival. 'B' Flight's first task was the production of mosaic of the King's Lynn-Colchester-Hemel Hempstead-Rugby area, required for army exercises by 22 March. It called for twenty runs of eighty exposures—and fine weather.

Flt. Lt. C. T. P. Stephenson, D.F.C., made the first high-level operational P.R. sortie by the Squadron on 20 March, when he flew MM313 close to the coast of France. Eight enemy fighters were seen below, but they did not interfere with his progress. It was a cloudy day, and he returned with shots only of Le Touquet from 30,000 feet. The fourth operation on 26th was the first to be successful, and in the course of which the Roubaix-Ath-Mons region was recorded on the cameras. There were three more successful P.R. flights before the squadron moved to Gatwick. With many obstructions on the approach, and runways barely long enough for the Mosquito, Gatwick was hardly an ideal choice. A day after their arrival the squadron received their first two paddle-blade-airscrew aircraft from Benson. During the first ten days of the month fourteen sorties had been flown over Amiens, Compeign, Cherbourg, Dieppe and the rivers in the Douai area, obtaining photographs for future use by the army, as well as further *Crossbow* records. Flt. Lt. L. W. Lowther and Sgt. J. White, flying MM273, went to Pointe d'Ailly and Bellencombe on 25th. During the flight they observed a large circle of foam eight miles west of Le Treport. They recorded on return that they were 'still puzzling over whether it was a submarine, a jettisoned bomb or even the Loch Ness monster gone astray'.

During the remainder of April and May the work over Northern France continued; then, unexpectedly, came news that the squadron was to cease Mosquito operations, the last of which was flown by MM309:U on 20 May, to the obscured Lille area. Conversion of 'B Flight to Spitfires had begun on 11 May.

Operations: Photographic Reconnaissance

400 Squadron received its first two Mosquitoes, Mk. XVIs, on 22 December, 1943. Due to the aforementioned troubles it was 26 March before the first operational flight, by Sqn. Ldr. P. Bissby in MM284, took place. Rivers in Northern France were the main objectives during forty-three Mosquito sorties, the last of which came on 2 May, when Flt. Lt. H. R. Pinsent photographed the Caen region. Brest, Denain and Reims were visited by the squadron's aircraft two of which, MM284 and MM356, took photographs for a mosaic of the Cherbourg Peninsula on 30 April. The Squadron's six Mosquitoes were taken off front-line status on 12 May and off establishment on 29 May.

Squadrons of the Tactical Air Force were now busily employed, and thus it was that MM251 flew over Ouistreham in Normandy photographing its beach on 6 April, and again on 11 April when Flg. Off. J. G. Bishop of 140 Squadron repeated the task—to obtain material for use in connection with the D-Day landings. 140 Squadron now concentrated on rivers, harbours, beaches and railways south of Paris. From 10 April Mosquitoes entirely replaced the squadron's Spitfires.

Night photography played a large part in 140 Squadron's operational role. Dr. Rohmer on 30 April on a night exercise in a Mosquito fitted with a K.19 camera and moving film, and, using a Mk. II photoflash, obtained the finest photographs so far, taken from 3,000–5,000 feet. On 4 May 140's first night operational sorties were flown. MM282, coned by searchlights over Abbeville, nevertheless dropped her flashes and photographed the area. Two other Mk. XVIs were less successful, and only visual reconnaissance was possible by the light of flares. However, Flg. Off. R. Batenburg brought MM305/G down to a hundred feet in order to observe roads in the Avranches-Vire region. A week later came tests with the Mk. III flash.

As D-Day approached the squadrons were given important responsibilities. 540 began operations as a whole squadron based at Benson on 1 March, many of their targets now lying in France or southern Europe. Enemy fighters were often seen, LR416 being twice intercepted over Denmark on 6 March. As LR435 approached Paris on 26 March bound for Munich she was intercepted—fortunately by a Mosquito. Two enemy fighters later approached and, caught out, the Mosquito was forced to head homewards.

Mingled with deep penetration flights were such sorties as Sqn. Ldr. Merifield's to Limoges at only 1,000 feet on 10 March in LR422, and a reconnaissance of the coastline near Caen by LR433 on 20th. Merifield was near Vienna on 24 May when six Mustangs intercepted him. Three broke away after flying close for twenty minutes, and two more after he had fired the correct colour Verey lights. The last, still untrusting, refused to leave for some time. 540 Squadron first used the Mk. XVI on operations on 31 May.

Bad weather immediately prior to D-Day almost precluded P.R. operations, 540 Squadron flying none between 2 and 7 June. Then

Operations: Photographic Reconnaissance

LR415 photographed the Saumur tunnel region prior to the night's attack by Lancasters of 617 Squadron. On the night of 5-6 June MM305/G of 140 Squadron, equipped with a 10-inch Fairchild camera, dropped flashes in the Châteaudun/Freteval area revealing traffic movements, and four others of the squadron were similarly employed elsewhere. On D-Day MM279, the first Mk. XVI to fail to return, was lost in the Montdidier-Cambrai region. To check on troop movements on D-Day four of 140 Squadron's aircraft made a dawn get-away to the Abbeville-Amiens-Pontoise areas. A frightening experience befell Plt. Off. G. H. Ardley in MM250, for the hood came off his aircraft. Repaired, the machine was off again at 16.45 hours, photographing the Normandy beachhead between Trouville and St. Vaart. Shortly before midnight four more sorties were despatched, to light, observe and photograph roads in the Liseaux and St. Lo regions. Throughout June the work was continued and 177 sorties flown.

To 544 Squadron went the high-priority task of twice daily obtaining coverage of the rail networks of Central and Southern France, using 5 or 6-inch front-lens cameras. Like 140 Squadron, its crews flew out at high-level, descended and flew home low. On D-Day eight sorties were flown and Flt. Lt. Hampson's aircraft, NS500, damaged by flak over Saumur, had its rudder cables cut, electrics put out of action, engine instruments rendered useless, radiators jammed shut and three other hits. Next day ten sorties were flown by 544 Squadron; included amongst them were the first two by Mk. XVIs of the unit, to Toulouse. Railway surveillance of Western Germany soon became the major task, until 1 July when the squadron took over deep-penetration flights to East and Central Germany.

Fear of special attacks by novel naval forces caused a patrol by 540 Squadron at 3,000 feet between Cherbourg and Le Havre on 10 June. Flt. Sgt. McGoldrick and Flt. Sgt. Inskeep of 540 Squadron, were on a reconnaissance of Coswig on 19 June when at 29,000 feet their starboard engine was hit. They switched off their electrics and headed for Den Helder, by which time the engine fire was out. They were losing 400 feet per minute, and after an hour's flying had been forced down to 18,000 feet and an I.A.S. of 140 m.p.h. To escape from view they skimmed across a 12,000-foot layer of strato-cumulus. LR422 was nursed to Terschelling and then to Southwold. Their return track on one engine was 640 miles. Eventually they landed safely at Benson, bearing wonderful testimony to the qualities of the Mosquito—and its crew.

A second interception by friendly American aircraft took place on 5 August, after LR433 had set out for Gdynia and Stettin. Three Fw 190s intercepted the Mosquito thirty miles south-west of Bremen, but the aircraft soon escaped. Five minutes later six American P-47s closed in from port. They jettisoned their drop tanks and positioned themselves for the attack, staved off only when the Mosquito's crew fired the colour of the day. A little later two Fw 190s were seen, but they did not engage.

Operations: Photographic Reconnaissance

Installation of K.19 cameras in 140 Squadron's aircraft was carried out during July, 1944, and *Rebecca H* fitted in an aircraft which, by the close of the month, also carried *Gee*. Using a beacon situated on Beachy Head very accurate navigation was now possible to a distance of fifty miles. 80/88 of 140 Squadron's night sorties were judged effective this month and 88/113 by day. Surveillance of roads around Caen occupied much of the squadron's effort.

Much of the time both strategic reconnaissance squadrons were photographing troop concentrations and movements, many of which were by rail. Between 2,000 and 3,000 miles of track were daily surveyed, so that trains and their loading could be regularly and continuously analysed, and daily estimates of army strength along and behind the fronts compiled. In fair weather the high-altitude photographs supplied the information required, but cloud frequently drove the aircraft very low. It was not easy for Spitfire pilots to photograph particular railway lines, whereas the navigator in the nose of the Mosquito could readily direct his pilot. Since much of the enemy traffic moved at night, photographic missions employing photoflash equipment were now of great value. 140 Squadron flew 164 daylight and 66/67 successful night sorties in August, 1944. Only four of its aircraft were damaged, in all cases when they flew low over vital areas.

An exceptional flight by NS504 is worthy of mention here. It left for the detachment of 544 Squadron at Leuchars on 9 July and, flown by Flt. Lt. F. L. Dodd and Flt. Sgt. E. Hill, proceeded on a daylight visual reconnaissance to Statlandet, Narvik and the Lofoten Islands, a return flight of 7 hours 45 minutes ranging in height from 6,000 to 24,000 feet. Weather precluded a refuelling stop on the flight home. This proved to be one of the longest-ever flights by a Mk. IX or XVI in the European conflict.

An equally remarkable experience befell Flg. Off. H. R. Vickers on 19 August after he was intercepted by two Bf 109s over Western Germany. He nursed MM354 to the Châteaudun area, where sudden engine failure forced him down. He crashed in a wood, and the machine ricocheted into a field. Miraculously he was able to walk from the aircraft, and immediately joined a retreating American patrol!

When the U.S. 8th Air Force formed a meteorological reconnaissance flight it took advice from Bomber Command. The Americans were particularly interested in cloud level observations by the light of flares. For weather scouting, photography (especially at night) and precise navigation at long range, the Americans found their Mosquito PR. XVIs, supplied under reverse Lend-Lease, a great change from their B-17s and B-24s, and very different from the Lightnings they had used. Mosquitoes were employed where brushes with the enemy were likely and high speed necessary, leaving the large bombers to reconnoitre the Atlantic.

Twelve American M46 photoflashes each of 700,000 candlepower could be carried in a Mosquito bomb bay, and the R.A.F. developed

Two-stage Merlin Reconnaissance Mosquitoes

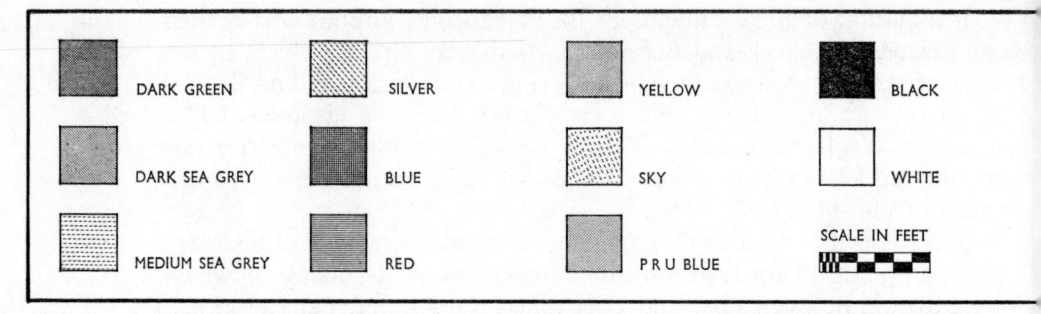

■ DARK GREEN	▨ SILVER	■ YELLOW	■ BLACK
■ DARK SEA GREY	■ BLUE	▨ SKY	□ WHITE
▨ MEDIUM SEA GREY	■ RED	▨ PRU BLUE	SCALE IN FEET

Mosquito PR. XVI with A.E.A.F. identity markings, in June, 1944. Black and white stripes were applied to many Allied aircraft between 4 June and 6 June, 1944, rendering them easily identifiable to Allied forces. Stripes were retained in full during the summer of 1944, but later often appeared only on the undersides of aircraft involved in tactical operations. Early in 1945 a narrow white ring was added to the roundels above the wings, and often to the fuselage roundels on reconnaissance aircraft. No roundels were painted beneath the wings of reconnaissance Mosquitoes. Mosquitoes used by the U.S. 8th A.A.F. had markings similar to those on R.A.F. aircraft. On 16 August, 1944, an order was given as a result of which both surfaces of vertical and horizontal controls, also the fin, were painted 'crimson red' on the American weather reconnaissance Mosquito XVIs. From 23 September the entire tail surfaces were red.

Mosquito PR. 34 (Merlin 114) PF662 was delivered 28.2.46. Used by 58 Squadron then 540 Squadron (as depicted; recorded July, 1950). Modified into PR. 34A and then used by 81 Squadron, F.E.A.F., 4.54 to 8.55.

Mosquito PR. IX (Merlin 72) LR416 reached 540 Sqn. 4.7.43, and first operated 10.8.43. The intended sortie to Northern Italy was abandoned due to trouble with drop tanks. Next day it photographed Friedrichshafen, flown by Wg. Cdr. Douglas-Hamilton who over-flew Europe and landed at La Marsa and made a return sortie on 15 August. LR416 flew ten sorties in 1943, targets photographed including Toulouse (30.8.43), Trento and Venice (16.9.43), Modane (18.9.43) and Leipzig (23.12.43). On 5 January, 1944, it attempted to photograph Berlin, and on 25.3.44 secured photographs of the Siegfried Line. It took-off on its eighth 1944 sortie on 13.4.44 leaving Benson for Munich at 09.50 hrs. As a result of trouble it made an early return, and was destroyed in a crash at Abingdon, near Oxford.

Mosquito PR. XVI (Merlin 72) NS502 reached 544 Squadron 23.5.44 and operated over Europe. Crashed 21.2.45 and passed to Martin Hearn for repair. To Royal Navy, Fleetlands, 7.11.47.

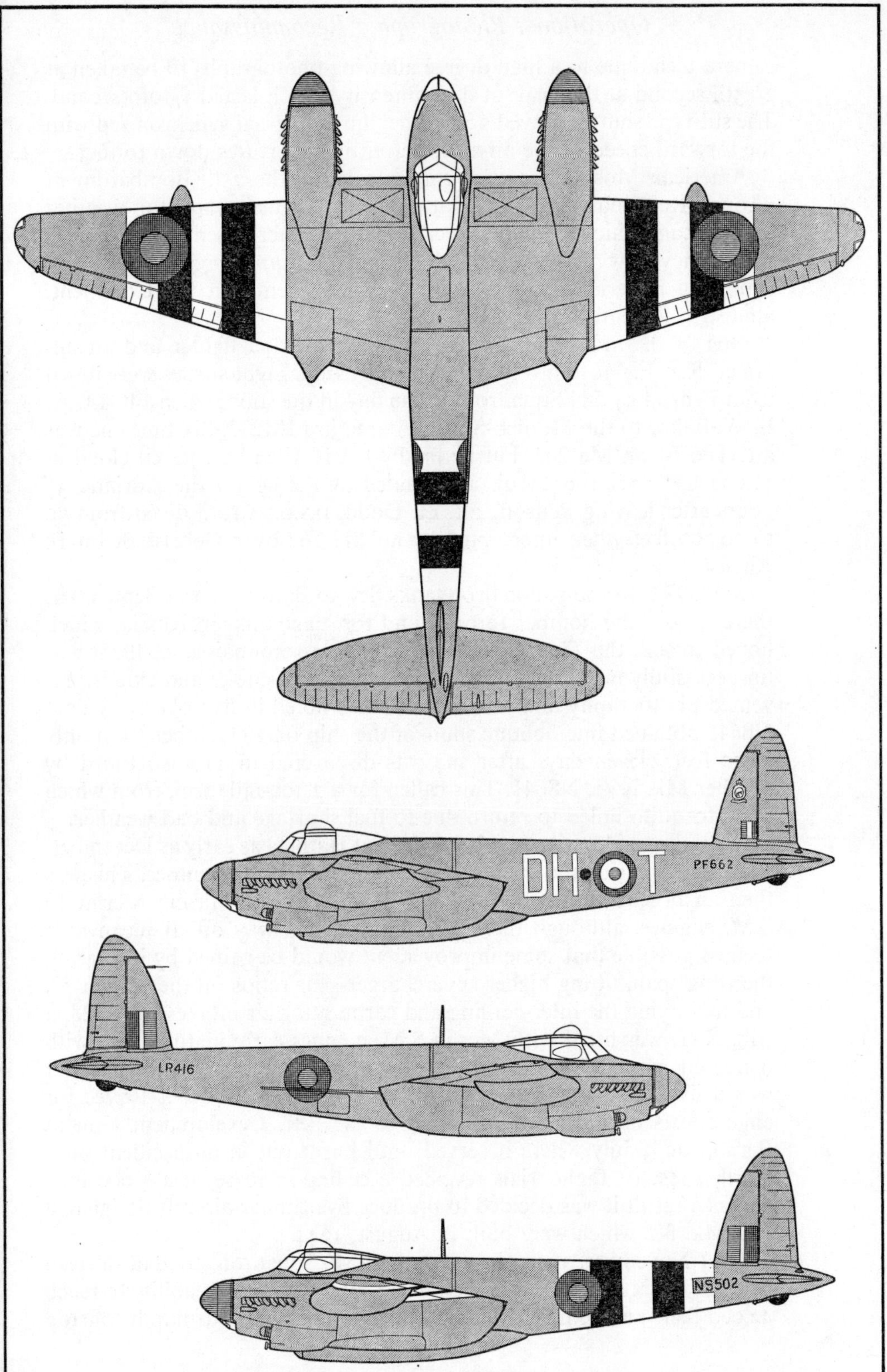
DH T
PF662
LR416
NS502
GB

Operations: Photographic Reconnaissance

camera technique to a high degree allowing photographs to be taken at 1/25th second at the peak of flash intensity which lasted 1/10th second. The slit-type shutter moved across the film at a speed synchronized with the forward speed of the aircraft, permitting apertures down to f6·3.

American Mosquitoes were concentrated in the 25th Bombardment Group (Reconnaissance) at Watton from August, 1944, for weather and photographic reconnaissance. Some were later based at Harrington where they flew with the 492nd Group on *Carpetbagger* operations, supplying material and messages to Allied agents on the Continent, almost to the end of the war.

One of the first combats between a German jet fighter and an unarmed R.A.F. Mosquito took place on 25 July. Eight sorties were flown from Benson by 544 Squadron on the day in question, when Flt. Lt. A. E. Wall flew to the Munich-Stuttgart area in MM273. Six times he was attacked by an Me 262, but he finally left it when he entered cloud at 16,000 feet over the Tyrol, and landed at Fermo on the Adriatic 3½ hours after leaving Benson. Flt. Lt. Dodd, D.S.O., A.F.C., dived from 30 to 20,000 feet when intercepted by an Me 262 over Gebelstadt on 18 August.

MM397 with 100-gallon drop tanks flew to Bardney on 11 September, there to join the bomber force bound for Yagodnik, in Russia, which hoped to sink the *Tirpitz* on return. The first reconnaissance flight was unsuccessfully made on 14 September. Next day smoke and clouds prevented photography and on 16th '397 was holed in five places by flak. NS643 obtained fine oblique shots of the ship on 29 October from only 4,500 feet, eleven days after she was discovered in Tromsö Fiord by another Mk. XVI, NS641. This called for a 2,100-mile trip, from which one Mosquito failed to return due to fuel shortage and bad weather.

Special consideration had been given at Hatfield as early as December, 1942, to increasing the operational ceiling of P.R. Mosquitoes. This was then an urgent requirement which might be met by fitting Merlin 15 S.M. engines, although these were then a long way off. It meanwhile seemed possible that some improvement would be gained by increasing the wing span, fitting higher supercharger gear ratios on the Merlin 72, and modifying the inter-cooling and carburettor air intakes. MM328, a Mk. XVI, was fitted with Merlin S.M. engines early in 1944 and, with increased span and decreased tare weight, underwent trials for which it was delivered to Benson on 20 May, then passed to Rolls-Royce for engine tests on 4 June. It returned to the P.R. Development Unit at Benson on 3 July where it served until burnt out in an accident on 1 April, 1945. Its flight trials revealed a ceiling increase of 4–5,000 feet, and as a result it was decided to produce five similar aircraft designated PR. Mk. 32s which were built in August, 1944.

On 1 November, 1944, the strength of 540 Squadron stood at thirteen Mk. XVIs, six Mk. IXs and one PR. 32. With the latter's ability to reach 42,000 feet, and reduced weight conferring an extra 30 m.p.h. on top

Operations: Photographic Reconnaissance

speed, the PR. 32 fortunately appeared when the enemy jet and rocket fighters were becoming more of a nuisance.

A typical interception by an Me 262 was described by Flt. Lt. Watson, R.A.A.F., who with his observer, Flg. Off. Pickup, was flying in the Munich area in September. At 29,000 feet they detected two Me 262s coming up behind them from the Messerschmitt factory. For fifteen minutes the jets attacked independently in wide circles and sweeps, the Mosquito avoiding every burst of fire by tight turns until it was down to ground level. The jets continued to attack, pulling out of their firing dive each time at about 1,000 feet, and the Mosquito, by weaving tightly amongst the tree tops (and collecting a piece of pine tree which smashed through the observer's nose window to become a treasured souvenir) avoided destruction. Going south, Watson crossed the Brenner Pass at nought feet and then, having shaken off the pursuers, who were probably running low on fuel, climbed to check his state and eventually landed safely in Italy. The total attack had lasted thirty minutes. Flt. Lt. Dodd and Flt. Sgt. Hill in NS639 had a similar experience near Munich on 16 September, when two Me 262s attacked them eight times until they found cloud cover at 6,000 feet.

Wg. Cdr. H. W. Ball, D.S.O., D.F.C., flying with Flt. Lt. E. G. Leatham, D.F.C., was first to fly a Mk. 32 on operations. He took off in NS589 at 11.30 a.m. on 5 December to photograph Darmstadt and Mannheim through cloud breaks. Two unidentified fighters intercepted, forcing him to dive from 38,500 feet to 22,000 feet.

During her fourteen operational sorties NS589 was used exclusively to photograph targets connected with the railway system in Western Germany and on 14 March, 1945, during a mission to Munster, Hamm, Gladbach and Rheine, secured photographs of the broken viaduct at Bielefeld following the initial use of the 22,000 lb. bomb. Fifteen sorties were flown by NS588 and only one by 540's other PR. 32, NS582, which flew to Nürnberg and Heilbronn at 40,000 feet on 25 December, 1944, in the hands of Sqn. Ldr. G. Watson. A run over Worms and another over Karlsruhe were included.

The original intention was that 544 Squadron should use all the Mk. 32s. Eventually only one was allotted, NS587, exhaustively trial-flown in November and found to be wanting, its engines suffering from over heating. There was a lack of cockpit heating, and icing troubles as encountered with the early Mk. XVIs. Her first operational flight came on 23 December, when oil targets at Hamburg and Magdeburg were photographed. Over the latter target the starboard engine had to be stopped and the propeller feathered. An Fw 190 attacked over Emden at 19,000 feet, but the PR. 32 escaped into cloud cover. During the engagement the propeller succeeded in unfeathering itself. NS587 flew only two other sorties, to Denmark on 2 April and to the Lake Constance area on 9 April, when, flown by Flt. Lt. G. E. Grover, 42,500 feet was reached.

Operations: Photographic Reconnaissance

Mosquito versatility was ever evident. In contrast to the high flying of the PR. 32s there were some very-low-level flights. These occurred in the early days of the reconnaissance Mosquito and again came to the fore on 4 August, 1944, when a forward-facing camera was installed in each drop tank of MM235. Flt. Lt. D. Adcock took the aeroplane to Watten V-weapon site on a low-level damage-assessment evening flight, but found the approaches to the target difficult and the flak too intense to permit photography.

The aircraft's second sortie to Forêt de Nieppe on 11 August resulted in its loss. Another Mk. IX, LR432, was then suitably modified and Wg. Cdr. D. W. Steventon, D.S.O., D.F.C., first used it on 21 August to secure low-level photographs of a radio station near Ijmuiden. A Mk. VI, PZ345, able to shoot at defenders, proved more suitable to the task. It made several sorties, thus employed by 544 Squadron. On 24 September Wg. Cdr. Steventon and Flt. Lt. A. M. Cross, D.F.M., flew in '345', now christened 'Dicer', to a radar station at Tybyaert, East Denmark, where they also shot up a factory and collected two flak holes and a dent resulting from bird collision.

Shrinkage of enemy territory brought about by the Allied advance meant that flights could now begin in France or Belgium, and penetrate over the remaining region held by the enemy. 140 Squadron maintained their tactical role, rarely finding enemy fighters yet operating by day and night. In December, 1944, they flew 47 night sorties, and a further 33 in January, 1945, in addition to 40/51 successful day sorties in December. On one of the few occasions when their aircraft saw an enemy fighter MM349 in the hands of Flt. Lt. L. T. Butt, D.F.C., sighted an Me 163 and a '262 at the end of November, but no attack materialized.

Operation *Frugal* began on 9 October, 1944. It called for flights by 544 Squadron direct to the Soviet Union by way of Memel and thence to Ramenskoye, a journey on average of $6\frac{1}{2}$ hours. Later, Vnukovo/Moscow was chosen as the terminus. Twenty-seven single flights in fourteen days were made carrying mail between Northolt and Moscow during the Summit Conference, the average speeds from take-off to landing being 314 m.p.h. east-bound and 300 m.p.h. west-bound. On his return the Prime Minister sent a message of thanks for the punctuality of his mail. These flights were not without incident, for upon one a Mosquito encountered three enemy fighters and shortly after an Me 262; another was intercepted by an Me 163 near Dortmund, but escaped. Operation *Frugal* also called for flights to Cairo and Naples. A second use of the squadron's Mosquitoes as mail courier aircraft came between 31 January and 20 February, 1945. PR. Mk. XVIs were used fitted with 50-gallon drop tanks, but they carried no P.R. gear. For this, Operation *Haycock*, the supplying of a daily air-letter service between Benson and the Crimea or Cairo in connection with the Argonaut Summit Conference at Yalta, many flights were made. Sometimes the Mosquitoes flew to San Severo, then to Saki in the Crimea; alternatively the route lay by way of Malta.

67. High-level P.R. photograph of Gdynia, 2 June, 1942

68. Gnôme Rhône Limoges factory, 10 March, 1944

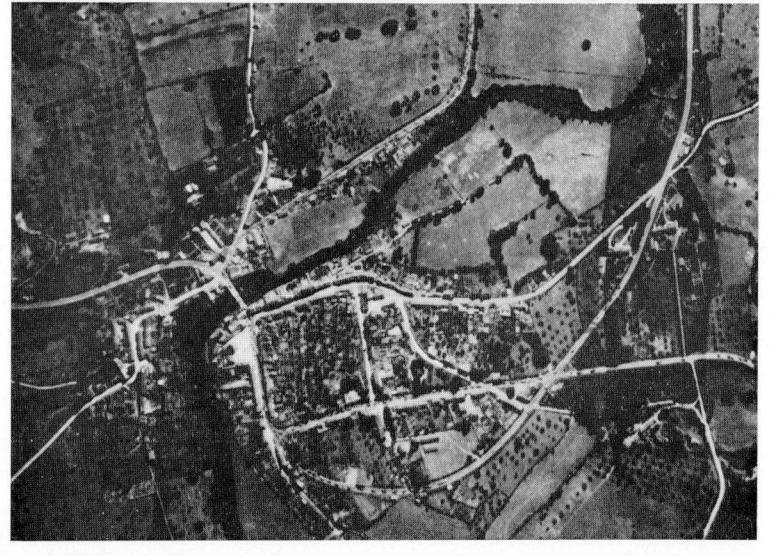

69. Enemy night movement in 1944, photographed using photo-flash

70.
RDF station at Bergen-Am-Zee taken using forward facing camera

71.
Geoffrey de Havilland checks his flight test log after flying a Mk. VI

72.
La Pallice photographed during the first Mosquito operational sortie

73.
The Phillips works, Eindhoven, after 2 Group's raid in December, 1942

74.
Two Mk. IV bombers of 105 Squadron, DZ353 and DZ367

87. Crews back from Berlin on 31 January, 1943
88. Railway workshops and power station under attack Trier, 1 April, 1943

Operations: Photographic Reconnaissance

Needless to say the flights were made without loss, the Mosquito having been chosen for its likely immunity to interception, and justifying the faith bestowed upon it for its important duty. 544 Squadron performed similar work during the later Potsdam Conference.

March, 1945, was a busy month for the reconnaissance squadrons as they eyed the fall of the Third Reich. On 19 March a Mosquito of 140 Squadron was given close escort by Spitfires of 127 Wing as it reconnoitred an autobahn in the Ruhr, a most rare event. To expand its night activities 140 Squadron was now experimenting with external wing racks carrying eight extra photo-flashes. During 1945 the squadron was flying from dawn to dusk, and frequently during the night and at all levels, gathering material for the Allied Armies. In the closing days of the war there were searches for enemy shipping. Fighter interceptions still occurred and three Me 262s chased MM283 of 544 Squadron into clouds near Peenemünde on 9 March.

On 16 March came an exciting engagement between Flg. Off. R. M. Hays in NS795 and two Me 163s. Over Leipzig the two rocket fighters hurtled skywards to either side of the Mosquito flying at 30,000 feet, and simultaneously delivered beam attacks. Hays did a half roll and dived vertically to 12,000 feet, raising his I.A.S. to 480 m.p.h. The 163s were not easily shaken off and made further beam and stern passes, so the Mosquito peeled away to starboard to fly at tree top height. By this time its starboard engine was smoking, and there was little power available. A climb to 2,000 feet was accomplished, by which time the enemy fighters had given up the battle. Then a Bf 109 came upon the crippled machine, which now threaded its way along valleys near Kassel until the enemy fighter was shaken off. Over a small town it encountered flak, but in spite of its severe damage the Mosquito landed at Lille. For his exploit Flg. Off. Hays was awarded an immediate D.F.C.

Although the need for long-range flights over Europe was diminishing, in the Far East, whither the Mosquito squadrons might have been sent, distances to be flown were immense. Very long range was a feature of the PR. Mk. 34, which did not see wartime operational service in Europe. Two-hundred-gallon drop tanks designed for the PR. 34 were experimentally fitted to PR. XVI MM283 by 544 Squadron. When fuelled they resulted in a 49-minute dihedral change. The weight caused sagging on the rear section of the wings, so the scheme was abandoned on 5 April, the day it was born, but not until fifty gallons of fuel had been pumped into each tank and the configuration flight-tested. A PR. 34, RG179, reached the squadron in the third week of April and was first flight tested on 21st. Wing drop tanks carrying 200 gallons were fitted, making the total tankage 1,256 gallons. The squadron was most impressed with its new variant.

Final operational flights by Mk. IXs of 540 and 544 Squadrons came on 2 and 5 March respectively, the aircraft involved being LR426 and MM276. The former set out for Hamburg and Hanover but, finding

Operations: Photographic Reconnaissance

10/10th cloud, photographed Cuxhaven, Nienburg, Deventer, Amersfoort, Amsterdam and Rotterdam instead. MM276 despatched to Neustadt, Tarnewitz and Ludwigshaven encountered cloud and came home empty handed. 540 Squadron moved to France on 29 March, taking with it nine Mk. XVIs, a Mk. VI and a PR. 32. In the first four months of 1945 the squadron had flown 393 sorties, losing one aircraft. Their final operational flight occurred on 26 April, when RF970, flown by Flg. Off. Browne and W. Off. Lodge, set out for the railway network between Prague and Budejoice, only to be thwarted by radio trouble, causing an early return. 544's last sortie left Benson at 10.05 for the airfields at Sola, Forus, Lista and Kjevik in the hands of Flt. Lt. H. C. L. Leech who, as he brought RG115 into land, completed one of the Mosquito's most vital contributions to the war against Germany.

Unarmed photo-reconnaissance Mosquitoes had roamed the length and breadth of Europe almost immune to successful enemy action for four years. Adaptations led to range being extended, ceiling heightened and speed increased. Engine changes and detail modifications were made, but no major structural alteration was required due to the soundness of the original design, which proved to have exceptional adaptability. The success achieved in the vital realm of reconnaissance makes it amazing, on reflection, that the enemy barely attempted to emulate the British success, particularly as the build-up for the Second Front took place. Reconnaissance by Mosquitoes over twenty-four hour periods contributed much to the success of the Allied offensive, protection of which was, in large measure, in the hands of the Mosquito nightfighters, as will now be seen.

Operations: Photographic Reconnaissance

Analysis of strength and operations of No. 1 PRU Mosquitoes September, 1941–October, 1942.

Acft.	Sept.	Oct.	Nov.	Dec.	Jan.	Feb.	Mar.	Apr.	May	June	July	Aug.	Sept.	Oct.
W4055	2	9; lab	2	FTR	—	—	3	2; 1w	12	9	—	5	1; 2w	3; 1w, FTR
W4058	—	3	4	4; 1w	2; 1w	1m	3; 3w	1; FTR	—	—	—	—	—	1
W4056	—	4; 1w	4	2; 2w 1w	5; 3w 2m	9; 2w 2m	5	1; FA	—	—	—	2	5	—
W4061		—	—											
W4059		—	—	2; 1m	6; 1w 1m	5; 2w	6; 1w 3m	5; 1m	10; 1m	6	4	2; 1m, 1w	—	—
W4051					.1	1w	3; 1m	2; 1w	2; 1w	1	—	1	To 521 Sqn.	
W4060							6; 1m	5; 2w lab	9	5; 1w 1m	1m; FA	—	—	—
W4067									1; 1w	4; 2w	FTR 27th			
DK284								1	2; 1m	—	5; 2m	2		
DD615									3	4				
DD620										1				
W4054									1	1; 1m	4; 1m	4; 1w	1	1
DD659											2; FTR	1		1
W4089												4; FTR		
DK310												2	4	1
DK311													2	1
DK320														3w
DK319													1	2
DK314													1	
DK315														
Totals	2	16; 1w lab	10	8; 4w 1m 1 FTR	14; 5w 3m	14; 5w 3m	26; 4w 6m	17; 4w 1m lab 1 FTR	40; 2w 2m	31; 3w 2m	15; 4w. 2m 2 FTR	23; 3w 1m 1.FTR	15; 2w	11; 2w 1 FTR

Operations survey: 212 successful operations, 41 abandoned due to weather at target. 20 failures due to mechanical faults. 3 abortive before target. 6 aircraft FTR.

Legend:— Total sorties per month successful listed, followed by number abandoned due to weather (e.g. 1w) followed by sorties unsuccessful due to mechanical failure (e.g. 4m) followed by abortives (e.g. lab.) and missing aircraft (e.g. 1 FTR). FA indicates seriously damaged in flying accident.

147

Chapter 12

Operations: Fighters

Wing Commander Gordon Slade, who flew the Mosquito prototype during its official trials at Boscombe Down, was given command of the first Mosquito fighter squadron, 157, formed at Debden on 13 December, 1941. A nucleus of ground crews, Hatfield trained, talked to the squadron about its future equipment on 16 December. Five days later Wg. Cdr. Slade arrived from 604 Squadron, bringing Plt. Off. Truscott, his observer. On 26 January, 1942, he delivered the squadron's first aircraft, a 'dual control Mk. II' W4073 to Castle Camps, then Debden's satellite station. Poor accommodation, lack of equipment, rudimentary workshops and bitterly cold weather made the bleak airfield, far from any sizeable town, a miserable place. Yet, to this station, in the south of Cambridgeshire, the entire squadron moved in January.

Seventeen Mk. IIs were delivered to Maintenance Units for the fitting of service equipment and, in particular, A.I. Mk. V radar. Several of 157 Squadron's pilots assisted in the deliveries to the M.Us. The first Mk. IIs, W4087 and W4098, neither fully equipped, were delivered to 157 Squadron from 32 M.U. on 9 March. Squadron returns on 28th showed fourteen Mosquitoes on strength, and that one night training flight had been made, on 13 March.

By mid-April 157 Squadron had a T. Mk. III and nineteen NF. Mk. IIs, three of them without radar. Modified gun-sight brackets were being busily installed, a blind being fitted to shut out glare from aerodrome floodlights; and lamps and switches were being repositioned, as suggested by the first few hours of Mosquito night flying. Only seven crews were yet available, but there was terrific keenness to get after the enemy bombers in the Mosquito.

Flt. Lt. Darling delivered to 151 Squadron its first Mosquito II, DD608, on 6 April. Two days later 'A' Flight were told that they would be converting to Mosquitoes, leaving 'B' Flight to carry on with Defiants. 151 Squadron received sixteen Mosquito IIs during April, also the second dual-control aircraft to enter service, W4077. Of particular interest to the squadron were German minelaying aircraft operating 20-30 miles off shore, also daylight shipping reconnaissance bombers. 151 Squadron's Commanding Officer was Wg. Cdr. Smith, a New Zealander, and Wittering its base was under the command of Grp. Capt. Basil Embry,

Operations: Fighters

with a D.S.O. and two bars, who had had some remarkable experiences. He spent much time flying his pet Mosquito, DD628.

Throughout April, 1942, both squadrons built up their night-flying hours, and practised interceptions with one Mosquito 'stooging' for another. Co-operation between crews and ground controllers also was built up. Night gun-firing practice over the North Sea showed a need for flash eliminators on the ·303-in. Browning guns, without which they blinded the pilot to anything outside the cockpit, including the brightest stars. Flt. Lt. Stoneman, Engineering Officer of 157 Squadron, helped by Flt. Sgt. Burge, set about designing and making a device which incorporated a venturi to suck the flame downward, and which was successfully adopted. From this it emerged that the ·303 guns had never been fired in the air at night during the official gunnery trials at Boscombe Down. The flash from the 20-mm cannon was not troublesome, but caused some reflection from the propeller discs if the blades were wet.

The most serious defect so far encountered with the Mosquito was the unserviceability of the exhaust manifold (with its rather cumbersome flare-damping shroud) and the cowling side. Quite apart from causing about 10 m.p.h. speed loss, this exhaust system was a nightmare to designers and maintenance men in these early days, and on 11 April 157 Squadron grounded its Mosquitoes after a cowling had burnt through during a flight by Slade to Boscombe Down. John E. Walker and Hereward de Havilland hurried to Castle Camps to investigate, and returned to Hatfield with two sets of cowlings for scrutiny. One aircraft was permitted to fly again on 14 April, four more on 17th; and, following careful inspections and repairs, and modifications, the trouble was slowly overcome. Tailwheel shimmy was cured, by the use of the Marstrand double-track tyre. Modifications entailing about 500 man-hours per aircraft were carried out at Hatfield, 157's aircraft arriving in batches of three or four. All had been modified by 27 May. About sixty aircraft required the cowling modifications, and it was mid-July before 151 Squadron's machines featured the improvements.

There was keen rivalry between the squadrons to become operational and draw first blood. Hitler sanctioned the Baedeker raids on 14 April. Norwich was bombed by 47 aircraft early on 27th and Flg. Off. Graham-Little and Flt. Sgt. Walters took off from Castle Camps at 01.10 hours in DD603 on the first Mosquito patrol, to the target area. They had two radar contacts, both on friendly aircraft. An hour later Flg. Off. Babington and Plt. Off. Bowman in DD627 were away, obtained a brief contact, but failed to intercept. Wg. Cdr. Slade and Plt. Off. Truscott made the night's third patrol, but had no success. During an attack on York and East Anglian targets the following night four sorties were flown, but no enemy aircraft were located. On 29 April Ipswich and Norwich attacked by 75 aircraft were the targets. Sqn. Ldr. Ashfield contacted a Do 217, but it saw him in the moonlight and escaped.

Operations: Fighters

Shortly afterwards he contacted—a Halifax. One had to be extremely careful. . . .

Flt. Lt. Pennington in DD613 made 151 Squadron's first patrol on 30 April. Two other sorties were flown by the squadron that night. Although the Luftwaffe suffered casualties in the short period of the attacks on East Anglia, Mosquitoes made no score. The enemy had some lucky escapes—and so did the Mosquitoes. On the night of 1 May Grp. Capt. Embry narrowly missed a head-on collision with a Heinkel 111, and Sqn. Ldr. Darling almost hit a Ju 88. Another pilot followed a Hun to the Dutch coast, but the chase ended in cloud. By 10 May more than half the squadron's aircraft were unserviceable with cowling trouble, and ten civilian workers left Martin Hearn to assist in modifying them.

8 May brought success nearer to 157, when Sqn. Ldr. Ashfield was directed to an enemy aircraft under Trimley radar control. This, and two later contacts by Flg. Off. Stevens, were also lost. Sqn. Ldr. Ashfield, Senior Flight Commander on 157, was a tough and enthusiastic officer with eighteen months night fighting experience. He, like others, found that the Mosquito, for all its speed, was none too fast, for the Do 217s were in the habit of going home at over 300 m.p.h., so that a speed advantage of even 60 m.p.h. meant a twenty-minute hundred-mile chase to catch a Dornier with only a twenty-mile start. Many enemy aircraft crossed the coast at less than 3,000 feet, making it difficult for ground radar stations to plot them.

The first Mosquito day patrol was made on 12 May, 1942, by Flg. Off. Stevens of 157 Squadron. On the 19th the squadron witnessed the tragic loss of Flg. Off. Babington and Plt. Off. Reeves, when DD601, which had been acting as a target aircraft for Wg. Cdr. Slade, crashed on the north-east side of Castle Camps following an engine defect. Another accident befell the station when DD604 overshot the runway in heavy rain on 26 May careering into a coach load of men, four days after the squadron became fully operational. A third crash followed after Sgts. Heath and McIlvenny took off from Debden on 29 May for firing practice over the sea. During take-off the cockpit emergency hatch blew off, but the pilot proceeded with the exercise. As he turned towards land after firing, his port engine revs. jumped to 3,500 and, finding that he could not feather the propeller or maintain height, he was forced to ditch the aircraft, which was found to float well and level in spite of losing its tail on impact. The crew were rescued. In the absence of the enemy, training continued until the end of the month, when the Luftwaffe at last came face-to-face with the Mosquito fighter.

The first successful interceptions were made by 151 Squadron on 29 May when Grimsby was attacked, and by 157 Squadron on 30 May. On 29th Flt. Lt. Pennington took off at 04.30 in DD628 and contacted a bomber, possibly an He 111, over the North Sea at dawn, heading for home and trailing black smoke. Pennington opened fire at 400 yards with cannon. The bomber returned a short burst putting bullets through

Operations: Fighters

both starboard main spars and tailplane spars of the Mosquito, also the port engine coolant system and elsewhere. Pennington closed to 80 yards without difficulty and the enemy, badly hit and with his port engine apparently on fire, spiralled smoking towards the sea. It was lost in the haze and is now listed as damaged although was then claimed as a probable. Pennington feathered his port propeller and flew 140 miles to base on one engine. DD608 also engaged in combat, Plt. Off. Wain damaging a Do 217 E.

On 30 May, Sqn. Ldr. Ashfield of 157 Squadron flying W4099 was vectored on to a Do 217 E4 south of Dover at 04.45, flying at 4,000 feet. He moved in to 250 yards and fired as he closed to 150 yards. The enemy aircraft went into cloud. Ashfield fired again at 200 yards until he had used 450 cannon shells. He attacked once more, this time using his machine-guns from 100 to 60 yards, and the Dornier dived from 2,500 feet vertically into cloud. It was last plotted by radar ten miles from Dover, and recorded as a probable. German records list four Do 217 Es missing from that night's operations. 68 Squadron claimed to destroy one and damage another, one fell to an anti-aircraft site in the Humber area and 157 Squadron claimed a probable. In the light of all information now available it seems almost certain that Sqn. Ldr. Ashfield destroyed a Do 217.

30–31 May, 1942, the night after Sqn. Ldr. Ashfield's success, was also the night of the 1,000 bomber raid on Cologne, and Embry, patrolling in a Mosquito, had to climb to 20,000 feet to get out of the traffic—after making more than 40 radar contacts and eight visuals in the moonlight with bombers engaged on the raid. Thus it was a historic week-end for the Mosquitoes, for it was also on the morning of the 31st that the bombers first went into action.

On 3 May 'B' Flight of 264 Squadron at Colerne received their first Mosquito, W4086, two days after they had flown their last operational sorties on Defiants. Soon after, W4053 arrived from Debden. Originally the first turret fighter, it had been returned to de Havilland from R.A.E. and modified into a T. Mk. III. Unfortunately 264's first three aircraft had unmodified cowlings presenting an unfavourable impression of the Mosquito. The Commanding Officer of 264 Squadron was Wg. Cdr. Hamish Kerr, who converted his pilots to the Mosquito after a couple of hours on Oxfords, and in some instances directly from Defiants without any twin-engined experience.

There wasn't much night-fighter business in June, 1942, although on the 1st, Flg. Off. Little of 157 found trade at 9,000 feet, unexpectedly identified at 50 yards as a Messerschmitt Bf 109 F in day camouflage yet flying at night, and the enemy escaped. Nine nights later Little and his observer were killed in a night-flying training accident near Bishops Stortford. Wg. Cdr. Kerr and Flt. Lt. Lesk in DD642 made 264's first night patrol on 13 June, with W4081 closely following them.

Enemy activity increased at the close of the month, and Mosquitoes

Operations: Fighters

of 151 Squadron drew first definite blood, scoring four confirmed and three damaged—and they added a further five confirmed during July. Their Commanding Officer, Wg. Cdr. Smith had a great night on 24–25 June. Flying with Flt. Lt. Sheppard in W4097, he scored one damaged and two confirmed in less than half an hour. He sighted an He 111 against the Northern Lights at 8,000 feet at 23.30, but was spotted, for the Heinkel dived just as Smith opened fire from 300 yards. Its port tanks were set on fire, then the Heinkel was pulled out of its dive and into a stalled turn to port jettisoning what resembled a torpedo. Smith fired another burst and large pieces fell off as the aircraft went steeply into cloud to limp home damaged, although at the time it was claimed as a probable.

At 23.40 ground control told Smith to search four miles due east, where he immediately picked up a Do 217 E4 F8+AC:5454 of KG 40. He closed unseen to 100 yards, fired about four rounds from each cannon, and then the Dornier dived into the sea and exploded. Eight minutes later Smith contacted a Do 217 of I/KG 2 by radar and fired a long burst from 200 yards, after which the bomber's wings were almost completely enveloped in flame. Even so the enemy fired a short burst, without hitting the fighter. Smith closed in and fired again and finished off his foe. His windscreen was soon covered with German oil.

Meanwhile Sqn. Ldr. Darling and Plt. Off. Wright engaged a Do 217 at 5,000 feet, but it dived away from them. Similarly, they located a Ju 88 but lost it at sea level. The following night it was the turn of Plt. Off. Wain and Flt. Sgt. Grieve, who caught a bomber well out at sea carrying externally two large bombs. It was claimed as shot down in flames. In the early hours of 26 June Flt. Lt. Moody and Plt. Off. Marsh, flying DD609, were hit by return fire from an unidentified enemy, then to their surprise found themselves alongside a Do 217. This they promptly shot into the sea. These successes were scored mostly 60 to 80 miles out from the East Coast.

Despite having topped the list for hours flown per month and serviceability, 157 Squadron still had no proven interception success. It maintained patrols well out to sea and over land, but German aircraft consistently entered North or South of their sector. Minelayers frequented 157's area, but they were usually so low that the radar then in use was not very effective against them. Success during June came nearest to Wg. Cdr. Slade, who chased a Do 217 E4 over Norwich in DD612 on 26–27 June. Next night Sgt. Roe of 264, flying DD643, damaged a Do 217 E near Weston-Super-Mare in the squadron's first Mosquito engagement.

Six enemy aircraft were claimed by Mosquitoes at night over or near the United Kingdom in July, 1942. Plt. Off. Taylor damaged a Do 217 off Orfordness on 12–13, Sqn. Ldr. Ashfield claimed a probable, later confirmed, on 27–28 and Flt. Lt. Worthington in DD627 damaged a 217 but yet another month passed without confirmed success coming to 157

Operations: Fighters

Squadron. Meanwhile 151 rapidly increased their score, first when Plt. Off. Fisher found a Dornier twenty miles East of Skegness and chased it to a point 50 miles east of the Humber Estuary, there destroying it. Sqn. Ldr. Pennington and Plt. Off. Fielding each shot down a bomber at sea on 27–28 July, and two days later another fell to Flt. Lt. Ritchie.

So far the highest aircraft attacked had been Wain's —at 16,000 feet. The Northern Lights were making sighting conditions quite difficult especially when, after silhouetted against the light Northerly sky, an enemy aircraft manoeuvred to the south side of his pursuer.

There had been three cases of overshooting the enemy in combat by mid-July, this leading to the question of fitting airbrakes to Mosquitoes. More speed at low levels was clearly demanded for night work, and higher speeds between 15 and 20,000 feet. It was clear also that the performance of airborne interception radar varied between units and was due in part to the varying quality of installation work, its design, accessibility and maintenance—all factors to be considered in future. In their only action during fifteen July sorties 264 Squadron had first success on 30th, when Sqn. Ldr. L. A. Cooke destroyed a Ju 88 near Malvern.[1] On the previous day 85 Squadron had taken delivery of their first T. Mk. III, HJ 857, and they crashed it next day. Wg. Cdr. Raphael, a Canadian of long operational experience, was their Commanding Officer. DD718, their first Mk. II, arrived on 15 August, and V-Victor flew the squadron's first Mosquito patrol at dusk on the 18th of the month.

August, 1942, began with low cloud and poor visibility, 151 Squadron was put at day readiness to use its radar and catch such raiders as came across under cloud cover. In the event only a few ventured to Britain in August and September, relying upon weather conditions for protection. Their flame damping was not good, but their skill in locating targets was fairly high, bearing in mind the erratic courses they were forced to fly. There was evidence that some of the raiders had means of interfering with the radar reception in our fighters. On the other hand they were only able to detect interceptors visually.

When on 22 August Gordon Slade with Plt. Off. Truscott as observer at last opened the 'Confirmed as Destroyed' list for 157 Squadron, two of the crew of the Do 217 E4 Nr. 1152: U5+LP of II/KG2 which he shot down at Worlingworth, twenty miles away from his base, baled out and said afterwards they had not known of the Mosquito's presence until they felt it. In the moonlight our fighters were occasionally attacked by Bf 110 and Ju 88 intruders and needed to keep a good look out aft.

---

[1] Ju 88 A4 M2 + AK:2124 of unit 2/106 crashed at Hornyeld Farm, Malvern Wells at 02.05. It set out from a French base to bomb Birmingham and when at about 10,000 ft. was attacked by a fighter. The first attack started a fire in the port wing and soon this began to break off. Bombs were jettisoned and the crew baled out from 3,500 ft.

Operations: Fighters

By the end of September the four newly equipped night-fighter squadrons had flown about 11,000 hours with their Mosquitoes.

There was now further experimentation with the night-fighter colour scheme. Early F. Mk. IIs had a velvet black finish supposed to be glisten-proof in searchlights or moonlight. It looked rather like black suede, and was about as thick! It was found not to be as effective as had been expected, and when tests showed that it caused something like a 15 m.p.h. speed loss, it was very quickly substituted by plain matt black. Now, after six months' experience in all sky conditions, there was a growing opinion that black produced too much silhouette effect, especially on moonlight nights and against cloud, which could be avoided by using a modified day camouflage on top with lighter shades on the under surfaces. This was approved by Fighter Command and was retained for the rest of the war.

On 8 September DD669, flown by Flt. Lt. Ritchie, shot down a Do 217. According to the Squadron's Operations Record Book the Mosquito 'was camouflaged white, pale blue and grey . . . in a special manner devised by 151 Squadron'. The Dornier 217 E4 5502 F8+AP of I/KG 2 crashed in flames on Rectory Farm, Orwell, Cambridgeshire, its wreckage being strewn over a wide area. The same night Flt. Lt. Bodien, also of 151, chased an enemy aircraft from N.W. of Bedford to Clacton, and down to 10,000 feet. He shot up the target, but lost it in searchlight dazzle at 6,000 feet when his A.I. gear went unserviceable.

Ten enemy aircraft had crossed the coast that night, between Southwold and Yarmouth, and operated mainly in the Cambridge area, where 157 Squadron flew six patrols. The next major activity was an attack on King's Lynn on 17–18 September when Sgt. Walters of 157 damaged a Do 217 twenty miles north-west of Hamstede, and Flt. Lt. Bodien destroyed one, the crew of which became prisoners.

In the first Mosquito day combat, on 30 September, Wg. Cdr. R. F. H. Clerke of 157 in DD607 shot down a Ju 88[1] thirty miles off the Dutch coast. Trimley radar station guided him to the bomber flying at 1,100 feet off Orford. He lost it, but it re-appeared head-on fifty miles out to sea. The enemy dived taking violent evasive action, but was eventually shot down seventy miles East of Orfordness. Later 'F' of 85 Squadron made its squadron's first radar contact with the enemy, but no result came from two bursts fired.

During the morning of 19 October thirty-five sneak raiders ventured singly over East Anglia. Mosquitoes were scrambled, as a result of which 85 Squadron fought its first combats. Flt. Lt. Bunting took-off at 10.05 and, forty-five miles north-west of Foreness at 7–8,000 feet, fired at a Ju 88, which retired smoking. Sgt. D. I. Chimes, away thirty-five minutes later, fired three bursts into a Do 217 off Clacton. The burning bomber dived steeply to port, but was not seen to hit the water. Chimes then patrolled Bradwell and saw a Ju 88, which he chased without

[1] Ju 88 A4 144181:3E + AH of I/KG6.

Operations: Fighters

success before returning to base at 13.40. More successful was Flt. Sgt. N. Munro of 157 Squadron, who took off at 07.10 and shot down a Ju 88 A4 of KG 6 at Southwold, then damaged a Do 217 off Clacton.

A Ju 88[1] fell to Flg. Off. E. H. Cave of 157 Squadron, shot down during a daylight convoy patrol thirty miles south of Beachy Head on 26 October. Return fire from the bomber's rear gunner damaged the Mosquito's starboard spinner and tailplane. Two days later Sgt. Sullivan of 85 Squadron fired on a Do 217 during an afternoon patrol. Despite bad weather his squadron flew 51 hours on day and 199 hours on night patrols in October, making 81 sorties. On 31 October 157 Squadron flew eight patrols during the last Baedeker raid, directed against Canterbury, but the squadron's seventeen day patrols during the month constituted their principal effort.

Although contact with the enemy was rare the period October to December was not without incident since, at this time, four squadrons were rapidly converted to Mosquito fighters. 25 Squadron at Church Fenton received their first Mk. II on 21 October. 'To the delight of us all,' wrote the squadron historian, 'it was made known this morning that we were to be re-equipped with Mosquitoes. A few minutes later the first three arrived on the aerodrome. The American signals officer was perspicacious—"Gee! they must be fast ships, they've nearly beaten the rumour".' 25's first battle operation on the new mount came on 17 December, when Sqn. Ldr. J. L. Shaw and Plt. Off. Guthrie were scrambled in DD755 and operated under Patrington radar station. Enemy aircraft were still approaching at low levels, but on this occasion no contact was established.

410 Squadron at Acklington, whose personnel were nearly all Canadians, received their first Mosquito T. III, HJ 865, on 24 October. Squadron enthusiasm was not all that it might have been, partly because the unit had no enemy aircraft to its credit after 16 months' existence. A local brains trust to interest the squadron in the new aircraft was organized by the squadron commander, engineering officer, the D.H. liaison officer Carter, and Cruikshank of the Propeller Division. Fifteen hours flying in two days with seven pilots going solo resulted, and on 6 December Wg. Cdr. Hillock and Plt. Off. O'Neille Dunne, in DZ249, made the unit's first operational Mosquito sortie. A second patrol the same night was flown by Flg. Off. S. J. Fulton with Flg. Off. R. N. Rivers in DZ251. On 8th, 9th and 12th, Mosquitoes of the squadron set off unsuccessfully to intercept enemy weather reconnaissance flights. By the end of the month 41 sorties had been flown, and Beaufighter II operations ceased on 18th.

307 (Polish) Squadron was equipped with Mosquitoes at Exeter during December, 1942. Intelligent and highly disciplined, the Poles were extremely eager to learn all they could about the Mosquito. The engineer officer, Gatonoski, even translated into Polish the pilot's notes

[1] Ju 88 D1 1685:4U + IL of 3(F)123.

Operations: Fighters

and part of the maintenance handbook. During a practice flight the starboard engine of one of 307's aircraft caught fire owing to an internal structural failure. The pilot feathered the airscrew, but had to make another circuit because Typhoons were taking off at the time. He landed successfully with the engine on fire, yet although the installation was burnt out practically no damage occurred aft of the fireproof bulkhead. Sqn. Ldr. Ronszek and Plt. Off. Krawiecki made the squadron's first night defensive sortie on 14 January, 1943, in DZ271. Only one item was missing on their inventory—the enemy.

Three Mk. IIs had been received by the end of 1942 by 456 Squadron at Valley in Anglesey, operating under 9 Group. Sqn. Ldr. S. P. Richards made the first patrol with the new type on 22 January, 1943, using DZ297. By the end of February the squadron had only twice been scrambled. 456 was an Anglo-Australian unit previously using Beaufighters, the sleeve engines of which had required filters because there was so much sand blowing about at Valley. Two small lakes were drained and mud spread over the airfield. Grass seed was sown in an attempt to improve matters. The wind was frequently very strong, so much so that it was impossible even to fit ends to the blister hangars.

At Valley, as elsewhere, it was feared that water soakage around fuselage doors would cause a good deal of unserviceability. Hereward de Havilland inspected three aircraft of 85 Squadron at Hunsdon on 20 December for evidence of this trouble and of the effects of protective treatment. DD737, he recorded, was in particularly deplorable condition, but until the protective treatment could be improved, clearly the trouble would continue. Fortunately there was no great demand for aircraft at this time, but improved weather-proofing was introduced with all possible speed because some heavy night attacks were expected against our south-western airfields congested with aircraft in connection with the Allied landing in North Africa. Because of this Fighter Command took precedence over Bomber Command where Mosquito deliveries were concerned. By the end of January, 1943, there were eight night-fighter squadrons holding 159 aircraft, 60 per cent of all the Mosquitoes then in squadron service.

On 3 January, 1943, the Meteorological Office reported a north wind of 110 m.p.h. at 21,000 feet, so Hamish Kerr, O.C. of 264 Squadron, hurriedly flew to Wick in DD625 with the intention of establishing a speed record. Just after take-off an exhaust gasket blew and he had to reduce power to +2 boost and 2,300 r.p.m. His time from take-off to landing was 1 hour 30 minutes, equivalent to an average speed of about 335 m.p.h. 264 Squadron was particularly unlucky at this period, for on 3 December twelve of their twenty-five aircraft were undergoing repairs due to water soakage. Personnel at Colerne had witnessed a particularly unpleasant accident on 2 October, when DD639 ran into the corner of a hangar following a bad landing. Although the winter had little to offer the night fighters by way of action other events punctuated it.

Operations: Fighters

The most noteworthy night attack of January, 1943, came on the 17th when four raiders, a Do 217 E4, and three Ju 88 A14s of I/KG 6, were destroyed. Most of the Mosquito squadrons were cheated of action, simply because the raiders avoided their sectors. London was the target during this reprisal raid for the bombing of Berlin, and it was the first major attack on the City since 1941. Mainly concerned was 85 Squadron, which flew 55 hours on patrols between 17.15 and 07.20, a record so far. During the course of the two-phase raid, in which the enemy despatched 118 sorties, Wg. Cdr. Wight-Boycott destroyed a Ju 88 A14. He was no newcomer to the de Havilland sphere, since he had been a flying instructor at Hatfield in the E.F.T.S., under Wg. Cdr. C. A. Pike.

The Turbinlite Mosquito arrived on 151 Squadron for trials on 16 January, 1943. W4087, an unarmed Mk. II modified by Alan Muntz of Heston between 14 October and 31 December, 1941, at a cost of £620, had an airborne searchlight installed in its nose, the intention being that it would illuminate an enemy aircraft, which would then be fired upon by another fighter. Many trials with these searchlights had been conducted by Havocs and Bostons, which equipped ten squadrons, but few operations were carried out and the development of advanced radar interception gear outdated the searchlight. Favour rested on the Turbinlite in some quarters and as late as 3 September, 1942, a four cannon Turbinlite Mosquito was considered. As W4087 arrived on 151 Squadron the unit heard that it had been selected to be the first to operate the Mk. XII Mosquito with A.I. VIII radar. The Turbinlite Mosquito was flown by a crew from 532 Squadron whilst at Wittering. On 8 February Flg. Off. Hester of 151 Squadron took it to 85 Squadron, where Wg. Cdr. John Cunningham (who took over 85 Squadron on 27 January, 1943), after an adverse report on its capabilities returned the aircraft to 1422 Flight at Heston, from whence it had come.

Enemy activity increased during March, on the first night of which 25 Squadron engaged a Do 217 E4 being held by searchlights. The main action of the month fell on the night of the 3rd/4th, when about twenty-five enemy aircraft crossed the coast between 20.00 and 21.00 hours, and a further fifteen at 05.30. Some managed to reach London, and although 85 and 157 were scrambled, neither achieved more than contacts. At 04.30 Wg. Cdr. Cunningham took off, and fifteen minutes later he established contact with a Do 217 which he held for two frustrating minutes in his sights. His repeated attempts to open fire were of no avail for, it later transpired, the electric lead to his firing solenoid had been pulled off the terminal, probably after being caught on the overalls of a ground crew member. The occasion was all the more disappointing since he later held a second Dornier in his line of fire. The Luftwaffe despatched 117 bombers on the night's raids, and lost 6. It was a despicable performance, during which many of the bombers couldn't find London.

Two Beaufighters fitted with A.I. Mk. VIII came to 85 Squadron on

Mosquito Fighters

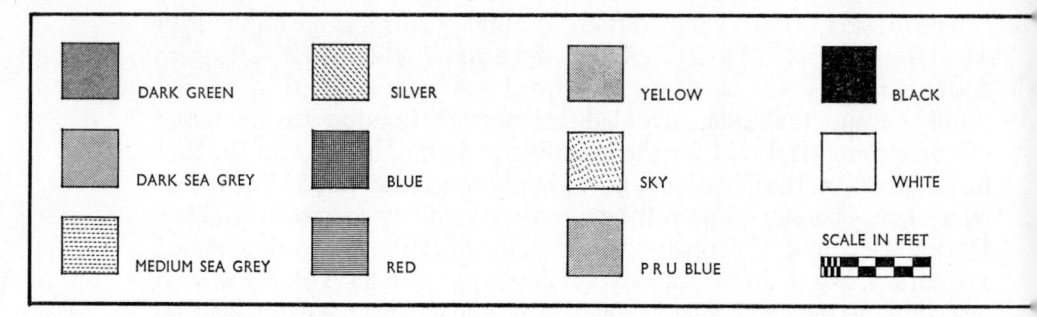

■ DARK GREEN	▦ SILVER	■ YELLOW	■ BLACK
■ DARK SEA GREY	■ BLUE	▒ SKY	□ WHITE
▦ MEDIUM SEA GREY	▦ RED	■ PRU BLUE	SCALE IN FEET

Mosquito NF. II (Merlin 21) W4079 in RDM2 finish. To 157 Squadron 12.6.42, shot down Do 217 E4 F8+MP: 4345 of II/KG40 off Lowestoft 28.3.43. To D.H. 25.9.43, to RAE 2.1.44, to 10 MU 12.3.44, 218 MU 22.5.44, to 51 OTU 5.8.44 (still all black), crashed 29.11.44 and destroyed.

Mosquito NF. II (Merlin 22) DZ726. Reached 410 Squadron 24.3.43. Slightly damaged when flown through H.T. cables at Apeldoorn 15.4.43, repaired in works. To 141 Squadron 18.10.43 and shot down an He 177 15.2.44. Markings as recorded on March, 1944.

Mosquito NF. II (Merlin 21) DD636 delivered 7.5.42 and used by Handling Squadron, Hullavington. Reached 264 Squadron 16.4.43. Engaged and damaged by Fw 190s on *Day Ranger* 11.5.43. To 307 Squadron 7.8.43, to 157 Squadron 9.11.43 and ditched on an *Instep* patrol 19.11.43 (engine failure).

Mosquito NF. XII (Merlin 21) HK119 reached 85 Squadron 17.4.43 and shot down a Ju 88 S on 29–30.5.43. On 22.1.44 it joined 307 Squadron and as EW-D probably destroyed an He 177 on 19.2.44. It was damaged in July, 1944, and saw no more front-line service.

Mosquito T. 3 (Merlin 23) VP351 reached the R.A.F. on 9.5.47. It appears in 1949 trainer markings, when it served with 19 Squadron Church Fenton. Wartime T. Mk. IIIs were usually camouflaged dark green and dark sea grey, with medium sea grey under surfaces. Like standard fighter aircraft they wore a Sky Type S rear fuselage band, and had yellow wing leading edge striping. Although unarmed they resembled armed single-engined fighters in a way that few armed Mosquitoes did!

Mosquito FB. XVIII (Merlin 25) NT224 was delivered direct to 248 Squadron on 2.6.44, and after many sorties was shot down 7.12.44 during an engagement over Norway with enemy fighters.

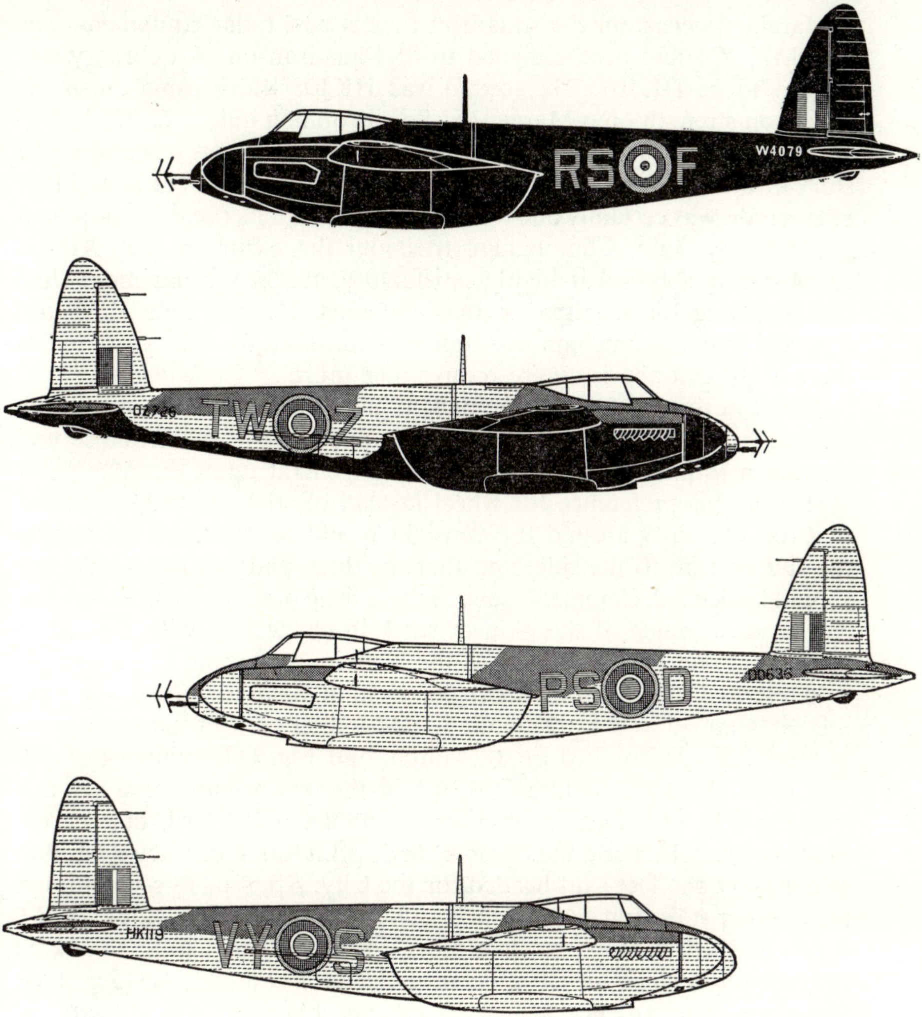
RS F
W4079
TW Z
PS D
VY S
HK119
VP351
NT 224
OE

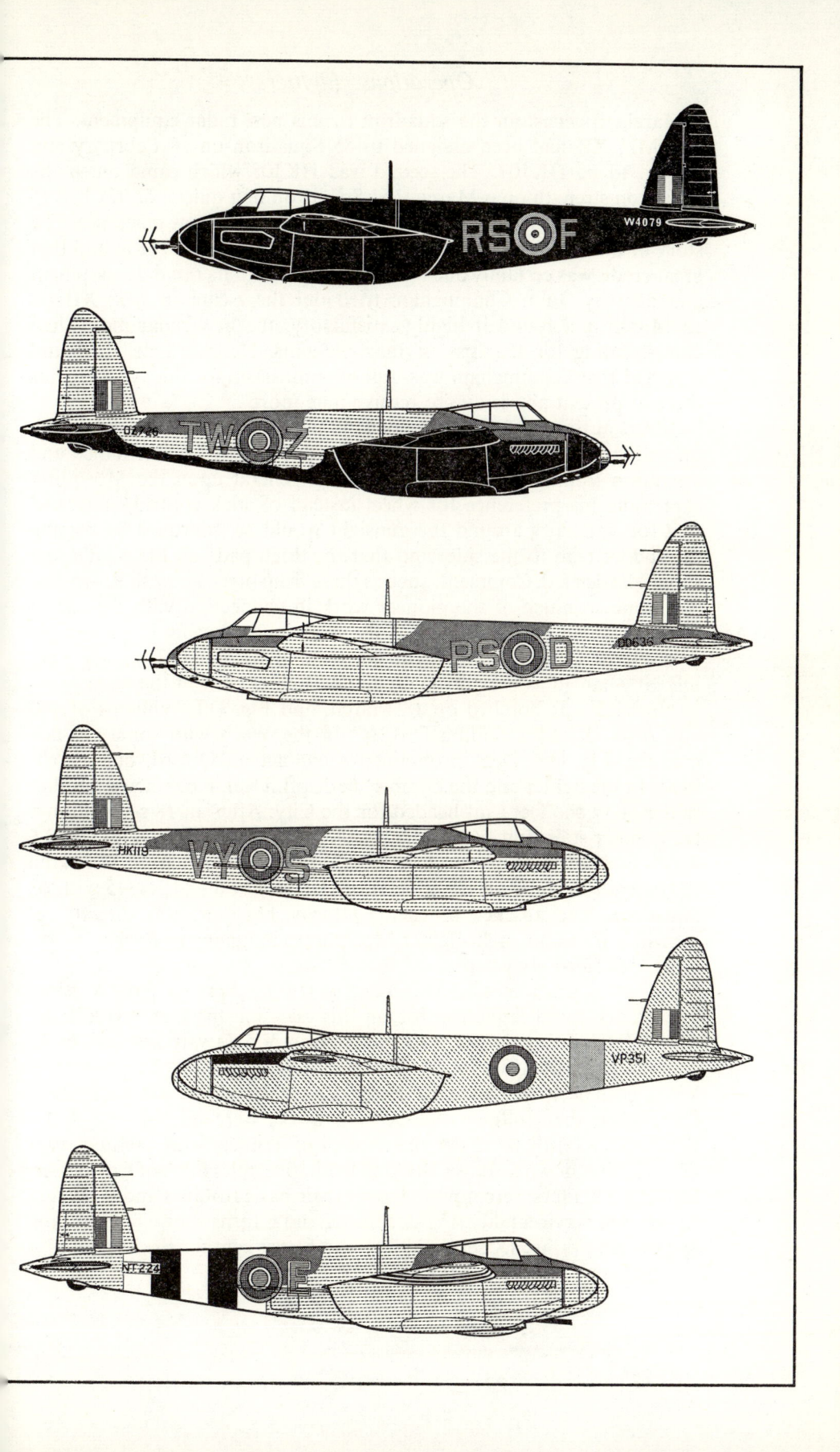

Operations: Fighters

5 March to accustom the squadron to this new radar equipment. The first Mk. XII had been assigned to 85 Squadron on 28 February and proved to be HK107. The second was HK108 which came on to the squadron strength on 9 March. 151 Squadron in a quieter sector had to accept being now placed second in line to receive the new machines. The story of the Mk. XII is elsewhere detailed, but it is pertinent to add that great credit was certainly due to all concerned with its rapid development and delivery. John Cunningham tried out the 4-cannon Mk. XII on 24 March and found it highly satisfactory, the new radar more than compensating for the loss of machine-guns. Hereward de Havilland reported that Cunningham was 'full of enthusiasm for this type. He has three at present and is due to receive nine more'.

It was at this time that Cunningham had also commented on the low operating efficiency of the A.I. Mk. V, due to poor maintenance and operation training. He also remarked in general upon the Mosquito, expressing his preference for wheel instead of stick control, suggested that forward view around the gunsight would be improved by placing the rudder trim to the side, and that the thigh pads on the pilot's seat should be deleted. Comments such as these from one who, with Rawnsley, had done so much of the pioneer work in connection with the use of A.I. radar and the perfection of the technique of using it, proved invaluable to Hereward de Havilland. Cunningham and Rawnsley were officially named as the outstanding night-fighter crew of the war.

Norwich was bombed on 18 March, and Flg. Off. Williams of 410 Squadron shared a raiding Do 217[1] in the Wash with some ack-ack gunners. Flg. Off. Deakin obtained a contact off Orford, but searchlights lit the fighter and the bomber made off. Deakin contacted another raider at 12,000 feet as it headed for the City. After an 18-second burst the bomber stalled; it proved to be Ju 88 A14 3E+AK:4322 of I/KG 6 which crashed into the sea. In a second raid on Norwich on 28 March 157 Squadron shared a Do 217 E4 of II/KG 40 F8+MP:4345 with 68 Squadron. The absence of enemy attacks, however, was directly responsible for many night-fighter Mosquitoes indulging in *Ranger* operations, described elsewhere.

Enemy aircraft crossed the coast on 14–15 April, a date of some importance to 85 Squadron for on this occasion the first two kills by Mk. XIIs were made. Plt. Off. Sutcliffe in VY:D was first to obtain contacts, and then Sqn. Ldr. Green in VY:F and Flt. Lt. Howitt in VY:L each destroyed a Do 217 E4 of II/KG 40 during a raid on Chelmsford, when the Hoffman ball-bearing works were hit. Flt. Lt. J. G. Benson, D.F.C., flying under the control of Trimley, was vectored to a Do 217, shot down at Layer Breton Heath, five miles S.W. of Colchester. Mosquito fighters were now well over their basic troubles and achieving 74 per cent serviceability. Fighter Command returns showed at this time 85 per cent Spitfire IX serviceability and 61 per cent for the Typhoon.

[1] Do 217 E4 5523:U5 + AH of I/KG 2.

Operations: Fighters

On 16–17 April the enemy introduced a new form of attack on Britain, using fast Fw 190 A4 and A5 fighters carrying a bomb beneath the fuselage and fitted with drop tanks beneath the wings, mainly to attack London, emulating fast raids by Mosquitoes on Germany. The first operation by thirty aircraft of SKG10 was a fiasco, for only two dropped bombs on London and three of the enemy landed in ignominy at West Malling through faulty navigation, whilst a fourth crashed nearby. Nevertheless, these attacks were continued for some time and it looked as if life was to become more interesting for the night-fighter squadrons, as the Mosquito had little or no speed advantage over these newcomers. German tactics were closely studied; the pilots chattered a great deal over their R.T., asking for fixes, and saying whether they were being chased and occasionally gave the international distress signal.

To be in a better position to catch the 190s 85 Squadron moved to West Malling and took over the quarters of 29 Squadron, leaving 157 Squadron to take up residence at Hunsdon the same day. The following night Plt. Off. R. L. Watts, of 157 Squadron, destroyed a Do 217 over Orfordness, one of thirty bombers in the sector when Chelmsford was again the target. On 16–17 May 85 Squadron was responsible for one of the most noteworthy night-fighter successes of the war, when it claimed five Fw 190s between Beachy Head and the Foreland. 11 Group was told that Typhoons could catch the enemy fighter-bombers and that Fighter Command believed Mosquitoes were not fast enough to destroy them. This was accepted, against the advice of experts. A few 190s came in after 23.00 hours and the Typhoons took off. After an hour they clearly could not cope, so the 11 Group Controller told them to land, 'and risked his bowler hat'. The Mk. XIIs of 85 Squadron were then sent into action. Between 23.09 and 04.25 seventeen 190s operated over Kent, Essex, Herts and Bucks, Surrey and Sussex, scattering their bombs. Sqn. Ldr. Green with Flt. Sgt. Grimstone destroyed one over Dover, the first Mosquito crew to do so over Britain. As a reward they were given £5 by the redoubtable Flt. Lt. Molony, a bottle of gin from John Cunningham, a bottle of Champagne from Sqn. Ldr. Crew and a silver model of a Mosquito for the squadron's silver by Sqn. Ldr. Bradshaw-Jones, Senior Controller at Hunsdon. Flt. Lt. Howitt destroyed his '190 at 4,000 feet over the sea near Hastings, Flg. Off. Thwaites downed one in the Channel and another near Ashford, and Flg. Off. Shaw shot one down near Gravesend, subsequently flying through its debris and getting his windscreen covered with soot and his rudder badly damaged. With news of victory the singing of 'Yip I Addy 85' invaded the sanctity of the Adjutant's bedroom in the old parsonage nearby!

First to claim an Fw 190 at night, on 14 May near Evreux, had been Flt. Lt. H. E. Tappin of 157 Squadron and, two nights after 85 Squadron's spectacular, Tappin came close to destroying one of the thirteen that ventured over the south-east. One of them fell to Flg. Off. Lintott of 85 Squadron. On the 19th ten 190s headed for London and the following

Operations: Fighters

night two of 157's aircraft chased several of the fighter-bombers then roaming over Essex, Hertfordshire and Kent. Twenty-two flew over the South-East in two phases on 21–22 May, one being shot down into the sea twenty-five miles north-west of Hardelot by Sqn. Ldr. Crew of 85 Squadron. On 29–30th Flg. Off. Lintott in S-Sugar of 85 Squadron was vectored on to a bandit flying along the South Coast at 29,000 feet. At first it was believed to be a Do 217 or Ju 88B from Schiphol but later was discovered to be a Ju 88S shot down at Isfield, five miles north-west of Lewes.[1]

During May 157 Squadron had flown 92 patrols and 85 Squadron recorded 264 hours of operational night flying.

Daylight attacks on Britain were rare now, but at breakfast time on 8 May six Ju 88 A14s of III/KG 6 made a low-level sortie into Essex. W4092 of 157 Squadron was scrambled, but Spitfires reached the enemy first, destroying two.

A further feature of the month was the continuing re-equipment of the night-fighter squadrons. Although 9 Group told 256 Squadron on 23 January that they would be receiving Mosquitoes, it was not until May, 1943, that instruction in the use of A.I. VIII radar began and the squadron started to receive Mk. XIIs at Ford. On the squadron's first sortie on 21st–22nd Flg. Off. Robinson and Sgt. Midgley contacted an Fw 190, but it outclimbed them. 29 Squadron at Bradwell received Mk. XIIs in May, Sqn. Ldr. P. W. Arbon making their first patrol on 31st in HK164 under the control of Trimley radar. On 20 May the last two NF. IIs had been accepted at Hatfield and despatched to 60 O.T.U., bringing delivery of the first fighter variant to completion.

June's first success fell to Flg. Off. Burnett of 256 Squadron, who shot down a Do 217 off the south coast. The Plymouth attack on 12 June was too far from the Mosquito squadrons to concern them, but two of 157's aircraft fitted with A.I. IV patrolled the Abbeville area in an attempt to shoot down any Fw 190s as they set out for Britain. Meanwhile four or five 190s had penetrated to the London area, where 'A' Flight of 85 Squadron searched unsuccessfully for them. Next night one of the half dozen that risked the journey to Britain was chased by Flg. Off. J. O. P. Turner as far back as Calais. Two 190s entered 85 Squadron's sector and Wg. Cdr. Cunningham engaged one at 23,000 feet, shooting it down for the cost of twenty rounds. It crashed at Wrotham, its pilot being taken prisoner. This was John Cunningham's seventeenth victory, his first on the Mosquito. June 14–15 brought seven 190s over the south-east, and 85, 157 and 256 Squadrons sought them without success. On 15th 29 Squadron had its first all-Mosquito

[1] It set out from Chartres at 00.49 to attack London. Still in its factory markings NL + EX, Nr. 550 is believed to have come from I/KG 66, and was the first Ju 88 S1 to come down in Britain. It was attacked over the coast from behind and below by the Mosquito. Hits on the starboard engine led to a wing fire. A steep dive developed, the starboard wing fell off and the crew bailed out at 25,000 feet.

erations: Fighters

ntil 26 June that the last Beaufighter patrol
bed London on 17–18 June, but it was very
or Lt. Rad of 85 Squadron claimed it as a
worked out by Cunningham and Rawnsley
d to 85 Squadron's Mosquitoes, were bring-
sporadic nuisance raids continued almost
merely illustrated the low ebb to which the
ad sunk since the days of the great blitzes.
ame over the South-East, yet only two
d the capital on 21–22—and one was shot
he City on 22–23rd and one narrowly
d by Wg. Cdr. Wheeler of 157 Squadron.
s did not mount to the expected degree,
ito crews rapidly increased with the June
appearance of the Me 410 over Britain. Powerful and fast—it had a top speed of 388 m.p.h. at 21,980 feet—the 410 had a 13 mm. MG 131 gun on each side of the rear fuselage mounted in a remotely controlled barbette. In addition to two forward-firing 20 mm. cannon and two 7·9 mm. machine-guns it carried a bomb load of 500 kg.

Mosquito XII VY:T of 85 Squadron, flown by Flt. Lt. Bunting, was on 13–14 July first to shoot down one of the new raiders. It came in over Dover and flew northwards. Bunting, on a night interception patrol, followed it to the Essex coast at 10,000 feet and saw the glow from its exhausts when a mile and a half away. Climbing at full boost he chased it for about fifteen minutes at 220 I.A.S. The enemy was climbing, and when the Mosquito was 1,800 feet away had reached 25,000 feet. Bunting closed to 200 yards as soon as he had identified the new shape, and gave it a two-second burst from below. The 410[1] turned on to its back and dived vertically into the sea five miles off Felixstowe. A second Me 410[2] was shot down by Flt. Lt. Thwaites of 85 Squadron, two nights later off Dunkirk from 20,000 feet, and a third fell on 29 July to Wg. Cdr. Park flying a Mk. XII of 256 Squadron some twenty miles south of Beachy Head.

Me 410 activity increased in August and 29 Squadron had spectacular success at intercepting intended airfield attacks on 23–24 August, the night of their first Mosquito successes. HK197 destroyed one east of Manston, HK175 shot one down near Dunkirk and another fell to HK164 near Knocke. It was a bad night for KG 2, during which bombs had been dropped on the airfield at Shipdham, around Walsham and near Cromer. Another six raiders had been turned back by the defenders, and Lt. Rad of 85 Squadron had also destroyed an Me 410. The previous night, when ten enemy aircraft came in over the south-east, Flt. Lt. Howitt shot down an Me 410[3] at Chelmondiston, near Ipswich,

[1] Me 410 A1 Nr. 238:U5 + KG of V/KG 2.
[2] Me 410 A1 Nr. 237:U5 + CJ of V/KG 2.
[3] Me 410 A1 Nr. 274:U5 + DG of 16/KG 2.

Operations: Fighters

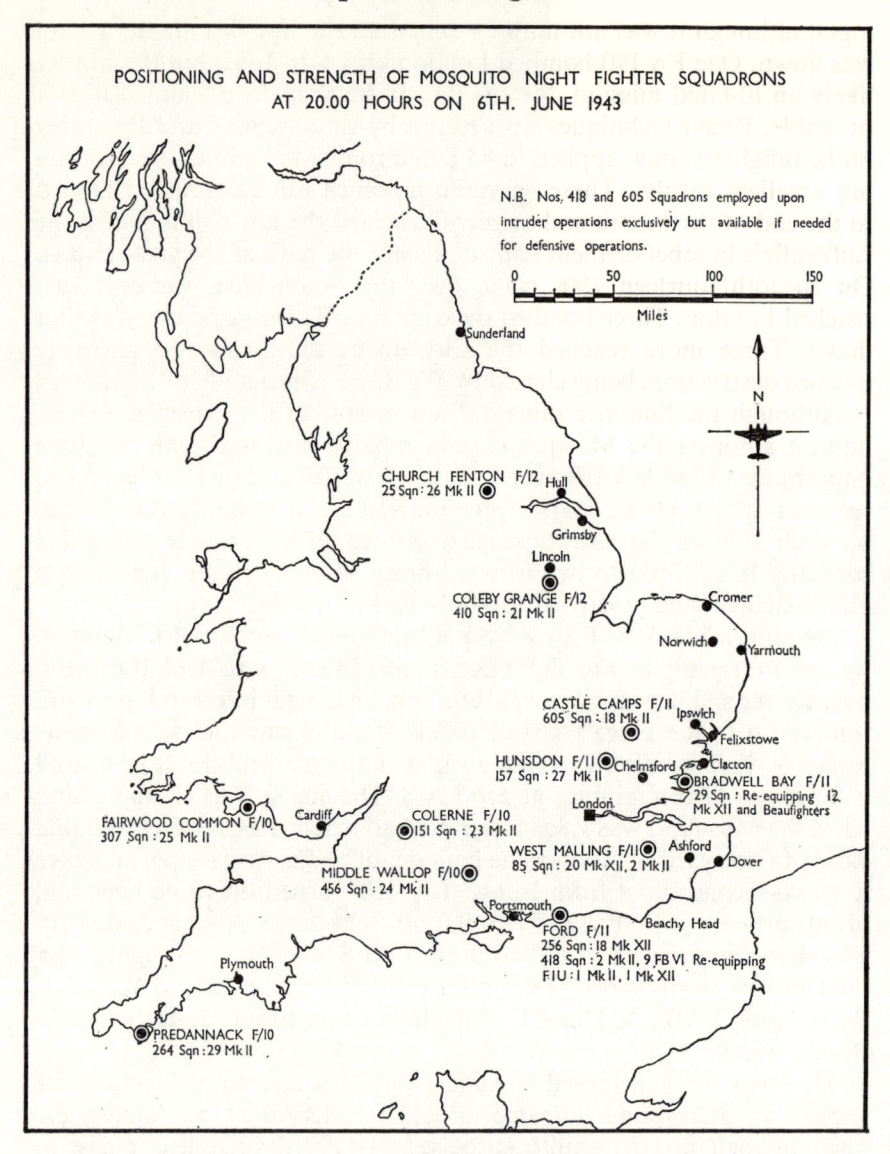

POSITIONING AND STRENGTH OF MOSQUITO NIGHT FIGHTER SQUADRONS AT 20.00 HOURS ON 6TH. JUNE 1943

N.B. Nos. 418 and 605 Squadrons employed upon intruder operations exclusively but available if needed for defensive operations.

CHURCH FENTON F/12
25 Sqn : 26 Mk II

COLEBY GRANGE F/12
410 Sqn : 21 Mk II

CASTLE CAMPS F/11
605 Sqn : 18 Mk II

HUNSDON F/11
157 Sqn : 27 Mk II

BRADWELL BAY F/11
29 Sqn : Re-equipping 12 Mk XII and Beaufighters

FAIRWOOD COMMON F/10
307 Sqn : 25 Mk II

COLERNE F/10
151 Sqn : 23 Mk II

WEST MALLING F/11
85 Sqn : 20 Mk XII, 2 Mk II

MIDDLE WALLOP F/10
456 Sqn : 24 Mk II

FORD F/11
256 Sqn : 18 Mk XII
418 Sqn : 2 Mk II, 9 FB VI Re-equipping
FIU : 1 Mk II, 1 Mk XII

PREDANNACK F/10
264 Sqn : 29 Mk II

after it had crossed the coast at Clacton. This was the first to be brought down in England, and its observer was taken prisoner. On 24 August yet another of 16/KG 2[1] was shot down off East Anglia. At 01.20 hours the brand new aircraft started from near Lille intending to bomb an airfield south of Cambridge, and intercept British bombers returning from Dusseldorf. It was set on fire by the Mosquito, crashed and was burnt out.

Learning from the successful ruse of the Mosquito, enemy intruders were now coming in with our heavy bombers returning from Germany.

[1] Me 410 287:U5+EG.

164

Mingling with the 'heavies' they were difficult t(
upon arrival practised several of the tricks Mosq(
finding profitable over the Continent. They attemp
so keeping the bombers orbiting and running shor
Me 410s attacked them. These resembled the Mosq
and care was accordingly needed before opening fire
was the annoyance being caused that by September F
had a detachment of NF. XIIs from 151 Squadron at (
with the intruders

Two of KG 2's Dornier 217s—one a Do 217 M1—on
patrol were destroyed, and another damaged on 15 August
Park, Officer Commanding 256 Squadron. The first was
steep climb and after a two-second burst was fired at it four
fell away. The second 217 was shot down immediately after
visual was obtained, but as range was closed the bomber fi..d back
and commenced violent evasive action. A running fight ensued during
which several bursts were exchanged, one of which caused an explosion
in the belly of the Dornier, but no fire resulted. The enemy's final burst
of fire put two bullets through the Mosquito's nose scanner and armour-
plated bulkhead, wounding the hand of the navigator. Large pieces of
the perspex nose punctured both radiators, speed was lost and the port
engine caught fire. After another long burst from 200 yards which caused
another explosion in the Dornier the combat was broken off, 55 miles
south of Ford. Park just managed to reach base on the starboard engine,
which was wrecked in the process.

Of seventeen aircraft claimed as shot down by Mosquitoes in August
and September, 1943, over or around Britain, seven were Me 410s and
six Fw 190s. Three of the latter, Fw 190 A5s from I/SKG 10, were
destroyed on 8–9 September by 85 Squadron, one by Wg. Cdr. Cunningham who returned on one engine in VY:R after being hit by debris
from his foe shot down off Aldeburgh. Flt. Lt. Thwaites shot one into
the sea east of the Foreland, and another was destroyed nearby. On
6–7 September a dozen 190s came in over Essex, Kent and Suffolk. A
100 kg. bomb fell 50 yards away from a Mosquito at Castle Camps and
severely damaged it. 85 Squadron shot down two of the raiders. With a
bomb beneath the fuselage and two long-range wing tanks the 190s
made ideal targets—provided the wreckage was avoided after the engagement.

October brought the Ju 188, an improved version of the ubiquitous
Ju 88. Night raiders operated on 21 nights in October, making just over
500 sorties, all aimed at London. The largest night attack for many
months came on 7–8 October. Fifteen enemy aircraft flew into Kent and
Surrey and one reached Oxford and another Chelmsford. In a second
phase 30 flew to Yarmouth and then inland at between 10 and 15,000
feet, over Norfolk, Cambridgeshire and Huntingdonshire. One went to
Woburn and Bedford, and probably left Britain via Kent. In the third

Operations: Fighters

phase a dozen came in from Lille and flew along the Thames Estuary between 9 and 20,000 feet, eight reaching London. In all, three were shot down and three damaged. Eight defensive sorties were flown by 'A' Flight of 85 Squadron, and Sqn. Ldr. Maguire had a terrific dog fight with an Me 410 of KG 2.[1] He took off at 19.10 and clambered to 25,000 feet then dived on the raider 9,000 feet below. At 20.55 he lost it on climbing over Hastings, but another crew saw it fall into the sea. 85 Squadron laid a half claim to another Me 410 shot into the sea off Dungeness, and DZ260 a Mk. II of 157 Squadron, in the hands of Flt. Sgt. Robertson, damaged another. Amongst the raiders were some of the new Ju 188s one of which, 260160:Z6+IK, crashed undetected in the sea.

85 Squadron claimed their first Ju 188 on 8–9 October south east of Dover, one of ten raiders that came in from Holland. It was a Ju 188 E1 260204:3E+KF of Erprob. St/KG 6. Absolute proof that the 188 was in use arrived on 15–16 October, when 3E+BL came down at Woodbridge and 3E+HH at Birchington.[2] The former, operated by I/KG 6 was shot down from 20,000 feet by Sqn. Ldr. Maguire after twice being attacked since it had set out from Rheine. The other aircraft was brought down by Flg. Off. Thomas in VY:K. It was one of three of I/KG 6 operating Ju 188s for the first time which, with four of III/KG 6, had flown to Munster/Handorf to refuel prior to the operation. Due to insufficient fuel some of the aircraft had to land at Rheine and so arrived late for the raid. The crew of 3E+HH believed they had bombed London, and flying alone they were an easy target. On their attack they had relied upon *Knickebein*, the radio beam used by the Germans since 1940. They were carrying *Duppel*, the German version of *Window*, but, strangely, did not use it. Despite skilful use of the high flying Me 410, the metal *Duppel* strips and tail warning radar in their bombers, the enemy lost 28 aircraft in three weeks.

Defending night-fighters were now benefiting from the new 150 mm. radar directed searchlights, which could illuminate, for the fighter crews, targets at 35,000 feet. It was at this time that the Germans began coming in at heights up to 30,000 feet so as to be able to lose height with high speed on the journey home. Unlike our bombers, the enemy aircraft were over unfriendly territory for but a few minutes. They relied upon somewhat violent evasive tactics, particularly the Me 410s which frequently fired at their pursuers. The margin in performance superiority between the Mosquitoes and the German raiders was small. Skilful handling and maximum use of the slender advantage of performance and A.I. radar was called for from the Mosquito crews. German losses averaged about 7 per cent while losses amongst Mosquito bombers at this time were about 1·75 per cent despite their being over enemy

---

[1] Possibly Me 410 A1 10185:U5+KG of 16/KG2. German records also list Me 410 A1:103 as crashing near Ghent due to enemy action.
[2] 3E+BL W.Nr. 260179 and 3E+HH W.Nr. 260177.

Operations: Fighters

territory for about 1½ hours compared with the German's 30 minutes or so. Whereas the Mosquito bombers were normally delivering 3 × 500 lb. and 1 × 250 lb. bombs with precision, the Luftwaffe operated with nervous abandon, scattering bombs far from prescribed targets. The enemy load was often 10 × 50 kg. or 1 × 1,000 kg. The results of the attacks at this period may be compared thus:

	Mosquito night raids	German night raids
Time spent over enemy territory per aircraft	178 hours	7 hours
Wt. of bombs dropped by each aircraft lost	103,000 lb.	40,000 lb.

It must be remembered that, in addition to bombs, the Mosquitoes raiding Germany carried five times as much fuel on many raids as the Luftwaffe raiders did.

Development of the Mosquito fighter was proceeding along three lines, giving improved A.I. radar, range and performance. The NF. XII stemmed from the Mk. II, with superior radar fitted in the nose in place of four machine-guns. The standard fighter-bomber wing had come on to the production lines early 1943, able to carry external bombs or drop tanks. Incorporating this, the NF. XIII could have an endurance of about 5¾ hours. Meanwhile American radar developed with British aid was coming into production and Lend-Lease supplies reaching England. The Mk. II was modified to carry it in the manner of the Mk. XII by Marshalls of Cambridge, and later the NF. XIII with American radar became the Mk. XIX. This mark acquired a performance advantage at low levels by being fitted with Merlin 25 engines.

Small-scale nuisance night raids continued throughout October. On 17–18th a two-wave attack from Lille and the Somme Estuary developed at 01.30, the enemy spreading out over the Home Counties and East Anglia. 85 Squadron shot down an Me 410 A1 (10176:U5+LF of 15/KG2) at Hornchurch, and two others were intercepted. Two days later 410 Squadron, using Mk. IIs fitted with A.I. Mk. IV or V, was brought south from Coleby to West Malling to increase the strength of the defenders and on 22 October it was decided to allow searchlight aided night-fighters over London, provided they kept above 20,000 feet, thereby leaving the guns to fire in the I.A. Zone to 18,000 feet. Recently German bombers had been flying straight in a slight dive over the Capital ignoring the barrage, at anything over 22,000 feet. On 23–24 October eighteen enemy aircraft entered East Anglian skies and eight reached London. Next night the small raids continued and on 31 October–1 November fourteen enemy aircraft approached the coast between Sidmouth and North Foreland, but only six flew inland.

488 Squadron's destruction of an Me 410 of KG2 off Clacton on 8 November was noteworthy, being the first occasion when a Mosquito NF. XIII (HK367) destroyed an enemy aircraft. Their next success with

Operations: Fighters

the new type came on 20 December, when an Me 410 A1 (420085: U5+HE of 14/KG2) was shot down near Rye, by which time a Mk. XIII of 29 Squadron, HK403, had destroyed an Fw 190 near Horsham.

In November ten night raiders were destroyed by Mosquitoes, five of them 410s. Enemy activity over regions other than the east and south-east was rare. When 307 Squadron moved to Drem in Scotland on 9 November fresh from *Instep* patrols they assumed there would be little to watch but shipping. It was during a patrol on 22 November that Flt. Sgt. Jaworski in HJ651 claimed to come across one of the large He 177 bombers 120 miles north-east of the Shetlands and shot it down. The Luftwaffe really lost a Fw 200. Four days later 307 Squadron despatched a Ju 88 to a watery grave, a success repeated on 9 December. But the most remarkable success that month came to Flg. Off. R. Schultz in DZ292, a Mk. II of 410 Squadron, who on 10th brought down three Do 217 M1s of I/KG2 raiding Chelmsford, all of them crashing into the sea off Clacton, and set up a new Mosquito record. By the end of the month Mosquito night-fighters had claimed $16\frac{1}{2}$ Me 410s as having been shot down, one probably destroyed and three damaged.

Paramount needs at this time in the Mosquito were steeper climb, more speed around 20,000 feet and quick clearance of ice from windscreens, a problem currently being investigated. Steps were meanwhile taken to ensure the best possible performance from existing aircraft fitted with Merlin 23s, by cleaning up the cowling and sealing oil leaks, especially in the carburettor air intake ducts fitting badly on some machines. Snowguards were left off when weather conditions permitted. Stub exhaust pipes were tried at night, but their conspicuousness proved them unsuitable.

Two aircraft were at Farnborough being fitted with nitrous oxide injection for trials with 85 Squadron and the Fighter Interception Unit. By the end of 1943 the first had made successful test flights during which R.A.E. reported an increase of 47 m.p.h. at 28,000 feet using 4·85 lb. of gas per hundred h.p. per engine, sufficient being carried for six minutes' use. It was subsequently decided that Heston Aircraft should equip fifty Mk. XIIIs with N_2O installations for 96 and 410 Squadrons. Cunningham flew one, HK374:L of 85 Squadron, on operations for the first time on 2–3 January, and bagged an Me 410 off Le Touquet.

Closing weeks of 1943 found Fighter Command bisecting itself. One half under Air Marshal Roderic Hill, known as 'the Home Guard' but officially as Air Defence of Great Britain, controlled $10\frac{1}{2}$ Mosquito night interceptor squadrons and the intruders. Other squadrons came under the control of 85 Group, Allied Expeditionary Air Force, in readiness for the assault on Fortress Europe. Two fears in the minds of British Intelligence were that the enemy would resume an offensive against London and, that he would launch an attack on the invasion build up.

On the night of 21–22 January, 1944, the expected Operation *Stein-*

Operations: Fighters

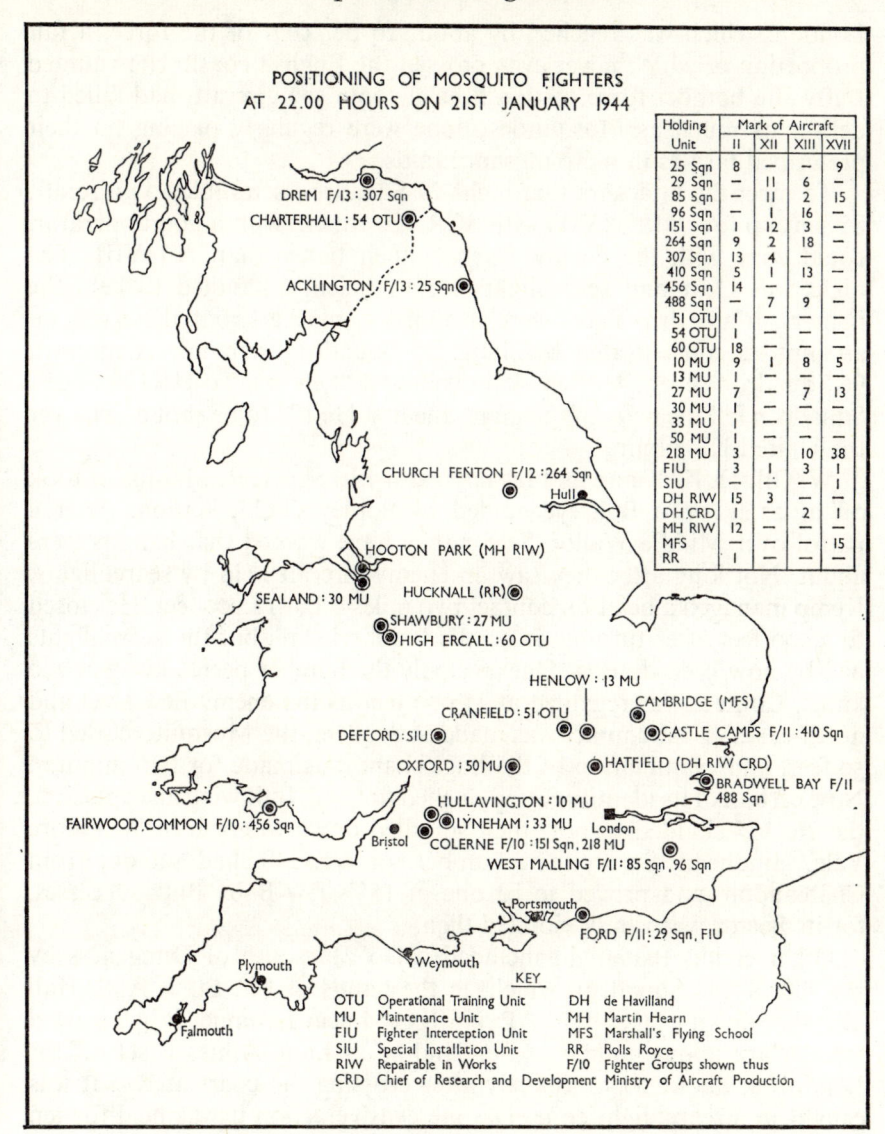

POSITIONING OF MOSQUITO FIGHTERS AT 22.00 HOURS ON 21ST JANUARY 1944

Holding Unit	II	XII	XIII	XVII
25 Sqn	8	—	—	9
29 Sqn	—	11	6	—
85 Sqn	—	11	2	15
96 Sqn	—	1	16	—
151 Sqn	1	12	3	—
264 Sqn	9	2	18	—
307 Sqn	13	4	—	—
410 Sqn	5	1	13	—
456 Sqn	14	—	—	—
488 Sqn	—	7	9	—
51 OTU	1	—	—	—
54 OTU	1	—	—	—
60 OTU	18	—	—	—
10 MU	9	1	8	5
13 MU	1	—	—	—
27 MU	7	—	7	13
30 MU	1	—	—	—
33 MU	1	—	—	—
50 MU	1	—	—	—
218 MU	3	—	10	38
FIU	3	1	3	1
SIU	—	—	—	1
DH RIW	15	3	—	—
DH CRD	1	—	2	1
MH RIW	12	—	—	—
MFS	—	—	—	15
RR	1	—	—	—

Locations shown on map: DREM F/13 : 307 Sqn; CHARTERHALL : 54 OTU; ACKLINGTON F/13 : 25 Sqn; CHURCH FENTON F/12 : 264 Sqn; Hull; HOOTON PARK (MH RIW); SEALAND : 30 MU; HUCKNALL (RR); SHAWBURY : 27 MU; HIGH ERCALL : 60 OTU; HENLOW : 13 MU; CRANFIELD : 51 OTU; CAMBRIDGE (MFS); DEFFORD : SIU; CASTLE CAMPS F/11 : 410 Sqn; OXFORD : 50 MU; HATFIELD (DH RIW CRD); BRADWELL BAY F/11 488 Sqn; HULLAVINGTON : 10 MU; FAIRWOOD COMMON F/10 : 456 Sqn; LYNEHAM : 33 MU; London; Bristol; COLERNE F/10 : 151 Sqn, 218 MU; WEST MALLING F/11 : 85 Sqn, 96 Sqn; Portsmouth; FORD F/11 : 29 Sqn, FIU; Plymouth; Weymouth; Falmouth

KEY
OTU — Operational Training Unit
MU — Maintenance Unit
FIU — Fighter Interception Unit
SIU — Special Installation Unit
RIW — Repairable in Works
CRD — Chief of Research and Development
DH — de Havilland
MH — Martin Hearn
MFS — Marshall's Flying School
RR — Rolls Royce
F/10 — Fighter Groups shown thus
Ministry of Aircraft Production

bock offensive against London, intended to have opened at Christmas, began with the enemy despatching 447 sorties. Mainly used were Ju 88s and the latest versions of the Do 217. Only one Geschwader yet had Ju 188s and another used Me 410s, which, with Fw 190s, led the attack, dropping *Duppel*, which seriously affected the performance of our G.C.I. stations. Parts of two units also flew about thirty He 177s, extremely troublesome aircraft. The attack came in two waves about six hours apart. The nearest bombs to Hatfield fell at Potters Bar, but none exploded. Although the bomb load of the force was around 500 tons only about half of this fell on land, and a mere 30 or so on Greater

Operations: Fighters

London, which was reached by about 20 per cent of the force, a fair proportion of which never even crossed the English coast. The vaunted Luftwaffe bomber force, with a fleet of some 500 aircraft, had failed to deliver as much as Mosquitoes alone were regularly placing on their prescribed targets in mere nuisance raids.

To meet the attackers that night 29 Squadron scrambled 13 aircraft, 85 sent up four Mk. XVIIs with SCR 720 which, with a good operator, could cope with the enemy *Duppel* much better than A.I VIII. The night was clear and searchlights were efficiently handled to help the fighters. Many drip flares were dropped as markers, but there was no attempt at concentrated bombing. 25 Squadron sent up six aircraft, 96 Squadron seven, 410 ten, 456 two and 488 eleven. To HK193 of 151 Squadron fell the prize of the night, the first He 177 to be shot down over the United Kingdom.

W. Off. H. K. Kemp and his navigator Flt. Sgt. J. R. Maidment took off at 22.25 on a Bullseye guided by Sopley G.C.I. station. After a patrol over Middle Wallop Sector they were warned that bandits were about. Not long after they saw an enemy aircraft held by searchlights. Kemp managed a head-on contact two miles off at 14,000 feet. He closed to 3,000 feet after turning on to the bomber's tail, but the searchlights had by now been doused. Unexpectedly the bomber peeled away at 220 knots. Contact was regained at 11,000 feet as the enemy flew level and north-east. Visual contact was made at 800 feet, the Mosquito closed to 50 feet, and examination of the huge shape was made for two minutes. Now certain of its identity Kemp opened fire and, following an explosion, the He 177 skidded to port then spiralled down to crash in Whitmore Vale, Hindhead, Surrey. The bomber, of 1/KG40, had set out from Châteaudun, and proved to be one of 151's few Baby Blitz successes, for its Sector was rarely violated then.

Other claims that night included a Do 217 south of Dungeness by HK380 of 488 Squadron which, in the hands of Flt. Lt. J. A. S. Hall also shot down a Ju 88 B3+AP of 6/KG 54 near Lympne. It had started from Marx about midday. After landing at Laon/Athies it set off for London at about 20.30. Shortly after crossing the coast at Rye it was caught in a searchlight cone. Despite evasive action it was held for ten minutes. After it encountered A/A fire a night-fighter attacked, and set an engine on fire. Height was lost to about 6,000 feet before the crew baled out. Both enemy aircraft had been held in the searchlights. Flt. Lt. Head of 96 Squadron in HK372 claimed two Ju 88s as probables. He also was helped by searchlights. Sgt. Wakelin, also of 96 Squadron, guided by Wartling radar, shot down a Ju 88 A4 4D+EP:0414 of 6/KG30 which crashed at Paddock Wood Station, near Tonbridge. It had started from Eindhoven at 02.20 hours to attack the Charing Cross area of London, and was hit by flak when flying at 1,500 feet. HK425 destroyed a Ju 88 too. 25 Squadron had no success but flew three sorties with Mk. IIs (HJ654, HJ649 and HJ654 again on a second flight) as

Operations: Fighters

well as three using Mk. XVIIs. Mk. IIs were flying for the last time operationally with 25 Squadron during the night. 456 Squadron scrambled only HJ925, for the enemy kept clear of their sector. 410 Squadron had two aircraft on patrol and scrambled six. They estimated that forty bombers entered their sector in the first wave, sixty in the second. 29 Squadron claimed an Fw 190, making nine raiders destroyed by defending Mosquitoes. A second equally unsuccessful attempt was made to bomb London on 29 January.

Sizeable raids were directed against London on eight nights in February. Metal *Duppel* strips proved effective, blotting out large areas of the radar coverage of the South-East and making it difficult for the Mosquitoes to depend upon their A.I. Defences had to resort to the use of sound locating apparatus, and many of the Mosquito successes of the month came when searchlights illuminated the raiders.

Six squadrons faced the bombers on 3 February. 410 Squadron flew twelve patrols and scrambled six aircraft from Castle Camps between 20.15 and 20.20 hours. Flt. Lt. Geary in HK429 had a visual on an Fw 190 and nearly collided with it before he lost it. Flg. Off. W. G. Dinsdale contacted, and had a visual on, a Ju 88 in HK476. He grazed the enemy bomber with his starboard propeller. Neither engagement resulted in success, but a Do 217 fell to Flg. Off. E. S. P. Fox. Flt. Sgt. Vlotman in HK367 of 488 Squadron shot down a Do 217 forty miles east of Foreness, and another fell to 85 Squadron.

The next major attack came on 13–14 February, after sporadic raids on most nights, as on the 11th when four out of fifteen enemy aircraft which came over the south-east reached London. To combat a few enemy aircraft on 12 February, 456 Squadron scrambled Mk. XVIIs, making its first operational use of the new mark.

The raid of the 13–14th caused 410 Squadron to scramble five Mosquitoes to join three on standing patrols. Sqn. Ldr. Somerville in HK466 shot down a Ju 88 S of 1/KG66 and later engaged a Ju 88, which returned his fire. The 88 S was caught in searchlights, and the pilot's evasive action was so violent that a survivor of the incident believed control had been lost. There was a loud explosion as the Mosquito's shells rammed home, and the cabin seemed to collapse. Wreckage of the aircraft fell at Havering-atte-Bower, near Romford, Essex. Flg. Off. Schultz in HK429 destroyed a Ju 188, but his aircraft was badly shot up. Another '188 fell to Flt. Lt. Bunting, flying MM476 of 488 Squadron.

On 18 February came the most successful enemy attack of the series, when 175 tons of bombs fell on the Capital, partly because the enemy used a simplified pathfinder technique. At ten minutes past midnight ten of 410 Squadron's aircraft scrambled after ninety raiders crossing the coast, a third of which eventually reached London.

Two nights later eighty-two headed for London. Flg. Off. Dinsdale found a Ju 188, but W. Off. Miller, also of 410 Squadron, flew in his

Operations: Fighters

way preventing an attack. The bomber had a bright light on its tail, making it very difficult to attack. HK285 of 25 Squadron flown by Plt. Off. J. R. Brockbank shot down a Ju 188, the first enemy aircraft to fall to a Mk. XVII. Flt. Lt. Singleton in HK255 of 25 Squadron shot down a Do 217. Twelve enemy aircraft ventured past Dover around 03.30 on 21 February; then the next night came another raid on London. Two Me 410s fell to 85 Squadron, and Sqn. Ldr. C. A. S. Anderson, in HK521 of 410 Squadron, shot down Ju 88 A4 3E+GS of 8/KG6, at Earls Colne, which was flying from Melsbroek to bomb London, and Ju 188 3E+EK of 2/KG6 which fell in flames, at high speed, on Bullers Farm, Shopland, near Rochford. Sixty enemy aircraft over East Anglia had prompted sixteen sorties by the squadron. HK283 of 25 Squadron claimed a Do 217.

A raid by about sixty enemy bombers was mounted from the direction of East Anglia on 23-24th, and 25 Squadron shot down an He 177 of 1/KG100 at Wolseley, near Yoxford, and destroyed a Do 217. An Me 410 was brought down by 96 Squadron, and Sqn. Ldr. Thwaites, D.F.C., destroyed one of the few Fw 190s now being used at night. On 24th about 150 aircraft came over England and Flt. Lt. Ward of 96 Squadron shot down an Me 410 from 28,000 feet, while Sqn. Ldr. Parker-Rees claimed an unidentified raider. Two hours before midnight Flt. Lt. P. F. Hall of 488 Squadron went on a free-lance patrol in HK228 and eventually found a Do 217 held by searchlights. This he set afire north-east of Brighton; he then shot down what he believed was a Ju 188. Examination of the wreckage, which fell at Lamberhurst, near Tonbridge, Kent, showed it to be an He 177 of 3/KG100. It had allegedly started from Châteaudun at about 21.00 hours to bomb London. Whilst approaching target at about 12,000 feet the night-fighter attacked. Fire broke out in the aircraft's underside and it crashed in flames at 22.44 hours. Thirteen enemy aircraft were shot down that night, seven over Britain or close by. Five more were claimed as probables and three damaged. 29 Squadron shot down five of them, a night after its Commanding Officer, Wg. Cdr. Mack, had failed to return from chasing a Hun off Beachy Head. Flt. Lt. Cox of 29 Squadron destroyed an He 177 the same night, whilst his colleague, Flg. Off. W. W. Provan, shot down a Ju 88 or 188, Flt. Lt. J. E. Barry a Do 217, and a probable Me 410, in the face of a shower of *Duppel* from each, and Flt. Lt. R. C. Pargeler destroyed a Ju 88. Flg. Off. Hedgecoe of 85 Squadron claimed a Ju 188, and Flg. Off. Burbridge a Ju 88. Seventy-two raiders or 5·2 per cent were claimed as destroyed during February in attacks on Britain. Additionally there were those which failed to return to their bases due to low standards of airmanship. For the raiders there was the ever-present hazard of meeting intruding Mosquitoes on the return journey and over home bases.

Forces under Peltz attempted four major raids on London in March in addition to one on Hull and another on Bristol, during both of which no bombs fell on the cities themselves. All but two of the night-fighter

Operations: Fighters

squadrons in England had, or were receiving, Mosquitoes. Both squadrons scheduled to have N_2O gear were having their aircraft modified on site.

Sixty enemy aircraft crossed 410's sector on their way to London on 1 March, yet only six are recorded as reaching London. Incendiaries fell near West Malling, whence HK499 of 96 Squadron had taken off flown by Flg. Off. Gough, who closed in on an He 177 only to find that his guns would not fire. Later he had a visual on an Me 410 at 9,000 feet over Northern France. He opened fire from 700 feet and saw a glare as the enemy fell towards the ground. 151 Squadron was more successful, for Wg. Cdr. G. H. Goodman in HK377 destroyed two Ju 188s and damaged a Ju 88. A Ju 188 fell to Sqn. Ldr. Harrison and a Ju 88 to Flt. Lt. Stevens, but soon after the squadron concentrated on flying *Insteps* and figured no more in the Baby Blitz.

On 14 March seven of 410 Squadron were scrambled to join three already on patrols. Six contacts were achieved, helped by searchlights. A Ju 188 was shot down by Lt. Harrington, a U.S.A.A.F. officer, Sqn. Ldr. Green destroyed a Ju 88 and Flt. Lt. Bunting of 488 Squadron shot down Ju 188 U5+B of 5/KG2 which crashed at White House Farm, Great Leighs, Essex. One hundred and forty enemy aircraft were estimated to have crossed the coast, in four waves, three of which came across between Cromer and Shoeburyness. The fourth formation, from the Rouen-Somme region, crossed Sussex and met the main force as it headed for home, and was engaged by 96 Squadron, which had scrambled nine Mosquitoes. As a result a Ju 188 was seen heading into cloud producing showers of sparks, and a Ju 88 fell in flames to Flg. Off. Gough. The next major raids on London developed on 21 and 24 March, when again 85, 96, 410, 456 and 488 Squadrons engaged them.

Small-scale attacks occurred on most nights, but already the enemy was turning, perhaps hopefully, to what might appear to be less well defended targets. On 19 March fifty raiders were reported over the Wash, and attacks on Hull and Norwich were intended. Flt. Lt. Singleton of 25 Squadron shot down no fewer than three Ju 188s heading for Hull; then his Mosquito HK255 developed engine trouble and, with the starboard engine on fire and his port engine cut, he crash landed the aircraft—both of whose engines were now afire—in a ploughed field to the south of Coltishall. Out clambered the crew and Flt. Lt. Singleton, unable to grab a fire extinguisher from the cockpit, did the next best thing. He simply threw handfuls of dirt on to the burning engine, but the aircraft had to be written off. 25 Squadron, which had arrived at Coltishall on 5 February, had been particularly successful during the night, for to HK278 had fallen an He 177 and a Do 217. HK285 had contacted another 177, but it had escaped. The following day the C-in-C visited Coltishall to congratulate the squadron. 307 Squadron also scrambled to meet the Hull raid, and Plt. Off. J. Brochocki in a Mk. XII HK119:J claimed an He 177, 307's only probable success in the Baby

Operations: Fighters

Blitz. 264 Squadron also claimed a share of the kills, a Do 217, a Ju 188 and one unidentified bomber. A two-wave raid on Hull developed on 20th, as a result of which 264 Squadron engaged an He 177 and 25 Squadron a Ju 188. Next night it was the turn of London and Flt. Lt. L. R. Davies of 25 Squadron shot two Ju 188s into the North Sea from 20,000 feet, bringing 25's total since moving to Coltishall to eleven.

Squadrons of 10 Group had their chance for battle on 27 March, when Bristol was target for more than 100 aircraft, eight of which fell to the night-fighters. One shot down by HK260 was Ju 88 B3+BL of 3/KG54, which fell at Ifle Brewers, near Ilminster, the first enemy aircraft to be destroyed by a Mosquito of 219 Squadron now operating Mk. XVIIs from Colerne. Target for the Ju 88 was Bristol docks but it was attacked soon after it had crossed the coast. After jettisoning the bombs the crew baled out.

The last attack of the series directed against London took place on 18 April. Flg. Off. S. B. Huppert of 410 Squadron, now at Hunsdon, shot down an He 177, one of fifty enemy aircraft over the Sector. At 23.18 hours the He 177 6N+AK of 2/KG100 took off from Rheine to bomb the Tower Bridge area of London. After crossing the coast in the Orfordness area it was surprised by Mosquito MM456, which made two damaging attacks. At 01.03 the crew jettisoned bombs at Little Walden, Essex, then baled out. Wreckage fell at Cole End, near Saffron Walden. 410 Squadron flew eight patrols, had nine contacts and seven visuals, and damaged a Ju 188 seven miles from base. Plt. Off. Allen of 96 Squadron shot down a Ju 88, the descent of which was watched by the squadron ground crew. Wg. Cdr. Crew, D.F.C., and Bar, destroyed an Me 410 of 1/KG51, and its wreckage fell amongst the tomb stones in St. Nicholas churchyard, Brighton. He almost collided with a Do 217 a few minutes later. Sqn. Ldr. Green, D.F.C., shot down a Ju 88 ten miles north of Margate, and 488 Squadron had their last success of the period when Flt. Lt. J. A. S. Hall and W. Off. Bourke each destroyed a Ju 88. A third Ju 88 landed at the squadron's base during the night and the crew surrendered.

Sporadic enemy activity continued during May, falling off towards its end. On the 15th 106 enemy aircraft operated, sixty of which congregated around Portsmouth in the last major night raid of the war on Britain. Sqn. Ldr. Gill of 125 Squadron shot down a Ju 88 near Cherbourg and an Me 410 was damaged North of Portland Bill. A Ju 88 B3+DT of 9/KG54 shot down by 456 Squadron after being held in searchlights, fell at Medstead, near Alton, and a Ju 188 was brought down by 604 Squadron. A week later 125 Squadron claimed a Ju 88 and a '188 in the Southampton area, where two '88s were added to 456 Squadron's score.

At this critical time, as the Allies concentrated their forces for the D-Day assault, the Luftwaffe's offensive petered out, and the Baby Blitz ended on 29 May, with scattered attacks by fast aircraft on south-east harbours now filling with shipping, and destined nevertheless to be

Operations: Fighters

almost completely ignored by the enemy bomber force by day and by night.

Defensive Mosquito operations acquired a new look from D-Day, and fell into three main categories. Six squadrons forming part of 85 Group, Tactical Air Force, added night protection to Allied forces, from the invasion until VE-Day. Seven squadrons and the Fighter Interception Unit stood by for possible enemy attacks on Britain, and flew patrols over the beachhead until the flying bombs appeared. Then they were committed to the night defence of Southern England, supplemented by two squadrons of 100 Group. Thirdly, there was defence against air-launched flying bombs released from Heinkel 111s over the North Sea.

All squadrons of 85 Group were alerted for operations on the night of 5–6 June, 1944, covering Allied forces landing in or approaching France. Flt. Lt. Fox of 264 Squadron took off at 22.45 hours on 5 June in HK480 to be first over the beachhead, his squadron flying jamming patrols between St. Martin and St. Pierre, and defensive patrols later. About forty-five enemy aircraft were plotted during the night, approaching the Channel from the Paris area, but they turned back avoiding battle. Only to 409 Squadron did this come, when Flg. Off. Pearce destroyed a Ju 188 near the English coast. Next night Mosquitoes searched for German aircraft attempting to interfere with Allied shipping. Ju 88 bombers of Fliegerkorps IX and Ju 88 torpedo-bombers with He 177 and Do 217 glider bomb carriers of Fliegerkorps X flew about 150 sorties, repeating their efforts on 7–8 June. They carried Hs 293 glider bombs and Fx 1400 radio-controlled bombs, and attempted low-level attacks in the Seine Bay. On 8–9 and 9–10 June they flew about eighty sorties and on the next two nights raised their efforts to around a hundred. Soon these fell to sixty to seventy nightly sorties, during which oyster mines, circling torpedoes and the Mistel combination were used. Mines caused some embarrassment to the Allies, yet on 4 July the millionth man stepped ashore and 300,000 vehicles had safely landed.

Early on 7 June engagements by Mosquito night-fighters took place between Le Havre and Cherbourg, and four He 177s were destroyed by 456 Squadron. 219 destroyed a Ju 188, and, inland an Me 410, another of which fell to 604 Squadron. During this, their first night over France after the landings, the Middlesex Squadron shot down five of the enemy. Next night they made five more kills, then trade slackened. Two He 177s were destroyed when flying very low, another when silhouetted against a bank of white cloud. Early on the 8th 456 Squadron accounted for three more 177s, Sqn. Ldr. D. Howard in HK323 claiming two after cunningly stalking them so as not to reveal himself until firing. The other fell to HK302, and D/406 Squadron claimed a Do 217. Off Southwold 25 Squadron destroyed an intruding Me 410. Bad weather then made operations difficult and when 264 Squadron's six aircraft landed early on 9 June they did so with the aid of FIDO.

Operations: Fighters

It was too much to hope that the success rate would continue. Nevertheless 456 Squadron added to its claims when on 9–10 June HK353 destroyed an He 177 carrying Hs 293s off Cape Levy and a Do 217 near Cap de la Hague. At 02.00 hours MM460 of 409 Squadron shot down a Ju 188 40 miles S.E. of Le Havre. Plt. Off. Sanderson in HK249 intercepted an He 177 on 10–11 June. Before he could open fire gunners in the bomber damaged his starboard wing, stripping six feet of the covering. One-legged Flt. Sgt. S. H. J. Elliott of 409 Squadron with Flt. Lt. R. A. Miller shot down an Fw 190 and severely damaged another.

Throughout June operations to prevent enemy night bombers reaching Allied shipping continued. June 14–15 was a busy night, which brought Flg. Off. Dinsdale in HK476:O face to face with a new foe—to quote the ORB of 410 Squadron a 'Ju 88 with a glider bomb on top' a Mistel combination,[1] which he destroyed off Normandy. Another was shot down by Flt. Lt. Corre of 264 Squadron flying HK502. During June Mosquitoes of 2 T.A.F. and A.D.G.B. accounted for a greater number of enemy aircraft than in any preceding month. Fifty-three[2] were claimed over the beachhead area, one over England and the remainder over France, Belgium, Holland and Germany. Ten Mosquitoes were lost due to enemy action, nine to accidents.

As they returned on 14 June shortly before 04.00 hours from an intruder patrol Flt. Lt. D. MacFayden and Flg. Off. J. Wright of 418 Squadron saw what they reported as 'a rocket projectile heading northwards and leaving a red trail'. A new menace had revealed itself, the V-1 flying-bomb. Ten were fired towards Britain in the early hours of 14 June and the first fell four miles from Gravesend airfield, base for 140 Wing of Mosquito fighter-bombers. Another reached Bethnal Green and two fell in Southern England. Planned Luftwaffe support resolved itself as merely one Me 410 over London, and this was promptly destroyed. Although it was a weak effort compared with the intended launch of 500 missiles, British authorities realized that it was prelude to a large-scale offensive directed mainly against London. It was resumed on 15–16th when 244 V-1s were fired, of which 144 crossed our coast and seventy-three reached Greater London. Many guns were now in position to deal with them, and fighters went into action.

To a Mosquito VI of 605 Squadron flown by Flt. Lt. J. G. Musgrave and Flt. Sgt. Sanewell went the distinction of being first crew to destroy one. This was on 15 June and the bomb, caught in low and level flight, exploded in mid air. Soon after midnight they had taken off from Manston on receiving warning of the approach of V-1s, and they shot their prey to pieces over the Channel. Musgrave reported that 'it was like chasing a ball of fire across the sky. It flashed by our starboard side

---

[1] Un-manned Ju 88 A with warhead in the nose, launched from beneath a Bf 109 or Fw 190.

[2] These and subsequent claims accepted in good faith are unlikely now to be proven from German records.

Operations: Fighters

a few thousand feet away at the same height as we were flying. I quickly turned to port and gave chase. It was going pretty fast, but I caught up with it and opened fire from astern. At first, there was no effect so I closed in another hundred yards and gave it another burst. Then I went closer still and pressed the button again. This time, there was a terrific flash and explosion and the thing fell down in a vertical dive into the sea. The whole show was over in about three minutes.'

96 Squadron at West Malling was also in at the start of the V-1 offensive. Its diarist wrote on 13 June: 'Well, well, whatever will happen next? At 03.42 the air raid warning was sounded, all clear went at 04.00. At 04.15 the red was sounded again and this time the news came from Biggin that the Hun was sending over pilotless aircraft of some sort, his secret weapon perhaps?' Two were seen to pass over West Malling as Flt. Lt. Mellersh, who later destroyed seven in MM577, was landing. At a loss to describe them 96 coined the appellation 'chuff bombs'. On 17 June HK415 flown by Rees was first on the squadron to claim a V-1. By 23rd the squadron had twenty-four to its credit. The nose of the C.O.'s aircraft split open on 25th as he chased one at high speed, and he had to abandon MM499. By 20 June about half the bombs were being destroyed, or crashing, before they reached London.

Four Mosquito squadrons—96, 219, 409 and 418—were almost immediately engaged upon the destruction of flying-bombs, each putting up about ten three-hour patrols per night as the enemy offensive came under way. Others worked part time. On 27 June 85 and 157 Mk. X A.I. Squadrons of 100 Group were applied to anti-diver work, as described in Chapter 19. Mr. Duncan Sandys, Chairman of the Anti-Diver Committee, went on patrol with Wg. Cdr. Crew of 96 Squadron; they shot at a V-1 but missed; a Tempest roared past them and attacked —and also missed it. Patrols were made along the French coast at about 8,000 feet in clear weather, in an attempt to spot launchings, seen as bright flashes followed soon after by flames from the propulsion unit, visible from about fifteen miles in good conditions. Bad weather during June, however, kept the number of successes against the V-1s at night rather low.

To catch the bombs in about thirty miles necessitated flying much faster than the foe, but with the speed gained in the dive from patrol height Mosquitoes seldom had difficulty in catching V-1s over the Channel when they were about 2,500 feet high, and travelling at 320–350 m.p.h. Judging the vector and distance from which to open fire at these small aeroplanes was not easy by day; at night the distance from the flame was even more deceptive. Various ranging devices were tried including photo-cell indicators which proved unsuccessful, apparently due to vibration. Long bursts of cannon fire were at first needed, and even then the bombs often flew straight on after receiving many hits. There were three problems facing the intercepting fighter. It had first to get on to course behind the V-1, then close for attack. Thirdly it

Operations: Fighters

needed to be within range either before it reached the gun belt at the coast, or between this and the balloons, situated immediately to the south of London. A few V-1s were observed at 6,000 feet, and one was attacked by a Mosquito at 12,000 feet.

Nine Mosquito squadrons were fully committed to the anti-diver role by mid-July. 418 Squadron began anti-diver patrols on 17 June and off Beachy Head Sqn. Ldr. D. A. MacFadyen, D.S.O., D.F.C. and Bar, twice engaged a V-1 before destroying it. An hour later off Dungeness he had a second to his credit. Two others also fell to the squadron, which scored three kills and one 'probable' in five sorties. More than two-thirds of 418s many sorties in the next fourteen weeks were against V-1s. In the next ten days ten different crews destroyed twenty flying-bombs and experienced all too many nerve-racking moments, as when Flg. Off. S. P. Seid closed to 50 yards, fired, and flew through the exploding bomb, his Mosquito being stripped of paint in the process.

456 Squadron first flew a diver patrol on 24–25 June when, although he tried for three, Flt. Lt. Houston was out of luck. Not until 9–10 July did one fall to the squadron. Great initiative was being shown by pilots, leading to unconventional methods of interception. One unusual way of obtaining success came when Flt. Lt. I. A. Dobie raised his score to thirteen by bringing down a flying-bomb with his slipstream. 'The V-1 was tearing along low down over the sea,' he said. 'We dived down to the same level and shot across in front of it at about 150 feet and approximately the same distance ahead. It was immediately tossed out of control in our slipstream, and dived full tilt into the sea.'

On 5 July notification was received that thereafter a V-1 shot down over the sea would count as one enemy aircraft destroyed; over land it would count as half. This gave fresh enticement to the defenders and one 418 crew destroyed four, and two others three each, on 6 July. Sqn. Ldr. Russ Bannock and Flg. Off. Bob Bruce scored the quadruple kill and, having had a triple kill three nights previously, they became acknowledged experts in the field.

On 7 July 96 Squadron recorded that out of eighty-eight V-1s which they claimed, only four had exploded in the air, two causing radiator damage to the Mosquitoes—which could be expected to be damaged if within 150 yards. On one such occasion Wg. Cdr. E. D. Crew and his observer had to bale out when, due to loss of coolant, both engine temperatures went 'off the clock', and one engine caught fire and both radiators had leaks. The aircraft crashed $4\frac{1}{2}$ miles from Dover.

An average of one V-1 per day fell to 418 Squadron in August. Shortly after 02.00 on 21 August Flg. Off. R. D. Thomas scored the unit's last V-1 kill, and their score stood at 83 V-1s destroyed, only seven over land. To achieve this they had flown 402 sorties, and the squadron had destroyed about 14 per cent of the total night V-1 kill. With $18\frac{1}{2}$ to his credit Sqn. Ldr. Bannock was top scorer, and Flt. Lt. Evans next

Operations: Fighters

with 7½. Four hundred and seventy-one flying-bombs were claimed by seven full-time Mosquito squadrons and 152 by the part timers.

Meanwhile night operations off Normandy continued. Near Fécamp on 22–23 June Flg. Off. W. J. Grey of 125 Squadron claimed three Ju 88s when flying HK238, and on 4–5 July HK356 and HK249 were each responsible for the destruction of an He 177 armed with FX 1400 radio controlled bombs. There were many other successes and between 6 and 30 June 409 Squadron flew 227 night sorties, claiming eleven bombers with two more as probables and five damaged. By the end of June sixty-five bombers had been claimed by 85 Group, ninety-seven by 31 July. Wg. Cdr. Maxwell's 604 Squadron intensively engaged against shipping raiders scored its 100th victory on 8 July.

From D-Day to 31 August 488 Squadron shot down thirty-four raiders for the loss of one, a record for 85 Group. Flt. Lt. G. E. Jameson in MM466 destroyed three Ju 88s at Caen and a Do 217 in twenty minutes on 29 July. By mid-August his eleven successes made him leading New Zealand night-fighter pilot. Usually it was a question of finding the enemy and shooting him down, but Flt. Lt. Huppert of 410 Squadron flying MM570, destroyed a Ju 188 then an Me 410, only to be shot down by defensive fire when engaging the latter. Although there were no large-scale enemy attacks, formations of from ten to forty Ju 88s and 188s had made accurate night attacks around Caen.

The first Mk. XXX to enter service arrived on 219 Squadron on 13 June, and in early July this and others began operations. Two Mk. XXXs reached 406 Squadron in mid-July and operations were begun on 10 August. After patrols over Biscay and the French coast 406 Squadron transferred to bomber support duties. Third to receive Mk. XXXs was 410 Squadron which operated MM760 and 762 only a few hours after their arrival. On 19–20 August MM744 of 410 Squadron shot down two Ju 88s, first enemy aircraft to fall to the new type. No sooner had the engagement taken place when, with an engine out of use, the Mosquito landed at B.5.

Early days of the Mk. XXX were dogged with snags. Despite more than a hundred hours of testing at Hatfield, the exhaust system was found to be completely unsuitable immediately the aircraft entered service, and all were soon grounded. An interim modification to the outer exhaust shroud permitted the Mk. XXXs to operate; then trouble was experienced with the inner shroud. Fifty cases of failure were reported by two squadrons in forty days. An inevitable problem had arisen when striving to reconcile the needs of efficient combustion (for range and speed) with suppression of flame brightness against the eyes of German fighters. Trouble persisted well into November, holding up issue to the squadrons. Louvred shrouds came into production in November and were fitted in retrospect. Seven night-fighter squadrons had Mk. XXXs before the end of 1944.

A V-1 approaching from the North Sea on 8–9 July, 1944, opened a

Two-Stage Merlin Mosquito Fighters

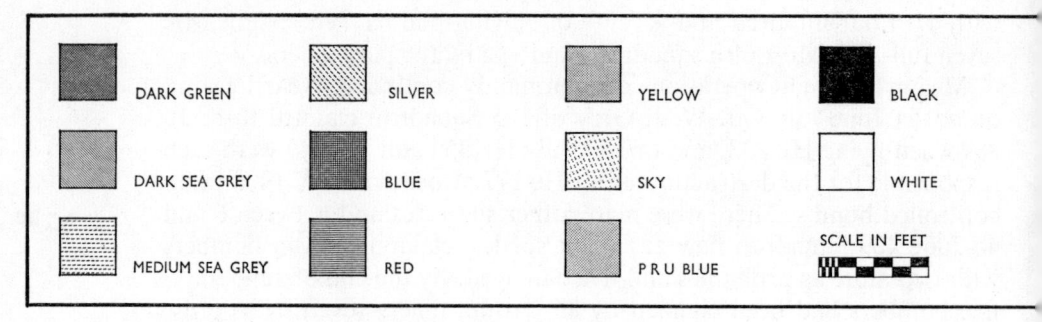

DARK GREEN	SILVER	YELLOW	BLACK
DARK SEA GREY	BLUE	SKY	WHITE
MEDIUM SEA GREY	RED	PRU BLUE	SCALE IN FEET

Mosquito NF. 30 in 1945-style camouflage. Dark green and medium sea grey overall camouflage was retained on Mosquito night-fighters until their withdrawal in the 1950s. Black serial numbers were painted beneath the wings from June, 1945, and the white ring in the upper surface roundels widened in 1948. Squadron badges appeared on the fins of fighters in 1949. In the summer of 1950 squadron identity letters on the fuselage sides began to be replaced by colourful squadron colours on the NF. Mk. 36s of the Fighter Command squadrons, Nos. 23, 25, 29, 85, 141 and 264. Black individual aircraft letters were painted above the fin stripes on some machines.

Mosquito NF. 30 (Merlin 76) NT283/G delivered 24.11.44, to 406 Squadron 18.12.44 and flew bomber support sorties. Destroyed a Bf 110 1.1.45 near Bonninghardt and Ju 88 on 14–15.4.45 near Prewzlau. Used by 609 Squadron from 30.4.46, with 616 Squadron 11.7.48 to 24.4.49. Markings as recorded 7.9.45.

Mosquito NF. 30 (Merlin 76) MB24 of the Belgian Air Force. Delivered to R.A.F. 25.5.45, stored in MUs, sold to Belgium 31.10.51.

Mosquito NF. 36 (Merlin 113) RL250 delivered to 51 M.U., R.A.F., 27.12.45. Issued to 264 Squadron 3.5.49 and passed to 27 M.U. 28.1.52. Sold as scrap 31.5.55. Markings as recorded June, 1949.

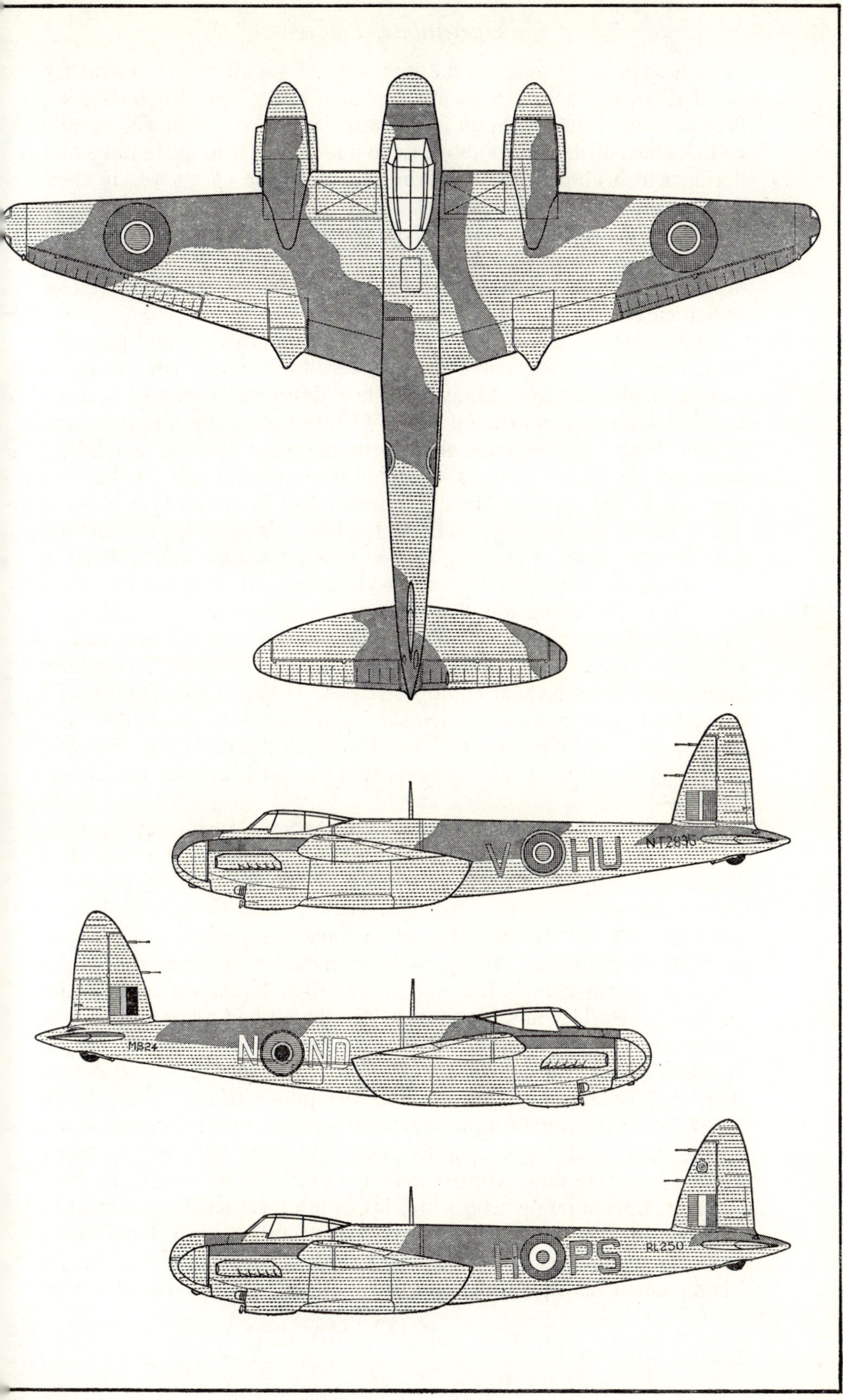

V HU
N ND
MB24
H PS
RL250

Operations: Fighters

new phase in the V-weapons battle, release of V-1s from He 111 carriers of III/KG3. Early on 5 September a batch of air-launched bombs was fired towards London. Soon after units concerned combined, forming KG53 which air-launched about 1,200 bombs. Early in the initial phase, during which about 130 V-1s were launched, the enemy wisely chose Portsmouth and Southampton as targets, fired a few towards Gloucester, then released bombs well out over the North Sea, towards London. Guns, balloons and fighter patrols were then re-arranged. First confirmed destruction of a carrier was by MM589 of 409 Squadron on 25 September.

Facing He 111 launchers off the East Coast in their second phase, 16 September, 1944 to 14 January, 1945, were Mosquitoes based at Castle Camps, Coltishall and Manston. They destroyed certainly fourteen Heinkels—six of them by the end of October. After the war it became known from the Germans that forty-one were lost on operations, seventy-seven from all causes. Catching them was difficult, for they flew low and slowly. Several Mosquitoes were lost in action against them, due to return fire, stalling and flying too low, then crashing into the sea. Patrols were flown at about 4,000 feet between Britain and Holland. It was difficult to locate the Heinkels with radar, and often the flash from an in-flight launching was first evidence that the enemy was at hand. Mosquitoes then dived on to the bomber which, as soon as it had released its load, would scurry back very low to its base. To improve interception rates a radar equipped frigate, H.M.S. *Caicos*, and later a radar Wellington, were used. Heinkels operated only on dark nights and, to have sufficient time to attack, the Mosquito needed to fly at a hundred or so feet above water at nearly its stalling speed, a hazardous manoeuvre.

25 Squadron began patrols on 24 September, to catch these Heinkels, four of which it damaged on the first night. Five nights later two fell to HK357 flown by Wg. Cdr. L. J. Mitchell. When 25 Squadron left Coltishall in October it at least had three Heinkels and twenty-two V-1s to its credit. Mk. XVIIs and XIXs of 68 Squadron patrolled off Holland in September. October 25 was a busy night for 68 Squadron, which engaged nine air-launched V-1s. Sgt. Neal of 68 Squadron flying TA389 claimed an He 111 on 5 November, just after it had released its bomb. The Fighter Interception Unit also operated experimentally against Heinkels, using Mosquitoes based at Coltishall, a week after moving there. 125 Squadron scrambled three Mosquitoes to meet V-1s. Beadle in HK310 of the squadron was vectored on to an enemy bomber at 060 degrees and contacted it $4\frac{1}{2}$ miles ahead. At 2,000 feet range he saw a V-1 released. He reduced his speed to 120–130 m.p.h. and fired from 900 feet. Encountering return fire, his cannon chattered again, and he obtained hits on the enemy's fuselage. He finished the 111 off with a two-second burst as he closed to 30 yards. Five nights later Flt. Lt. Thompson of 125 Squadron destroyed another He 111. W. Off. Brooking

Operations: Fighters

of 68 Squadron in HK296 claimed an He 111 on 6 January, but he and his navigator were lost without trace after calling Greyfriars and saying that they were in trouble. 307 had entered the campaign on 22 September by despatching HK231:O to patrol Ardorf airfield, base for some Heinkels. On 28 November the last Mk. XII home defence patrol was flown, using HK165:A of 307 Squadron.

Enemy air activity over France diminished after August, when fifty-five aircraft were claimed by Mosquitoes. Rivalry to be top scoring squadron on the Continent was now keen. On 9 September 409 Squadron moved into B17 (Caen/Carpiquet), resumed patrols, but found little trade. Only three German aircraft were claimed in September, including a Ju 87 which fell to 219 Squadron, which destroyed another three in the Nijmegen region on 2 October. October 6 brought much activity to Le Culot airfield; first a 409 Squadron Mosquito belly landed, this to be followed by another attempting a single-engined landing after destroying a Ju 88. It was diverted, then its other engine stopped so the crew baled out. Another Mosquito returned with a Bf 110 to its credit, then a '410' machine force-landed after having destroyed a Ju 88. It was a busy night.

Early in October 488 Squadron began operations with Mk. XXXs. On 4 November, flying MM820, W. Off. J. W. Marshall made the first claim, a Bf 110. In destroying two Ju 88s on 29–30 November a Canadian crew brought the total of enemy aircraft claimed as destroyed since D-Day at night by 85 Group to 200. Since D-Day the Cougars (410 Squadron) and Night Hawks (409) had each claimed thirty-eight of the enemy, their nearest rivals being 488 Squadron.

Although the Arnhem landings generated additional night defensive patrols it was the Ardennes attack that brought the final period of intense activity to 2 T.A.F. night fighters. Cold, cloud and fog made interceptions difficult but on 18–19 December the Canadian Squadrons claimed three Ju 88s and a Bf 110, probably protecting enemy troops. Weather clamped down for the next few nights, although a Ju 88 was destroyed. On the 23rd–24th Mosquitoes really came into their own, claiming six Ju 88s, Ju 188s and an Me 410. Of these 488 Squadron claimed two 88s, a Ju 188 and the '410'. Six others were shot down on Christmas Eve, including two 188s and two Ju 87s. Five more were destroyed during the next three nights, then the weather deteriorated. Leading now with 46 destroyed were the Cougars, followed by the Night Hawks (44) and then 488 Squadron (39). Before von Rundstedt's advance was halted six Ju 87s were shot down, and in the first ten nights of the counter attack twenty-five enemy aircraft were claimed by 85 Group. Successes included nine in one night, bringing the total claim since D-Day to 227 destroyed at night, all by Mosquitoes.

In January, 1945, Continental-based Mosquitoes were as busy as ever over the Allied lines. On 1–2 January night-fighters shot down five of the enemy trying to deliver low-level attacks on Allied positions. On 6 January Flt. Lt. F. A. Campbell and W. Off. G. H. Lawrence of 488

Operations: Fighters

Squadron had a thirty-minute running fight with a Bf 110 over Holland before shooting it down. One of the first Mosquito night-fighter—versus jet combats had taken place on 23 December, 1944, when Flg. Off. Taylor of 29 Squadron engaged a jet over Hespe. MT470 of 25 Squadron inconclusively battled with an Me 262 over Stuttgart on 28-29 January, 1945.

Mosquitoes went into action against manned enemy aircraft attacking targets in Britain on 3-4 March. That night 140 intruders came over, bombed fourteen airfields and shot down nineteen British bombers. The crew of Ju 88 G6 C9+RR of 7/N J G 5 set out from Lubeck/Blankensee to intrude on airfields in Eastern England. Near Scampton they attacked a motor car whose headlights, they probably believed, indicated activity on the airfield. While diving to attack the aircraft struck telegraph wires and crashed, burning parts being widely scattered. Wg. Cdr. Griffiths of 125 Squadron flying NT415 destroyed a Ju 188 at sea, and another fell to NT381 of 68 Squadron. Ju 188s were also destroyed by Flt. Lt. D. B. Wills and Flt. Lt. R. B. Miles. A Liberator was landing at Metfield, Suffolk, when a Ju 88 G6 of 5/NJG4 approached from starboard and opened fire, just as a Mosquito shot down the intruder. The raiders caused chaos, but they were not entirely unexpected. There had been activity off the coast on 2-3 March when 125 Squadron scrambled to intercept. March 7 was the last occasion upon which the Mk. XVII flew operationally, HK283 being despatched by 125 Squadron to deal with more intruders. 68 Squadron was among those scrambled on the 17-18th, after intruders which shot up Carnaby, Coltishall and other airfields. Out of a small force that crossed the coast on 20-21 March, a Ju 188, was claimed by Flt. Lt. Kennedy who chased it out to sea and sent it spinning into the water. About a score of enemy aircraft were claimed during these final operations against Britain.

Mosquitoes stood by during March to face a new assault by long-range flying-bombs launched against Britain from Holland between 3 and 29 March, 1945. Mk. XXX Mosquitoes were used, on anti-diver operations for the first time. Between October, 1944, and March, 1945, more than 6,500 V-1s were fired into the Continental Allied lines, and Mosquitoes intercepted a number whilst on patrols, intruder operations and rangers. Shortly before 9 a.m. on 29 March, 1945, the last flying bomb to elude the defences landed at Datchworth, near Hatfield, bringing an end to attacks on the British Isles.

Mosquito fighters of 85 Group patrolled around central Germany looking for escaping Nazi officers, who used an assortment of transports. 477 added two Ju 52s and a Fw 189 to its score and 409 Squadron found two Ju 87s and a Fw 190, not to mention a huge Ju 290 brought down by MM517. Even with such a tractable aeroplane as the Mosquito there were times when combat difficulties arose, particularly when the enemy was flying slowly. Such cases arose for 264 Squadron on 1 May. HK528 found an Me 108 flying too slowly to be shot down, and a Fi 156 dawdling along at 65 m.p.h., the last enemy machine the squadron found. On

Operations: Fighters

26 April Berlin was completely surrounded by the Red Army. Escape could only be by air, which, at night, the Mosquitoes tried to prevent. On VE-Day 85 Group's claim stood at 299 enemy aircraft destroyed since March, 1944, the last confirmed successes being the destruction on 25–26 April of a Fw 189 and a Fw 190 near Brandenburg by HK466 of 264 Squadron, by chance one of the high scoring Mosquitoes with five confirmed victories.

The number of German aircraft destroyed by Mosquito night-fighters will never be precisely known due to the uncertainties involved. It is positively true to say that the Mosquito shield was almost entirely responsible for the prevention of large-scale night attacks on Britain, and the Allied forces on the Continent, from 1943. Its part was vital to the build-up of the Allied Expeditionary Force. Against the flying-bombs the one-time fastest aircraft in the world proved that it had an edge—barely so. Without the defensive protection Mosquito squadrons provided, the Allies would have been far less secure during the long and the short nights.

First Patrols by various Mosquito Marks
(Squadrons of Fighter, Bomber Command, 2nd T.A.F., M.A.C.)

Sqn.	II	VI	XII	XIII	XVII	XIX	XXX
23	5–6.7.42	17.7.43	—	—	—	—	—
23[1]	—	5–6.7.44	—	—	—	—	—
25	17.11.42	12–13.8.43	—	—	4.1.44	—	4.10.44
29	—	—	3.5.43	14. 5.44	—	—	14. 4.45
68	—	—	—	—	9.7.44	26.7.44	21. 2.45
85	20. 8.42	—	24.3.43	—	2.44	5–6.6.44[1]	.44[1]
96	—	—	—	13.11.43	—	—	—
108	—	—	10.4.44	5. 4.44	—	—	—
125	—	—	—	—	18.2.44	—	27. 2.45
141[1]	14. 1.44	16.8.44	—	—	—	—	4. 4.45
151	30. 4.42	17.8.43	22.7.43	—	—	—	11.10.44
		7.7.44	—	—	—	—	—
157	27. 4.42	8.43	—	—	—	5–6.6.44[1]	2. 3.45[1]
169[1]	20. 1.44	18.6.44	—	—	—	—	—
219	—	—	—	—	19.3.44	—	21. 6.44
239[1]	20. 1.44	11–12.9.44	—	—	—	—	21–22. 1.45
255	—	—	—	—	—	26–27.2.45	11–12. 4.45
256	—	18.4.45	21.5.43	11.43	—	—	—
264	13. 6.42	28.8.43	—	25. 2.44	—	—	—
307	14. 1.43	2.9.43	10.2.44	—	—	—	23.11.44
406	—	—	29–30.4.44	—	—	—	10. 8.44
409	—	—	—	4. 5.44	—	—	—
410	6.12.42	—	—	19.12.43	—	—	17–18. 8.44
418	7. 5.43	.43	—	—	—	—	—
456	22. 1.43	25.7.43	—	—	12.2.44	—	3. 1.45
488	—	—	30.8.43	8.10.43	—	—	11.10.44
515[1]	—	5.3.44	—	—	—	—	—
	—	(605 Sq. a/c)	—	—	—	—	—
600	—	—	—	—	—	22.1.45	—
604	—	—	28.4.44	28. 4.44	—	—	—
605	18. 2.43	3.7.43	—	—	—	—	—

[1] Bomber Command Squadron.

Chapter 13

Operations: Day Bombers

High-speed runs and a shattering aerobatic performance by Geoffrey de Havilland heralded the arrival of the first Mosquito W4064 at Swanton Morley, on 15 November, 1941. The theory of a bomber relying on speed for safety was now to be tested in the hands of 105 Squadron, mostly in low-level attacks on heavily defended targets, before major policy decisions, in the event, left daylight strategic bombing in the hands of the Americans. For a year Mosquitoes were to blaze a trail outstanding in its effectiveness and dramatic appeal. To the Occupied Countries the sight of the fast little bombers roaring overhead brought hope; to the enemy, despair.

Two bombers W4066 and '67 were on flight tests and engine runs and thirteen on the floor at Hatfield on 5 November, 1941, when Grp. Capt. C. Williamson Jones, Bomber Command Engineering Officer, and Wg. Cdr. Jordan visited Hatfield with the Commander of 105 Squadron. He was eager that two or three Mosquitoes should reach his squadron even lacking operational equipment, so that crew training could commence. On 10 November, 1941, conversion for the squadron's observers began. Some redundant air gunners of the squadron's Blenheim days opted to become navigators. Immediately, if the bomber proved a success, there was to be a saving in crew training. Both flights of the squadron began to prepare for Mosquitoes, an unknown quantity yet one about which cheerful rumours were flying; in part, rumour was surpassed by reality. On receiving W4064 the squadron recorded: 'Another great day in the history of 105 Squadron. All crews watched with great enthusiasm the performance in the air. Even the Spitfire pilots of 152 Squadron were impressed. Wg. Cdr. P. H. A. Simmons, the Squadron Commander, flew with Mr. Geoffrey de Havilland in the Mosquito and was greatly impressed with its capabilities. Even in the mess he couldn't stop talking about it.' Next day there was bitter disappointment, the aeroplane developed hydraulic leaks and oil feed trouble; Geoffrey took it back to Hatfield for adjustments. On the squadron there was nothing but immense enthusiasm and crews willingly underwent bends tests in the Station Sick Quarters for four hours apiece, simulating operations at 35,000 feet, a height that was rarely known. When W4066 reached the

Operations: Day Bombers

squadron on 17 November the A.O.C. 2 Group and other officers had arrived to visit the station—and the amazing newcomer.

On 22 November positioning of the bombers was thus: W4064 at Hatfield was having snags cleared, and was due on 105 Squadron the next week; 4065 was on trial at Duxford; 4066 was with 105; 4067 was at R.A.E. for radio checks; 4068 was officially delivered yet still at Hatfield. Seventeen Mosquitoes had in all been delivered. A cavity window screen was being tried on W4072 and tests of W4057 with short vanes on her bombs were proceeding at A.A.E.E. It had been decided on 8 November that the first ten bombers should have short nacelles, initially at least.

Flights were few at this time and for special purposes, but on 25 November, on the third trip of the day by newly arrived W4068, Simmons took Sqn. Ldr. Darwen, C.O. of 152 Squadron, for a ride and wrote, 'It is thought the Spitfire C.O. was considerably shaken by the capabilities of the Mosquito.' A few more flights were made before 105 Squadron moved to Horsham St. Faith, taking four Mosquitoes. Soon, fast round-Britain flights were being flown, fuel-consumption tests and high-altitude trials were being made. Realizing the value of dual-control for training purposes Flt. Lt. Houston visited Hatfield to discuss the requirement, on 20 December. On 27 November a Mosquito reached 30,000 feet, then bad weather interfered and as Sgt. Swann landed W4070 from a diversion to Portreath he overshot through a hedge, damaging the aircraft but not its occupants. Hereward de Havilland hastened to see the wreckage, for this was the first time a Mosquito had crashed in such circumstances. In April he was to point out that the machine looked like taking five months to repair, 'deplorable since we have boosted the ease of repair of wooden aircraft'. A boost was then given to the planned repair organization.

On 11 February two low-flying Mosquitoes flown by Wg. Cdr. Simmons and Sqn. Ldr. Oakeshott raced across the countryside to St. Eval in Cornwall, averaging 280 m.p.h.; on 9 March the first high-level bombing run was made, followed by one at low-level later in the day. A week later 105 Squadron engaged in mock combat with a Spitfire of A.F.D.U. Duxford and considered it out-performed. March 27 brought the excitement of a mock Mosquito attack on Horsham. Plans now called for the commencement of operations in May with about seven aircraft. A tour of 2 Group airfields came on 2 April; cameras, whirring away, provided a new aspect of Mosquito use by the squadron. This was to have valuable application on 30 April, when photographs taken of Norwich showed the extent of the previous night's bomb damage.

Twenty crews and seven aircraft were 105's complement by 7 April. Lorenz beam apparatus was fitted in one at Wg. Cdr. Simmons' wish. Four days later delivery of the first Mk. IV DK288 was made—to the P.R.U. 105's first Mk. IV did not arrive until mid-May. With the longer

Operations: Day Bombers

nacelles and flame dampers these were reckoned to be faster than the P.R.U./Bomber Conversion Type Mosquito I, but on tests the first showed a top speed of 2 m.p.h. less, 380 m.p.h. at 22,000 feet in F.S. gear, and 105 Squadron maintained that this was general.

There were still only seven machines with 105 Squadron and some Masters for navigation training on 9 May. Fearing a fall in morale now that Wg. Cdr. Simmons was unwell the Station Commander, Grp. Capt. G. R. C. Spencer, had in mind a day sortie, but nothing came of this.

Bomber Command was still inclined to be sceptical about the unarmed bomber, even if 105 Squadron showered it with enthusiasm. Experienced officers with keenness for the idea considered that by the time sufficient Mosquitoes were received to make a really effective force, due to the demands already on production for fighting and reconnaissance, the superior speed—their sole means of defence—would disappear unless de Havilland could step up performance. Meanwhile the Company looked into ways of making each sortie more effective by increasing the bomb load. On 23 May DK290 began stability tests carrying 4×500 lb. bombs at an all-up weight of 20,700 lb. The planned Mk. V bomber would carry this load, plus 2×250 lb. bombs on wing racks at an a.u.w. of 21,300 lb. Low-level attack was not preconceived as the best method of using the Mosquito, in fact much bombing practice at this time was delivered from over 20,000 feet. From 26,000 feet in daylight mean errors were in the region of 400 yards, but there were hopes of reducing this to about 150 yards when the Mk. XIV bombsight, now on test in DK286, was used.

Eight Mosquito IVs were on 105 Squadron's strength when the Cologne 'Thousand Bomber' raid took place on 30/31 May, 1942. Before the heterogeneous cavalcade returned two Mosquito bombers were being prepared for the first bombing operation. At 04.00 Sqn. Ldr. Oakeshott took off in W4072-D Dog for Cologne taking 2×500 lb. and 2×250 lb. bombs. An hour and a half later Plt. Offs. Kennard and Johnson left in W4064. Oakeshott flew across at 24,000 feet and found smoke, billowing to 14,000 feet, completely covering Cologne. Of Kennard no further news was heard. Three Mosquitoes fruitlessly searched for him. Plt. Off. Costello-Bowen and W. Off. Broom were away at 11.40 in W4065:N followed five minutes later by Flt. Lt. Houlston and Flt. Sgt. Armitage in W4071:L. Clouds and smoke obscured the city still, so bombs were dropped by dead reckoning. In the late afternoon Sqn. Ldr. Channer left Horsham in W4069:M to make a low-level reconnaissance and photographic flight over the city, making first use of the Mosquito's high speed for such passage. He flew in cloud until 60 miles from the target, then shallow dived to nearly 380 m.p.h. A.S.I. Passing a large marshalling yard he noticed that nobody looked up at him. Cattle in fields failed to react until he was well past. He said afterwards that his belief in the soundness of Mosquito attacks on selected

Operations: Day Bombers

targets, then under discussion as a means of employing the new bombers, was greatly strengthened. On the evening of 1 June two Mosquitoes again visited Cologne to bomb and photograph the city; W4068:B failed to return, its crew having baled out. Early next day another sortie was flown to the stricken city, some hours after a mission to Essen. Sixteen successful sorties took place in June for the loss of one aircraft, all being attacks by single machines from high altitudes and with varying success, on clear and cloudy days.

The German High Command claimed to have shot down a Mosquito on 31 May. Had they not known of the aircraft through intelligence sources they would have known of its existence from the September, 1941, issue of *Commercial Aviation* announcing that de Havilland, Toronto, were to build the Mosquito. A week later *The Times* mentioned the item. Soon after, a house magazine *The Mosquito* was brought out in Toronto, and matters came to a head when the *Sunday Express* of 14 June referred to ferry pilots delivering Mosquitoes. M.A.P. the same day considered prosecution, but a warning to news editors was instead sent. After all, 197,000 A.T.C. cadets had access to photographs and silhouettes of the aircraft. The ways of 'security' were as ever weird, and one sympathized with the department in its problems.

A second Mosquito Squadron formed at Horsham on 8 June, 1942, and used 105 Squadron's aircraft until December. It speaks much for the ground crews of 105 Squadron, not to mention the aircraft, that the scale of operations was considerable. 139's first sortie flown by Sqn. Ldr. Houlston, A.F.C., in DK296, was a low-level attack on the airfield at Stade, Wilhelmshaven, on 25–26 June. It was noteworthy because landing took place after nightfall. Thus Mosquitoes were almost round-the-clock bombers already. The same evening Flt. Lt. Bagguley flew W4072 on a low-level raid on Dorum. He lowered his flaps instead of bomb doors and they were torn away, so on return he had to make a fast landing, overshooting the flare path and catching his undercarriage in a trench.

An important attack affecting policy formulation occurred on July 2 when Mosquitoes[1] left shortly before noon for the submarine yards at Flensburg. It cost two very fine pilots, Grp. Capt. MacDonald and Sqn. Ldr. Oakeshott. During his sortie Sqn. Ldr. Houlston was chased by three Fw 190s, intercepting Mosquitoes for the first time. He comfortably drew away from them, using plus $12\frac{1}{2}$ lb. boost at sea level. Flt. Lt. Hughes was chased by two fighters for twenty minutes yet escaped without being hit. He reckoned his bombs hit the target, then bounced to about a quarter of a mile away. His machine, DK298, was hit by flak in the centre fuel tank and fuselage. Fortunately tank sealing proved effective and very little fuel was lost. As a result of pilots' suggestions following the raid a perspex blister was fitted on canopy roofs allowing the navigator to kneel on his seat for a clearer view aft looking

[1] DK294, DK295:G, DK296, DK298:H, DK299:S, W4069:M.

Operations: Day Bombers

for fighters. Working on A.F.D.U.[1] trials it was felt that unless the enemy was higher he would not catch the Mosquitoes—provided he was kept far behind. Casualties showed that there was nevertheless no room for complacency, nor for conclusions.

At 19.00 hours on 11 July six Mosquitoes[2] of 105 Squadron made a diversionary attack on Flensburg for the Lancaster raid on Danzig. Five carried 4 × 500 lb. bombs, one had 2 × 250 lb. HE and incendiaries. Over the target DK300 (Plt. Off. Laston) was hit on the fin by light flak which blew both ply skins away for about two square feet. From the second formation one of the three aborted, after falling behind. Twenty miles from the target Sgt. Rowlands in DK296 felt a slight jar, then his observer found pieces of chimney pot on his lap and a large hole in the fuselage side. When the port engine started to vibrate badly it was closed to 2,300 r.p.m. Rowlands couldn't keep up with Hughes so found some factories, dropped his bombs and turned back. Clear of the coast he feathered his port propeller. After landing he found that thick cables must have gashed the spinner and radiator, but they could not have been balloon cables and the mystery remained unsolved. The chimney pot was the souvenir of the day: the hole by which it entered was 4 feet long—18 inches at its deepest part. Ralston said that when Rowlands hit the chimney he was looking down on Hughes, leading, who failed to return. The German radio claimed that a Mosquito came down near Flensburg. Such were the early adventures of the dauntless few who worked out the tactical use of the Mosquito, paying for every scrap of knowledge with casualties as high as 16 per cent. Some of the crews who had flown on earlier day raids had experienced much higher casualty rates and were less daunted now. Now it was decided to operate Mosquitoes at high levels on clear days and low levels when clouds permitted. New tactics followed because of losses, and individual lists of targets were given for attack when weather conditions permitted individual sorties. Already the volume of repairs required was increasing rapidly due to accidents and operational flying.

Major H. L. Armstrong, Director General of Aircraft Production, visited Hatfield and Leavesden on 24 July to get an impression of Mos-

[1] Report No. 39 Air Fighting Development Unit Duxford summarized results of trials with W4065, commenting on its excellent manoeuvrability irrespective of load and positive controls. With operational load the Mosquito was about 5 m.p.h. faster than the Spitfire V when the latter used emergency boost, although above 24,000 feet the Mosquito's speed fell away. It took nine minutes to reach 20,000 feet and could maintain height and good climb on one engine, and turn easily. When the Spitfire V came within close range of the bomber the latter was unable to shake it off due to its lower safety factor, but when the Mosquito was flying fast the only way in which a fighter could catch it was to dive upon it. If the Mosquito operated at about 21,000 feet it could itself dive and escape. 'If a fighter manages to get into range without being observed,' the report stated, 'the Mosquito should accelerate away with a diving turn or cork-screwing movement (down to the left and up to the right) which upsets the fighter's aim at long range and high speed.'

[2] W4070:C, DK300:F, DK297:O, DK299:S, DK295:P, DK296:G.

Operations: Day Bombers

quito production, and to tell de Havilland in very strong terms that the Mosquito was now highly thought of throughout the Ministry and R.A.F., and that every fibre of the D.H. organization must be strained to build up production. It so happened that this week rumours had spread around the factory that all was not well with the Mosquitoes in service, so at 12.15 on the 25th Lee Murray, General Production Manager, broadcast to the workpeople on the success of the aeroplane, telling them of the Flensburg raid. He said, 'It is really only in the past few weeks that Mosquito night-fighters and intruders and day bombers have been going right into battle, and what the Director General told us yesterday puts the seal upon the favourable reports of the Mosquito which we have been hearing daily and nightly from the pilots. My purpose in telling you of yesterday's visit is firstly to ask you for your whole-hearted effort in building up production, and secondly to give you the satisfaction of knowing that the aeroplane you are making is one of the outstanding successes of this war up to the present.' Earlier in the week at Farnborough, Air Ministry staged a display of the Fw 190, and a contest between the recently captured machine and a Merlin 61-engined Spitfire IX. Subsequently tests proved the Mosquito faster than the Fw 190A3, and W4050 now fitted with Merlin 61s was even faster. It seemed certain now that the Mosquito was the fastest aeroplane in the world.

Many a wing tip was damaged in these early days and A. J. Brant, de Havilland Service Manager, introduced in conjunction with the design office and others a simple procedure for wing repairs. It was needed because the wing was in one completed piece from tip to tip. Instead of traditional scarfes, or feather-edge joints, meticulously applied to every stringer and skin that had to be married up, a plain butt jointed was substituted. The damaged wing end was sawn off and a standard end from stock attached by overlapping patches, cemented and screwed inboard and outboard of the joint. This greatly simplified wing repairs throughout the war.

Sgt. Smith had bombed Essen and been intercepted by two Fw 190s which scored hits and sent the Mosquito DK313:M-GB into a spin. When the navigator prepared to abandon the machine he found petrol swirling around the floor, but as the aircraft seemed to be flying normally he closed the hatch. They flew home above cloud and belly landed at the first airfield they found. Flt. Lt. Parry and Plt. Off. Robinson in DK292 took a diplomatic bag to Stockholm on 6 August and stayed the night, their sortie presaging those by B.O.A.C. All roundels, numbers, letters were removed from the aircraft and the crew wore civilian clothing. They landed at Stockholm a few minutes before a German Ju 52 used by Goebbels and his staff. The Mosquito was put under joint Swedish and British guard for the night.

Wg. Cdr. Hughie Edwards, v.c., d.f.c., was returning from a sortie on 29 August in DK323 when about twelve Fw 190s attacked 40 miles

Operations: Day Bombers

from inside enemy territory. Initially they came head on, and this was the only method effective. Apparently the 190s were unaware that they were after a Mosquito. They lost ground as they swept round for a stern attack and were unable to close. Over the sea the port engine of the Mosquito had to be shut due to a bullet through a coolant pipe. Edwards decided to land at Lympne and told his navigator to lower the undercarriage by hand pump, but the aircraft performed an unexpected belly landing. Another Mosquito DK330 made a belly landing after encountering Fw 190s. Local ground crew used to Oakington's metal Stirlings loaded the Mosquito on a trolley, causing more damage. Wooden aeroplanes to them were a thing of the past.

Six Mosquitoes on 19 September attempted the most audacious act yet when they left Horsham at 12.30 for a high-level daylight raid on Berlin, the first of the war. Luck was certainly out. The city was cloud clad and enemy fighters responded to the intrusion. Sgt. Booth in DZ312:U, unable to maintain height, turned about after a short while. Flt. Sgt. Monaghan in DK336:P was forced home, as he came across the others already battling with enemy fighters. The intrepid Ralston reached Berlin but dense cloud over the city caused him to head for Hamburg—his secondary target—which, to his chagrin, he found equally cloud covered. Four times Flt. Lt. Parry was intercepted in DK 339:C and his aircraft sustained damage from two Fw 190s. He jettisoned his bombs near Hamburg. Sqn. Ldr. Messervy in DK326:M was shot down between Wesermunde and Stade. Only to W. Off. Bools in DK337: N fell the distinction of flying a successful bomber sortie to the city, and to Sgt. Jackson the satisfaction of being first supposedly to bomb it in daylight, albeit on dead reckoning. Had the raid been successful it would have won wide acclaim, such as was shortly to come.

On 25 September, 22nd birthday of the de Havilland organization, four Mosquitoes flown by Sqn. Ldr. D. A. G. Parry, D.S.O., D.F.C., (DK296:G), Plt. Off. Rowlands (DK313:U), Flg. Off. Bristow (DK328: V) and Flt. Sgt. Carter (DK325:S) left Leuchars for a low-level assault on the Gestapo H.Q. in Oslo. As they attacked in pairs—Parry and Rowlands leading—three Fw 190s found themselves above the bombers racing along Oslo Fiord. One dived for the leaders, two for the second pair going in to attack at 280 I.A.S. and 100 feet. Carter's aircraft fell a trifle behind and was hit. He turned towards Sweden, a fighter on his tail and an engine afire, and crashed in a lake. Others placed their bombs with precision that captured public imagination in Oslo, and in Britain where the existence of the Mosquito was revealed next day on the six o'clock news. At least four bombs entered the roof of the building, one stayed inside and failed to explode and the three others careered out of the opposite wall before exploding. By then the Mosquitoes were streaking for Sumburgh at 330 I.A.S. as low along the Norwegian valleys as they dared, steadily drawing away from the fighters. One of the Fw 190s crashed near Oslo and the pilot was killed. Either the

Operations: Day Bombers

Mosquito's slipstream forced it out of control or it was affected by bomb blast. In Oslo rumours circulated that it had been shot down by a Mosquito. When the first Mosquito photograph was released on 27 October, the accompanying caption stated 'Armament may consist of four 20 mm. cannon and four ·303 inch guns'; this was an attempt to keep the unarmed state of the bombers from enemy knowledge. There can be little doubt that throughout the war enemy fighter pilots expected that, sooner or later, a Mosquito would fire back. Accent was given in the first public revelation of the Mosquito on its production in Canada and Australia and in a widely dispersed manner in Britain, so as to discourage another low-level German raid on Hatfield.

The flight to Oslo and back took $4\frac{3}{4}$ hours. Fuel consumption on the 1,100-mile flight averaged 2·75 m.p.g. leaving about 140 gallons in the tanks. The Mosquitoes flew out at sea level and expected 10/10th cloud at 2,000 feet over Oslo, and no fighters. They found a cloudless sky, and Fw 190s. After the operation it was concluded that the Mosquito had an operational range of about 900 miles low-level and 1,350 high-level, but after further experience an ultimate range of 1,220 miles was accepted.

By the end of September 105 Squadron had nineteen Mosquitoes, all with 14 lb. boost. They had experimentally polished a Mosquito but reached the same conclusion as de Havilland engineers who had given the treatment to DK290—now converted to carry $1 \times 1,000$ lb. and 2×500 lb. bombs—that the gain was only about 5 m.p.h. Removal of the side blister windows added 3 m.p.h. A more hopeful modification, recognised at Hatfield from early tests, was the use of stub exhausts in place of the shrouded saxophone flame-damping type. Trials at Marham with DK336 fitted with open stubs indicated that these gave too much glare at night, revealing the position of the aircraft and making night landings difficult. Closing the stub ends to oval section of slightly less area reduced the glare in tests on 25 November and increased the jet propulsion effect without reducing engine efficiency, giving a worthwhile net gain of from 10 to 13 m.p.h. depending on altitude. Since they operated from dawn to dusk 105 Squadron called for fifteen aircraft with flame dampers and three with stubs. Various other tricks to increase speed included reducing the minimum radiator opening, wing-root leading-edge fillets and aerial mast removal; only changes to the exhaust manifolds were worth the effort. The shrouded exhausts—never popular with ground crews—were retained for dusk and dawn attacks, and others were fitted with oval stubs for day raids, as on reconnaissance aircraft.

While any means of gathering an extra 10 m.p.h. were eagerly explored the nagging old question of rear guns was repeatedly raised, despite likely performance losses. On 13 September, 1942, a discussion between Major de Havilland and Air Vice-Marshal Lees, Grp. Capt. Hesketh, Grp. Capt. Kyle and Wg. Cdr. Edwards was held at 2 Group H.Q., Huntingdon. Edwards was not in favour of scare guns if they

Operations: Day Bombers

reduced top speed by more than 5 m.p.h. De Havilland considered the pros and cons and four weeks later another meeting was held. A fixed pan-fed scare gun in the tail cone, a gun in the tail of each nacelle adjustable for elevation on the ground, another with limited traverse in the cockpit canopy manually worked by the navigator, were three possibilities. All, of doubtful value, presented difficulties. Crews considered there was a psychological difference in attacking an aircraft which had good, or at any rate movable guns, and one which had none or only fixed scare guns. Nevertheless, it was decided to fit a Browning gun inside the rear of each nacelle to fire aft. Rose Brothers at Scampton prepared a trial installation, which underwent successful ground trials there, but it was soon after decided to drop the idea.

An alternative proposal was to fly a few Mosquito fighters painted as bombers within bomber formations to confuse the interceptor and remove the 'take-your-time-they-can't-hit-me-back' complex. A request for six fighters was made to No. 2 Group in September, 1942, but the idea was dropped as performance improvements made the risk of interception less. One ruse employed was to paint a 'Sky' band round the rear fuselage, and decorate spinners this colour, associated with British fighter aircraft. Two Spitfires from A.F.D.U. assisted in training the pilots in fighter evasion tactics in September, 1942, and the nucleus of the Mosquito Training Unit, No. 1655 Flight, formed at Horsham using three T. Mk. IIIs to train crews in special operating techniques. Already 109 Squadron at Wyton was receiving Mosquitoes, but Fighter Command now had more than twice as many as Bomber Command.

Wing Commander Hughie Edwards, V.C., D.F.C., took over command of 105 Squadron on 3 August, 1942. This Australian officer earned the supreme award on a Blenheim raid calling for the highest courage—a remarkable low-level day raid on Bremen. He led many Mosquito raids in the same tradition and did much to build up the squadron's techniques and spirit, although he would be the last to admit it. It is of course difficult to single out individuals without making unfair omissions where team work was so essential.

Operating tactics were assuming great importance. In hilly country it was sometimes preferable to bomb from 2,000 feet in a shallow dive, which gave an average error of about 60 yards. 105 Squadron raided the steel works at Liège in this manner on 2 October. A combination of shallow dive and low-level was next worked out, and frequently used. While ground defences were preoccupied with the roar and explosions of a tree-top attack a second formation would approach in a swift climb to 1,500 or 2,000 feet and get steadily settled on to their shallow dives without attracting much flak. Their bomb bursts often came as a surprise to the enemy, and they were favourably placed for a high-speed getaway at hedge-top level a mile or two after the first formation. At medium levels the German coastal radar would detect the raiders about 80 miles away; at sea level it was ineffective, and so the Mosquitoes

Operations: Day Bombers

hugged the wave tops. Dawn and dusk attacks gained favour. Providing poor light for either of the runs, it was nevertheless sufficient for viewing the target.

One feature of low flying was accentuated at these times—the risk of bird collision. Leading aircraft roused the birds which would fly into those following, often with serious results. Wing leading edges needed to be reinforced, for on more than one occasion a bird penetrated both webs of the front spar and the bullet-proof windscreens later fitted. One of the most dreadful bird experiences befell Alec Bristow and Plt. Off. Marshall returning in DK296 from a low-level dusk attack on Hengelo led by Wg. Cdr. Edwards. Crews of three other Mosquitoes were surprised to see Bristow's aircraft nearly hit the ground then shoot up into the sky before falling behind. A curlew which rose from a meadow had penetrated the windscreen, crashing into Bristow's face. It momentarily knocked him unconscious and he fell forward on to the control column, bleeding badly from cuts. Marshall, cut in the forehead, grabbed the stick and yanked it back in time to avoid a crash. This almost stood the Mosquito on its tail, so Marshall pushed the stick forward again rather heartily and both engines cut under negative G. Bristow came round and took over, but could see little because blood was streaming into his eyes. It was impossible to see through either windscreen. After trying the direct vision panel for a while Marshall went to the bomb aimer's window and told Bristow how to steer by kicking him once for right twice for left. When they reached Marham it was dark and Marshall continued to prompt Bristow during approach and landing. After three attempts they finally bounced to a standstill. Bristow's injuries proved not to be serious.

Another danger of low flying, this time over the sea, was the risk when the surface was glassy and the weather hazy. On 27 August, 1942, Sqn. Ldr. Collins and Plt. Off. May in W4070:C were lost in this way heading for Vegesack. Raid leaders thereafter used to insist, whenever conditions were dangerous, that nobody was to get below them while crossing the sea. Fortunately not all of the accidents arising from low flying ended in tragedy, as witness the low-level return of DK297:O on 25 August, 1942, following an attack on the Brauweiler power station near Cologne by Flt. Lt. Costello-Bowen and W. Off. Broome. Near Rotterdam they struck a pylon which wrecked an engine and tossed the wingless aeroplane into a wood. The pilot was knocked out but Broome managed to lower him from a tree, after which came an adventurous journey—back to England.

High-level Mosquito attacks were all but abandoned in September, 1942. For large cloud covered targets and Mosquito speed they were acceptable. Smoky factories, although not typical targets, were difficult to attack at low levels. Flying high and out of cloud it was important that no contrails should be left. A good look-out behind was essential and variations in altitude were beneficial to safety. Sometimes radar-directed fighters would jump upon the Mosquito as it left cloud cover.

Operations: Day Bombers

For example Flg. Off. Downs and Flg. Off. Graves in DK313 were starting a run up to Essen at 27,000 feet when they were set upon by Fw 190s diving out of the sun, and their aircraft was shot up. Thanks to cloud they escaped, but their bomb doors were so badly damaged that they would not open and the port undercarriage dangled. In this condition with bombs aboard Downs landed at base without flaps, and with little aileron control. As the leg chassis collapsed the aircraft swung violently, extensively damaging the port wing. But the crew stepped out—and the aircraft was repaired. When saying that the Mosquito attack techniques were worked out and perfected in 1942 it must be realized, that in terms of lives and aircraft, this was not without considerable loss in spite of the excellence of the machine and its relative immunity.

Early operations vindicated the faith Sir Geoffrey and C. C. Walker had from the start in their idea; it *was* possible to fly low in daylight without excessive risk, achieve surprise and accuracy and destroy a defended target of medium size. Furthermore this *could* be done by an unarmed high-speed bomber. Many of the Mosquito attacks were on targets in Occupied Countries where those in the surrounding districts were friends, who must not be injured in the raids. Quite a number of these targets were beyond the range of single-engined fighter-bombers, whereas the Mosquito could reach them with a sizeable load. Four Mosquitoes could place sixteen 500 pounders into a building and all get by before the first bomb exploded. The combination of Mosquito reconnaissance aircraft, effective intelligence methods and bombers to act accordingly made an extremely effective team.

'Mosquito' came to be a most appropriate name, as the little aeroplanes stung vital and unexpected places. For instance, Sqn. Ldr. J. R. G. Ralston, D.S.O., D.F.M., led six 105 Squadron Mosquitoes on to two large motor vessels entering the Gironde on 7 November, 1942; they approached from afar at wave-top height achieving complete surprise. At first there was no flak and seamen were walking about the ships. Nobody stayed to see the full results, but hits with 500 lb. bombs were made on one ship. At the roll-call on turning for home one Mosquito was missing, DK328:V flown by Bristow and Marshall who had recently been awarded D.F.C.s; they survived a crash in France to become P.O.W.s. By the end of November, twenty-four Mosquitoes had been lost from the 282 sorties flown, nine-tenths of which had been in broad daylight. The loss rate was about 8 per cent, whereas night bomber losses were currently about 5 per cent. Bomber Command considered that while valuable work was being done Mosquitoes were not available in sufficient numbers to make the raids really telling. By 30 November only 330 tons of bombs had been dropped. Had the Mosquito project been approved in 1938, and had production thus been advanced by a year, and entire production been for Bomber Command, then there can be no doubt that the effects would have been startling. Crew losses would have been less, and considerably fewer than occurred with four-engined

Operations: Day Bombers

bombers. When Major de Havilland lunched at Bomber Command on 4 November, however, he was disappointed to find so little enthusiasm there for the Mosquito. The Commander-in-Chief was clearly committed beyond return to the four-engined bomber concept and No. 2 Group, for all its good work and the tremendous courage of its crews throughout the early years of the war, was discouraged by the emphasis on the night offensive. There was tremendous enthusiasm for the Mosquito on the squadrons and when a batch earmarked for 139 Squadron was transferred to 109 Squadron there was strong resentment. It is interesting to surmise what might have been achieved had Bomber Command been equipped with large numbers of Mosquitoes and projected successors.

Apart from its importance in depriving the Germans of valuable production from a leading factory, the lunch-time raid on the Philips radio factory in Eindhoven was the first occasion when 139 Squadron operated its own aircraft, and marked the beginning of an intensive offensive of low-level attacks. On 20 November Flt. Lt. Patterson, D.F.C. flew DK338:O on the route planned for the attack, along the Scheldt Estuary to Woensdrecht, recording it on a cine camera. Beyond the range of fighter escort, situated well in the enemy fighter belt and in a friendly city, the target called for specialized training and skilful attack. Two large-scale practices for the raid occurred, the power station at St. Neots simulating the Philips works. At 11.22 the leading Mosquito DZ365 flown by Wg. Cdr. Hughie Edwards, V.C., D.F.C., took off to lead ten from 105[1] Squadron and 139 Squadron. Near the Dutch coast the Mosquitoes formated behind the leading Bostons. As they passed Woensdrecht airfield enemy fighters were taking off and, as Fw 190s attacked Sqn. Ldr. Parry's section, two Mosquitoes turned away attracting the assault upon themselves and meanwhile drawing away from the rest of the formation and the pursuing fighters. DZ367 aborted when intercepted but the remainder climbed to 2,000 feet and, as planned, dived through the Boston formation and placed their bombs in the blazing factory. Past Utrecht and south of Amsterdam the Mosquitoes roared home. DZ371 of 139 Squadron, hit by flak over the target, was flying home on one engine with smoke pouring from the other. It fell into the sea 30 miles off Den Helder. Following the raid DZ314:F left on a photographic reconnaissance sortie to Eindhoven over which it made two runs at 800 feet. High buildings were burning like a furnace. There was plenty of flak, but the Mosquito returned safely. The Eindhoven raid was the last full-scale attack of the year, the two squadrons despatching fifty-six mainly low-level sorties on individual industrial and railway targets before the bad weather of January brought operations to a halt.

On 20 January six of 105 Squadron's Mosquitoes[2] led by Sqn. Ldr.

[1] 139 Squadron used DZ373:B and DZ371:A. 105 Squadron used DZ365:V, DK296:G, DK338:O, DZ367:J, DZ372:C, DZ370:Z, DZ374:X and DK336:P.
[2] DZ353:E, DZ416:Q, DZ379:H, DZ408:F, DK302:D and DK337:N.

Operations: Day Bombers

Ralston in DZ353 delivered a low-level assault on the railway installation at Hengelo. Eight had set out but one had been hit by flak. Another lost the formation and bombed the docks at Lingen. Two attacked from 1,000 feet. to which height the smoke curled after the raid. On 27 January Wg. Cdr. Edwards led nine Mosquitoes[1] to the Burmeister Wain Diesel Works at Copenhagen. At 17.05 their bombs fell, after a fortnight's hard training for the raid, and following a course that took the formation well south of the target before it swept round in sight of the Swedish coast. As the bombers turned away flak came up, damaging DZ365 leading. North of Grimstrup DK338 flown by Flt. Lt. J. Gordon had aborted for, when flying at 50 feet, blue smoke had appeared around the starboard wing tip. He lifted his port wing and it scraped telegraph wires damaging the aileron. Since the formation had pulled far ahead he dropped his bombs and turned for home. DZ407 collided with high-tension cables and was destroyed, and DK336 crashed near Shipdham, its crew being killed. Some of the bombs dropped had delayed action, and in the Mess at Marham those who had dropped them drank their health, at the time when each one was timed to explode.

Parties, a vital part of Royal Air Force life, were also scheduled in Berlin on 30 January, 1943, on a day when both Goering and Goebbels were to address mass demonstrations of Nazi might. Interruption of these events was an attractive idea, and the only aeroplane that could conceivably get away with a daylight raid on the city at this time was of course the Mosquito. For 105 Squadron it was an ideal opportunity to improve upon their attempt of last September. First away was Sqn. Ldr. R. W. Reynolds in DZ413:K navigated by Plt. Off. Sismore, followed by Flg. Off. Wickham and Plt. Off. Makin in DZ408:F and Flt. Lt. J. Gordon and Flg. Off. R. G. Hayes in DZ372. To their surprise they received no welcome from flak or fighters until DZ372:C passed Bremen on its return. To their joy they reached Berlin a few minutes before 11.00 hours when Goering was to speak. Around their radio sets in Britain those in the know heard muffled noises and some cries followed by an hour of martial music as the prime boaster's speech was delayed. Reynolds and Wickham aimed their bombs at a railway junction to the north of the city and Gordon bombed on E.T.A. through cloud. All three landed safely back at Marham.

Next it was the turn of 139 Squadron which despatched DK337:N, DZ379:H and DZ367:J flown respectively by Flt. Sgt. P. J. McGeehan and Flg. Off. R. C. Morris, Sgt. J. Massey and Sgt. R. C. Fletcher and Sqn. Ldr. D. F. W. Darling and Flg. Off. Wright. Sqn. Ldr. Darling led them at low-level north of Heligoland, thence to Lübeck which they reached at 20,000 feet. Above cloud they flew to Schwerm and they found Berlin shortly before 16.00 hours local time, clear of cloud, as Goebbels was about to speak. One of 337's bombs fell half a mile from

[1] DZ365:V, DZ413:K, DZ415:A, DK302:D, DK338:O and DZ407:R of 105 Squadron. DZ379:H, DK336:P and DZ416:Q of 139 Squadron.

Operations: Day Bombers

the city centre, then it had to evade heavy flak and two fighters as a power descent home was made. Over the Frisians flak was encountered but 337 safely reached base. It was 15.52 when Sgt. Massey bombed Berlin's S.W. suburbs. He avoided a Fw 190 by dashing into cloud and after flying at 15,000 feet to Hanover flew home low for the rest of the journey. Darling—last to attack—encountered heavy flak over the city and was soon after shot down.

For the raid Sqn. Ldr. Reynolds was awarded the D.S.O. and all the other crews D.F.C.s or D.F.M.s. To the planners there was a reward of considerable value which was to have great importance for Berliners. Recorded with great care were timings and fuel consumption. 105's total track mileage averaged 1,145 miles. Times from take-off to landing, and fuel consumption for their aircraft were: DZ413—412 gallons in 4 hours 36 minutes, DZ372—450 gallons in 4 hours 42 minutes, DZ408—450 gallons in 5 hours 03 minutes. This gave ideas for flight profiles for possible future attacks on the city.

From February until the end of May, 1943, Mosquitoes of the two squadrons delivered spectacular, exciting and very effective day raids on targets as listed at the end of this chapter.

The armament works at Liège were attacked on 12 February by 139 Squadron. DZ386 was hit in the windscreen by a 37 mm. shell which injured Flg. Off. H. J. Brown and Flg. Off. F. E. Hay who were forced to turn for home. Route for the formation lay by way of Marham–Orfordness and low level to Furnes after which they turned into line abreast, entered an area of rain and made their shallow dive into the poor light at Liège. Usually the squadrons operated individually as on 14 February when ten of 139 led by Sqn. Ldr. Reynolds made a dusk shallow dive and low-level assault on the engine sheds at Tours. Two days later DZ463:O flew a reconnaissance flight over the target which next day suffered a combined attack, the squadrons despatching twenty aircraft. Wg. Cdr. Peter Shand in DZ421:G led 139 Squadron which encountered an assortment of the usual complications. DZ418:M had an electrical fault so turned back. DZ464:C had its bomb aimer's panel smashed and wing leading edges damaged when it collided with a flock of birds. DZ470:N narrowly missed enemy balloons and was damaged by flak. DZ422:R lost the formation and had to turn back and DZ420 was shot down. Low clouds and rain hampered the entire operation but at the target it allowed good attacks on the engine sheds, roundhouse and turntable. In three attacks at this period thirty locomotives suffered badly.

Unopposed, ten aircraft from each squadron raided the naval stores serving U-boats at Rennes on 26 February, with Sqn. Ldr. Bagguley and Wg. Cdr. Longfield leading 139 and 105 respectively. There were resounding explosions from ignited ammunition. On the route however disaster had struck when, turning sharply to avoid an airfield, DZ413:K with a glycol leak chopped DZ365:V in halves. 105's other aircraft

Operations: Day Bombers

arrived late, as 139 was making its shallow dive attack. So successful was the bombing that DZ468:E dropped her load on the railway yards at Vire. For 105 it was a tragic mission for its leader, Wg. Cdr. Longfield, was killed in the collision. Subsequently Wg. Cdr. John Wooldridge, D.F.C. and Bar, D.F.M., took over Command.

The outstanding raid of March came on the 3rd when the molybdenum mine and washing plant at Knaben in Norway was the target for which 139 Squadron despatched nine aircraft.[1] These constituted two waves for low-level and shallow dive on the plant supplying four-fifths of Germany's molybdenum. Wg. Cdr. Peter Shand, D.F.C., led the operation in DZ421 and Sqn. Ldr. Bagguley the low-level formation in DZ469. Navigation was excellent and no bombs were wasted on the highly rated yet small target for which, again, there had been special practices. Their course took them to Flamborough Head, then the force flew low across the North Sea and over the clear snow-clad Norwegian mountains. Smoke from the low-level attack usefully marked the target for the shallow divers. Two Fw 190s intercepted Flg. Off. A. N. Bulpitt's DZ463 which was shot down, the only aircraft to be lost in spite of attempts by other Focke-Wulfs to destroy the raiders as they sped along valleys homewards.

On 4 March 139 Squadron bombed the railway sheds at Aulnoye at dusk and 105 Squadron hit a similar target at Le Mans, led by Sqn. Ldr. Reynolds. On 9 March Bagguley and Ralston led fifteen Mosquitoes to bomb the engine sheds at Le Mans once again, and it was this raid that cost 139 Squadron Bagguley and Hadden flying DZ469. Flg. Off. Brown's Mosquito was hit over the target and for the second time in a week he made a forced landing, this time with his rudder all but useless. Six shallow divers of 139 and six low-level attackers of 105 raided the John Cockerill Armament works in Liège on the 12th, and evaded six intercepting Fw 190s. Flak was encountered on the return flight and DZ373 plunged into the Scheldt Estuary. The attack left the works totally idle for five weeks, and it was only partially reconstructed.

Engine sheds at Paderborn came next, on the deepest daylight penetration the low-level raiders had yet made into Germany. Six of 105[2] led by Flt. Lt. W. C. S. Blessing, D.F.C., and ten shallow divers of 139[3] under Sqn. Ldr. Berggren, D.F.C., set off for the raid on 16 March. Over the Zuyder Zee Sgt. Cummins in DZ423 was intercepted by three ducks, one of which crashed into his observer's stomach whilst the others hit the starboard engine nacelle. He turned for home followed by 2nd Lt. Wenger, who had radio trouble, and one of 105 Squadron. Eight leading machines climbed for attack some 25 miles from the target and six in

---

[1] DZ421:G, DZ469:J, DZ373:B, DZ423:T, DZ418:M, DZ428:K, DZ470:N, DZ422:R, DZ463:O.

[2] 105 Sqn. DZ489:B, DZ518:A, DZ408:F, DZ461:G, DZ351:L, DK302:D.

[3] 139 Sqn. DZ423:T, DZ465:E, DZ477:D, DZ478:V, DZ464:C, DZ497:Q, DZ491:F, DZ470:N, DZ476:S and DZ482:P.

Operations: Day Bombers

the rear raced in, first hugging the hilly ground skilfully and surprising the enemy gunners, who opened up on the second formation and damaged Flt. Sgt. McGeehan's aircraft, DZ497; this crashed on Texel. Low-lying ground mist fortunately covered the withdrawal. Sgt. Massey's machine DZ477 was hit by flak over Paderborn in the port wing root, and fuel leakage caused him to cut the port engine. He tried soon after to restart it without success and flying low his aircraft silhouetted against searchlight beams over the enemy coast was hit again. With controls not very responsive, and his remaining engine giving trouble, he had considerable difficulty in landing. Eventually his aircraft crashed into the windsock at Docking and was written-off.

Not always did the Mosquitoes bomb their target even upon reaching it, for in Occupied Countries no risk was taken which might harm the population. Thus on 20 March, 1943, when six of 139 Squadron went to attack the marshalling yards at Malines at dusk, flak over the coast crippled the leading aircraft, which sped for home. In its ensuing crash at Martlesham both Flt. Lt. Wayman, D.F.C., and Flg. Off. Clear, D.F.C., were killed. With hydraulics shot away Cussens crash landed DZ422:R at Stradishall. There was much flak over the target, reached in bad visibility and when it was almost dark, so the battered and depleted force flew home without attacking the main target.

A long sea trip around Brittany was called for when the St. Joseph locomotive works at Nantes were attended to on 23 March. Six enemy fighters were at one point seen and several marauding Ju 88s, but none approached the Mosquitoes. Eleven made the attack, led by Shand and Blessing bombing from between 50 and 1,200 feet. Only an office building escaped destruction—not one bomb landed outside the works. In six raids on railway workshops 114 engines were put out of action, many being totally destroyed.

A week later because smoke was again leaving chimneys of the Philips works at Eindhoven, it was treated by two formations of 139 Squadron led by Wg. Cdr. Shand and Handley in DZ421. Hit over Aalkmaar Flg. Off. Crampton's aircraft turned for home leaving four to make the low-level assault. The attack worked out very well and the works again took an enforced rest. Next day two squadrons called on the railway installations at Trier and Ehrang. In foul weather Talbot nursed DZ381 damaged by the exploding bombs and came home on one engine and without instruments, and Crampton was forced to make a belly landing in DZ428.

Mosquitoes had now dropped 2,000 tons of bombs. Raids mentioned represent a mere fraction of those undertaken by the two Squadrons, a complete listing of which appears at the end of this chapter, but they are typical examples and give a fair impression of the work. Losses averaged $7\frac{1}{2}$ per cent, and there was talk of developing specialized high-altitude operating units using two-stage supercharged Merlin aircraft. Photographs taken on the low-level raids proved beyond any doubt that the

Mosquito Mk. IV Bombers

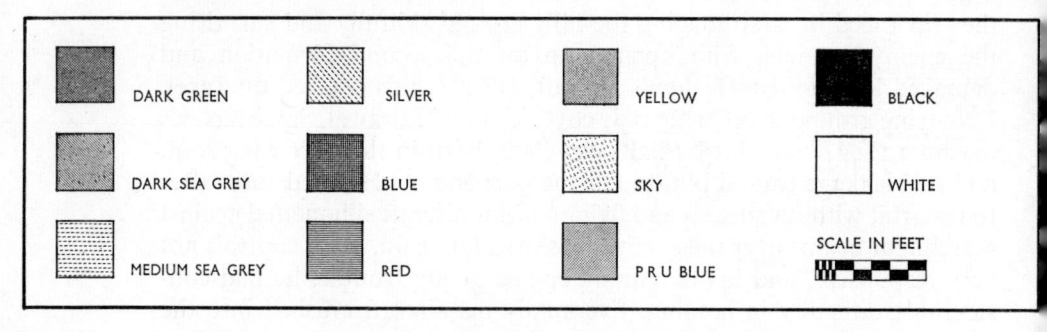

DARK GREEN	SILVER	YELLOW	BLACK
DARK SEA GREY	BLUE	SKY	WHITE
MEDIUM SEA GREY	RED	PRU BLUE	SCALE IN FEET

Mosquito B. IV Srs. ii plan view showing camouflage pattern *circa* 1943. In keeping with other day-bombers of 2 Group the Mosquito PRU/Bomber Conversion aircraft initially had dark green and dark earth upper surfaces and duck egg blue (officially called Sky) under surfaces. Spinners were black. During summer, 1942, these aircraft and the early Mk. IV Srs. ii had their spinners painted Sky, and an eighteen-inch wide band was painted around the rear fuselage. The markings somewhat resembled those of fighter aircraft. In late summer, 1942, the bomber Mosquitoes were re-painted dark green and dark sea grey, and had medium sea grey under surfaces. Sky fuselage bands were painted out, and spinners painted grey. But, as with all marking schemes, there were variations and combinations as one succeeded another.

Mosquito B. IV Srs. ii (Merlin 21) DK292 reached 105 Squadron 7.6.42 and flew 7 successful sorties before passing to 1655 MTU 3.10.42. After repair and overhaul joined 192 Squadron 4.10.44 and failed to return from Munich area 27.11.44. Markings depicted show Sky band and spinners to delude enemy fighters into thinking the machine an armed fighter.

Mosquito B. IV Srs. ii (Merlin 21) DZ601 reached 139 Squadron 17.5.43 and first operated 27.5.43 and flew many operations at night. Passed to 627 Squadron 24.5.44, damaged attacking Saumur marshalling yard 1.6.44. Repaired in works, but saw no further squadron use. SOC 16.10.46.

Mosquito B. IV Srs. ii (Merlin 23) DZ650 after conversion to carry 4000 lb. bomb reached 692 Squadron 15.5.44. Began operating 28–29.5.44 and raided Osnabruck 5–6.6.44. Passed to 627 Squadron 7.44, became AZ-Q. Damaged beyond repair 29.12.44.

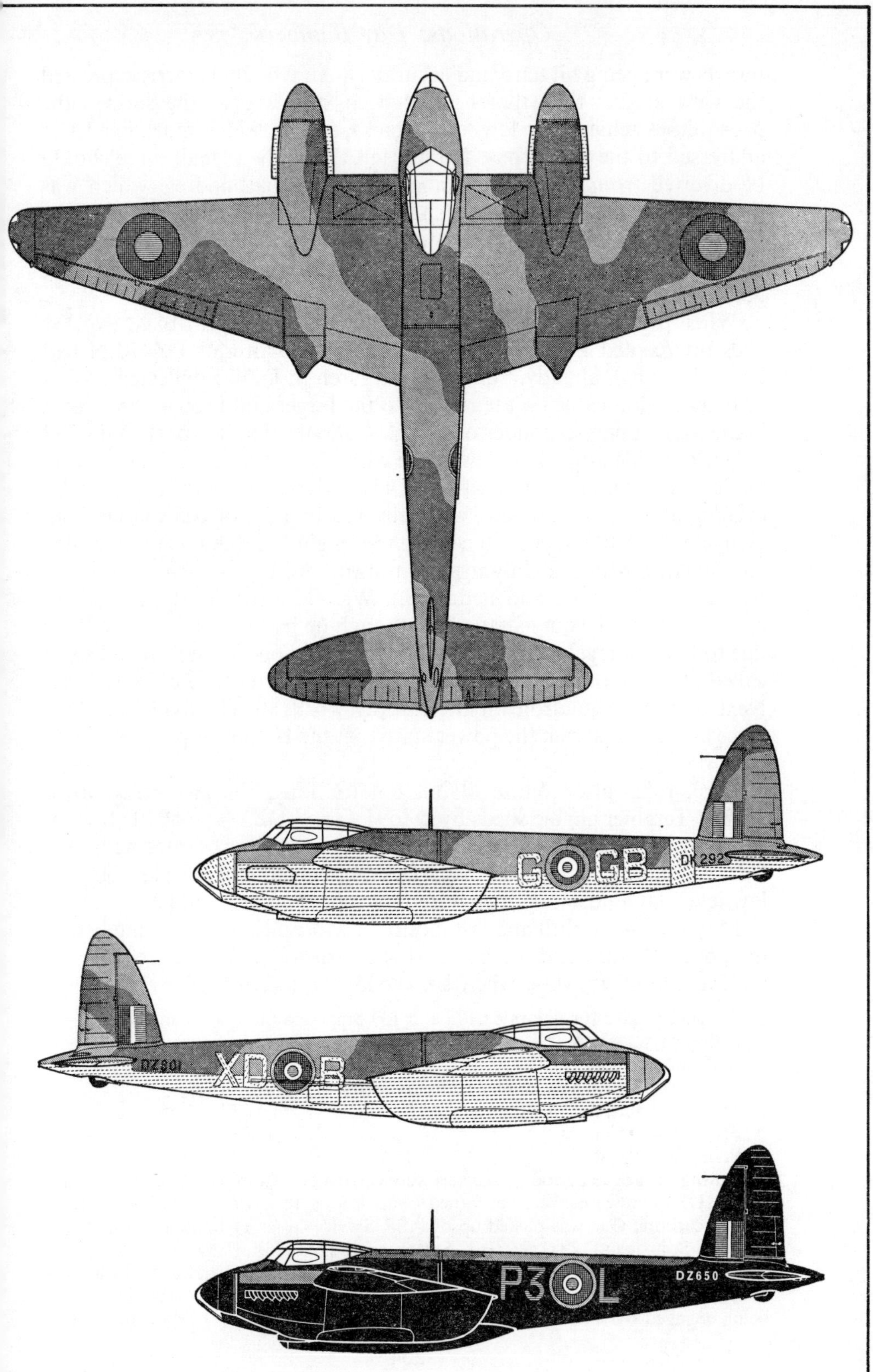
G GB
DK 292
XD B
DZ601
P3 L
DZ650

Operations: Day Bombers

targets were being hit hard and accurately. Air Marshal Harris expressed the view at this time that he had been surprised at the success the Mosquitoes achieved on low-level attacks, and he said as much in a letter addressed to the squadrons. He still felt that only a small force should be diverted from the main bomber effort; for pathfinding, which was then assuming great importance in the Command, he considered that the Mosquito was indispensable. It was concluded beyond any doubt that the Mosquito relying on speed had a good edge over the defending gunners and fighters. So, the job went on.

When four Mosquitoes attacked Malines on 11 April, 1943, two Fw 190s intercepted Flg. Off. J. H. Brown, whose aircraft DZ470:N had been hit by flak already. Sqn. Ldr. Berggren in DZ421 collected 20 feet of high-tension cable on the run up to the target and brought most of it home as a trophy to ponder over. Sgt. Cummins had his port engine set ablaze over the target and flew home with his port oleo leg dangling and made a crash landing in DZ482:P. Only DZ464:C escaped damage. An evening attack the same day was delivered by four of 105's aircraft on Hengelo, then 105 Squadron commenced night bombing. Day raids also continued, directed mainly against railway objectives such as those at Namur, Tours, Trier and Paderborn. Wg. Cdr. Wooldridge led a successful attack on Tours at 100 feet, though he had to belly land in Kent due to flak damage. An intended follow-up raid on Eindhoven had to be called off when one of the aircraft crashed soon after take-off on 1 May. Next day 105 Squadron hit the railway works at Thionville and sent three aircraft to attack the power station at the Hague, at precisely 12.24 on 4 May.

Dusk raids also continued but not for long, the last being flown against Tergnier engine sheds by DZ591:O and DZ374:X of 105 Squadron on 20 May, and when 105 Squadron set out for Nantes with Wg. Cdr. Wooldridge in the lead they were intending to make the last but one low-level day raid, when they were intercepted and turned back.[1]

May was the twelfth and last month of Mosquito day bombing before the policy change, and the last daylight raid of the series took place on the evening of 27 May, when six of 139 Squadron[2] led by Wg. Cdr.

[1] Casualty rates for day raids 105 and 139 Sqns. may now be summarised:—
Period 31.5.42–31.5.43.

Acft. missing/acft. sortie	6·7%
Acft. missing plus acft. totally destroyed/acft. sorties	8·15%
Total loss, acft. per month	18·6%
Flying hours/Cat. AC	138
Flying hours/Cat. B.	233

During the above period 726 sorties were carried out, from which 48 acft. did not return. Of 96 crew members concerned 12 are known to have become POWs and 3 evaded capture. One was picked up by ASR Services after 11 hours in a dinghy.

[2] 139 Sqn. lost DZ381:W (F/L Sutton) and DZ602:R (F/O Openshaw) who collided west of Kassel when avoiding flak. DZ598:N (F/O Pereira) had his starboard engine fail and diverted to bomb the railway near Kassel narrowly missing being engaged by six 190s. DZ605:D (F/L Sutherland and F/O Dean) hit a H.T.

Operations: Day Bombers

Reynolds, and navigated by Flt. Lt. Sismore in DZ593, set off at 19.15 hours for the Schott glass works at Jena. Five minutes later 105 Squadron despatched eight aircraft to the Zeiss optical works at Jena. Wg. Cdr. Reynolds' account of the trip partly draws this chapter to a close: 'I was leading in B for Beer, DZ601, a particularly fine Mosquito, and I was airborne with the rest of the formation at a certain hour in the early evening. We formed up in the circuit area of the airfield and then set course on what was to be the deepest low-level daylight penetration we have yet undertaken, and also what was to be one of the most eventful—for me anyway!

'The North Sea was crossed at wavetop height and on approaching the enemy coast, always the most tense moment, speed was increased and the formation closed up for the quick dash across. Once over the trip becomes more interesting. One sees such things as cyclists jumping off their cycles to have a look round to see what is coming, children look up at us then put their heads down and run as hard as they can. Mechanical vehicles are almost non-existent—everyone cycles. Startled horses, sheep and cows scatter in every direction. Frequently one has narrow squeaks with birds—an enemy of the low flying aircraft. If they do hit the kite serious damage is almost sure to be caused.

'There were no further incidents until we ran over one of Germany's reservoirs, when the Hun pushed up some accurate flak at us; fortunately no one was hit but shortly afterwards two of the formation collided and crashed.

'It was another fifty miles or so further on when bad weather was encountered. Firstly thick industrial haze and then heavy low clouds covering the tops of the hills over which we had to fly. Of course we had to enter cloud and climb to safety height, i.e. high enough to clear the hills with a margin of safety. As soon as we estimated that we should have cleared the high hills we descended into the gloom beneath the clouds.

'To enable the other chaps to pick me up again, a very difficult job under the existing weather conditions, I put on my navigation lights for a short while. This was also done by the Leader of the second section, Sqn. Ldr. Blessing. Crews afterwards stated that this was a great help for them to regain formation.

'At this stage we had arrived at a point approximately twenty miles from the target and we turned on to our run up, increasing speed then opening bomb doors. We picked up various land marks which we were looking for and knew from these that we were dead on track. The

---

cable when landing at Coltishall and the crew was killed. 105 Sqn. lost DZ467:P in the attack. DZ591:O and DZ595:C attacked the target. DK337:N bombed the town encountering much flak. DZ521:V raided a factory at Lobeda after losing formation, and DZ414:E also attacked this. DZ483:R crashed on her return and the crew was killed. DZ548:D diverted and bombed a train at Lapstrup. Return by most came after nightfall.

Operations: Day Bombers

visibility was now down to about 1,500 yards—not much when one is travelling at such high speed.

'The target was now only two miles away but not yet in sight. At 1,000 yards I picked up the tall chimneys and opened to full throttle. My observer pointed out the balloons and immediately the flak came up at us in bright red streams and unhealthily close. I could see one gun on a flak tower firing away as hard as he could at someone on my right. Now it was every man for himself; I picked out a tall building and went for it, releasing my bombs at point-blank range. I yanked the stick back to climb over the building and as I topped it the airscrew received a direct hit. There was a violent explosion in front of my eyes and I felt something tug at my hand and leg but took no notice for the time being—things were too hot.

'Now we were in a veritable hail of tracer shells, dodging and twisting for dear life. More balloons lay ahead which we missed by the Grace of God and now, apart from a few inaccurate bursts, we were clear and I was able to survey the damage.

'My left hand was bleeding freely as was my left leg. The kite was vibrating considerably and I could see holes in the fairing immediately in front of the radiator. Flak had pierced a hole just aft of the port radiator and close to one of the main tanks. There were two large holes in the fuselage close to the throttle box where some fittings had been blown away. My intercom had packed up and I discovered later that a splinter had severed the lead just below my starboard ear. The collar of my battle dress was torn also, this wasn't noticed until I arrived back at base when Flt. Lt. Sismore, my navigator, asked me what I had done to get that!

'However, to continue; after that one violent explosion it seemed a miracle that the aircraft could keep in the air. I was especially anxious about my port radiator with that hole so near and constantly checked the temps. to watch for any rise. Fortunately it remained constant at 97°C. and the vibrating got no worse so the need to feather the damaged propeller never arose.

'We were now returning individually and so I nipped into low cloud for safety—to clear hill and avoid any flak that may have been put up. My observer bound up my hand and then we settled down to the long journey home with a frequent apprehensive glance at the engine instruments and fuel. I, personally, felt satisfied that I got the target with my bombs and later one of the boys said that he saw them go in followed by a sheet of flame a hundred feet high—and it would be some time before they actually exploded.

'On the way back we ran into more trouble by entering two more defended areas. The second one was very hot and it was with luck that we escaped by means of violent evasive action combined with full throttle and fine pitch.

'From now on the return journey was uneventful. Petrol was checked

Operations: Day Bombers

and we decided that we should have sufficient. On arrival back at base we discovered that there was not much left but our worries were over now so we did not bother any more.

'Even though "B-Beer" was badly damaged she behaved magnificently throughout and is now being repaired, which will take a few days. One of the port engine bearers had a large hole through the middle of it but the vital parts of the engine were sound.'

Wing Commander Wooldridge, D.F.C., commanding 105 Squadron wrote to de Havilland about the Mosquito IV in service:

'The bomber is, in every way, an outstanding aeroplane—easy to fly, highly manoeuvrable, fast and completely free from vices of any sort. From our point of view it has a further quality, a highly important one in wartime—and that is the extraordinary capacity for taking a knocking about. I myself had an experience of this, a short time ago; while approaching a target at approximately 100 feet, with bomb doors open, my aircraft was hit by three Bofors shells. Apart from the distinct thuds as the shells exploded and a rather unpleasant smell of petrol, the behaviour of the aircraft after impact appeared to be normal and the bombs were dropped successfully. Actual damage was :— The first shell entered the lower surface of the port mainplane, approximately four feet from the wing tip, and burst inside removing three square feet of the upper wing surface. The aileron was fortunately undamaged. The second shell hit the port engine nacelle fairly far back, wrecking the undercarriage retraction gear, severing the main oil pipe line, damaging the airscrew pitch control and putting the instruments of the blind flying panel out of action. The third shell entered the fuselage just in front of the tailplane and severed the tailwheel hydraulic line and the pressure head line, rendering the A.S.I. useless. After awhile, on the way home, port engine began to give trouble and eventually it failed. Although the airscrew could not be feathered a ground speed of almost 200 m.p.h. was maintained on the return journey, and the aircraft was landed in pitch darkness on its belly without the assistance of flaps.

'On one occasion a Mosquito went through a set of high-tension cables which appeared unexpectedly in the target area, but returned to its base slightly bent, and was landed on the wheels in the dark, without further damage. Another machine had its elevator controls severed, but was brought home controlled fore and aft purely by means of the flaps and throttles! The aeroplanes fly so well on one engine that it is the opinion of this squadron that de Havilland must have originally designed it as a single-engined aeroplane, and then stuck another one in for luck. It is entirely free from unpleasant vices at all times, which is a great factor when making night landings when damaged, and owing to the clean design of the underside, it can in emergency be landed on its belly with very little damage, an important factor when considering serviceability. All round, it is a sturdy pugnacious little brute, but thoroughly friendly to its pilot.

Operations: Day Bombers

'In conclusion the Mosquito represents all that is finest in aeronautical design. It is an aeroplane that could only have been conceived in this country, and combines the British genius for building a practical and straightforward machine with the typical de Havilland flair for producing a first rate aeroplane that looks right and IS right.'

Writing of his experience on Mosquitoes the late Wg. Cdr. John Wooldridge in his epic account *Low Attack* (Sampson Low) said:— 'For those of us who flew the Mosquitoes on these attacks the memory of their versatility and their achievements will always remain. It would be impossible to forget such experience as the thunderous din of twenty aircraft sweeping across the hangars as low as possible, setting course like bullets in tight formation for the enemy coast. The whole station would be out watching, and each leader would vie for the honour of bringing his formation lower across the aerodrome than anyone else. Nor would it be possible to forget the sensation of looking back over enemy territory and seeing your formation behind you, wing-tip to wing-tip, their racing shadows moving only a few feet below them across the earth's surface; or that feeling of sudden exhilaration when the target was definitely located and the whole pack were following you on to it with their bomb doors open, while people below scattered in every direction and the long streams of flak came swinging up; or the sudden jerk of consternation of the German soldiers lounging on the coast, their moment of indecision, and then their mad scramble for the guns; or the memory of racing across The Hague at midday on a bright spring morning, while the Dutchmen below hurled their hats in the air and beat each other over the back. All these are unforgettable memories. Many of them will be recalled also by the peoples of Europe long after peace has been declared, for to them the Mosquito came to be an ambassador during their darkest hours.'

89.
Trier on 1 April, 1943—a few moments later

90.
A line-up of 139 Squadron at Marham

91.
Wg. Cdr. Peter Shand briefing 139 Squadron crews

92. One of 60 Squadron's **PR. II**s, DD744
93. Lt. Archie Lockhart with the Mosquito engaged by an Me 262 in August, 1944
94. 'Lovely Lady', a PR. Mk. IX of 60 Squadron, S.A.A.F.

95. An early production Mk. VI fighter-bomber
96. Amiens prison under attack
97. The prison photographed later from DZ414

104. Mosquito VI PZ446 of 143 Squadron being re-armed

105. A ship under attack at Sandshavn, 23 March, 1945

106. Rockets, long-range tank, paddle-bladed airscrews and nose camera on a Mk. VI RS625

107.
235 Squadron attacking shipping on 19 September, 1944

108.
A shipping strike at Nordgalen, 5 December, 1944

109.
Banff Wing attack on U-boats, 9 April, 1945

110. The first Mk. XVIII 'Tsetse'
111. A Mk. IV during *Highball* trials at Loch Striven
112. Mk. 30 NT585 of the type used in later stages of Bomber Support

Operations: Day Bombers

Summary of Operations flown in daylight by Numbers 105 and 139 Squadrons May, 1942 to 31 May, 1943

Date	Target	Role	105 Sqn. a	b	c	d	e	139 Sqn. a	b	c	d	e
31. 5	Cologne	HBPR	5	3	–	1	1					
1. 6	Cologne	HBPR	3	2	–	–	1					
1. 6	Essen	HBPR	1	1	–	–	–					
5. 6	Schiphol	PR	1	1	–	–	–					
18. 6	Bremen	BPR	2	–	2	–	–					
18. 6	Bremerhaven	BPR	1	–	1	–	–					
20. 6	Schutorff	BPR	1	1	–	–	–					
20. 6	Emden	BPR	1	1	–	–	–					
25. 6	Schleswig	BPR	2	–	2	–	–					
26. 6	Bremen	BPR	4	2	1	1	–					
1. 7	Kiel	HBPR	1	1	–	–	–					
2. 7	Flensburg	LL	4	3	–	1	–					
2. 7	Flensburg	HB	1	–	–	–	1	2	–	–	1	1
2. 7	Bremen	MR	1	1	–	–	–					
9. 7	Wilhelmshaven	HB	1	1	–	–	–					
11. 7	Flensburg	LL	6	4	–	1	–					
16. 7	Ijmuiden	LL	2	2	–	–	–					
16. 7	Wilhelmshaven	HB	2	–	–	1	1					
21. 7	N.W. Germany	HB	7	5	–	2	–					
22. 7	Munster	HB	1	–	–	1	–					
23. 7	Cologne	CCB	4	–	–	4	–					
23. 7	Ijmuiden	CCB	1	–	–	1	–					
25. 7	Mannheim, Frankfurt	BPR	2	2	–	–	–					
25. 7	Bremen	MR	1	–	–	1	–					
26. 7	Cologne, Essen, Duisburg	HB	3	3	–	–	–					
28. 7	Lubeck, Flensburg	BPR	2	2	–	–	–					
28. 7	W. Germany	HB	4	2	–	1	1					
29. 7	W. Germany	HB	3	1	2	–	–					
30. 7	W. Germany	HB	5	2	1	2	–					
1. 8	W. Germany	HB	5	3	–	1	1					
3. 8	Hagen	HB	1	–	–	1	–					
4. 8	W. Germany	HB	3	–	–	3	–					
5. 8	Brauweiler/Stuttgart	CCB	2	–	1	1	–					
5. 8	Ijmuiden	CCB	1	–	–	1	–					
6. 8	Kiel	HB	1	–	–	1	–					
6. 8	Essen, Hanover	HB	2	1	1	–	–					
7. 8	W. Germany	HB	3	1	–	2	–					
9. 8	Cologne, Frankfurt	HB	2	1	–	1	–					
9. 8	Kiel	MR	1	–	–	1	–					
10. 8	Essen, Cologne	HB	2	1	–	1	–					
10. 8	Osnabruck	LL	1	1	–	–	–					
12. 8	Wiesbaden	HB	1	1	–	–	–					
13. 8	Essen	HB	1	–	1	–	–					
14. 8	Mannheim	HB	1	–	–	1	–					

209

Operations: Day Bombers

Date	Target	Role	105 Sqn. a b c d e	139 Sqn. a b c d e
15. 8	Mainz	HB	1 – – – 1	
16. 8	Vegesack	CCB	1 – 1 – –	
17. 8	Kiel	HB	1 – 1 – –	
18. 8	Hamburg	HB	1 1 – – –	
19. 8	Bremen	HB	1 – – – 1	
21. 8	Lille	PR	1 1 – – –	
23. 8	Flensburg	CCB	1 – 1 – –	
25. 8	W. Germany	LLB	4 1 2 – 1	
27. 8	N.W. Germany	LL	4 3 – – 1	
29. 8	Pont à Vendin	LL	2 2 – – 1	
1. 9	Thionville	MR	1 1 – – –	
2. 9	Holland—no details	CCB	2 – – – 2	
2. 9	Sas van Gant	DB	1 1 – – –	
2. 9	W. Germany	HB	3 3 – – –	
4. 9	Holland	CCB	3 – – 3 –	
4. 9	W. Germany	HB	3 3 – – –	
6. 9	W. Germany	HB	4 – 2 1 1	
6. 9	Ijmuiden	LL	1 1 – – –	
7. 9	W/N.W. Germany	HB	5 3 2 – –	
9. 9	W. Germany	HB	6 5 1 – –	
13. 9	N.W. Germany	MR	1 – – 1 –	
14. 9	N.W. Germany	HL	5 3 2 – –	
14. 9	N.W. Germany	MR	1 1 – – –	
18. 9	N.W. Germany	CCB	3 – 1 2 –	
19. 9	Berlin	HB	6 1 3 1 1	1 – 1 – –
22. 9	Ijmuiden	LL	6 5 – 1 –	
25. 9	Oslo	LL	4 4 – – 1	
26. 9	N.W. Germany	MR	1 1 – – –	
1.10	Dutch & Belgian targets	LL	3 2 – 1 –	
2.10	Liège	SD	6 5 1 – –	
5.10	Frankfurt	HB		1 – 1 – –
6.10	Essen, Bremen, Saarbrucken	HB	2 2 – – –	1 – 1 – –
6.10	Munster	MR	1 1 – – –	
6.10	Hengelo	LL	4 3 – 1 –	
8.10	Saarlouis	BPR	1 1 – – –	
9.10	W. Germany	BPR	5 – 3 1 1	
11.10	W. Germany	HB	5 3 1 – –	2 – 1 – 1
11.10	Sluiskil	LL		1 1 – – –
12.10	N.W. Germany	MR	1 1 – – –	
13.10	N.W. Germany	MR	1 1 – – –	
14.10	N.W. Germany	MR	1 1 – – –	
15.10	Hengelo	SD	4 4 – – –	
15.10	Den Helder	LL		1 1 – – –
16.10	Hengelo	LL	5 5 – – –	1 1 – – –
20.10	W. Germany	HB	5 1 1 2 1	1 – – 1 –
21.10	N.W. Germany	RC	4 4 – – –	
22.10	Le Creusot	PR		1 1 – – –
23.10	Hengelo	LL	5 3 1 – 1	

210

Operations: Day Bombers

Date	Target	Role	105 Sqn. a	b	c	d	e	139 Sqn. a	b	c	d	e
24.10	Genoa (cloud covered)	PR	1	1	–	–	–					
25.10	Munster, Bremen	CCB	2	–	–	2	–					
27.10	Flensburg	LL	3	2	1	–	–					
27.10	Den Helder/Ghent	LL	2	1	–	1	–					
27.10	Antwerp	LL	2	–	1	1	–					
29.10	Ijmuiden, Borkum	CCB	2	–	1	1	–					
29.10	Langroog airfield	CCB	1	1	–	–	–					
29.10	Hanover	HB	1	–	–	1	–					
29.10	Flushing	LL						1	–	–	1	–
29.10	Wangeroog	LL						2	2	–	–	–
29.10	Osnabruck	CCB						1	–	–	1	–
30.10	Leeuwarden airfield	LL	2	2	–	–	–					
30.10	Deelen/Jever	CCB	2	–	2	–	–					
30.10	Lingen	CCB	1	–	–	–	1					
30.10	Saarlouis	PR						1	–	–	1	–
30.10	Flushing/Woensdrecht	CCB						2	–	1	1	–
7.11	Gironde Estuary, ships	LL	6	6	–	–	1					
13.11	Flushing	LL	2	–	–	–	2					
16.11	Rly, Lingen & Julich	LL	6	6	–	–	–					
20.11	Eindhoven	PR	1	1	–	–	–					
29.11	Rlys in Belgium	LL	4	4	–	–	–					
6.12	Eindhoven, Philips works.	SD	8	6	–	2	–	3	2	–	–	1
8.12	Rlys in Holland	LL	6	4	2	–	1					
9.12	Hengelo	LL						1	1	–	–	–
9.12	Rlys in France	LL	3	3	–	–	–	1	1	–	–	–
13.12	Rlys in France	LL						4	3	–	1	–
14.12	Ghent, Courtrai, Rlys	LL	2	1	–	1	–					
14.12	Eindhoven	PR						1	–	–	1	–
14.12	Roosendaal M.Y.	LL						2	–	–	2	–
17.12	Rlys in the Low Countries	SD						4	4	–	–	–
20.12	Rlys in N.W. Germany	RC	6	5	–	1	–	5	4	–	–	1
22.12	Rlys. France & N.W. Germany	LL	6	3	–	3	1					
29.12	Rlys in France	LL	4	4	–	–	–	3	3	–	–	–
31.12	Rlys in France	LL	5	4	1	–	–					
1943												
1.1	Raismes, Mons	PR	1	–	–	1	–					
1.1	Mons, Rly M.Y.	LL						1	1	–	–	–
3.1	Tergnier	LL	3	3	–	–	–					
3.1	Amiens	LL	3	3	–	–	–					
9.1	Tergnier, Mons, Raismes	PR	1	1	–	–	–					
9.1	Rouen E.S.	LL	6	5	–	1	1					
13.1	Aulnoye, Laon, Tergnier	LL	6	6	–	–	–					
20.1	Hengelo	LL	8	6	1	1	–					
21.1	N.W. Germany	MR						1	1	–	–	–
23.1	Rlys. W. Germany	CCB	4	2	–	1	1					
26.1	Rlys N.W. Germany	MR						1	1	–	–	–
27.1	Copenhagen	LL	6	5	–	1	1	3	3	–	–	–

Operations: Day Bombers

Date	Target	Role	105 a	105 b	105 c	105 d	105 e	139 a	139 b	139 c	139 d	139 e
30. 1	Berlin	HB	3	3	–	–	–	3	2	–	–	1
8. 2	Rheine M.Y.	LL	4	3	–	1	–					
8. 2	Lingen Rly works	LL	4	4	–	–	–					
12. 2	Arm. works Liège	SD						6	4	–	2	–
12. 2	Tergnier	LL						2	1	–	1	–
14. 2	Tours E.S.	LL						10	6	1	3	–
17. 2	Tours	PR						1	1	–	–	–
18. 2	Tours	SD						13	10	–	2	1
18. 2	Liège	LL	5	–	–	5	–					
26. 2	Rennes	LL/SD	10	7	–	1	2	10	9	–	–	1
28. 2	Liège	LL	6	6	–	–	–					
28. 2	Hengelo	LL	4	4	–	–	–					
3. 3	Knaben	SD/LL						9	8	–	–	1
4. 3	Le Mans E.S.	LL	6	6	–	–	–					
6. 3	Low Countries	MR	1	1	–	–	–					
8. 3	Aulnoye E.S.	SD/LL						9	9	–	–	–
8. 3	Lingen Rly sheds	SD	3	3	–	–	–					
8. 3	Tergnier	SD	3	3	–	–	–					
9. 3	Le Mans (Renault works)	LL	5	5	–	–	–	10	10	–	–	1
12. 3	Liège arms factory	SD/LL	6	6	–	–	1	6	5	–	1	–
16. 3	Paderborn E.S.	LL	6	5	–	1	–	10	8	–	2	–
18. 3	N.W. Germany	MR	1	1	–	–	–					
20. 3	Louvain	LL	6	6	–	–	–					
20. 3	Malines	LL						6	–	–	6	–
23. 3	Nantes	LL	4	3	–	1	–	10	8	–	2	–
24. 3	Trains, N. France	SD	3	3	–	–	–					
27. 3	Liège	LL	6	–	–	6	–					
27. 3	Hengelo	LL						6	6	–	–	–
28. 3	Liège	LL	6	4	–	–	2					
30. 3	Eindhoven	SD/LL						10	10	–	–	–
1. 4	Ehrang	LL						6	6	–	–	–
1. 4	Trier	LL	6	6	–	–	–					
3. 4	Tergnier E.S.	LL						2	1	–	1	–
3. 4	Aulnoye E.S.	LL						2	1	–	–	1
3. 4	Malines	LL	2	–	2	–	–					
3. 4	Namur	LL	2	2	–	–	–					
6. 4	Namur	LL	6	5	1	–	–	2	2	–	–	–
9. 4	Julich R.W.	LL	4	2	2	–	–					
9. 4	Orléans E.S.	LL						4	–	–	4	–
11. 4	Hengelo	LL	4	2	–	1	1					
11. 4	Malines	LL						4	3	–	–	1
14. 4	N.W. Germany	HB	6	6	–	–	–					
19. 4	Namur	LL	4	–	3	1	–	2	1	–	1	–
24. 4	Tours/Trier	LL	3	–	–	3	–					
24. 4	Paderborn	LL	6	5	–	1	–	2	1	1	–	–
26. 4	Tours, Julich, Lingen	LL	3	2	–	1	–					
27. 4	Duisburg	HB	3	2	–	1	–	2	1	–	1	–
1. 5	Eindhoven	LL	6	–	–	6	–					

Operations: Day Bombers

Date	Target	Role	105 Sqn. a	b	c	d	e	139 Sqn. a	b	c	d	e
2. 5	Thionville	LL	7	6	–	1	–					
4. 5	Hague, power station	LL	3	3	–	–	–					
4. 5	Haarlem	LL	3	–	–	3	–					
5. 5	Tubize R.W.	LL	6	1	–	5	–					
20. 5	Tergnier	LL	2	2	–	–	–					
22. 5	Nantes	LL	7	–	–	7	–					
27. 5	Jena: Zeiss (105) Schott (139)	LL	8	3	5	–	1	6	3	1	–	2
Sortie totals:			524	331	57	108	35	202	146	8	36	13

Legend.
Roles: HBPR—High-level bombing and photo reconnaissance
          BPR—Bombing & photo reconnaissance
            PR—Photo reconnaissance
            LL—Low-level bombing
          CCB—Cloud-cover bombing
           MR—Meteorological reconnaissance
            SD—Shallow dive
            HB—High-level bombing
            RC—Railways under cloud cover

Sortie Columns. (a) Total despatched by Sqn.
            (b) Attacked primary target
            (c) Attacked another target
            (d) Aborted; no attack
            (e) Failed to return

Abbreviations: ARM.: Armamens works
              E.S.: Engine sheds
              R.W./M.Y.: Railway works/Marshalling yards

Chapter 14

Operations: Meteorological Reconnaissance

As the tempo of the bomber offensive against Germany increased it was decided that regular meteorological flights should be flown over enemy territory, in order to reduce losses and abortive sorties attributable to adverse weather conditions. Two Spitfires were delivered to 1401 Flight in August, 1941, for the task, and the first operation, code name *PAMPA*, was undertaken on 7 November. The aircraft concerned failed to return and no more such sorties were flown until crews had been specifically trained for the task. It was realized that the Mosquito with a pilot and navigator/observer would be more suitable for the work, and on 15 April, 1942, DK285 was delivered to the Flight, complete with externally fitted thermometers. DK289 followed five weeks later.

The first Mosquito *PAMPA* sortie was flown by DK289 on 2 July, 1942, in the hands of Flg. Off. C. F. Rose, who set off at 17.20 hours with Sgt. Prag as his navigator. Nine more *PAMPAS* were then successfully flown from Bircham Newton by Flg. Off. Rose, Flt. Lt. Braithwaite and Sqn. Ldr. Wellings, including two on 19 July.

On 26 July Flt. Lt. De Jace, a Belgian, and Sgt. Prag flying DK289 failed to return, leaving the Flight without a serviceable Mosquito for DK285 was having her cowling modified.

On 1 August Squadron status was awarded to the Flight which remained at Bircham Newton as 521 Squadron continuing long-range sorties over Europe, usually below contrail height. The third aircraft, DK329, was the first of the squadron's to fly an operational sortie, when Flg. Off. C. F. Rose with Plt. Off. N. W. F. Green as observer took off at 11.00 hr. on 1 September, 1942, and flew to Hamm, Kahle, Asten, Epinal, Charleroi and back to base. Two more sorties followed on 2 September. Then came a memorable engagement.

On 4 September Sqn. Ldr. Braithwaite, outward bound in DK329 and still climbing, was attacked over the Zuyder Zee at 28,500 feet by two Fw 190s which approached head on at the same height as the Mosquito. They turned outwards, and Braithwaite turned to port under one of them. During the ensuing battle from 28,000 to 22,000 feet he found he could readily turn inside the Fws, but they appeared 'very much faster' than the Mosquito. Braithwaite, in the heat of the battle, had forgotten to close the radiator flaps and his speed was reduced by 16 m.p.h. Also he

Operations: Meteorological Reconnaissance

failed to pull the boost cut-out, which would have increased his speed below 22,000 feet. His observer Davis, an experienced air gunner, knelt facing aft, holding on to the observer's armour plate and giving helpful suggestions to his pilot. After going into a tight turn at about 22,000 feet the Mosquito started to shudder mildly then flicked on to its back. The first time this happened Braithwaite was rather bewildered, but when he found he could quickly roll out he several times used the ruse intentionally as an evasive one.

Davis found it impossible to retain his grasp of the armour plate during the turns and was flung to the floor. On one turn he injured his back and was violently sick. About fifteen attacks, mainly from the port quarter and over a period of 35 minutes, were made, during which the Mosquito progressed homewards. The encounter ended about 50 miles out to sea at 3,000 feet the only damage to the Mosquito being, amazingly, a solitary bullet hole in its tailplane.

From her sixth sortie, which occurred on 8 September, to Saarbrucken and Dortmund, the aircraft failed to return. Consideration was then given to equipping the squadron with long-range escorting Mosquito fighters.

The prime purpose of *PAMPA* was the provision of material upon which forecasts of the weather over enemy territory could be made. Special forecasts were needed to provide details of weather conditions before Bomber Command raids. To cope with these requirements extra Mosquito IVs were supplied.

Two flights were made in October, to Falkenburg and Flensburg. On 3 November DZ316 was selected to be the first and only Mosquito of the squadron to fly a *RHOMBUS*, a met. sortie across the North Sea and back to the north of Scotland. Once in November enemy fighters were sighted, and several times they were evaded. On one such sortie Sqn. Ldr. Braithwaite's observer found himself looking at eleven Bf 109s just after the Mosquito had emerged from cloud at 18,000 feet! Luckily there was plenty of cloud about. Mosquitoes flew deep into enemy territory, to Emden, Hamburg, Hanover and Rostock.

DZ359 went to the Mont Blanc region and Turin on both 13 and 15 November, and visited the Turin area four more times before the end of the month. The Marseilles area was the target on 11 and 13 December, but it was a five-hour mission by DZ406 to Alessandria over the cloud-clad Alps that proved to be the longest journey of the year. During December DZ359 undertook two sorties to Munich, and in 1942 there were 43 successful sorties; and in the first quarter of 1943 91 *PAMPA* sorties were undertaken by Mosquitoes.

Early 1943 it was decided that these two tasks could better be undertaken if the Mosquitoes operated under Bomber instead of Coastal Command. Thus, after DZ406 had flown the squadron's last Mosquito sortie on 31 March, 1943, the Mosquitoes were taken to 1409 Flight, formed next day at Oakington with an Establishment for 8+2 aircraft

Operations: Meteorological Reconnaissance

under the Command of 8 Group. Flt. Lt. P. Cunliffe-Lister began the unit's operational career the following day with a long-range flight to Plymouth, Idouessant and Lorient. Again the Mosquitoes soon ranged far and wide across Europe, usually returning to Wyton to supply information directly to Group H.Q. Mk. IVs continued operations until 13 October, when DZ479 made the last, a two-hour twenty-minute sortie over the Irish Sea.

There was 'pleasurable excitement' at Oakington on 21 May, when the Flight was told that LR502, a Mk. IX, was soon to reach them. The following day to quote the unit's diary, 'a Mosquito in black camouflage arrived and all and sundry poured out to inspect it as it taxied in after a rather rapid landing. It proved to be LR498 but on the theory that possession is nine points of the law it was quickly whipped into the hangar, and before you could say whatever you would say under the circumstances the mechanics had all the cowlings off for a routine inspection'. At 17.00 hours a signal arrived to say the Mosquito was for 1409 Flight. This was indeed fortunate, for 1409 was so keen to possess it that they had already half repainted it!

Two more Mk. IXs arrived two days later and on 27 May LR502 made the first sortie to Helmond and Vianen. Two days later '498 went to Doren but when LR501 failed to return from a sortie to St. Quentin it was still obvious that even for the superior Mosquito these operations were not without risks.

The Flight moved to Wyton on 8 January. Three days before it had attempted to fly its first Mk. XVI operation, one shortened to fifteen minutes, for the navigator found his maps missing. Consequently it was from Wyton, on the 10th, that the first Mk. XVI operation was undertaken, by ML928. The last operation by a Mk. IX MM229 was a long-range *PAMPA* to Kassel, Bonn and Gladbach.

For nearly three years Mosquitoes flew over enemy territory ahead of each major Bomber Command attack and, to the closing stages of the war, by the Americans. Often the Mosquitoes set out in weather so bad that flying seemed impossible. Sometimes the aircraft broke cloud a few hundred feet above enemy country in order to report accurately cloud base conditions. It was common practice for them to fly for long periods in the icing layer to bring back reliable information of its extent. For these purposes the Mosquito was the ideal aircraft.

The sorties preceding the great attacks on Germany needed to cover the region windward of the target area, and often the target itself. Since the Mosquito was probably being plotted throughout its course it was necessary to mislead the enemy by careful flight planning, yet ensure that the aircraft reached home as long before the bomber force set out as possible. Special training in meteorological theory was necessary to enable crews to bring back really useful and comprehensive material. As several cameras were carried on the aircraft photographic training, particularly of clouds, needed to be given. Often the aircraft's

Operations: Meteorological Reconnaissance

cameras were turned on to objects other than clouds, and on one occasion photographs taken of Peenemünde were rapidly followed by a raid by American bombers.

The crews compiled reports as they flew home. In cases of urgency findings were radioed home in code, otherwise the material was presented in concise form on landing and passed by a special telephone link to Command and Group Meteorological officers.

On night flights flares were dropped to assist observation of cloud heights and formations. Night flying became important when the Americans embarked on their day bombing policy in 1943, and it was necessary to evolve special methods of observing weather in darkness. The Mosquitoes needed to go out well before dawn to be back in time to provide the vital material for the Fortress and Liberator Groups.

Crews invariably had completed one tour of duty before they were accepted in 1409 Flight. The atmosphere at Oakington was quite unlike that of a bomber station, for the crews had little chance to be away from the scene of operations, being ever on call by day or night. Because the work was all the time concerned with future operations security was of special importance throughout the affairs of the Flight, as for example in the days immediately preceding D-Day when Mosquitoes were constantly over the Atlantic keeping track of the movement of the weather systems in readiness. It was a Meteorological Flight Mosquito which helped to seal the fate of the *Tirpitz*, and when King George VI flew to Italy a Mosquito went ahead to keep a watch on the weather. Mosquitoes always preceded the overseas flights of Sir Winston Churchill.

The Meteorological Flight shared with the Photographic Reconnaissance Unit the lessons of evading flak and interception, contrails and high patrolling fighters.

A check made late 1944 showed that the average number of operational sorties made by each member of the Meteorological Flight was eighty-seven, and that they had among them as many awards as there were men in the Flight. With about twenty-five trips per month per crew this was hardly surprising.

1409's operations did not cease with the end of the war, by which time it had flown 1,359 sorties, for the Unit found useful employment for many months as a weather reconnaissance formation and eventually settled at Lyneham, operating under Transport Command.

Chapter 15

Operations: Middle East

An initial intention was that the Mosquito should be suitable for overseas service, including operation in the tropics and sub-tropics. Problems of changing over to a formaldehyde glue (from casein glue), to resist tropical and humid conditions, made early Ministry decisions on overseas operations very difficult. The first Mosquitoes to fly overseas went to the Middle East, from British bases spending a few hours there before returning. Several such flights were made by W4055 from Benson in November, 1941, and about 4 November Sqn. Ldr. Clerke reached Malta in $4\tfrac{3}{4}$ hours, turning on his cameras as he crossed Italy. An important aspect of these long-range flights at the time was the opportunity to check fuel consumption.

At Hatfield the first two Mosquitoes produced for overseas service were meanwhile being prepared, long-range P.R.1s, tropicalized and with lowered radiator flaps. W4062, the first, reached Benson on 19 October, 1942, and was flown to Malta in the hands of the P.R.U. early in January. There disaster befell, for it was badly damaged in soft ground, injuring its observer. W4063 faired little better for, after flying to Malta via Gibraltar, it was seriously damaged by bombing. Adrian Maule, de Havilland representative then in Egypt, left for Malta and supervised repair of W4063, but it was then destroyed in the severe bombing of Malta around the end of March. Overflights from Benson, and outbase refuelling were serving many of the purposes for which the Mosquito Mediterranean detachment was intended. Thus, whilst there were often Mosquitoes in the Middle East during 1942, they were nearly all home-based machines.

After five months of successful intruder flights over Europe 23 Squadron was unexpectedly withdrawn from operations on the afternoon of 6 December, 1942, and next day began collecting new aircraft from St. Athan and Shawbury. Hereward de Havilland was summoned to Fighter Command and told that 23 Squadron would shortly be leaving for Malta, and that eighteen aircraft needed to be fitted with long-range tanks before they went. A de Havilland working party was busy at Bradwell next day, and the installations were completed by 13 December. Wg. Cdr. Wykeham Barnes (now Air Marshal Sir Peter Wykeham) in the days of his youth recorded these events at the time:—

Operations: Middle East

'We knew Winter was coming when Flt. Lt. Mattingley appeared one morning with a great tin of de-icing paste and announced that this was to spread over our aeroplanes, and next morning he was seen covering the leading edges of the main and tail planes with a messy mixture which looked like soft soap. A few days later the C.O. left for Northolt and vanished into the wilds of Bentley Priory, Headquarters, Fighter Command; he returned at the end of the day with a smug and secretative air and disappeared into his office. A few days later the news came out, 23 Squadron was going Overseas. A fine crop of healthy rumours posted us to Iceland, North Africa, Russia and Burma. To ensure the utmost secrecy Digger Aitken placed the whole of his security staff on the job, and an unlucky Orderly Room Corporal who opened a 'Most Secret' signal inadvertently read the squadron's destination and was put into solitary confinement until after we had left.

'The station's resources were mobilized for the gigantic task of making ready eighteen Mosquitoes for the big trip. All needed overload petrol tanks, tropical equipment, Marstrand tailwheels; and aircraft had to be recamouflaged. When the work had been completed each of the aircraft received a fuel consumption test consisting of a cross-country embracing Land's End and Edinburgh. On the day that six set out the weather unexpectedly closed in and visibility was much reduced, giving the management some anxious moments as six beautiful new aircraft full of picked crews groped their way successfully homewards.

'Exactly ten days from the issue of orders all was ready. The Ground Party had made an inauspicious getaway, where they were led aboard the ancient aircraft carrier *Argus*. On the morning of 20 December, when the red sails of the barges in the Blackwater were reflecting the sun across the river, the executive signal came through to leave at once for Portreath, and engines were started at 10 o'clock. Around the perimeter track they came, eighteen Mosquitoes, their propellers twirling in the winter sunshine. Digger Aitken stood on the end of the runway to see the boys off, and with a last call to Flying Control that had brought us home so many times from France and Holland, YP-A taxied on to the runway and rose into the air, tucking in her legs like a modest spinster. The squadron formed up over the Estuary and flew past the control tower setting noses to the West. Portreath, Cornwall, was reached by lunchtime. The despatching staff then briefed us, blandly announcing that we must be airborne by 4 a.m., would receive our final briefing at 3 a.m., would get breakfast at 2 a.m., and must therefore get up at 1 a.m. This whole procedure went through on the first evening, with the exception of the take-off. However, next night we were more fortunate.

'Jack Starr and I were awakened at 1 a.m. We looked out on to a brilliant moon and a clear sky. At 4 a.m. DZ230 "A Apple" taxied out to the take-off point, followed by the rest of the squadron, and rose into the starlit sky, then headed out towards the Scilly Isles, lying black

Operations: Middle East

in the path of the moon. Two hours later the sun rose on the left out of the mountains of Spain showing the Bay of Biscay lying calm and milky 15,000 feet below. As the early rays glinted off the airscrews, Geoff Palmer, my navigator, tuned into the 9 o'clock news. The tension of the night now relaxed, the pilot and navigator toasted each other in hot coffee and barley sugar, and watched with interest the appearance of Spain. The coastline passed into Portugal, when it was noticed that only the port wing tank was drawing petrol and speculation arose as to whether the starboard tank was going to draw at all. After a short cut Gibraltar was sighted four hours ten minutes after take-off.

'One by one the squadron came in on the tricky race-course runway, and the crews assembled for a late breakfast in the nissen-hut mess in the old grandstand. Tiger Tym and Babe Barkell, their eyes as big as saucers, sat one each side of a great plate of fruit, tentatively feeling the bananas to see whether they were real. A touching reunion took place with the sea party that had arrived two days previously. In *Argus* they had ridden out one of the worst storms ever known in the Bay, and the Intelligence Officer in particular had reached a high state of operational fatigue, culminating in his permanent retirement for the rest of the voyage. On 27 December orders were received to move into Malta six aircraft at a time, and to go at once. This was interpreted by the squadron as meaning "start yesterday if possible". The first party made one circuit of the Rock and without waiting even to form up set course for Algiers, refuelled, and left at dusk for Malta. Fifty miles away the four searchlights that marked the island were seen standing up into the night like one pillar. One circuit and A-Apple touched down on Luqa runway, the first Mosquito fighter in Malta.

'Having heard all the stories of the fearsome conditions at Malta we taxied A-Apple quickly off the runway and opened the door of the cabin for the Station Commander, who appeared out of the night in his car. "Do you want us to disperse very widely?" we whispered. "No, put them there and we'll disperse them in the morning." A few hours later we were taking stock of our new situation, and after calling loudly for a Squadron H.Q., and Intelligence Room, Briefing Room and pilots' dispersal and locker room we were given a stone-walled cell about twelve feet square and informed that it was to fill all of these functions. The roof, of fine old-world corrugated iron, was precariously attached by wire and string. In this palace of luxury a table and two chairs were set and, as a crowning glory, a small safe was moved in. Next we taxied the aircraft to their dispersals down a country lane, round a small clump of trees, through a flock of goats, past a farmyard, over the brow of a hill, and across the bed of a stream; finally we were beckoned into a dispersal in a hillside cranny, from which could be seen no sign of an aerodrome. Back in the Engineering Headquarters Flt. Lt. Mattingley was expanding upon the beauties of his office when his eulogies were cut short smartly by a shower of rain that reduced his text books and records

Operations: Middle East

to a sodden pulp. Since the next night was to be the first for operations I soon after left for Air Headquarters in an antique Morris van.

'Penetration to Charles Riley, SASO, was made via a deep dungeon at Valetta. Dealing with our business at lightning speed he simultaneously, or so it seemed, told two gross stories which no one had heard before, disposed of a great deal of business, interviewed two squadron commanders who dropped in, and at the same time dispensed cups of Bovril with his free hand to anyone who came in. Finally the AOC admitted us to his office, summed up the situation in a few words and stressed Malta's role in disrupting Axis shipping and air transport. The dominating position of his Command, and the high state of efficiency to which the island had been brought, were well explained and the squadron's operational role was provisionally planned out.

'There seemed to be disappointment among the Malta staff that the Mosquito could not drop torpedoes, lay mines or carry paratroops. The impression conveyed was that the type of aeroplane required at Malta should be able to do all these things with a little day fighting and night bombing thrown in, whilst the initial enthusiasm of the SASO for immediate attacks on Rumania, Athens and Budapest had all to be modified when he learnt the Mosquito's range was less than 3,000 miles.

'The sky was clear, the moon in its last quarter, when early next morning YP-A took off with a great whirr and turned her nose towards Sicily. This first patrol, although uneventful, was very instructive. The initial objective, which was Catania, was reached just as all aerodrome lights were switched on including a Visual Lorenz, four miles long. A searchlight waved in a friendly fashion and then went out, and a green Verey light was fired from the lee end of the flarepath. This cordial invitation to land had to be regretfully refused, but the thought that it might be for someone else kept the Mosquito patrolling until the grey light of dawn crept up and Apple had to find her way back to Malta through a virulent thunderstorm. Similar experiences came to the other crew, but no guns were fired.

'In a matter of days all the crews arrived, and we had our plans before us. We were to harass and as far as possible neutralize enemy air bases in Sicily by night. There were torpedo bombers operating from Malta, and the protection of these against enemy night-fighters could best be achieved by attacking night-fighter aerodromes. The Army in Africa was meanwhile gathering itself for a culminating offensive against Tripoli, and the long open roads of enemy communications offered fine harassing targets for ground-strafing Mosquitoes by moonlight. For the first few weeks we contented ourselves by visiting each of the enemy airfields in turn, going out at two-hourly intervals and flying round Sicily clockwise or anti-clockwise.

'Before the operations against Sicily could be expanded to full strength the big attack on Tripoli began and attention had to be diverted almost entirely to the mainland. To reach the African coast meant a full hour's sea crossing, and it seemed a long way to go for little return. In the third week of January, with the armies pressing into Tripoli, the Germans began to evacuate westwards along the road to Sfax, and on that night maximum effort was called for and turned on. Each crew made two trips and found the road packed with transports, easily visible in the moonlight, and attacked as they struggled towards Gabes. A cannon-hit on a tanker and trailer carrying petrol blocked the road with flaming fuel at a point where it passed through a date-palm plantation, and about fifteen miles on another petrol tanker was set afire and overturned, blocking the whole road. Between these two blocks developed the most terrific traffic jam. Up and down went the Mosquitoes, while fires, tracer and explosions lit the sand dunes, pale beneath the African moon. Beaufighter bombers joined the tumult of destruction, and resulting damage was very heavy. As these attacks increased so did the light flak guarding the highway. It became the practice of the more wily pilots to fly about a mile away from the road, fire off a few rounds of tracer, and watch for the enemy gunners to open fire thereby revealing their positions. The crew would then attack transport where it was unprotected. A ground-level get-away sometimes made the enemy guns plaster their own trucks, while trying to get their sights on the snaky Mosquitoes.

'None of the expected aircraft spares had yet reached Malta and the unlucky plumber found himself without even plywood to make his repairs. Low-level attacks brought a considerable amount of damage and woodwork repairs were carried out with cigar box wood, old tea chests and bits of bomb doors. Aid was also obtained from the local coffin maker, whose woodworking was of the highest order. His presence spread a slight gloom over the pilots, one of whom was heard to remark "to watch that blighter working on my aeroplane makes me expect to find it fitted with brass handles when I next go to fly". Wal Williams, our Rhodesian, stopped a 20 mm. shell with his tailplane. Passing between the front and rear spars it made a hole big enough to put your head and shoulders through, but apparently had little effect on the aeroplane's flying characteristics.

'Early in February there was still no sign of spares, yet on one night serviceability was 100 per cent, with every aircraft making a sortie. With the African theatre comparatively quiet attention was again turned to Sicily. The score was opened by Jack Starr who, at Castel Vetrano, caught up with a Ju 52 coming in to land, and gave it a squirt. The Hun took evasive action by blowing up into a thousand pieces and E-Evelyn (DD687) flew through the bits with the crew temporarily blinded by the explosion. This caused "A" Flight to wear a slightly superior look, but Rusty evened the score two nights later when he found a Ju 88 and set it on fire after a short running fight, and watched it crash into the sea.

'As the moon came up and the nightly Sicilian round began to pay fewer dividends the squadron turned to the old stand-by, train busting. The Mosquitoes now would fly to the foot of Italy, and would creep

Operations: Middle East

into the naval base of Taranto, lying screened by its balloon barrage. A quick look at the harbour, then a turn south would take the aircraft to where the railway line clung close to the beach. A little added employment on these trips was the production by Ted Lewis of rubber-banded bundles of letters written by Italian POWs and addressed to a specific locality. These we were asked to deliver ("what, personally?" asked Babe) and it was usually easy, for the localities were often sleepy little towns. This was service indeed, and we liked to think that the letters had only to be collected and delivered next morning.

'But all this time a steady mounting toll of damage too was coming into the squadron's bag. John Striebel blew up an anti-submarine vessel by setting off depth charges on its after deck, and explorations were being made further up the Italian coast railways in search of steam engines. Permission was granted for attacks on electric trains, and during one a pilot managed to cause a short circuit that melted the rails beneath the train. Never underequipped imaginationwise Ted Lewis visualized this engine welded to the rails and having to be cut adrift with a hack-saw.

'It was never easy to ground strafe at night, and as soon as the squadron arrived in Malta it was decided that the standard night-fighter reflector sight was not going to be satisfactory. An idea stolen from 151 Squadron was tried out, and a G.M.2 day reflector sight minus its glass screen was mounted, throwing its reflection directly on to the windscreen of the Mosquito. Brackets and fittings were made in the Naval Dockyard, and with a little careful adjustment the sight could be arranged so that the whole windscreen was completely clear, except for the red image of the sight which floated on the glass in front of you. This was doubly valuable, for the pilot could sight and judge his distance from the ground simultaneously. Practised teams adopted the system of the pilot concentrating on aiming, whilst his observer read the altimeter readings over the intercomm.

'The battle for Africa was now reaching its climax and the stranglehold which the Mosquitoes had upon the enemy's bases at night was strengthened. Enemy night movements on land and in the air had to be reduced to a minimum. Tripoli passed to AHQ the news of any raid together with its outgoing direction. Mosquitoes were then sent to the bases from which the enemy had set out and covered these, keeping them closed until the German bombers ran out of fuel and had to land. If the intruder did not get them on the circuit the danger still occurred of the bomber being shot up on its own runway. This tactic was a favourite with Wal Williams, who provoked pandemonium at Comisso and Catania, being prepared even to follow the returning bomber to its dispersal point, or perhaps into its hangar. Harrying and driving them from place to place, giving no rest, the Intruders ranged far and wide over Sicilian and Italian airfields. No sooner would an aerodrome open for night flying when, winging out from Malta, travelling at great speed

Operations: Middle East

low on the sea, a Mosquito would arrive. Every night-flying German who made the smallest mistake would be rewarded with a burst of cannon fire.

'The squadron's casualties were not light when compared with the sortie totals, but were negligible compared to the damage done. None the less one by one the crews fell away. Flt. Lt. Paterson and his navigator were shot down and taken prisoners, Flt. Sgt. Clunes and his navigator failed to return, young Cave-Brown was shot down over Trapani, W. Off. Woodman, navigator to Flt. Lt. Hodgkinson was killed. In addition many Mosquitoes came home heavily damaged, and Johnny Striebel came to be known as the "one-motor-king", having three times brought his machine home on one engine. Jack Starr's E-Evelyn, savaged by flak and hit in the radiator and wings, came home successfully and flew again in a very short time, while another machine had a 20 mm. shell clean through the fuselage, apparently with no effect on performance. Small flak holes from rifles wielded by Italian Home Guards, and other small-arms fire, became a matter of no importance, and were repaired as a matter of course during the daily inspection. By May the squadron had destroyed 15 enemy aircraft and had 3 probables and 11 damaged to its credit. 200 trains and an enormous collection of miscellaneous road vehicles were also claimed.'

Late in 1942 60 Squadron, S.A.A.F., operating Baltimores and Marylands in the Western Desert on photographic reconnaissance, was largely employed providing photographs from which the 8th Army Survey Directorate produced attack maps for General Montgomery's drive to sweep enemy forces from North Africa. This was short-range work, for Rommel would retreat a little way then make a stand. Before attacking, Montgomery would demand full photo coverage. Unfortunately German fighter airfields were within a few miles of targets such as Zem-Zem and the Mareth Line at El Agheila. Day after day 60's aircraft were being seriously damaged, or shot down. General Montgomery was frustrated at the delays, and sent for O. G. Davies to say that if his squadron was unable to provide photographs he would have to ask the Americans to undertake the task. Davies explained the difficulties, then General Montgomery asked what aircraft were required to ensure success. 'Mosquitoes,' Davies replied.

Two days later news was received that the R.A.F. had refused to release any Mosquitoes, and following further enquiry from Montgomery, Davies suggested that a Flying Fortress be obtained and, relying on its offensive fire power, photographs might be taken before enemy fighters destroyed it. Montgomery sent a personal request to General Eisenhower, then in Algiers, for the loan of a few B-17s, but this was somewhat curtly refused.

So 60 Squadron battled on with Baltimores with little success and finally, after a somewhat explosive interview at Montgomery's H.Q., the General dictated a signal to Mr. Winston Churchill saying that

Operations: Middle East

unless 60 Squadron S.A.A.F. was equipped with Mosquitoes forthwith, he would be unable to attack the Mareth Line. Within twenty-four hours news came that two Mk. II Mosquitoes would be arriving on the squadron!

The original intention had been that these Mosquitoes would be picketed on a desert airfield to discover how they would withstand sand and climatic conditions, but now they were to be diverted to 60 Squadron. As soon as they arrived at Kasfareet in the Nile Delta area, Davies and Lt. Oliver Martin flew there to collect them. Orders were that the Mosquitoes were not to be touched until full cockpit briefing had been given, but the R.A.F. pilots who had delivered them were away in Cairo. The C.O. of the station was then persuaded of the urgency of the need for the aircraft and relented, allowing Davies to fly the machine. He recalls the experience in these words:— 'My first flight in a Mosquito is still engraved in my memory. Engine covers were removed and I spent some time studying the cockpit layout, controls etc., but no one remembered the pitot head cover. It was not until I was airborne that Lt. Martin enquired whether it was not unusual for a Mosquito to climb at 60 knots. Fortunately the runway at Kasfareet was a good 1,700 yards, for I used up most of it in landing!'

The M.U. did a rush job, fitting a camera beneath the seat of each aircraft and, after one flight by Lt. Martin, both were flown to Castel Benito on 4 February, 1943. Here Davies had difficulty with his tailwheel refusing to extend. 'My attempts to get it down were not made easier', he afterwards recalled, 'by constant messages on the R/T from the Station Commander, that I was delaying the parade of Squadron C.O.s waiting to be introduced to the Prime Minister, Mr. Winston Churchill. Eventually the application of much "G" got the tailwheel down and locked, and I landed safely and hurried to the honour of a meeting with the great man himself.'

Thereafter, without experience of Merlin engines, ground crews miraculously kept the aircraft serviceable for carrying out up to four sorties a day, week after week. On his first operational flight Davies climbed to about 14,000 feet for the first time when the high blowers cut out. He thought he had been hit by flak. This indicates how little was known of the aircraft, yet the Mosquitoes gave hours of trouble-free flying without untoward incidents.

Meanwhile 683 P.R. Squadron had formed at Luqa in February, 1943, and on 13 May began operating Mk. IIs over Sicily and Southern Italy, the first three sorties being made by 1st Lt. Marec. PR. 4 DZ553 was briefly used and twenty sorties were flown by Mosquitoes before the squadron relinquished its aircraft.

60 Squadron's first operation was flown on 15 February, when DD744 made a coastal reconnaissance flight. With two Mk. IIs 60 Squadron managed fifteen sorties by the end of the month. They were used until November, 1943, and supplemented in June by a PR. IV DZ553 which

began operating on 11 July, when it reconnoitred Venice, Pola and Trieste from Ariana. On its seventh sortie it was shot down. A Mk. VI also joined the squadron, but it was acquisition of the PR. IX in August that boosted the importance of 60 Squadron, S.A.A.F., for this was the latest version of the Mosquito in the area. Operating began on 28 August, with a flight to Milan. September saw a further extension of operations, as the Mosquitoes reached Spezia and Turin. Next month Bolzano and Trento in the Dolomites, Budapest and, on 25 October, Klagenfurt in Austria, were all photographed. But 60's greatest honour was its selection for the photography prior to the Sicily invasion. Following the assault General Montgomery signalled to the Squadron:— 'The success of the Campaign was largely due to the magnificent air photography with which I was provided.'

Sorties by 23 Squadron, hectic as ever, totalled 233 in May, 175 of them intruder patrols. Six enemy aircraft were destroyed, sixty-five locomotives attacked and twelve seriously damaged. Near Marettime an escort vessel was blown up and a 1,500-ton ship left blazing by the stern. Defences were stronger, and seven Mosquitoes were lost. A new phase in the activities of 23 Squadron began in July, for on 1st of that month the Mk. VI fighter-bomber, which began to arrive in Malta in May, 1943,[1] reached the unit. Flg. Off. Menkes in HJ716 attacked three airfields near Rome on 17 July, becoming the first 23 Squadron pilot to use this version.

Long-range tanks were fitted to these aircraft, but two nights later wing bombs were initially carried by the seven aircraft detailed to attack road targets in the South of Italy. Such operations increased following the invasion of Sicily, when 23 Squadron intruded over Sicilian airfields. Unexpectedly, Malta was raided on 19–20 July and the squadron was ordered immediately to remove wing bomb racks, in case another attack came and the Mosquitoes were needed for night fighting.

In the July moonlight period 126 night sorties were flown by 23 Squadron. Drop tanks allowed missions north of Rome and Foggia, record distances for intruder fighters. Sqn. Ldr. Rabone damaged three Ca 506 floatplanes on Lake Bracciano on 20 July, but destruction of enemy aircraft was now less common. Flg. Sgt. P. Rudd, however, destroyed an Me 210 when returning from the Taranto area on 26 July.

On 13–14 August Sgt. Dawson successfully force-landed his aircraft at Palermo by the lamplight from two jeeps, after it had been hit over Foggia. Next day Sqn. Ldr. Rabone borrowed a Spitfire in which to fly spares for the grounded Mosquito. He turned his mission to good account, by shooting down a Ju 88 off Trapani on his return. Delivery of P.O.W. mail continued in the Rome and Naples areas. The last Mk. II sortie came on 17 August using DZ238:Z, the first major inspection done on a Mosquito in Malta was made on 24 August—previously

[1] The first three Mk. VIs to arrive were HJ672 on 17.5.43, HJ653, and thirdly HJ688 despatched there on 31.5.43.

Operations: Middle East

these were done in Britain—and the 1,000th sortie was flown to Taranto by Flg. Off. C. J. Barber and Flg. Off. J. A. le Rossignil in HJ737:R on 30 August. Congratulations from the A.O.C. followed, for in eight months 23 Squadron had claimed twenty-four enemy aircraft, hit 172 locomotives and attacked twenty-five ships.

Rain and mud now beset the squadron, but insufficiently to prevent nine crews from operating on 8 September, the day of Italy's collapse, when a Ju 88 and an He 111 were destroyed near Rome and another '111 was damaged. Night fighting, however, was primarily the task of 256 Squadron.

Ford in Sussex received 256 Squadron in April, 1943. Using Mk. XIIs the squadron flew defensive patrols off the South Coast. To give support for the Sicilian invasion the squadron was notified on 1 July that six crews were to prepare for detachment to Luqa. Overnight long-range tanks were fitted and six Mosquitoes left Ford next evening, delayed only by the unfortunate release of a parachute in an aircraft. Their route lay via Gibraltar and North Africa, where they refuelled taking lunch at Rasel Ma. Five safely landed at Luqa and one crashed on arrival at Krendi.

256 was attached to 108 Squadron and, immediately following the invasion, trade was hectic. In one amazing week sixteen enemy aircraft were claimed including Ju 88s, Cant 1007s and He 111s. Ten fell to Sqn. Ldr. J. W. Allan. Following a raid on Malta by Kampfgeschwader 1 the squadron stood by, but it never repeated its success. HK133 flown by Flt. Sgt. Jenkins on 25 July failed to return after one German bomber was destroyed and another damaged. Five days later a Cant Z1007 floatplane was set ablaze.

Then the remainder of 256 Squadron prepared to move from England to Malta, where ground crews went ashore on 13 October. First call on shipping was currently given to the Army, and the squadron's stores had only reached Algiers by this time, whereas the Mosquito XIIs had landed at Luqa on 29 September. A month later re-equipment with Mk. XIIIs began. The first, HK339 and '400, touched down on 20 November in the hands of two new crews. 256 Squadron now had a major commitment, the defence of Malta by day and night with four and three crews respectively standing by. Major action first came on 21 December when six Ju 88s attacked a convoy and lost one of their number to two intercepting Mosquitoes.

Two or three Mosquito IV reconnaissance aircraft were operating from Malta in May. Another was in American hands at Algiers, and on its nose was inscribed 'PILOT—COLONEL ROOSEVELT'. The President's son thought very highly of the aeroplane, but it was bedogged by engine trouble. Another Mk. IV in American hands in Algeria was lost on a flight to Britain for major overhaul.

Changes in reconnaissance units occurred during the summer, 1943. The first three Mosquito PR. IXs were detached from 540 Squadron to

Operations: Middle East

the Theatre in June, under the command of Wg. Cdr. Douglas-Hamilton. The intention was to see how two-stage Merlin engines reacted to the climate. MM292, a PR. XVI, was the first of its mark to be allocated to the Middle East, and reached Gibraltar on 29 January, 1944. Another arrival, this time from Italy, was a PR. IX LR444 which Wg. Cdr. Hughes, D.S.O., D.F.C., brought to LG219 on 16 February. Here it was received with great excitement by 680 P.R. Squadron which had been relying mainly upon Spitfires. Next day their first PR. XVI MM297 arrived from Britain, and by the end of April, 1944, the squadron had but one PR. IX and nine PR. XVIs, with which came a plentiful supply of window icing troubles and, unfortunately, no spares.

Mosquito reconnaissance operations commenced in May and twenty-five sorties had been flown by the end of the month. In its diary the squadron recorded that 'the speed and manoeuvrability of this twin continue to amaze the most experienced pilots'. Under Wg. Cdr. J. R. Whelan operations were flown from Tocra, where the 1st Detachment 680 Squadron was based. Greater range, accurate sighting of targets and useful radio equipment were aspects of the Mosquito appreciated by crews. Flt. Lt. A. M. Yelland, in MM333 was, on 7 May, first to fly on operations, over the ports and airfields of Crete and the Cyclades. By mid-May sorties had covered Navarino, Pilos, Koroni, Araxos, Patrus, Salonika and Larissa which previously were largely unphotographed. A remarkable flight was made by Flt. Sgt. W. S. G. Chandler on 26 June, when in MM297 he photographed the coast around Gavdos at only 1,300 feet. Concentration was at this time on the ports and airfields of Greece.

As 680 Squadron received Mosquitoes 108 Squadron at Luqa likewise re-equipped. Again the story was of great delight. The first machine arrived on 16 February; three only were on strength, placed in 'B' Flight where ex-256 crews flew them. Soon there was great disappointment. It became known that the squadron's Beaufighters would continue to operate. Many crew changes followed, and an unfortunate series of Beaufighter crashes. Not until 5 April was the first Mosquito operation flown. Night convoy patrols were then begun, ending 24 July. Mosquitoes made intruder flights in June, but at no time did they entirely replace Beaufighters, and the squadron surrendered them on 28 July. Although unaware of it, disbandment had officially taken place, but to the squadron 'the unkindest cut of all was losing our Mosquitoes'.

Most active Mosquito squadron in the Middle East, 23 Squadron, partly stationed at Signella, was forced to move to Gerbini when rain washed away its runway. Previously Gerbini had served as an advance base. On 27 October 23 Squadron proceeded north to Pomigliano by Mount Vesuvius. Although their H.Q. remained in Malta 23 Squadron's Mosquitoes operated night intruder flights now from advanced bases, because targets lay too far from Luqa.

Operations from Pomigliano began 1 November, Wg. Cdr. Burton-

Operations: Middle East

Gyles intruding on the Viterbo area. Fifteen Fw 190s bore down upon Pomigliano on 30 November, but 23 Squadron survived the onslaught without loss. Typical target areas were the rail connections between Rome and Florence, Ancona and Bologna, routes leading to the port of Leghorn, and railways around Venice, Padua and Vicenza. To permit deeper penetration and longer time in target areas a detachment moved to Alghero, Sardinia, on 7 December, 1943. Trains and road transport in the Genoa-Turin-Milan regions now felt the Mosquito's sting. So far advanced was the new base that Flt. Sgt. T. Griffiths was able to strafe two trains near Bordeaux on 7-8 January, 1944, in HX804:P Peter, and the coastline from Toulon to Rome could be covered. Night attacks around Lyons were now made. Wing tanks were replaced by racks for 250 lb. bombs on 17 January, 1944. Next night, operations were concentrated on roads and railways radiating from Florence, Rome and Trasimeno, using machine-gun and cannon fire supported by more bombs. At the end of January 23 Squadron had destroyed thirty-three enemy aircraft in flight and thirty-nine on the ground, also 331 locomotives. All squadrons were badly off for spares and equipment, and their airfields were in a shocking state, with cracked up runways amid seas of mud.

23 Squadron was always high spirited and saw to it that the finer side of life was catered for in spite of mud, heat, rain and the enemy. 'There was an unexpected appearance of two charming FANY's escorted by a member of the Senior Service in the mess at Luqa', records the squadron diary. 'A brilliant interception on the part of the C.O. rescued these attractive members of H.M. Forces from the clutches of the Navy. The C.O.'s entry for the squadron's line book read "I only brought them down for you chaps." There were no operations that night.' This was indeed exceptional, for the squadron pestered the enemy almost nightly, and from 9 April by day, for this date saw the commencement of *Day Ranger* flights. The first two such sorties were flown by Sqn. Ldr. P. Russell and Flg. Off. D. Badley who damaged two Do 217s and destroyed one at Perpignan. Two days later two Mosquitoes ranged in the Marseilles-Beziers region whilst another two, one of which was shot down, attacked six trains in Southern France. For the most part intruders concentrated on roads and railways north of Rome. On 1 May, Sqn. Ldr. P. Russell damaged a Ju 188. It was the last enemy machine to be attacked by 23 Squadron, for on 2 May four sorties to the Rome-Trasimeno area were the last the squadron flew in the Middle East. It then withdrew to Britain.

Plans to extend the radio war in the Middle East during 1944 called for a handful of Mosquitoes in 'A' Flight of 162 Squadron, based at LG 91, where Mk. VI HJ671 arrived in January. Its first operation, flown on 13 January, was an investigation in daylight of the use of defensive radar by the enemy. Unserviceability, and the fitting of more equipment, then grounded the Mosquito. Not until 18 May was the

Operations: Middle East

second operation flown, by which time the machine was shorn of guns. During June it was used for GCI exercises and calibration of radar. Then, unexpectedly, it was decided to attempt interception of high-flying night intruders with '671, for which it flew to 168 M.U. for the fitting of A.I. radar. Before its return 162 Squadron was disbanded.

By the start of 1944 60 Squadron was covering Southern Europe, leaving the remainder to British-based P.R. squadrons. 680 concentrated on Greece and the Balkans. Deep penetration flights by 60 Squadron from San Severo became commonplace in the early months of 1944, the range of the Mosquito allowing flights even into Poland. During these the V-2 testing ground at Blizna was watched, and on 31 May, 60 Squadron flew its 464th sortie. Of this number half had been with Mosquitoes.

In June and July, 1944, Stuttgart, Augsburg to view Messerschmitt production, and Leipzig were principal targets. It was a flight to Gunzburg and Leipheim that resolved itself as one of the most memorable Mosquito sorties of the war. Capt. Pienaar of 60 Squadron took-off from San Severo on 15 August, 1944. In the target area at 30,000 feet he was intercepted by an Me 262 jet fighter from the Messerschmitt factory. He jettisoned his drop tanks, turned to starboard and put the Mosquito into a spin. As the enemy made a first firing pass Pienaar found his throttle jammed open. He reached 420 I.A.S. before he could pull out at 19,000 feet. The fighter had shot away his port flaps, port aileron, and elevator and damaged the main spar. Despite the damage Pienaar endured a second attack and turned to port as the enemy shot past. In all the jet attacked twelve times with varying success, and for no less than forty minutes. At last the out-manoeuvred Messerschmitt gave up the battle ninety miles south of Leipheim, its starting point. When it landed in Italy, the Mosquito, NS520, was seen to have lost the whole of its port flaps and the rear of the nacelle. The fuselage tail cone had disappeared with much of the port tailplane. Both Captain Pienaar and his navigator Lt. Archie Lockhart-Ross were immediately awarded the D.F.C.

Against the new jet fighters the Mosquito could rely only upon its manoeuvrability, and eleven days after Pienaar's interception another Mosquito, NS521, was lost on a flight to Munich. It seemed likely that a jet had engaged it.

Flying at 30,000 feet over Bavaria on 13 September a single-engined fighter intercepted a Mosquito of 60 Squadron, flying at 290 m.p.h. For once a Mosquito found itself quite unable to escape, although fortunately the German fighter gave up the chase. Rather than risk further interceptions it was decided to give close escort to P.R. Mosquitoes, using P-51 Mustangs of the 31st Fighter Group, U.S.A.A.F. Already this had been experimentally tried on 10 September, when sixteen P-51s escorted three Mosquitoes photographing the Bunzlau-Prague area. Just how fast the Mosquito still was when compared even with the best

Operations: Middle East

Allied fighters was at once apparent, for the American fighters—especially P-38 Lightnings—found it difficult to keep up with the Mosquitoes, and operations required carefully planned rendezvous. Throughout October American long-range fighters escorted the P.R. Mosquitoes into Europe and only one was lost, over Central Czechoslovakia.

As the bad weather in Italy took charge of the operational situation the number of P.R. sorties fell. Tricky runways made the swing on take-off increasingly difficult to master. Icing conditions, vapour trails, poor light and cloud-covered targets all combined to reduce operations, but over four days at Christmas twenty-eight sorties were flown. Eight times in December there were escorts of four or six P-38s or P-51s. On Boxing Day an Me 262 appeared and was promptly set upon by five P-51s. To the South Africans a new name now appealed, 'No. 60 S.A.A.F. P.R. and Decoy Squadron'.

680 Squadron had fared similarly in the closing months of 1944, when it extended its sphere of operations to Central and Southern Europe, in addition to covering the Balkans. Between 20 and 23 August part of the squadron moved into San Severo, and sorties followed to Sarajevo and the Budapest area. On 4 October Lt. W. B. Tilley broke new ground by flying to Austria and Bavaria. Both regions were cloud-clad. Nine days later MM287 was lost over Austria. Escorts were then provided for 680 Squadron, but this alone could not prevent all losses. On 20 November Flt. Lt. K. R. Booth, escorted by four P-51s to Prague, encountered heavy flak and had to make a hazardous flapless landing.

At the end of December, 1944, the squadron's equipment included ten Mk. XVIs and one Mk. IX, and it consisted of three Flights. 'A' Flight covered targets in Czechoslovakia, Breslau, Bratislava, Vienna, the oil refinery at Brux, Munich and Stuttgart. 'B' Flight watched over Crete, Athens, Rhodes and the Dodecanese, and the Spitfires equipping 'C' Flight made shorter-range flights.

Throughout 1944 256 Squadron busily used Mosquitoes. Although Luqa was the home base many operations started from Catania, mainly convoy patrols. A long chase across the sea brought the first success of 1944, a Ju 88 off Augusta. For a fortnight in February four crews joined 23 Squadron at Alghero for moonlight intruder missions. Convoys to Italy attracted most of the squadron's patrols, but in March a detachment moved forward to Pomigliano, from where defensive sorties were flown over Naples and the Anzio beachhead. On 1 April the squadron, now assembled at Malta, sailed in H.M.S. *Royal Scotsman* for Algiers. There were valuable convoys needing night protection as they entered the Mediterranean and in May two Ju 88s were shot down and another seriously damaged. Mosquito XIIs and XIIIs still flew with the squadron, which returned to Alghero in August and soon after installed itself at Foggia there to begin patrols, scrambles and intruder flights.

256's Mk. XIIIs now ranged far over the Balkans, and on 4 October two Ju 52s and He 111 were shot down near Salonika. Mosquitoes were

Operations: Middle East

a familiar sight over a wide area now. HK435 which intruded on the Padua–Verona region on 4 October two days later was searching for the enemy near Athens. *Day* and *Night Rangers* became commonplace, and flights over Turkey were included. Many of the November sorties were to Yugoslavia. A weather reconnaissance flight by MM580, a night fighter, to Sarajevo and Brad on 26 December opened a new phase, and illustrated yet again the versatility of the Mosquito where roles often for which it was not precisely intended were demanded of it. Mud and rain then interrupted 256's operations, but it managed to destroy a Ju 188 on 3 January.

During October, 1944, the first NF. XXXs were delivered to the Middle East. A month later Mk. XIXs were despatched for two new night-fighter squadrons. For the Americans in North-West Africa flying Beaufighters more Mosquito XXXs arrived, but they saw little service with the American squadrons. At the end of December, 1944, 600 Squadron began receiving Mosquito XIXs at Foggia. Wg. Cdr. A. H. Drummond made the first operational use of one, TA448, on 22 January. First contact with the enemy came to TA426:G near Venice on 16 February. 255 Squadron converted to Mosquito XIXs at Rosignano in January, 1945, and such Beaufighters and Mosquitoes as the squadron had were all standing by for an expected very-large-scale daylight raid on Rome, which never materialized. 255's Mk. XIXs became operational on 26–27 February, when the only activity Wg. Cdr. J. Kempe found was an E-boat in the Ligurian Sea. A reconnaissance Ju 188 showered *Duppel* on a Mosquito near Florence on 1 March, but this was almost the only contact the new squadrons had this month. Chases, patrols, weather reconnaissance flights—all were made.

At this time Ju 87 dive bombers appeared in small numbers, but there were no night engagements with them. Higher night losses of Bostons in 232 Wing led to the belief that the enemy was operating night-fighters, so each night 255 Squadron despatched two Mosquitoes to watch fighter airfields and support the Boston Wing. Mosquitoes also patrolled at 2,000 feet off the mouth of the Po in an attempt to catch German reconnaissance aircraft. Hereabouts 255 Squadron successfully engaged a Ju 188 in the area, when Flt. Lt. Pertwee operated there on 22–23 March.

Sorties by reconnaissance squadrons centred on Germany's oil industry in the first weeks of 1945; next month troop movements and the Austrian border defences attracted sixty-three sorties from 60 Squadron. In the Vienna, Prague and Munich areas the frequent presence of enemy jets demanded continuing P-51 escort for the Mosquitoes. North-West Italy and the Alpine areas, then standing ahead of the advancing Allied troops, were the subject for photographs until the end of the war. A very long-distance flight of the period was made by Major P. P. Daphne, S.A.A.F., on 21 March, when he flew to the Ruhland oil refinery fifty miles north of Dresden, a 2,000-mile journey, and also spent forty minutes in the Prague area. By this time the likelihood of interception

Operations: Middle East

was less. An outstanding Mosquito IX in the Middle East was LR444, the first Mosquito of 680 Squadron. Her last flight took place on 28 February when, over Crete, engine trouble occurred. The port airscrew ran away and the crew baled out. The navigator landed safely, but the pilot's parachute roman-candled.

This was almost the end of the PR. IX in the Middle East, although Owen Davies and Peter Hingeston flew one from Cairo in an attempt to establish a fast time to Pretoria. 'Lovely Lady', as this famous Mosquito which had made so many flights with 60 Squadron was known, was bedevilled by refuelling delays *en route*, and lack of weather information. The flight was given up after eleven hours and in last light, as it came into Que-Que, Southern Rhodesia, the gallant aircraft slid to an ignominious halt. It is now displayed in the South African War Museum.

Shooting up barges on the Villa Canale leading from Cavarzere became an increasingly appealing pastime for 256 Squadron in March, 1945. On 16 March three Mosquitoes were used in an attempted army liaison operation employing the cab rank technique, but this was merely a brief flirtation with the soldiers. A Mk. IX LR461 was acquired at this time for the squadron's meteorological reconnaissance work, terminated near the end of March. Diversity became the order of the day, for from late March bomber-support operations were billed, 256 Squadron emulating the duties of the many Mosquitoes in 100 Group. The first of these sorties were flown on 30 March, and consisted of anti-flak and anti-night-fighter patrols by Mks. XII and XIII in the Polesella region. To assist with navigation all the aircraft in use were fitted with *Gee*.

Only on three occasions at night in April, 1945, did enemy aircraft cross Allied lines in Italy. 255 Squadron's first Mk. 30 began operating on 11–12 April patrolling the Leghorn area, but neither used in Italy scored any successes. Flg. Off. D. S. Denby of 600 Squadron inconclusively fought a Ju 87 on 11 April, and Sqn. Ldr. G. W. Hammond tangled with Fw 190s on 13 April, destroying one; it was the last enemy aircraft to fall to a Mosquito in the Middle East. 256 Squadron received a few Mk. VIs for ground strafing in April, but on 17th, the day when they commenced operations, both received damage in their radiators when strafing barges.

Only one NF. Mk. XIII is known ever to have been fitted with wing bombs, HK508 A—Apple of 256 Squadron. Engineered by the Squadron, the machine was thus modified on 10 April. Early next day Sqn. Ldr. G. M. Smith flew it to the Rimini area. Seeing a ground target he pressed the wrong tit, and nothing happened. He landed back at base at 06.50 after jettisoning his two bombs. At 05.20 on 12 April he set off again, bombed a staff car south of Ariano and had a near miss on a pylon near Taglio Di Po. Attacking a canal near Copparo HK508 was shot down, and the unique machine destroyed. Another noteworthy date in April was 29th when Mk. XII HK185 touched down with Plt. Off. J. D. Fox at the controls, for this, an armed reconnaissance flight to the Pordenone

Operations: Middle East

area, was the last operational sortie by this version. Next day MM531 became the last of 256 Squadron's Mk. XIIIs to operate.

After the war Mosquitoes remained in the Middle East for eight years. PR. XVIs and later Mk. 34s operated on photographic and reconnaissance duties in connection with civilian mapping and survey, also operations connected with the Palestine emergency. 39 Squadron was responsible for night defence in the Canal Zone for four years, using NF. 36s. Mk. XXs were brought into service with the Royal Navy for second-line purposes between April and June, 1945, at R.N.A.S., H.M.S. *Goldfinch*, and other Canadian built Mosquitoes, F.B. 26s, served with 249 Squadron. Some long-nosed target-tug T.T. 39s served in the Mediterranean region.

From small beginnings the Mosquito complement in the Middle East gradually expanded along specialized lines. Long range, long endurance and high speed were all used with measured purpose. Only the specialized bomber versions were neglected, but barely so for 614 Squadron had plans at the end of the war to operate B. XXVs in a pathfinder role. Approximately one hundred enemy aircraft are believed to have been shot down in the region by Mosquito fighters.

Chapter 16

Operations: Fighter-Bombers

From October, 1943, to November, 1944, almost the only Mosquito day bombing was by the fighter-bombers of 2 T.A.F. Day raids on industrial targets were followed by night intruder attacks, and an intensive campaign against the V-weapon sites. From D-Day close tactical support to the Army became the prime task of these aircraft. As the enemy retreated across France, movement by day was prohibited by Allied fighter-bombers. By night the Mosquito VI was the only aircraft that could really effectively carry on the task, for its speed, offensive load and manoeuvrability suited it ideally to the role enabling it to work with the slower flare dropping Mitchells which illuminated woods and other hide-outs. Effective destruction of transport columns was thus achieved by night throughout the retreat.

Precision attacks by day and night against small targets like individual buildings proved the more spectacular of the Mosquito operations. Great accuracy was the hall-mark of these raids, as when a suspected Gestapo Headquarters in a school at Egletons near Liège was smashed without killing any patriots amongst the surrounding trees. Occasionally straight night bombing was undertaken, as on the night of 3 January, 1945, against von Rundstedt's forces around Houffalize and St. Vith, when reliance had to be made on instruments in severe cloud and icing. As the final rout of the Wehrmacht followed, that once great army was mercilessly harried almost nightly, and to the end of the war, by the Mosquito fighter-bombers of 2 Tactical Air Force. More Mosquito Mk. VIs were built than any other version, and it was fortuitous that early in the war this variant was conceived.

To proceed with a fighter-bomber version of the Mosquito was first decided on 11 July, 1941. During the week following this was projected as a Mosquito bomber armed with four cannon. In the Autumn these ideas became formulated in the Mk. VI, planned as the Mk. VIA standard night-fighter and Mk. VIB fighter-bomber, intruder and long-range fighter. By early 1942 the night-fighter version had been superseded by later marks, leaving the Mk. VIB to be developed as the F.B. VI, the three versions of which differed in respect of weights and loads. DZ434, the prototype, appeared in May, 1942, and was delivered to the

Operations: Fighter-Bombers

A. and A.E.E. on 13 June, 1942, renumbered HJ662/G. It was wrecked on 10 July due to an engine failure when taking off at Boscombe Down. About ten feet above the ground it swung to port, its undercarriage crashing through a parked Beaufighter. The pilot tried to lift it over a second Beaufighter which the Mosquito's tail practically cut in half. The observer was slightly injured, and the pilot escaped. A considerable delay followed due to production planning decisions, the first production FB. Mk. VI emerging in February, 1943. 418 intruder Squadron at Ford was in May, 1943, equipped with Mk. VIs, fitted with Merlin 25s to increase low-altitude speed. The other intruder squadron, 605 at Castle Camps, received Mk. VIs by the end of July. 23 Squadron in Malta was similarly equipped.

Under A.V.M., D'Albiac, with Basil Embry providing much of the drive as A.O.C. 2 Group, the 2nd Tactical Air Force was formed in June, 1943. 464 (R.A.A.F.) and 487 (R.N.Z.A.F.) Squadrons, flying Venturas at Methwold, Norfolk, were then moved to Sculthorpe where, on 21 August, each received two Mosquito VIs. Next day a Conversion Flight was formed to provide crews for 464 and then 487 Squadrons, after which it converted crews of 21 Squadron. A month later Venturas left these squadrons, and celebrating the event 464 arranged a fly-past by twelve Mosquitoes. 21 Squadron arrived from Oulton to begin conversion, and complete the Sculthorpe Wing. On 27 September 464 Squadron recorded 'there are over seventy Mosquitoes here now, which is a very impressive sight'.

Their station commander, Grp. Capt. P. C. Pickard, D.S.O., D.F.C., well known to the public as star of the film *Target for Tonight*, decided to learn about the new aircraft so that he might fly as Wing Leader, and managed ten hours flying in Mosquitoes during a three-day visit to Hatfield. David Atcherley was Embry's S.A.S.O., and the two planned a vigorous offensive of low-level day and night operations, for which the squadrons were pronounced ready and enthusiastic on 2 October, 1943.

Next day 464 and 487 Squadrons each despatched, around 08.30 hours, a dozen aircraft and two reserves to Exeter to refuel. Shortly before 13.00 hours they were airborne in groups of six led by Grp. Capt. Pickard in 'Freddie' of 487 Squadron, with Wg. Cdr. Meakin leading 464 Squadron and Basil Embry as rear end Charlie to his box, with David Atcherley as his navigator. Two dozen Mosquitoes, each carrying 4 × 500 lb. bombs, headed for power stations at Pont-Château and Guerleden. The leading boxes made runs at tree-top height, dropping 11-second-delay bombs, leaving the others to deliver shallow dive attacks from 2,000 feet and dropping bombs from 800 feet with ·025 second delay. Several hits on target were obtained, and covering Typhoons destroyed two Fw 190s. On the way in Wg. Cdr. Wilson shot up a train. Pickard landed at Predannack having returned on one engine after a lump of flak lodged in a radiator. Flt. Lt. Patterson, flying DZ414 (*see* Appendix 11), a Mk. IV with a ciné camera and extra nose glazing, had most of the cockpit cover

Operations: Fighter-Bombers

of his machine blown off by a shell. Four other aircraft, including Embry's, were damaged.

Six days later a second raid was unsuccessful due to atrocious weather —and to its ambitious nature. An aero engine works at Woippy, near Metz, was the target, for which twenty-four aircraft took off around 11.05 with 464 Squadron leading, to face 10/10th cloud with a 500-foot base and poor visibility on the complicated 600-mile route. About eight miles out the formation had quickly to change course to avoid unexpected balloons over a British convoy. Near the Dutch coast the squadrons lost contact in the mist. Wg. Cdr. Wilson, leading 487 Squadron, soon after became detached from his formation of six aircraft, and others finding themselves lost returned to Sculthorpe. Wilson, however, found the target. Shortly after, his navigator was killed by flak, which meant a difficult return journey and a forced landing at Manston. Over Holland an explosion occurred beneath the leader of the second formation, who crashed. South of Antwerp another aircraft inadvertently dropped its bombs and blew itself up. Intercom switches on some aircraft had been altered to press buttons on the control column close to the bomb release switch, causing confusion. The remainder of the formation returned individually. Only one of 464 Squadron's aircraft attacked the factory, and three others bombed minor targets. Sqn. Ldr. Villiers saw Wg. Cdr. Meakin damaging a Bf 110 wearing desert camouflage. 464 Squadron lost two aircraft, one near Metz and one in Belgium. Two others were badly damaged by flak and unable to reach base. Five others landed away from home, due to weather conditions which enthusiastic crews and an ideal aeroplane were unable to combat. There was need for further training. Hereward de Havilland recorded, 'I drove round with Pickard to each aircraft as it came in to Sculthorpe; most crews were roundly cursing the weather, the convoy and instantaneous bombs; all windscreens were plastered thick with flies, which didn't improve some of the landings; four leading edges and two main spars were damaged by birds.' A few weeks later, after flying through a flock of birds a Mosquito returned with its bullet-proof windscreen rendered opaque by cracks and a 2-inch hole through it, the nose below the Browning guns bashed in, also the starboard spinner, the starboard-wing bomb fairing half carried away and the radiator duct badly damaged.

Concentrated training in low-level attack techniques followed. When on 23 October three Mosquitoes crashed, and were seriously damaged, there was no loss of life. Confidence in the Mosquito increased. Operations were resumed on 5 November, six 487 Squadron aircraft attacking Hengelo. The same day the unit was notified that its Merlin 23 aircraft were to be replaced by others powered by superior Merlin 25s. Six Mosquitoes were despatched on 21 Squadron's first operation on 10 November, but bad weather caused their recall. On 16th they set off for the same target, leaving Sculthorpe at 07.45 and refuelling at Hartford Bridge. *En route* 'K' left the formation after cannon fire hit the starboard

Operations: Fighter-Bombers

engine, leaving Sqn. Ldr. A. M. L. Alderton to take the lead. The target, the tetra-ethyl plant at Paimboeuf near Evian, was not seen, and the formation returned to Predannack to refuel. Over France LR257:Z flew so low that she hit a tree and was badly damaged.

464 Squadron commenced night-flying training during October. On 16 November, 1943, 'A' Flight was detached to Ford and 'B' Flight to Bradwell Bay, in both cases to learn night-intruder technique alongside 418 and 605 Squadrons, experienced in this trade. 'A' Flight operated first, on 19 November, when three *Flower* sorties were flown over enemy airfields. Eight more were flown during the month, by the end of which 'B' Flight had flown fifteen using *Gee* to aid navigation.

An armed motor vessel was located lying off Groix near Lorient. Three Mosquitoes of 487 Squadron were despatched to Predannack on 1 December there to refuel, after which they were to set off to attack the ship. Only HX962:C flown by Sqn. Ldr. A. S. Cussens, with Flg. Off. H. M. Mackay as navigator eventually left, and it was escorted to the target by twelve Typhoons. Following a most courageous attack the Mosquito crashed into the sea and sank immediately.

Establishment of a second Mosquito fighter-bomber wing commenced on 14 October, 1943, when 613 Squadron based at Lasham detached six crews to Sculthorpe for Mosquito experience. A T. Mk. III trainer was flown to Lasham by an A.T.A. woman pilot on 16 October and flying training began there. Three days later Sqn. Ldr. R. N. Bateson, with Flg. Off. B. J. Standish as navigator, arrived to command the squadron. Conversion training continued during November and Wg. Cdr. K. H. Blair, D.F.C., flying LR271 with Flg. Off. J. E. Majer made the unit's first operational sortie, a weather reconnaissance off Cherbourg on 19 December.

21 Squadron began *Day Rangers* on 24 November, when HX958 and HX959 visited the transformer station at Vannes. Two others planned to call at Vezins, but bird damage to the aircraft prevented the attack. Next day three set out for Aube-sur-Rile but only Sqn. Ldr. Alderton in HX958 attacked. Two others operating on *Gee* bombed a transformer station at Falaise. La Roche was raided on the 26th, while two Mosquitoes headed for engine sheds at Lingen. HX957 hit the sea *en route* and turned back; of HX906 no further word was heard. Three days later two Mosquitoes of 21 Squadron set out for an oil depot at Cleve, but failed to locate it, and so set out again to find it next day. On 4 December two of 21's aircraft flew along the Rhine looking for targets. Bad weather interferred with this and a planned *Ranger* to Coesfeld rail junction. When these were repeated on 10 December, ending the series, an interfering Ju 88 was shot up, also a train and barges. Meanwhile 464 Squadron was engaged almost exclusively on *Flowers* from Ford and Bradwell, six aircraft for instance supporting the Leipzig raid of 3 December. 487 Squadron despatched six aircraft to rail targets in North-West Germany on 4 December, and two formations of three on the 10th after

Operations: Fighter-Bombers

which vitally important tasks required assistance from the Mosquito squadrons.

Discovery of the V-1 flying bomb sites in which Mosquitoes played an important part called for urgent action, taken for the first time by Mosquito VIs on 21 December, 1943, when twenty-nine aircraft of the Sculthorpe Wing set out to attack a launching site at St. Agathe, but turned back when the aircraft flying ahead reported the site covered by cloud. Next morning the force took off again, led by Grp. Capt. Pickard. Thirteen, drawn from 21 and 487 Squadrons, repeated the attack during late afternoon. Next day ten Mosquitoes of each squadron attacked a site at Pommereval, near Dieppe. They were greeted by heavy flak which damaged three of 21 Squadron and caused Wg. Cdr. R. M. North to crash-land HX954 at West Malling. HX959 was forced to land at Friston, and Flt. Lt. Wickham brought his Mosquito home with a shattered rudder post.

The other V-weapon attack of the year came on 31 December. It was somewhat unusual in that most of the Mosquitoes were bombing Le Ploy as they moved to Hunsdon, their new base chosen months previously. Sculthorpe was too far from the scene of operations, northern France. Amongst the thirty-five aircraft were seven of 613 Squadron, making that unit's first bombing raid. They took off from Lasham at 09.35 to rendezvous with seven flights of the Sculthorpe Wing. The formation headed for St. Quentin, with 1½-minute intervals between the boxes and 613 Squadron in the rear. Their reserve aircraft pulled away 20 miles from the coast as the Mosquitoes raced towards Maintenay and a *Noball* target close by. Flg. Off. Hunter's LR260 of 613 Squadron was damaged by debris, and he was forced to fly home at low-level. Flt. Lt. Bodington's machine, LR297, also was hit by flak.

During January, 1944, forty-one raids were despatched against V-1 sites by Mosquitoes drawn from four squadrons. Six operations had to be abandoned due to weather. Sometimes DZ414 accompanied the raiders to photograph the attacks. Small targets difficult to locate, the sites were frequently surrounded by anti-aircraft guns, and the nature of the targets called for precise attacks which could be predicted by the defenders. 21, 487 and 613 Squadrons each supplied six aircraft for a raid on the Longuemont area on 14 January, 1944. They dived on their objective from 2,000 to 100 feet, in the course of which LR390 of 21 Squadron lurched away with engines afire. LR276 also of 21 Squadron nosed over on landing, the navigator Flg. Off. G. W. Williams, D.F.M., being injured. Ten days earlier HX921 of 21 Squadron crashed in sand dunes on the French coast, Flt. Lt. Pierce and his navigator becoming P.O.Ws. following a sortie to Puisseauville, where their aircraft was hit by flak. Whilst the day raiders concentrated on the V-weapon sites 464 Squadron commenced *Night Rangers* on 2 January, operating over France, and attacking airfields.

Great care was taken in smashing the V-sites in northern France.

Operations: Fighter-Bombers

They were more difficult to hit than targets hitherto tackled, and they were in territory friendly to Britain. During December, 1943, a technique was developed of approaching the French coast very low, pulling up to five or 6,000 feet and bombing with instantaneous fuses in a shallow dive from about 1,500 feet and flying flat out. If the horizontal low attack was employed the pilot rarely spotted the site soon enough, so dive bombing at 35° from 11,000 feet was practised, bombs falling away at 6,000 feet. This gave an average bombing error of 150 yards. Weather seldom allowed small sites to be seen from anything like 11,000 feet—and in any case only wing bombs could be dropped from a Mosquito in a steep dive.

It was exceedingly difficult to distinguish a genuine from a decoy site, and many sites were hidden amongst trees. The Germans ringed them with guns, installing twenty-six heavy and fifty-six light guns at one site within forty-eight hours of its completion. Although there were many cases of flak damage amongst the Mosquitoes, they were not engaged by enemy fighters during their first 150 sorties to France.

By the end of February, 2 Group had six Mosquito squadrons operating, and they had made 717 sorties against the V-1 sites, 628 effectively. Nearly 150 Mosquitoes had been damaged by flak. Eleven, or 1·53 per cent, had failed to return, and three were written off due to battle damage. Photo-reconnaissance suggested that as many as one hundred Mosquito sorties were needed to knock out just one site.

Usually the Mosquitoes operated in groups of six to eight at low levels, flying at about 260 I.A.S. in the target area, but from the beginning of March it was decided to despatch groups—pairs or fours—enabling them to fly at 280–290 I.A.S. and avoid crowding on the final run. There was not much to choose in efficiency between Mosquitoes and heavy bombers, simply because the sites were so small and mostly in woods. Basil Embry was at this time trying to obtain some 4,000 lb. bomber Mosquitoes for use against the V-weapon sites, and for a time it seemed likely he would be successful.

Two new Mosquito squadrons were 305 and 107. The former, a Polish Squadron, joined 2 Group on 5 September, 1943, afterwards flying five operations using Mitchells. On 3 December, 1943, 305 ceased flying them and it began ground training on Mosquitoes the following day. The first forty-five minutes ended with the collapse of the undercarriage of the squadron's T. III but the Poles, with their usual relish, hastily repaired it. Operational flying training took place during January, LR313 and NS823 setting out on the squadron's first operation against a V-site on 25 January. Unfortunately the attempt proved abortive. Three days later NS823 dropped four 500 lb. MC bombs 50 yards off target, to be the first 305 machine to deliver an attack. 107 Squadron based at Lasham with 613 and 305 equipped with Mosquitoes during February, 1944, launching its first operation on 15 March.

2 Group's squadrons were each spending a month on night attacks,

Operations: Fighter-Bombers

487 under the command of Wg. Cdr. I. S. Smith (late of 151 Squadron) taking over from 464 on 1 February, to fly thirty-three sorties during its tour in the course of eight nights. Usually airfields were attacked, but V-site workers also were given restless nights. Sometimes the squadrons and wings combined in various formations for day raids, but usually half a dozen aircraft from one squadron now comprised each force. In spite of wintry weather twenty-two operations (139 sorties) took place during the first fifteen days of February. No operations were undertaken on 16th or 17th, then on 18 February, 1944, came the epic attack on Amiens prison.

Information received in London during January, 1944, indicated that more than one hundred French patriots in Amiens jail were awaiting execution, some for assisting Allied airmen to escape. The French requested that the R.A.F. break down the prison walls, for this offered the only chance of escape for the prisoners. The task was accepted and put in the hands of Basil Embry. Provisionally the raid was fixed for 17 February, but dense cloud and snowstorms caused its postponement for twenty-four hours, by which time it was felt that the weather would have improved. To have delayed further would have been to jeopardize the whole purpose of Operation *Jericho*.

February 18, 1944, dawned misty and snowy as crews of the three Hunsdon squadrons rose for early briefing on what was still a secret target. 487 Squadron was given the lead part, to boost morale after a tecent distressing and serious accident on the airfield—and following the toss of a coin. Six crews were also chosen from 21 and 464 Squadrons, in addition to which Tony Wickham, flying DZ414 of the Film Production Unit, was to record the raid. In charge of the assault was Grp. Capt. P. C. Pickard, D.S.O. and two bars, D.F.C., with Flt. Lt. J. A. Broadley, D.S.O., D.F.C., D.F.M., as his navigator. Escort and cover was to be provided by Typhoons of 198 Squadron.

Detail planning of the raid, on the prison situated alongside the Amiens–Albert road, had been conducted in great secrecy, and at the briefing a model of the cruciform building was revealed and closely studied. Strict security precautions were in force in the operations block, each man being carefully checked as he entered. Crews experienced unusual emotions, for they were now to attack to save life rather than to destroy. Recalling the briefing an R.A.A.F. officer said, 'It was not a time for long speeches, but there was no mistaking the air of determination that was abroad that morning.'

The prison stood in a compound surrounded by a wall 20 feet high and 3 feet thick. Six aircraft of 487 Squadron forming the first wave were to breach the wall on its North and East sides. A second wave was to open up either end of the prison, destroying the quarters of the German garrison. A third wave was to stand by, and the filming aircraft to take ciné and still photographs. To avoid collisions over target a very exact timetable needed to be adhered to. Fears existed that slight

Operations: Fighter-Bombers

mistiming would cause a collision, as the Mosquitoes swept over the prison at right angles. To ensure maximum benefit from the assault the Resistance was to be informed of the precise time of the attack.

Nineteen Mosquitoes left Hunsdon in a swirling snowstorm a few minutes before 11 a.m., rapidly passed from view and headed for Littlehampton, there to meet escorting Typhoons. On the way four Mosquitoes became detached from the formation, two from 21 Squadron and two from 464 Squadron, and returned to base in a blizzard. Conditions were difficult for the Typhoons and only eight of the planned twelve met the Mosquitoes. The formation crossed the Channel at sea level with the photographic Mk. IV, which made a late take-off, tagging behind, and climbed to 5,000 feet as it crossed the coast 10 miles north-east of Dieppe. Over France the weather was better, the winter sun casting shadows across the snow-clad landscape as the force swept accurately round to the north of Amiens to approach the prison along the straight poplar-lined road towards the target, almost at tree-top height. Working to plan, the Mosquitoes arranged themselves in sections of three, the leading section of '487' hurling 500 lb. bombs at the eastern wall. They were flying in close formation, with Wg. Cdr. I. S. Smith, possibly the first to destroy an enemy aircraft in a Mosquito, in the lead. His bombs hit the wall. Others fell close, and one in a field to the north. 487's remaining two aircraft attacked the northern wall, just managing to clear it as they scampered away. One of their bombs fell on the prison building. Four of 464 Squadron—in pairs, since two of the Mosquitoes had already headed for home—attacked that part of the prison holding German guards. To the north the Film Unit Mosquito orbited from 12.03 to 12.10 and then photographs were taken by the fixed nose cameras and the observer's on three runs. Seeing that the walls of the prison and compound were pierced, Grp. Capt. Pickard flying with 464 Squadron, ordered 21 Squadron over the VHF radio not to attack, using code words 'Daddy daddy, red, red, red', and their four aircraft left the area. Pickard was meanwhile orbiting the prison at 500 feet, and Typhoons were engaging some Fw 190s which had raced to the scene. 464 Squadron reforming near Albert encountered light flak, which claimed MM404 and killed its navigator, Flt. Lt. R. W. Simpson. Sqn. Ldr. McRitchie, temporarily blinded and with his right arm paralysed, was flying at 300 m.p.h. at about 50 feet; he later managed a skilful landing in the snow.

Pickard, having observed the success of the raid, now left the scene at low-level. It seems likely that his aircraft was damaged by ground fire. A few moments later two Fw 190's attacked him, shooting the tail off HX922,[1] the wreckage of which fell near Montigny. Long after the raid it was felt certain that he and his navigator would somehow show up.

---

[1] HX922 received by 487 Squadron 13.9.43 had a flying accident 5.11.43. Was flying again 8.11.43 and was on 487 Squadron's strength when shot down, coded EG:F.

Operations: Fighter-Bombers

Reconnoitreing pilots saw the wreckage of his Mosquito, then information from the Resistance confirmed that both he and his navigator had died in the crash and were buried next day alongside the prison. In addition to the two Mosquitoes lost an escorting Typhoon was shot down near Amiens, another crashed in the Channel in a snowstorm and a third made a forced landing.

Flt. Lt. B. D. Hanafin had an engine fire on the way to the target, but by feathering the propeller managed to stay with his formation as they headed for Amiens. He restarted the engine, only to have trouble about ten miles from the target. He jettisoned his bombs and turned for home. Twice his aircraft was hit by flak and Hanafin was wounded in the neck following which his right side, arm and leg were paralysed. His navigator gave him morphine injections to relieve the pain, but by sheer determination and courage he brought his damaged aircraft back to an airfield in Sussex. Plt. Off. M. N. Sparks had some difficulty in keeping T-Tommy airborne, after an engine was hit by flak, also a wing tip from which the fabric was stripped. He managed to land at a south-coast airfield, where an undercarriage leg collapsed. Of the nineteen Mosquitoes[1] that set out for Amiens, three were damaged—one beyond repair —and two failed to return.

Subsequently it was learnt that following the attack 258 of the 700 prisoners had escaped, including half of those awaiting execution. One hundred and two of the inmates were killed during the bombing or by German machine-gun fire, some were recaptured and there were several civilian casualties outside the prison. Five days after the raid a message was received in London from the French Resistance which read:

'I thank you in the name of my comrades for bombardment of the prison. The delay was too short and we were not able to save all, but thanks to the admirable precision of attack the first bombs blew in nearly all the doors and many prisoners escaped with the help of the civilian

[1] Aircraft which participated and their pilots were:—
487 Squadron: T : Plt. Off. M. N. Sparks   R : Wg. Cdr. I. S. Smith
               C : Plt. Off. M. L. S. Darrall J : Plt. Off. D. R. Fowler
               H : Flt. Sgt. S. Jennings
               Q : Flt. Lt. B. D. Hanafin
HX856 and HX982 were used by 487; both damaged.
464 Squadron: F : LR334 Wg. Cdr. Iredale
               U : MM410 Flg. Off. K. L. Monaghan
               V : MM403 Flt. Lt. T. McPhee
               A : MM402 Sqn. Ldr. W. R. C. Sugden
               T : MM404 Sqn. Ldr. McRitchie (missing)
21 Squadron: C : HX950 Flt. Lt. A. E. C. Wheeler
               P : LR348 Flt. Lt. E. E. Hogan
               F : LR388 Flt. Sgt. Steadman
               D : LR385 Flt. Lt. D. A. Taylor
               J : MM398 Flt. Lt. M. J. Benn
               U : LR403 Wg. Cdr. I. G. Dale
F.P.U. aircraft DZ414:O.

Operations: Fighter-Bombers

population. Twelve of these prisoners were to have been shot the next day. . . .'

Doubts cast in recent years as to the ethics of the attack cannot dim the bravery of all concerned with it. The crew of many a Mosquito owed their life and freedom to the Frenchmen who risked their lives with the Resistance, and those who took part in the Amiens raid with such skill showed that the Royal Air Force realized its debt to its Allies under the ruthless German occupation. Commenting on the raid an Australian navigator said, 'This was the sort of operation that gave you the feeling that if you did nothing else in the war, you had done something.'

In the remainder of February the squadrons raided V-1 sites making sixteen-attacks and completing ninety-nine sorties. An important change of technique occurred on 1 March, when Sqn. Ldr. R. C. E. Law in ML905 and Flt. Lt. F. W. Walton flying LR500 both of 109 P.F.F. Squadron led a *Noball* by six of 21 Squadron, marking the target using *Oboe* on this operation for the first time. Bombing was from above cloud on sky markers. A second similar high-level attack came on 6 March, when the Forêt de Helles was marked, again by 109 Squadron, under fighter escort, and bombing was undertaken by six of 21 Squadron. Four other high-level raids were made in March with P.F.F. Mosquitoes marking.

A spectacular attack, more typical of a year before, came on 18 March. It was a low-level raid delivered by eleven Mosquitoes of 21, 464 and 487 Squadrons directed against Hazmeyer electrics factory at Hengelo. Led by Wg. Cdr. Iredale, the force attacked at 16.35 hours, encountering moderate flak and leaving a large fire. Sqn. Ldr. W. R. C. Sugden was flying MM402:SB-A whose starboard engine was set ablaze, and it crashed soon after.

464 and 487 operated by day and night in March, 487's effort amounting to twenty sorties on ten nights and 464 flying seven during nights of employment. Generally attacks continued on enemy airfields, but additionally some rather hopeful raids were made on V-sites.

The next special low-level daylight operation was undertaken by 613 Squadron on 11 April, 1944, when six aircraft[1] led by Wg. Cdr. Bateson with Flg. Off. Standish as navigator destroyed the Dutch Central Population Registry, situated in the Scheveningsche Wegg, den Haag. One house only was to be eliminated. It stood 95 feet high and had five storeys. On to it were showered 500 lb. delayed-action and incendiary bombs. The formation left at 13.05 hours and flew via Luton to Swanton Morley to refuel. Then it flew via Southwold, crossing the North Sea at 50 feet to achieve surprise. Twenty miles from the Dutch coast the Mosquitoes climbed to 4,000 feet crossing Over-Flakkee, then descended to a very low level. They made their way to Gouda then to Delft, where

---

[1] LR355H (Bateson's aircraft), LR376Q (Flt. Lt. P. C. Cobley), NS844A (Sqn. Ldr. C. W. M. Newman), HP927B (Flt. Lt. R. W. Smith), MM408F (Flg. Off. R. Cohen), LR366L (Flt. Lt. V. A. Hester).

Operations: Fighter-Bombers

flood water made map-reading difficult. The first two machines attacked whilst the other four circled Lake Gouda, allowing the 30-second D.A. bombs to explode. The third and fourth aircraft carried incendiaries, but by the time No. 4 reached the city the smoking target was difficult to locate. The fifth dropped two H.Es. and two incendiary bombs, but the sixth encountered trouble and failed to attack, and instead photographed the burning target. Spitfires escorted the Mosquitoes home. Only NS844 was damaged, by slight flak.

107 Squadron switched the communication targets in April, beginning on 12th when eight aircraft led by Wg. Cdr. Pollard joined six from 613 Squadron to attack locomotive sheds at Hirson. Rendezvous with escorting Typhoons was made over Littlehampton shortly before 13.00 hours. In good weather and little opposition the leading aircraft dived to attack from 3,600 feet, bombing and raking the target with cannon. Others attacked in pairs. LR384 of 107, although peppered by light flak, made its way safely back to base. Eight Mosquitoes of 487 Squadron were meanwhile attempting to bomb the railway works at Haine St. Pierre, also in Belgium. There were many slag heaps in the area, and the target was difficult to locate, two aircraft bringing their bombs back.

Another low-level raid was made by 464 Squadron on 15 April, when Sqn. Ldr. McPhee led six aircraft to railway installations at Ghislain, near Mons. Two Mosquitoes attacked the goods depot, two the wagon repair shops and two took a wrong course so returned with their bombs. 305, the Polish Squadron, made their first attack—other than a *Noball* —when twelve Mosquitoes led by Wg. Cdr. Konopasek raided heavy guns in the area of Ouville-la-Rivier, meeting intense light flak and machine-gun fire, which damaged LR328:G flown by Flt. Lt. Rayski. At this period 305 Squadron was flying almost exclusively night raids, begun on 9 April, when 500 lb. bombs dropped on the airfields at Chartres, Conches and Beaumont. True intruder attacks commenced two nights later.

Similarly 21 Squadron was flying mainly at night now, its intruder operations having commenced on 2–3 March, 1944, when two aircraft went to Montdidier airfield. On 26 March the first *Flower* was flown by Flt. Lt. Murray in LR292 and on ten nights in April *Flowers* were flown. An exception was a *Day Ranger* by LR388 and LR348 to the Panzer Division H.Q. near Laval. Although the target was located it was not bombed, for attention was directed towards three staff cars and two lorries near Mortain.

A series of experimental raids led by P.F.F. Mosquitoes began on 27 April. During these 21 and 464 Squadrons each usually despatched about six aircraft. On the first operation LR373, '383 and '402 led by ML907 and ML956 of 109 Squadron bombed the V-weapon working at Yvrench Conteville from 20,000 feet. Six of 464 and five of 21 Squadron made the first of a series of high-level raids on the railway installation at Abancourt on 29 April, again bombing from 20,000 feet. There was rain and poor

Mosquito Fighter-Bombers

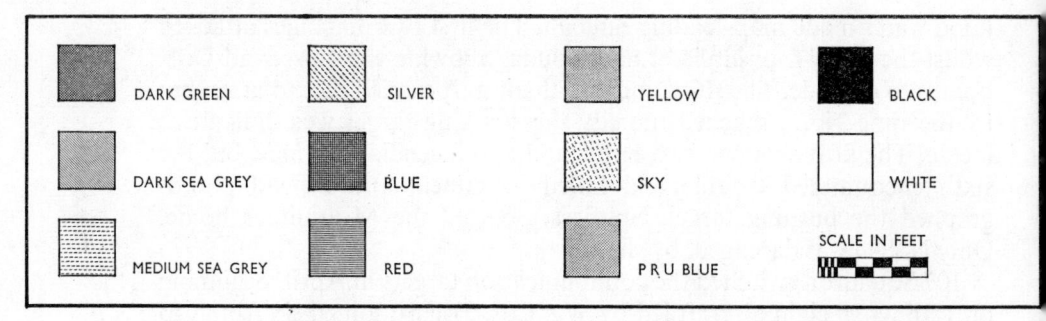

- DARK GREEN
- SILVER
- YELLOW
- BLACK
- DARK SEA GREY
- BLUE
- SKY
- WHITE
- MEDIUM SEA GREY
- RED
- PRU BLUE
- SCALE IN FEET

Mosquito FB. VI wearing standard camouflage markings. Initially 2 Group received Mk. VIs in two distinct schemes, standard night-fighter (dark green and medium sea grey) and day-bomber (dark green, dark sea grey and medium sea grey). The latter was not considered suitable for daylight fighter-bomber operations, thus the night-fighter scheme became standard. R.A.E. conducted experiments with a compromise scheme of light brown and dark green upper surfaces and light grey under surfaces. Fin and rudder were off-white, spinners light grey, but this scheme was not adopted.

In the Far East Hereward de Havilland on 20 January, 1944, recommended that all Mosquitoes for SEAC should have an overall aluminium finish. On 4 February, 1945, it was decided to give PR. IXs and XVIs PR. Blue under surfaces. That month Mk. VIs in SEAC had their aluminium dope plasticized with castor oil. Some Mk. VIs at this time had medium grey under surfaces. Six Mosquitoes arrived in India on 19 May, 1945, with an experimental infra-red reflecting finish.

Mosquito FB. VI (Merlin 25) NS898 reached 613 Squadron Spring, 1944, as SY-Z and operated over Normandy 5–6.6.44. It was damaged in action 5.8.44, operated again between 7–8 and 17.9.44. After overhaul it reached 613 Squadron 8.3.45 to be SY-E. Post-war 69 and 11 Squadrons used it.

Mosquito FB. VI (Merlin 25) PZ170 reached 141 Squadron 27.8.44, became TW-U and on 17.9.44 strafed Ju 88s at Steenwijk and was damaged. Passed to 239 Squadron 3.10.44. Leader of the Stuttgart raid of 7.1.45. Fitted with *ASH*, passed to 23 Squadron and, after overhauls, etc., was sold to France 7.10.46.

Mosquito FB. VI (Merlin 25) HP927 reached 613 Squadron 14.2.44. Flew *Noballs* and *Flowers* and was flown in the Hague raid of 11.4.44. Operated over Caen area 5–6.6.44 and was busily engaged in 1944 over France. Reached Marshalls of Cambridge 26.2.45, sold to Fairey Aviation 13.11.46.

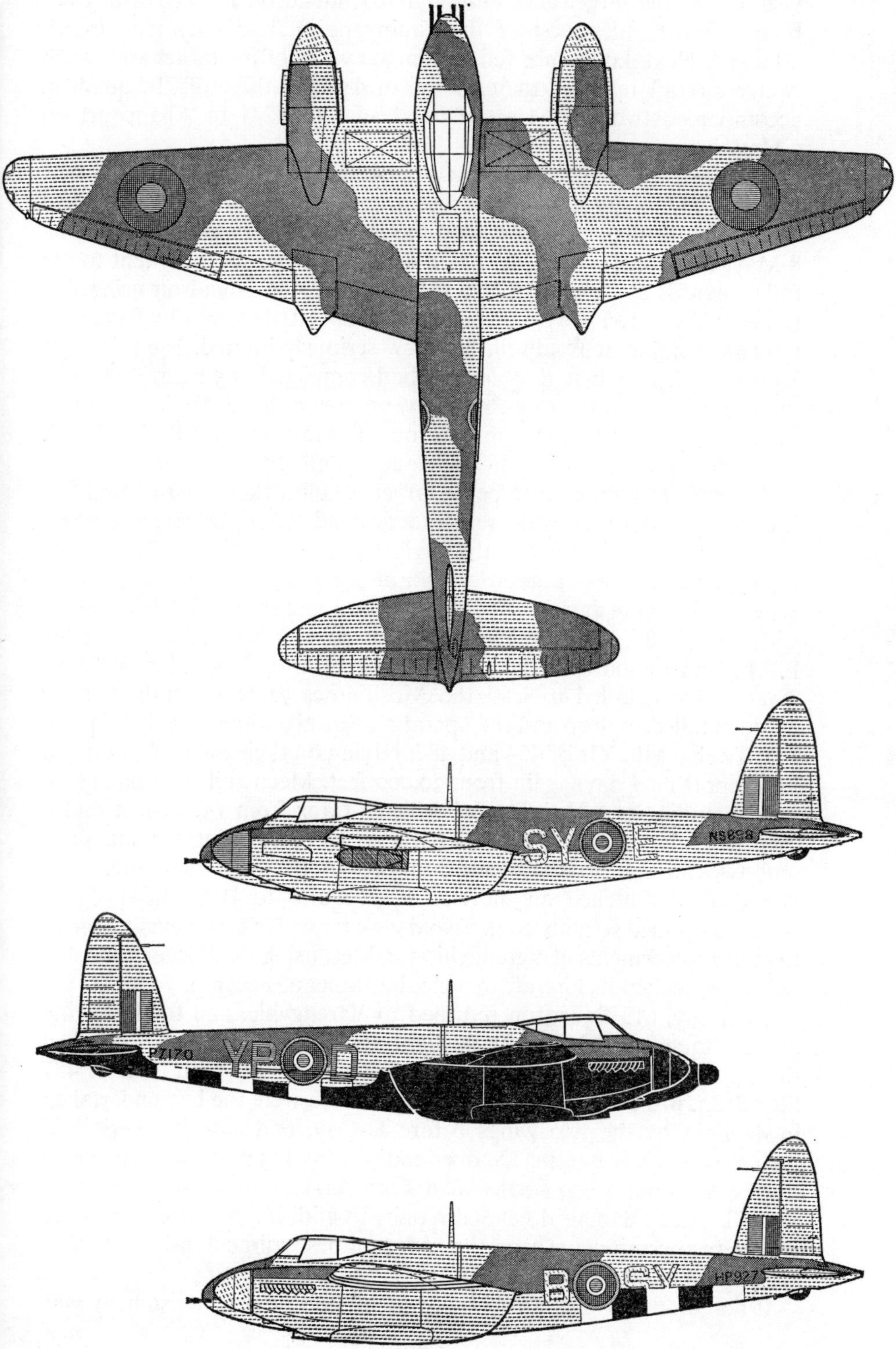
SY
E
PZ170
YP
D
B
SY
HP927

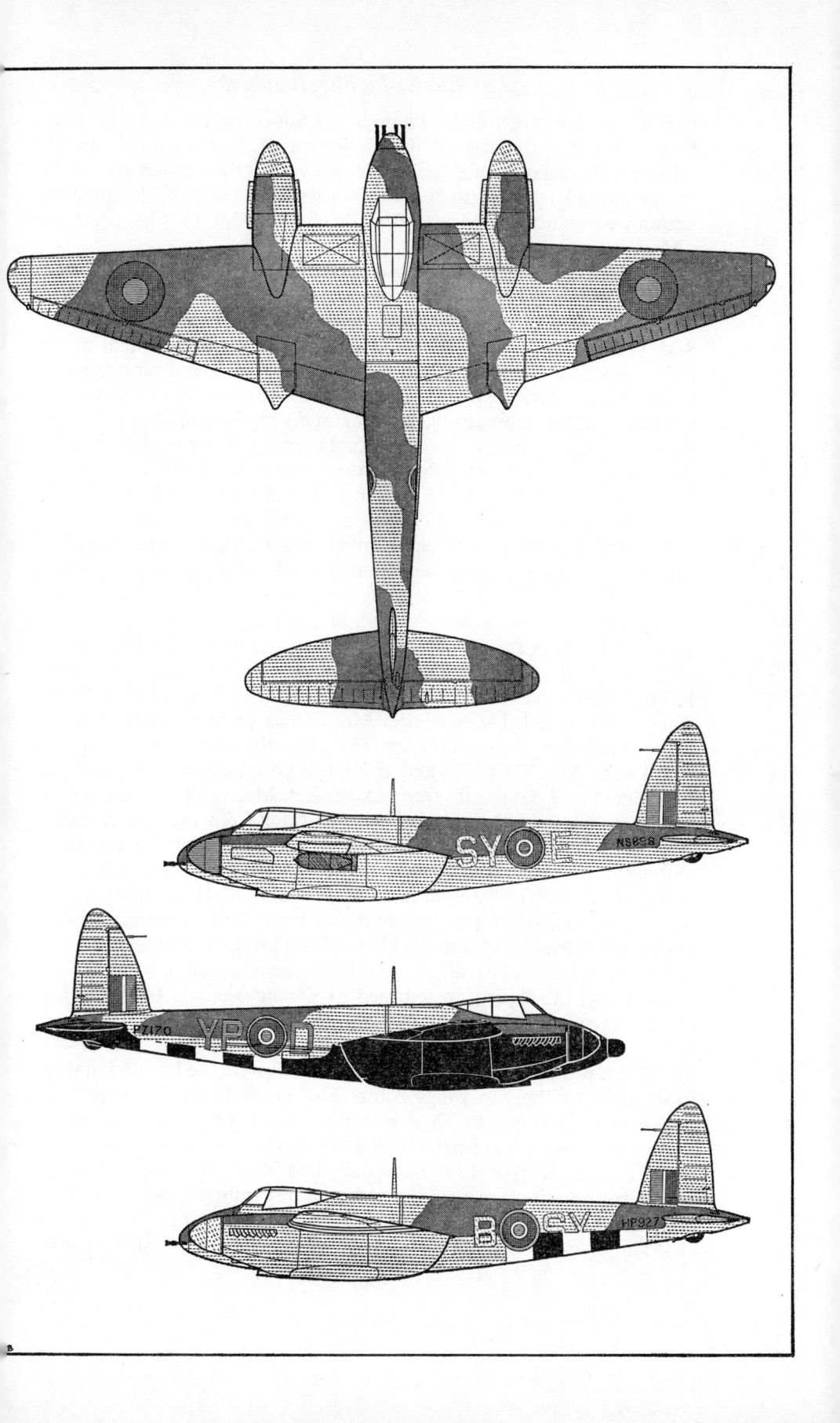

Operations: Fighter-Bombers

visibility at the target and, although 109 Squadron marked with care, bombs fell 7 miles west of the aiming point, and only ten aircraft attacked. Next day bombs fell within 300 yards of the target area when twelve aircraft returned there. Seven of 487 and three of 21 Squadron accompanied two of 464 and *Oboe* leader MM241 to Abancourt on 3 May.

A further attempt to employ fighter-bomber Mosquitoes in the new tactics of bombing on P.F.F. markers came on 6th, when eleven attacked a V-site. A similar number raided the marshalling yards at Serquez on 8 May. Next night six of 487 again led by an *Oboe* aircraft went to the railway yards at Juvincourt. On 9 May six of 464 Squadron using *Gee* raided a V-weapon site from 20,000 feet. MM410 damaged by flak made a forced landing at Bradwell, her crew seriously injured. Next day 305 Squadron flew its first *Rangers* and had completed 105 night sorties by the end of the month. *Day Rangers* were now occasionally flown by all the fighter-bomber squadrons, as on 16 May, when Flt. Lt. W. J. Bodington attacked targets in Châteaudun and Orleans areas.

A completely different type of target was attacked on 19 May, following the pattern set by the remainder of 2nd T.A.F. During a *Rhubarb* four of 613 Squadron raided the navigation beam station at Sortosville-en-Beaumont. It was a good attack, although the bombs hung up on HP927. The same day three of 21 Squadron and three of 464 Squadron with six of 487 attacked Yvrench from 20,000 feet, bombing on the P.F.F. leader's markers. The final occasion during May when a P.F.F. marker was detailed to lead the Mosquitoes came on 20th, but his markers failed to drop and the operation was abandoned by 464 Squadron. Twelve Mk. VIs of 464 and 487, relying on their own *GH*, bombed Le Treport next day, again from 20,000 feet. Meanwhile 613 Squadron were attending to a V-site at Belleville-en-Caux from 15,000 feet and a clear sky, also bombing on a *GH* leader. In the evening the squadron sent two aircraft to Florennes and two to the St. Trond area. 305 Squadron despatched an individual Mosquito to Belleville-en-Caux, but its *GH* failed so it aborted. Two days later 613 in two waves attended to gun emplacements at Varengville-sur-Mer, using a *GH* leader and 487 twice despatched its aircraft to a similar target at Fecamp.

Next day, 613 Squadron returned to Varengville, and the following day 21 Squadron, closing its daylight activity for some time, bombed three guns at St. Cecily also using *GH*, in partnership with three of 487 Squadron. 613's raids on guns at Ault on 26th were the last undertaken in daylight by the two wings before D-Day, and closed a period of intense activity. Between 3 October and 26 May 155 day bombing operations and about 1,600 sorties, had been carried out for the loss of 36 aircraft and 21 damaged beyond repair. In addition to these there were countless occasions when the Mosquitoes miraculously returned although shot up.

Proof of the Mosquito's ability to absorb heavy punishment was

Operations: Fighter-Bombers

typically illustrated on 18 April, 1944. LR328 of 305 Squadron entered a stream of flak bursting close by. The crew could hear the detonations, then suddenly felt their aircraft shudder. There was a great explosion and the machine banked on to its starboard wing. By the time the crew had righted it they were well into France. Great strain on the stick was necessary to keep the aircraft flying level. Instruments indicated that the engines were in order, but the navigator could see a large hole in the port wing. Nevertheless an attack was made from a shallow dive, very shallow in case the wing was loose. France was recrossed at low level and a normal landing made at the first airfield seen. LR328 was repaired on site.

On the same occasion the Wing Commander of 305 Squadron was crossing the French coast when a heavy detonation was heard, and his Mosquito became difficult to control. He suspected that the hydraulic system was damaged, and the bomb doors were immediately opened. Nevertheless the officer led the formation towards the target. Flak again was hurled at the machine as the coast was recrossed and over Lasham the undercarriage had to be lowered by the emergency method. When the Mosquito was on the ground the damage could be seen to be a starboard inner flap shot away, hydraulic installations of the undercarriage, flaps, tailwheel and bomb doors put out of action and thirty flak holes in the fuselage.

Conserving their strength, Mosquito fighter-bomber squadrons of 2nd T.A.F. flew only a few *Night Ranger* missions on the first four days of June, 613 commencing on 3-4 June. Several Mosquito squadrons of the Allied Expeditionary Air Force were now under canvas, prepared for a quick move to the Continent if required. Two de Havilland liaison engineers, Baker and Taylor, were already in battledress and getting periodically run in for failing to salute officers. 511 Forward Repair Unit of 85 Group at Odiham was ready to undertake all major Mosquito repairs on the Continent, under the control of Grp. Capt. Messiter. Banks of the D.H. Drawing Office was on hand and the Hatfield Design Office was investigating new means of effecting speedy repairs in the field, especially wooden components.

On 4 June black and white identity markings were applied to the A.E.A.F. Mosquitoes. As the transport aircraft and the seaborne forces set out for Normandy so did the Mosquitoes of the six squadrons. On the night of 5-6 June, 113 sorties were flown against cross-roads, small bridges, railway lines and stations, and many roads in the area around Caen. Busiest was 21 Squadron, which, using eighteen aircraft, flew twenty-five sorties. 464 Squadron—twenty sorties using eighteen aircraft —came a close second. The only loss was, sadly, Sqn. Ldr. A. G. Oxlade, shot down in NS897 of 464 Squadron. The presence of the Mosquitoes virtually put a stop to enemy troop movements, a process continued by other A.E.A.F. aircraft during the hours of daylight, whilst the Mosquitoes were prepared for intensive operations on the night of 6-7 June,

Operations: Fighter-Bombers

when 196 sorties were flown. Dropping their own flares, also being assisted by flare-dropping Mitchells, the six Mosquito squadrons harassed enemy troops in positions located the previous evening. Wherever they found trains, road transport, troops and armoured vehicles down went 500 lb. bombs and cannon fire. Often by radio they called upon colleagues when their discovery was too much to keep to themselves.

Several nights passed before troop movements were on any sizeable scale, then the Mosquitoes dropped flares over the targets and attacked, or relied upon the Mitchells of 98 and 180 Squadrons for 'nitelights'. On moonlight nights roads, railways and their traffic, woods and lakes were visible to a height of 4,000 feet, but in bad weather the aircraft flew below the cloud, a dangerous pursuit due to valleys and lack of sensitive equipment. These tactics, however, slowed enemy movement, even if no attacks were made. In the first two months following D-Day casualties were less than half the expected rate, and only twenty-six Mosquito VIs failed to return from just over 2,000 sorties. For the first fortnight after D-Day T.A.F. Mosquito squadrons maintained 93 per cent serviceability and subsequently 85 per cent, although ground crews were hard pressed to keep it so high. Part of the duty of the Mosquito fighter-bombers was the delivery of beer to the beachhead in wing drop tanks. Supplied to the right place at the right time this was reckoned to achieve a greater destruction of the enemy per sortie than the more direct method for which the Mosquito was intended.

Between 7-8 and 12-13 June, 464 Squadron flew seventy-five night sorties, twenty-seven more on 6-7, impeding road transport heading for Caen. 613 Squadron meanwhile made twenty-five sorties, and five aircraft of 487 Squadron each completed two trips. Twelve 487-Squadron aircraft delivered a concentrated night attack on the marshalling yards at Mezidon on 8-9 June, losing one of their number, bringing their total loss to nineteen Mosquitoes. Nine of 305's machines attacked Mezidon on 10-11th whilst three other aircraft using *GH* made a night attack from 16,000 feet on the St. Sauveur marshalling yards. 107 and 613 Squadrons each sent three aircraft on this raid. Although it was cloudy radar direction made the operation a success. This indicates the diverse forms of attack on which Mosquitoes were engaged. While this operation was progressing 'E' of 305 Squadron alone bombed railway yards at Le Haye de Puits, where its flares were the first to be dropped preceding any attack by a 305 Squadron aircraft.

At the request of the Army Wg. Cdr. Iredale, D.F.C., led a dusk attack on 10 June against petrol tankers in the marshalling yards at Chatellerault, 200 miles into France. This low-level operation called for bombing from 20 feet on to the target south of Tours. The force comprised three aircraft from 464 Squadron and three from 487, followed forty-five minutes later by six of 107 Squadron. On the way home the Mosquitoes shot up such trains on the line north as they could find.

Operations: Fighter-Bombers

During the night a road convoy near Coustances was set afire, and others bombed a seaplane base near Caudebac. In pairs on 12–13th 464 Squadron attacked the railway installation at Mesuil, and they conducted armed reconnaissance flights in the Tours–Alençon region the following night.

Operations by the Mk. VI squadrons were now interfered with by the flying-bomb attack, Gravesend being directly in the line of V-1 fire. From 16 June squadrons there began operating from Dunsfold, before moving to Thorney Island. Sparse comfort came to enemy troops, for 464 Squadron alone managed seventy-eight sorties between 21 and 30 June, sixteen of them to the Villers Bocage area on 29–30th.

Each squadron despatched nightly on average fifteen to twenty sorties, to range over regions close to the battle-grounds. Late in July 305 Squadron on three nights fired rockets, as well as dropping flares and bombs at road targets in the Amiens–Rouen region. Rockets were first fired by Mosquitoes on 25–26 July against trains in Normandy, and on the nights of 27–28th and 30–31st.

Precision bombing attacks were made during August on buildings chosen by Allied Intelligence, great care being taken not to kill civilians nearby. An enemy Divisional H.Q., barracks or Gestapo centre, picked out by French agents—and sometimes even marked by them with white sheets spread on the roof—would be scheduled for destruction at a specified time on a certain day. Only in rare circumstances could such attacks be made at night. More often daylight strikes with bombs horizontally flying released from roof-top height in traditional Mosquito style were required, and many a good job was done by Tactical Air Force crews practised in the technique. Both high-explosive and 500 lb. phosphorus bombs were used, the latter sometimes serving as target indicators.

Gestapo barracks at Bonneuil Matours in the north-east corner of the Forêt de Moulière were destroyed on 14 July. Its six buildings fitted into a rectangle only 170 by 100 feet, close to a village that must not be hit. Nine tons of bombs were delivered by shallow-dive Mosquitoes, led by Grp. Capt. Peter Wykeham Barnes, D.S.O. and Bar, D.F.C., and Flg. Off. Chaplin at the head of the 487 Squadron aircraft, with Wg. Cdr. R. H. Reynolds, D.S.O., D.F.C., leading four of 464 Squadron.

Twelve Mosquitoes of 487 Squadron and a dozen from 21 Squadron led by Wg. Cdr. I. S. Smith, and escorted by Mustangs, made a low-level attack on the barracks at Poitiers on 1 August. Next evening six of 305 Squadron and seventeen of 107 Squadron set off shortly before 20.00 to attack the saboteur school in the Château Maulny. Twenty-three Mk. 6s and a photographic Mk. IV flew to Selsey Bill, thence to Cap de la Heve where they crossed at low level before climbing to 3 to 4,000 feet. Three of the Polish squadron claimed hits on the Château and three shot it up. Four hours after take-off all the aircraft returned safely to Lasham.

Air Vice-Marshal Embry flying in PZ222, with Sqn. Ldr. Chapman

Operations: Fighter-Bombers

as his navigator, led fourteen aircraft of 613 Squadron to attack Egletons school, believed to be an S.S. barracks, in support of the Maquis. All but one of the Mosquitoes attacked, over twenty hits being obtained. Half of the formation carried 2×500 lb. 11-second delayed-action bombs, the remainder had each 2×500 lb. instantaneous bombs, and all were fitted with long-range tanks. Participating in the raid were Grp. Capt. L. W. C. Bower in HP923, Wg. Cdr. C. W. M. Newman in NS859 and Wg. Cdr. R. C. Porteous in PZ223. Accompanying the force was Mk. IV DZ383, from which the operation was photographed. NT230 was shot down at the target, but others all returned safely to base.

Six of 487 Squadron on 31 August practically destroyed the S.S., H.Q. at Vincey, near Metz, a building believed to be housing some 2,000 personnel. 305 Squadron flew the final raid of the series on 31 August, when six Mk. 6s set off to deliver a low-level attack on petrol storage tanks at Nomeny. Each carried 2×500 lb. bombs. Although there was 10/10th low cloud over the target the visibility was fair. 'A' released her bombs too soon, but the others succeeded in destroying four large tanks and several smaller ones.

Meanwhile the constant night offensive had been maintained by the six squadrons, and was aimed at movement generally in the direction of the front, until the disintegration began, leading to the terribly harassed and disorderly retreat of the Wehrmacht across France. No difficulty contributed more to the enemy's failure to resist the Allied invasion than air superiority which pinned under cover his moving forces during the long days, and barely permitted them to progress at night, due to the attacks, for which the Mosquito VI was proved the ideal weapon and the principal performer. With its speed, and the manoeuvrability necessary for protection, possession of navigational aids and a second crew member to work them, it could accurately rendezvous at night with other aircraft such as the flare-dropping Mitchells, at the precise cross-roads, woods, river bridges or rail centres pinpointed at the evening briefing, and was ready to go in and destroy them as soon as flares illuminated the ground. These burnt a few minutes only and what the Mosquito missed on its first run along a woodside road was, therefore, hidden in the trees by the time it had circled. Surprise was, therefore, essential. As well as its high performance the Mosquito also had good striking power, using bombs, cannon and machine-gun fire with a generous ammunition supply—though pilots always wanted more.

In that busy and exciting summer, with great troop pockets on the Seine and the helter-skelter movement of German forces further south and east, the liberation of Paris, and the drive into Belgium, 2 Group of the Second T.A.F. sometimes ordered as many as 140 Mosquito sorties in a night, and crews often made two trips between dusk and dawn, scoring well on both. By 17 August, seventy-two days after D-Day, one Mosquito Wing commanded by Grp. Capt. L. W. C. Bower, D.F.C., with Wg. Cdr. H. J. W. Meakin, D.F.C., as Wing Leader, had made 2,319

Operations: Fighter-Bombers

sorties, averaging thirty-two sorties daily irrespective of weather conditions. On one night alone, 1 September, T.A.F. Mosquitoes attacked 23 trains, 300 wagons, 300 road vehicles and 20 barges in France, Belgium and Germany.

At 07.00 hours on 17 September, 1944, the day of the airborne assault, thirty-two crews of 107 and 613 Squadrons assembled for briefing, during which models of barracks around Arnhem were unveiled, and plans for their destruction and the reason were detailed. Take-off was at 10.45, and 107's aircraft in waves of four attacked houses used by German troops from heights varying between 800 and 1,500 feet. Flak was intense, and two aircraft were shot down. 613's aircraft flew in two groups accompanied by a photographic Mk. IV flown by Flt. Lt. V. A. Hester, who was injured by ·303 in. ground fire as he flew low to deliver 4×500 lb. TD bombs, and secure photographs. Each of the other aircraft dropped 4×500 MC TD 025 sec. bombs, a frequent Mosquito load, on houses in which there were German troops. Nijmegen was meanwhile being attacked by 21 Squadron, seventeen of which also set off at 10.45. Five claimed to have bombed enemy barracks as detailed, three attacked secondary targets and seven other crews reported that there were so many aircraft in the area that they were unable to attack. One Mosquito was shot down. Three led by Wg. Cdr. Dennis bombed a barrack square at Cleve, where they machine-gunned troops, except for the Wing Commander, whose aircraft struck a bird, after which he had to jettison his load. Apart from these operations preceding the airborne landing in the area later that day, night interdiction sorties in large numbers comprised the entire effort by the FB. VI squadrons during September.

464 Squadron logged their 1,000th sortie on the night of 19–20 September, when attacking water crossing points and rail routes in the northern sector and supporting the Arnhem landings. Their aircraft returned safely, as now was customary. The squadron flew only forty sorties at night in support of the Arnhem landings, limited by bad weather. At night the 2 Group Mosquitoes attacked in an arc from Walcheren to Cologne, denying forces for Arnhem and preventing others from crossing the Rhine. Rarely did enemy fighters engage the Mosquitoes, although on 27–28 September, Flg. Off. G. M. L. Doube of 464 Squadron claimed that he was intercepted by two Me 410s, and the squadron lost an aircraft during the night.

Lasham's squadrons moved to Hartford Bridge in October to prepare themselves for a move to Continental bases. Eventually 107 Squadron, also 305 and 613, flew to Epinoy on 19–20 November. The move brought little respite to the Germans, for night operations were resumed on 21st, 305 Squadron being detailed for their 366th operation, the 140th using Mosquitoes. In the weeks ahead wet stormy weather, poor visibility and water-logged airfields disrupted operations. Even so, forty-six trains were attacked in one average night by Mosquitoes in western Holland.

Operations: Fighter-Bombers

On 31 October the Gestapo's Headquarters in Jutland, housed in two buildings of Aarhus University, was destroyed by twenty-four Mosquitoes of 21, 464 and 487 Squadrons,[1] led by Grp. Capt. Wykeham-Barnes and escorted by eight Mustangs. With resistance and sabotage increasing the Gestapo had been strengthening its grip on the country. Documents and dossiers in the buildings had been the basis for the persecution of patriots. Crews were briefed with the aid of a model, again in great secrecy. As in most attacks, a heavy responsibility fell on the navigator leading the formation on a long sea flight, and approach to a precise point on the enemy coast. Four sections, each of six, including a Mosquito of the Film Production Unit, comprised the formation. The 11-second fusing of the bombs, carried only in the fuselage for precision attacks, was arranged so that each section of six aircraft could bomb without risk of being blown up by bombs of their own section, yet leave as little time as possible between the attacks.

Immediately before the target the weather was bad, but it cleared somewhat at Aarhus, and from the ground-level attack bombs were well placed. Sqn. Ldr. Denton flew PZ332 so low that he returned with a damaged nacelle and only half the tailplane and tailwheel. His observer helped him pull the stick back to maintain control. PZ164, damaged, force landed in Sweden, but two others of 487 Squadron also damaged by bomb blast flew home. Two of 21 Squadron's aircraft collided with birds. Others of the squadrons shot up engines, troops and two trains. The round trip totalled 1,235 miles. The target area was covered by low cloud and although it was noon many lights were to be seen. It was such a surprise attack that it was some time before any anti-aircraft guns opened up. The ciné-equipped Mosquito was able to make three runs across the target. As the formation crossed Denmark many Danes waved, and one man near the target was seen to duck as the bombs whistled overhead. A farmer stopped ploughing, came to attention and saluted as the Mosquitoes swept across his field. While the Aarhus raid took place eight aircraft of 107 Squadron in foul weather were responsible for the final softening up, prior to the Canadian assault, of the island of Walcheren.

Whenever the weather permitted, night operations continued. 305 Squadron, now closer to the front line, flew 344 hours operationally on twelve nights in December and 464 Squadron managed 421 sorties between 1 October and 31 December. By then, von Rundstedt's unexpected thrust had plunged into the Allied line. His Ardennes offensive opened on 16 December, in the thick fog, during a period when the weather was adjudged the worst for fifty years. In the week following the fog scarcely lifted, severely hampering the Tactical Air Force. Nevertheless, away

[1] Aircraft used:— *21 Sqn.* PZ306, PZ314, PZ316, NT200, PZ304, NT170, LR353, plus filming aircraft; *464 Sqn.* NS890, NS896, NS943, LR256, HR185, HR352:S, NS994, HX920; *487 Sqn.* PZ332:A, NS981:B, NT171:F, PZ164:K, PZ195:L, HR182:O, PZ330:V, NS840:X.

Operations: Fighter-Bombers

from the battle-ground the Allies hammered rail targets and supply bases. On 17–18 December, 464 Squadron, after six nights of enforced inactivity, bombed Mayen and attacked motor convoys heading for the front. On 18–19th 2 Group despatched forty-six Mosquito sorties. Then for four nights the weather prevented operations. Dramatically on 23 December the weather changed. On 23–24th 107 Squadron resumed the offensive, attacking targets close to the German thrust, whilst 464 Squadron attacked Prum, forward base of the Fifth Panzer Army, then turned to shoot up vehicles in the region around. On Christmas Eve seventy-one Mosquitoes were despatched, one unit sending each aircraft out twice. Choke points reminiscent of those around Falaise in June were being created. Fog now hampered the operations of 140 Wing, still based at Thorney Island, but from Epinoy the other squadrons operated often in appalling weather.

Failing to secure Bastogne, the enemy began to fall back towards Houffalize, heavily raided by fifty-two Mosquitoes on the night of 3–4 January, 1945, along with St. Vith, through low blanketing cloud and with the aid of *Gee*. This offensive proved to be the last serious German counter-attack and the losses suffered were a further terrible blow. Plans of concentration which the Germans had made were thrown out of gear and by 14 January Mosquito pilots found difficulty in assessing the position of the bomb line ahead of the advancing Allied troops. On the night of 19–20 January many Mosquitoes made two sorties dealing with villages just behind the front line packed with German troops and stores. Heavy snowstorms made the task difficult, and a very low icing level was encountered. Next night, in continuing snowstorms, Mosquitoes were called upon to halt road and rail movements mainly north of the Ruhr, and had one of their best nights since the new drive had begun, although in the most severe weather they had yet flown through. Night intruder operations, under way before the offensive, were resumed on 16 January. Nine hundred and thirty-five Mosquito fighter-bomber sorties were flown during the Ardennes offensive.

Two operations per night per aircraft were now commonplace. 107 Squadron for instance managed thirty sorties by seventeen machines—four of which were damaged—on 19–20 January. On the night of 1–2 February, whilst attacking eastward-moving troop trains, Mosquitoes patrolled as far as 350 miles from base, striking at targets within 80 miles of Berlin. Then, on 22 February, came Operation *Clarion*, when German communications, factories and targets of opportunity were attacked in daylight. Mosquitoes flew deeper into Germany than usual, ranging well beyond Bremen and Hamburg. 140 Wing, which moved to Rosière-en-Santerre on 7 February, put on maximum effort. 21 Squadron despatched 19 and lost 1, 464 lost 2 aircraft out of the 16 flying, 487 5 out of 18. 138 Wing despatched 613 Squadron to the Wessel-Kiel-Kirchbarkau-Wessel region, where they damaged ten trains, signal boxes, an airfield and road transport, and lost three aircraft.

Operations: Fighter-Bombers

305 Squadron operated nineteen aircraft between 11.30 and 15.00 hours in the Hamburg-Wesermunde-Meldorf region, losing one aircraft shot down 8 miles S.W. of Bremen. HR346:K shot up barges, a radio station and a factory for good measure; NS909:L dealt with five trucks, a signal box and a level crossing; NS844:N destroyed eight trucks before being hit by flak. 107 Squadron despatched three boxes each of five Mosquitoes with another roving behind. All carried 2 × 500 lb. bombs and some had 5-inch vertically fitted cameras. Subsequently they hunted in pairs, and a Mk. IV photographic aircraft tagged along with them, filming some of the eleven trains attacked for the loss of only HR188. In all 143 Mosquitoes had been despatched and twenty-one failed to return. Flak damaged twenty-five others. It was decided not to repeat the costly daylight onslaught using Mosquitoes, which resumed their night offensive until 21 March, 1945, when a specialized daylight precision attack was mounted.

Eighteen Mosquitoes were assembled at Fersfield in Norfolk on 20 March. At 08.40 the following morning in glorious spring weather, and in high spirits, the crews departed after being briefed to attack from 100 feet the Shellhaus building in Copenhagen. This housed the Gestapo Headquarters where, Danish Resistance leaders feared, plans were being completed to make mass arrests so as to paralyse the Danish underground movement. Leading the formation was Grp. Capt. R. N. Bateson flying RS570:X.[1] Wg. Cdr. P. A. Kleboe in SZ977:T of 21 Sqn. had Flg. Off. K. Hall as navigator. Following came six of 464 Squadron and seven of 487 Squadron. With an escort of 64 Squadron's Mustangs they flew for two hours across the North Sea, making landfall as planned. It was noon when, in poor weather, they reached Copenhagen, where the streets were thick with traffic and people. Bombs from the first wave smacked into the target. Wg. Cdr. Kleboe flew so low that he crashed into a flag pole. One crew reported seeing the bombs of the aircraft ahead of them enter the Gestapo building between the first and second storeys. The six-storey building was set ablaze. Many Gestapo officials were killed, their records largely destroyed. Three other Mosquitoes failed to return, one of them crashing into a school building causing grievous casualties. Two missing crews of 464 Squadron were amongst its most experienced, their pilots being Plt. Off. R. G. Dawson in SZ999 and Flg. Off. J. H. Palmer in RS609. Nevertheless the attack had achieved its aim. The Mustangs, two of which later fell to flak, found no fighter opposition. Five hundred yards from the target could be seen a group of guns from which not even the tarpaulins had been removed, such was the surprise achieved. Wg. Cdr. Denton, whose aircraft, PZ402, was damaged by flak which removed the starboard flap,

[1] Aircraft used:— *21 Sqn.* SZ977:T (FTR), PZ306, LR388, HR162; *464 Sqn.* PZ353:G, PZ463, PZ309, RS609 (FTR), SZ968, SZ999 (FTR); *487 Sqn.* RS570:X, PZ402:A, PZ462:J, PZ339:T, SZ985:M, NT123:Z (FTR), PZ242:P and Film Production Unit Aircraft.

Operations: Fighter-Bombers

and whose hydraulics were rendered useless, flew 400 miles home on one engine and belly landed.

During Operation *Plunder* the six squadrons of 2 Group gave assistance to the heavy bombardment near Wesel preceding the crossing of the Rhine. As the enemy retreated from his last line of defence mobile anti-aircraft weapons were taken along. Frequently the army columns put up intense flak barrages as they came under attack. On 2 April, Mosquitoes claimed destruction of 136 transport vehicles. There were few targets now left that could not be called tactical, and on 6–7 April 2 Group Mosquitoes began bombing Berlin. During the night Flt. Lt. Campbell logged 464 Squadron's 2,000th sortie since D-Day. On 10 April the squadron's aircraft made an unusual tactical daylight attack and on twelve more occasions the squadron operated in April.

The last of the Mosquito daylight 'spectaculars' was mounted by six aircraft[1] of 140 Wing during the afternoon of 17 April, against a school building used by the Gestapo and located in the western outskirts of Odense, on the large island of Fyn. Mustangs again escorted them, primarily to deal with flak positions. All six Mosquitoes attacked including the Mk. IV photographic aircraft. Flt. Lt. W. R. M. McClelland's aircraft, PZ463, was damaged by bomb bursts. He flew back on one engine to make a safe landing at Eindhoven. Besides fuselage bombs, wing bombs were carried in place of long-range tanks unnecessary because the raid was delivered from Melsbroek in Belgium. Operations by 138 Wing ended on the night of 25–26 April.

107 Squadron had during the month flown 260 night sorties and on 24th attacked Bremen under radar control, sending a dozen aircraft each with 2×500 pounders. 305 Squadron had logged 977·10 hours flying time in the course of twenty-two night operations and 613 also had operated twenty-two times.

On the same night 21 Squadron also flew its last operation, twelve aircraft intruding upon road transport at Bad Oldesloe and the railway at Wittenburg. Thirteen sorties by 464 Squadron on 2–3 May, when twelve trains were assaulted in the Emden-Bremen area, represented the squadron's final effort, the last machines HR352 flown by Flt. Lt. H. G. Hobson and HR186 by Flt. Lt. Killingworth, landing back at 04.30 hours. On 3–4 May, 487 Squadron despatched a dozen Mosquitoes to Heide and Itzeloe. After attacking a train near Friedrichstadt a huge explosion damaged one Mosquito flying at 11,000 feet. The distinction of 'last man home' fell to TA119:A, which touched down at 02.35. For 464 Squadron one exciting task remained, for it flew wave-top patrols over a British force taking Crown Prince Olaf back to Norway on 13 May to the excitement of everyone.

With the war over 2 Group's squadrons engaged themselves on peacetime training duties until they became part of British Air Forces of

---

[1] Aircraft used included RS573 (21 Sqn.), PZ463 (464 Sqn.), RS570:X and PZ399:T (487 Sqn.).

Operations: Fighter-Bombers

Occupation established on 15 July, 1945. 21 Squadron, based at Gutersloh, operated courier services between Furth and Blackbushe during the Nuremburg trials and after training detachments, was disbanded on 7 November, 1947. 107 Squadron operated as part of B.A.F.O. until 4 October, 1948, when it was renumbered 11 Squadron. It had moved to Wahn in November, 1947, where its Mosquito 6s were based until September, 1949, its home becoming Celle for a year before it proceeded to Wunsdorf, there to bid farewell to its Mosquitoes.

The Polish Squadron, 305, was based in Holland, then Belgium, before moving to Wahn in March, 1946. It returned to Britain in October and gave up its aircraft soon after arrival. 464 Squadron was disbanded on 25 September, 1945 (the de Havilland Company's twenty-fifth birthday) and its crews—mainly Australians—headed for home. 487, the New Zealand Squadron, had a more varied existence, becoming first 16 then 268 Squadron soon after. As such it was based in Belgium, until it disbanded in March, 1946. 613 Squadron was renumbered 69 Squadron on 8 August, 1945, and was based at Cambrai until March, 1946, when it moved to Wahn. The following month it was equipped with Mosquito B. XVIs which it used until disbanding on 6 November, 1947.

It is impossible to record in complete detail the successes achieved in the wartime interdiction operations by the 2 Group Mosquitoes. In the course of nearly 1,000 sorties they slowed the building rate of the flying-bomb sites; precisely eliminated buildings to aid the Resistance movements of Europe, and tormented the enemy's communications network to an incredible extent. The cost? Fifty-seven Mosquitoes damaged beyond repair and 136 which failed to return.

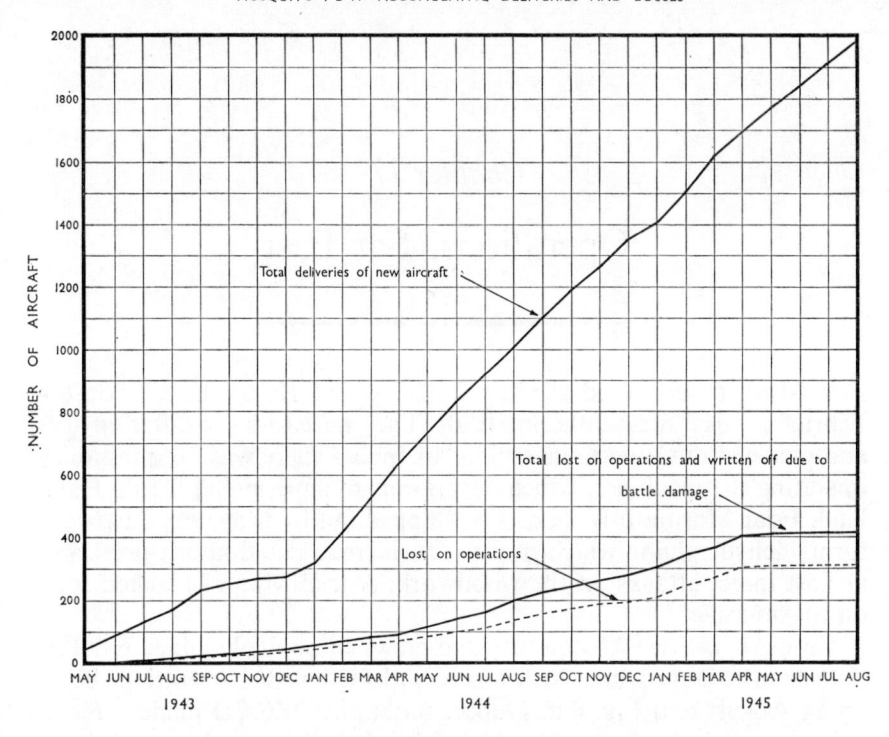

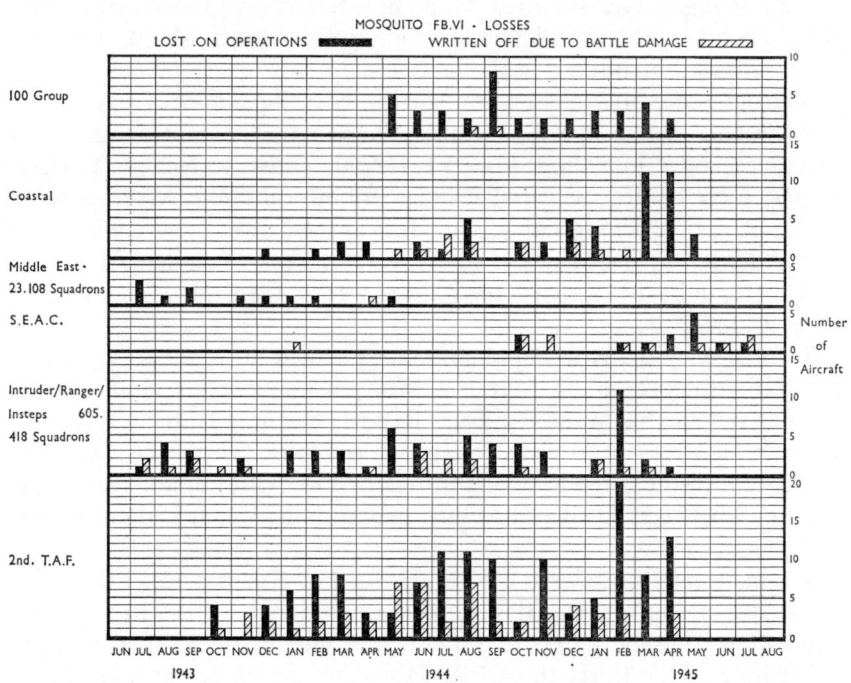

Deliveries and losses of fighter-bomber Mosquito, Mk. VI, between May, 1943 and August, 1945

Chapter 17

Operations: Far East

Air Ministry proposed sending six Mosquito IIs to the Far East in February, 1943, for weathering trials. They were to be of differing ages, and picketed out in various regions of India. There was no question of operating the machines, which Myers was to superintend for de Havilland. Four Mosquito IIs had been shipped out by May, 1943, two with formaldehyde gluing which, it was considered, would afford protection against insect attacks on the woodwork. A trial Mk. VI reached India on 10 August.

On 9 August two Mosquitoes and crews surprisingly joined 681 P.R. Squadron at Dum Dum. Such was its satisfaction that operations began on 23 August and Flg. Off. Dupee, D.F.M., in DZ697:J made a reconnaissance flight over the area Mandalay–Shewbo-ye u–Monywa–Wuntho. Next day Flt. Lt. Picknett reconnoitred Akyab Island. Thus the Mosquito was introduced to the Japanese.

During September eight more sorties, covering 11,000 square miles, were flown, Prome and Rangoon falling to the nose cameras of the machines, now including Mk. VI HJ730. Close to enemy territory DZ696 force landed, with bullet holes in both oil tanks. Myers assisted by an R.A.F. working party, with few spares, fitted new engines and undercarriage to the casualty, despite an outbreak of cholera, dysentery and torrential rain. Repairs took three weeks, then the Mosquito flew from a jungle strip.

Contingency plans in September, 1943, called for one PR. squadron, two NF. squadrons and four FB. Mosquito squadrons to operate as soon as practicable in the Far East. Reports reaching Hatfield regarding weathering were so good as to appear dubious, for the aircraft were being kept in the open in temperatures of 130° with humidity at 88 per cent; and between April and August the rainfall at Dum Dum measured 53·64 inches.

It was decided to release the PR. IX for Far East service. LR440 left for India 30 August, and flew its first operational sortie—to Rangoon and Magwe—in the hands of Flt. Lt. McCulloch, 21 October. PR. IXs made nine of the thirty-three PR. Mosquito sorties of the month. On the last day two IIs photographed the Akyab trail and Rangoon, then Mosquitoes were withdrawn from 681 Squadron.

Operations: Far East

684 Squadron formed at Dum Dum next day equipped with four Mitchells, two Mosquito IIs and three Mk. VIs. The first PR. IX arrived on 18 October, the second lost its way from Allahabad and was destroyed in a night landing at Ranchi and the third arrived on 23 October. 684 Squadron began operations on 1 November, Flt. Lt. C. R. Fenwick making a reconnaissance of the coastline from Port Campbell to Interview Isle, and a Mk. VI flew a tactical mission. Next day a Mk. VI surveyed Japanese airfields whilst DZ697 headed for Rangoon. It failed to return. According to Japanese radio a Mosquito shot down over Rangoon was 'no match for Japanese fighters'. Fighter Mosquitoes were used for low-level strafing missions as well as P.R. work. One was forced down over Allied territory by a Mitsubishi Zero, suffering a Glycol leak. Most of the flights were of long duration, extending past Rangoon and Moulmein, requiring refuelling stops at Chittagong. Build-up of Mosquito strength in the Far East was gradual, the monthly quota for the period being two Mk. VI and four P.R. machines. In its first month of operations 684 Squadron flew thirty-eight sorties, 68,000 miles. A further three sorties failed, two due to engine trouble and one to exhaust failings. There was no rain, but humidity averaged 78 per cent—and the Mosquitoes could fairly be said to be behaving well. Myers' contribution to the all-round efficiency was such that the C.T.O. India expressed a desire to keep his services at least until January, although impending despatch of pressurized Mk. XVIs made a refresher course more necessary, so he returned to Britain on 5 January.

On 9 December 684 Squadron moved to Comilla, whence very long-range flights began, to North Thailand, northern Burma and the Andamans. On 5 December LR443 secured pictures of the Burma–Siam railway and the 'Six Hundred' bridge, which was to be deeply engraved on the memories of so many, and photographs of Bangkok were taken a few days later. Reconnaissance flights were being made at around 28–33,000 feet. On 22 January the squadron strength had risen to twenty Mosquitoes—mainly Mk. VI and IX but with a sprinkling of tropicalized Mk. XVIs, the first three of which reached the squadron in February, and commenced survey work in March.

Such was the continuing success of the Mosquito in tropical conditions, and the high rate of production, that it was decided in January, 1944, that of the hundred squadrons planned for the Far East theatre twenty-two were to fly Mosquitoes. Beaufighters were to handle night defence and the strike role, until sufficient Mosquitoes reached the Far East. Spitfires would offer day defence and serve in a tactical role, and the bomber offensive would be mainly by American-built aircraft.

This called for elaborate preparations for servicing, undertaken now with high priority. Only very minor repairs were possible on the squadrons, and owing to the climate minimum stocks only of timber could be held in eastern India and Bengal. All repairs above a minor nature were to be dealt with by two R.A.F. Repair Depots within 50

Operations: Far East

miles of Calcutta, supplied by a stores depot close by. In Karachi the D.H. branch (the Company's first overseas establishment, opened 1927) would manufacture tailplanes, flaps, fins etc., which would be flown to Calcutta. D. S. K. Crosbie was in charge. D.H. would move later into better quarters in Mauripur. Interchangeability jigs were to be sent out from England and all raw material would be made free issue to D.H. at Karachi. Thus the decision to operate large numbers of Mosquitoes in the Far East can be seen to have required considerable planning. It was some time before the Mosquito served in quantity in that theatre.

Throughout 1944 684 Squadron was hectically engaged, using Mk. IXs and later almost entirely the Mk. XVI, on very deep penetration and mapping flights over Burma, Siam and the Nicobars. On 31 January they returned to Dum Dum, and settled at Alipore on 6 May for the rest of the year. Between 1 November, 1943, and 30 May, 1944, 3,000 hours' operational flying had been completed and some 232,100 square miles of enemy territory photographed. One flight was made over enemy territory carrying Earl Mountbatten. Typical examples of the length of the flights undertaken and return mileages are: Koh Si Cangs Is. (2,113), Mergui (2,084), Singapore and its railways (2,028), Bangkok (2,000), Car Nicobar (1,903), and Nancawry (2,256) this taking 7 hours 20 minutes, averaging 308 m.p.h. A need increasingly apparent was for a super-long-range Mosquito, a requirement attended to at Hatfield, and ultimately answered by the PR. Mk. 34.

By mid-1944 Australian built Mosquitoes were working as far west as Sourabaya but this left a gap of almost 2,000 miles of Japanese territory beyond the Mosquito's range. At Hatfield an in-flight-refuelled version of the PR. XVI was briefly considered in March, 1944, and was schemed with an all-up weight of 24,710 lb. More favour was won by the PR. XVI carrying 200 gallons in each long-range wing tank.

Steps implementing January's decisions were soon taken and 1672 Conversion Unit formed at Yellahanka to train crews to operate Mosquito VIs which were to replace Vultee Vengeance bombers in India. A motley collection of aircraft was first gathered, including two Mosquito III trainers, Blenheim Vs and a dual-control Anson. March 1 brought the first Mk. VIs for the unit which was initially to convert 45 Squadron, who greeted the news on 24 January with great enthusiasm, and received their first Mk. VI on 29 February. Three months' conversion training was undertaken, slowed by the few aircraft available and the pilots' need to convert via Blenheims to twin-engined aircraft. Several changes of bases followed before the squadron was ready for operations.

82 Squadron began conversion 4 July at Kolar. They moved to Ranchi to begin operations in October, by which time 45 Squadron were at Kumbhirgram. On 28 September two VIs of 45 Squadron flew a special reconnaissance flight for army officers, but the role was officially the destruction of roads, railways and associated bridges. These operations began on 1 October with three strikes by 45 Squadron, one by six aircraft,

Operations: Far East

on targets of opportunity. Such operations were flown by day and night against road, rail and river targets, huts and ferry boats. HP941 was lost on 16 October, and HP921 on 20th, the day when disaster of another sort struck.

Flg. Offs. A. E. Parker and M. D. Randall of 82 Squadron were on 20 October detailed to make a practice bombing attack on Random Range. During the shallow-dive run in, half of the starboard wing outboard of the nacelle crumpled and disintegrated. The Mosquito rolled into the ground. Immediately all others in India were grounded. Inspection of the aircraft revealed that heat and exposure were at last taking their toll, causing the glue to crack and skinning to lift from the spars. Fabric cracked, too. Accordingly all Mosquitoes were ordered to be flown to No. 1 C.M.U. Kanchrapara, if they had been in the Far East for three months or over, or had spent a month in the Middle East. Further investigation revealed that the aircraft in which formaldehyde glue had been used were in satisfactory state. At Hatfield all completed parts in which formaldehyde glue had not been used were ordered to be destroyed, or sold off as scrap.

47 Squadron were informed they were to convert to Mosquitoes on 19 September and moved to equip at Yelahanka early in October. On 5 October the last Beaufighter left and Mosquitoes were then received. An urgent signal of 25 October ordered all Mosquitoes grounded and by 20 November all had been found to be faulty. On 30 November Beaufighters returned.

Trouble could not have come at a worse time, for the campaign to reconquer Burma, in which the Mosquitoes played a long-range role, was about to begin. This, in effect, meant the neutralization of the railway network leading to Mandalay and Rangoon. Shorn of this, and lacking transport aircraft, the Japanese soon would be at a disadvantage, whereas General Slim's 14th Army could move steadily forward with plentiful air support. Thus, it was important to reintroduce Mosquito fighter-bombers as quickly as possible, especially for night intruder work.

45 Squadron resumed operations with cleared Mosquitoes for an attack on Meiktila airfield on 9 November. Eight Mosquitoes were involved. Near the target enemy fighters appeared, an Oscar II engaging and chasing Flg. Off. P. Ewing in HR368. Flt. Lt. C. Emery engaged another fighter, then his aircraft fell to flak. Lack of aircraft halted operations until 4 December, when a night attack re-opened the offensive. Airfields were bombed and strafed by day and night by small numbers of Mosquitoes. An order was received 15 December that all the squadron aircraft should have an F.24 camera installed in the nose.

On 19 December 82 Squadron commenced intruder operations. Sortie totals rose gradually and by 31 December ten aircraft were operating. Both squadrons were now working almost day and night on *Rhubarbs*, against trucks, trains, motor convoys and airfields and in January 45 Squadron completed 143 day and 64 night sorties. Since

Operations: Far East

September the squadron had dropped 1,183 × 500 lb. bombs, destroyed 3 enemy aircraft, 5 locomotives, 17 vehicles and 17 small boats—despite having been grounded three times. Low-level and shallow-glide attacks from 5,000 to 2,000 or 12,000 to 7,500 feet were usually flown.

Four Oscar fighters engaged three of 82's machines raiding Meiktila on 15 January, 1945. In the ensuing battle, around 8,000 feet, Flt. Lt. C. R. Goodwin in HR402 was shot down. After a spell of night operations 82 Squadron resumed day raids on 8 February despatching twelve crews to three targets in the Singu-Thila area, and from 15 February flew many dawn or dusk operations. There were a few more encounters with Oscars and on one occasion Sqn. Ldr. Tooth escaped from three by using superior speed. On another occasion an Oscar shot away his elevator cables, but he escaped. One hundred and seventy-one night sorties were made by 45 Squadron in February.

Conversion of two more squadrons commenced. 47 again received Mosquitoes from 18 February commencing operations with them five days later and flying with 45 and 82 Squadrons as part of 908 Wing. It was 22 April, 1945, however, before Beaufighter operations by No. 47 ceased. Mosquitoes bombed and strafed, whilst Beaufighters bombed and rocketed.

84 Squadron, like 47, progressed towards being equipped with Mosquito VIs the previous autumn using Oxford trainers for twin conversion and despatching two sergeants for training on Merlin engines at Kolar. The gluing trouble put an end to this programme and it was February before conversion was resumed. 110 Squadron at Yelahanka was also now converting, its first Mosquitoes having arrived on 28 January. During March 84 and 110 worked up, making low-level training assaults on the docks at Cochin. During April 84 Squadron moved away from the scene of action and 110 Squadron took its place at Joari, beginning operations on 31 March, its paired aircraft patrolling roads and waterways. These tactics continued until 24 April when two waves of six and five Mosquitoes attacked M.T.B.s at Thonga. One hundred and thirty-one sorties were completed by the end of the month.

Meanwhile 45 and 82 Squadrons had been busily engaged, 45 making 287 day and sixty-five night sorties in March, a record month. 82's 269 sorties represented for them a record, in spite of hazy weather making operations difficult. Many violent storms broke around their base. Accustomed to their aircraft, 82 Squadron made some highly successful attacks, during one of which on 8 March Wg. Cdr. F. W. Snell, D.F.C., made his run in so low against a bridge at Pyinmana that HR497 was damaged by an exploding land mine which blew a hole in its nose, and riddled it.

After the fall of Meiktila in April, 1945, the bomb line moved rapidly forward through the jungle, calling for fast long trips. Fewer sorties were flown, mainly concentrated on convoys proceeding south, as the weather continued to deteriorate. All was leading to the all-out attack

Operations: Far East

on Rangoon, Operation *Dracula*, made 1 May. For this the Mosquitoes of 45, 47 and 82 patrolled over the city mostly on a cab rank basis, attacking targets suggested by the Army. This was virtually the end of Mosquito fighter-bomber operations prior to VJ-Day, although 110 Squadron continued operations into August. May 12 was the last day of operations by the remainder.

In difficult conditions heroic deeds were many. Wg. Cdr. R. J. Walker made a remarkable return after being wounded when attacking a river bridge at Toungoo, 160 miles north of Rangoon in April, 1945. He made a trial run over the target at 50 feet, then followed another aircraft in at 75 feet. As he pulled away over trees he was hit by a shell splinter in the head and in one eye, then lost consciousness. Recovering he found his navigator, Flg. Off. F. J. Harler, in control of the aircraft. He took over so that they could regain sufficient height, and bale out. Then he decided they could make base, but the forward strip he flew towards was being attacked. In great pain he managed to fly another 100 miles, and put down. He collapsed on landing. In hospital he later said, 'During the flight back I couldn't talk to Harler because my helmet was damaged so we had to exchange notes. He bandaged my wound and his thoughtfulness and excellent behaviour were undoubtedly responsible for our getting back safely.'

As the Allied armies advanced into Burma 684 Squadron at Alipore maintained its watch upon enemy movements near and far. Rangoon, Bangkok and Victoria Point were frequently photographed. A Mosquito which gave excellent service was LR464, flown to India by Sqn. Ldr. K. J. Newman, D.F.C., and Bar, R.N.Z.A.F. and W. Off. A. K. Smith, D.F.M., in December, 1943. It made its first sortie on 2 January, 1944, to Sittang Bridge–Mesareing. Over Burma, Siam, South-West China, the Andamans, Nicobar Islands and N.W. Sumatra flying from the mainland or the long-range detachment base at China Bay, this aircraft logged 495 hours flying—320 on operations—before ending its operational career on 23 March, 1945, when it was returned to England for technical examination. It had covered some 90,000 miles behind enemy lines in the course of fifty-seven sorties. Only once did it turn back with a defect, and was noted for its consistently low fuel consumption.

Flt. Lt. Stoneham made a 5-hour flight home on one engine in NS657. Wg. Cdr. Lowry set up a long-distance record for S.E. Asia in January, 1945, with an 8-hour 20-minute flight of 2,350 miles. In the next two months it was thrice beaten, first by Sqn. Ldr. Newman and W. Off. Smith, then twice by Flt. Lt. Irvine and W. Off. Bannister, always in the same aircraft (NS675). The last trip of over 8 hours and 2,483 miles was through extremes of heat and cold, and in dangerous tropical storms. A thousand miles from base Jack Irvine circled the target after having spent two hours on the outward leg in cloud at 28,000 feet, when the nose and wings of the aircraft were covered in ice. This, he reported, did not worry him. But he was afraid of dangerous

Operations: Far East

winds and would have turned back, but for the importance of the target. On flying out into the sunshine he was only 20 miles off track. They landed back with fuel for 20 minutes. Flt. Lt. Irvine flew NS675 to Puket Sound on 9 March, obtaining excellent shoreline photographs on an 8-hour 48-minute flight. A flight of 8½ hours was made by NS479 too.

Still the watch on the Siam-Burma railway attracted the squadron, as on 14 February when Wg. Cdr. Lowry flew NS622 over the area. Irvine's record survived for about a month. In April, 1945, Flt. Lt. Robin Sinclair, son of the Secretary of State for Air, with Flg. Off. R. Stocks, flew 2,490 miles in 8 hours 45 minutes in NS688. But the ultimate record, 2,600 miles in just over 9 hours, came on a round trip over Penang and Taiping from the Cocos Islands on 20 August, 1945.

Sqn. Ldr. Clif Andrews R.N.Z.A.F. and Flt. Lt. Jack Irvine left Alipore at 10.00 on 16 June and headed north. At 27,000 feet they found clear weather and could see nearby the peak of Mt. Makalu, 27,800 feet, poking through the clouds. Ten miles to the East lay Everest obscured by a small cloud bank. Suddenly the cloud dispersed and the purpose of their mission, photographing the peak, could be achieved. 'We made a wide circuit and had a splendid view of the southern slopes', reported Andrews, 'of the South Peak, Lhotse, and the saddle between Lhotse and Everest. I was so dazzled with this scene of beauty that I think I just kept on circling.' He then flew closer to Makalu, but once clear Everest again absorbed attention. 'We circled the mountain for about 20 minutes, with ever changing beauty apparent in the scene.' As a farewell gesture they decided to fly as close as possible to the summit, flashing past about 30 feet away.

Still photography was done with 14-inch cameras mounted in wing drop tanks, taking a picture every 1½ seconds and 1/300th second exposure. Warm air was passed over the cameras to prevent the cold making the film brittle. A second flight to Everest was made by two aircraft—one with ciné cameras—on 1 July. Heavy snow was falling and gave only brief glimpses of the mountain.

The most important Mosquito event of 1945 was the arrival of the long-range PR. 34, which entered service June, 1945, with a special detachment of 684 Squadron established on the Cocos Islands for very-long-range missions. Wg. Cdr. W. E. M. Lowry, D.F.C., with Flt. Sgt. Pateman as navigator, flew first reconnaissance with the PR. 34 on 3 July, taking RG185:Z to Morib and making a run to Point Pinto. They flew to Gedong to cover the Port Swettenham area, and followed this with a run over Sumatra. This was the first sortie from the Cocos Islands, and was followed next day by a flight to Kuala Lumpur, over which RG186 made seven runs. Malacca, Singapore, Seletar and Kallang—all were photographed within the next few days and the PR. 34s had flown twenty-five sorties by the end of the month.

August 1 found the Cocos Detachment with three crews. Their

Operations: Far East

instructions were to cover three times in a fortnight all important targets. They encountered more opposition now than at any previous time, particularly over Palembang where the flak was intense. Even so, thunderstorms over Malaya created greater hazards. Thirteen sorties had been flown by VJ-Day, mainly over Malaya, and to Java, over which Flt. Lt. C. G. Andrews was the first to fly. By the end of August seven PR. 34s were based on the Cocos, but other long-range PR. XVIs equipped another detachment at China Bay which had concentrated on Akyab and the Andaman areas. Flt. Lt. J. R. Manners and W. Off. F. A. Burley in RG210:J made the longest-ever flight from the Cocos on 20 August. They took off at 06.10, flew 2,600 miles to Penang Island and Taiping, and landed back at 15.15, taking 9 hours 5 minutes for the trip.

Another version of the Mosquito appearing in large numbers in the Far East in the summer of 1945 was the NF. XIX. 89 Squadron stood down at Baigachi, where training for night fighting using A.I. Mk. X commenced. Wellingtons flew in from Italy, service personnel came from Naples by sea in April, four months after the plan to send the XIXs east had been made. Eleven Mosquitoes reached 89 Squadron in May and training was completed 18 July. Surplus aircraft were set aside for 176 Squadron, also at Biagachi. 176 began converting on 14 July. Progress of the war was such that neither squadron took its Mosquitoes into action, although 89 Squadron was ready for Operation *Zipper*. Both units disbanded in 1946, after comparatively little flying.

Plenty of Mosquito VIs in the Far East (351 in all) allowed 211 Squadron to convert in June, 1945. 211 left India for Siam October, 1945, but the order grounding Mosquitoes for spar inspection on 20 January, 1946, put paid to action. Apart from policing duties on the mainland of Asia, Mosquitoes of 684 Squadron found immediate post-war employment in a major role, a survey of Indo-China from Saigon begun in November, 1945. All Cambodia was photographed in January, 1946, the Mosquitoes operating from Bangkok. A few weeks later extensive surveying of India commenced, work that was to take many months.

For fighter-bomber squadrons offensive action had been resumed. A constant watch was maintained for pockets of resistance, following Japan's unconditional surrender. 110 Squadron's Mosquitoes made thirty-three sorties in the first eleven days of August, 1945; then the squadron was directed to neutralize Japanese resistance at Tikedo, and two other positions, with eight aircraft on 20 August, after the surrender. Thus, it fell to RF954:Y and HR562:T to make virtually the last two Mosquito strikes of the war, although there was soon a further campaign.

In the Netherlands East Indies there was an element that did not wish to come again under protection of the Dutch and, preferring independence, proclaimed the existence of Indonesia. Extremist elements

Operations: Far East

inflamed the situation, and a few Japanese aircraft on the islands were acquired. There were clashes with Allied troops rounding up Japanese, and repatriating prisoners. Following the murder of the Commander of the 49th (Indian) Infantry Brigade it was decided that action must be taken to curb the restive element. 84 and 110 Squadrons were called upon to detach Mosquitoes and from 9 November these were available for operations, making armed reconnaissance flights and ready to use their bombs and defensive armament. 82 Squadron briefly joined them on 15 November sending five rocket-armed aircraft to Java. This was somewhat surprising, for no one on the squadron had ever fired rockets and in the Far East this role so far had been restricted to Beaufighters! 47 Squadron, after the war, had worked up with rockets. On 17 November they despatched four R.P. aircraft to Kemajoran. That day they sought out a radio station playing a major part in the troubles. TacR flights continued. On 25 November, hits were obtained by four 47 Squadron Mosquitoes on the Soeramarta radio station. Next day three of 84's aircraft made a short strike on terrorists in the Ambarawa region. A shower of leaflets from a Beaufighter preceded an attack by four Mosquitoes of 47 and 84 Squadrons on the radio station at Tagjakarta.

During December 47 Squadron flew many patrols covering British troop movements. Twenty-four R.P.s were fired into a persistent road block in the Kampong area on 21 December. Although escorts and reconnaissance flights continued until 20 January there were no more strikes. Operations officially ended for the Mosquitoes of 47 Squadron on 12 March, 1946.

To the north France encountered similar trouble when she attempted to regain power over Indo-China. By the end of 1945 the French Air Force had been re-established. An assortment of aircraft were despatched to this troubled area, among them many surplus American types. With the war in Indo-China going increasingly badly, France began despatching aircraft from Europe. From Rabat/Sale, Morocco, 10/Groupe de Chasse 1/3 'CORSE' left on 3 January, 1947, for the Far East. A convoy of fifteen Mosquito VIs arrived at Saigon ten days later. Three days after the arrival of their ground equipment on 20 January, the Mosquitoes flew their first offensive sorties. By the end of May, 1947, 345 sorties had been flown and 169,000 pounds of bombs had been dropped during 748 hours 10 minutes of operational flying. The formation then returned to Rabat, to become Groupe de Chasse 1/6 Corse, and continued using Mosquitoes in Africa until July, 1949.

Whilst squadrons of Mk. 6s faced the Iron Curtain in Germany, in the Far East uneasy tension turned into the clearing of Communist terrorists from Malaya, during Operation *Firedog*. In conditions as difficult as any, where great expanses of jungle foliage gave plentiful cover to underground forces, aerial reconnaissance played a vital part, allowing the slightest evidence of enemy activity to be detected at a

Operations: Far East

distance and attacked by strike forces. By this time Brigands had taken over from the Mosquito 6s, leaving 81 Squadron's Mosquitoes the only representative of their breed in R.A.F. hands in Malaya.

From Singapore 81's Mosquito 34s went into action from the start of the clearing operation in July, 1949, daily making reconnaissance flights whenever the weather permitted. Over half of 81 Squadron was armed with Mosquitoes, the remainder using Spitfires. For seven years PR. 34s brought back vital information on terrorist movements, upon the evidence of which bombing raids were often despatched. On 15 December, 1955, it fell to RG314 to make the final operational flight by a Mosquito, and bring to a close the career of the Mosquito in the Far East.

Although it could not be so successful in jungle country as in other theatres, the Mosquito would have played an increasingly effective and large part in the defeat of Japan had it not been for the sudden end of hostilities. That the initial troubles (foreseen in 1940 as inevitable) which Mosquitoes encountered in the tropical climate were effectively cured is proven, beyond doubt, by the excellent service rendered over Malaya during the emergency there.

Chapter 18

Operations: Coastal Strike

The Mosquito fighter was envisaged in two forms, home defence and convoy escort, for which long range and endurance were essential. These features typified the Mk. VI with which squadrons of Coastal Command were equipped. 333 Norwegian Squadron, formed from 1477 Flight, was the first Coastal Command Mosquito Squadron other than P.R. squadrons. Its employment was on armed reconnaissance along the Norwegian coast, leaving strike duty to Beaufighters arming other squadrons. Another important task was destruction of enemy submarines creeping into Biscay ports, to oppose which the Mk. XVIII was developed.

On 19 March, 1943, R. E. Bishop received this letter from M.A.P.:—

'We want to investigate the possibility of making a six-pounder gun installation in a Mosquito; and the first stage is to know from you whether you are satisfied that the MOSQUITO structure is strong enough to carry a gun installation weighing—with ammunition and supporting bar—1,800 lb., with the trunnion reaction as high as 8,000 lb.

Would you please think this over quickly and let me know your views as soon as possible.

Yours sincerely,
J. E. Serby, D.D./R.D.A.

Bishop replied that de Havilland knew nothing of the gun, but from rough investigation it seemed there would be little difficulty in dealing with an 8,000 lb. reaction. In his reply of 28 April, J. Serby gave instructions to proceed with a prototype installation, promising contracting cover.

This was not the first time a heavy gun installation had been considered. Mounting a 3·7 in. anti-aircraft gun into the Mosquito, and on 4 December, 1942, a ground-attack version, with extra armour plating, were discussed and weight estimates made.

Rapid development characterized nearly every modification of the Mosquito. It speaks much for its design that modifications were invariably highly successful. So with the new machine, which became the Mk. XVIII. Working fast, de Havilland cut the nose portion from the fuselage of a crashed Mosquito, installed an ordinary six-pounder field gun and fired it on the Hatfield butts discovering the blast effect on the

Operations: Coastal Strike

nose. This happened on 29 April, 1943, and M.A.P. made a ciné recording showing a side view of the nose and recoil characteristics. In the first week of May the firm produced a mock-up of the proposed installation in another crashed Mosquito fuselage to see how the ammunition feed could be placed. That same week a Mk. VI HJ732 was wheeled into the Experimental Shop for the prototype installation.

On 10 May, 1943, M.A.P. wrote saying it had been decided thirty of the Mosquito VIs to be delivered in 1943 were to have the standard 6 lb. anti-tank gun.

Sunday, 6 June, saw the prototype aircraft taken to the butts and its gun fired. On 8 June it made its first flight, and three days later Air Vice Marshal Ralph Sorley, then Controller of Research and Development, came to see it, on the day before it flew to Boscombe Down. No difficulties with blast or recoil were encountered, and at A. and A.E.E. it began air firing trials. Armour plating within the cowling was fitted to another aircraft, also around the nose and over the cockpit floor. This added 900 lb. to the machine, but was necessary since it was known that the purpose of the Mk. XVIII was for destruction of heavily armed U-boats creeping into harbour.

At Boscombe Down the ammunition feed system proved troublesome. In the first week there one hundred rounds were fired at the butts, in an attempt to clear the trouble. This was not the fault of de Havilland, but the aircraft returned to Hatfield on 22 June for modifications to the feed unit and weapon bay doors. Thirty conversion sets were now underway in the shops, and in eight weeks the trial installation, prototype and production drawings had been produced. M.A.P. on 19 July advised de Havilland that because of the shortage of aircraft only three Mk. VIs were presently to be converted.

HJ732 spent July and August at Boscombe Down and Exeter for air firing trials, during which about eight hundred rounds were fired. The rifling on the gun wore at anywhere between three hundred and five hundred rounds per barrel, but it was found to be a good accurate weapon. By October three production aircraft had reached Boscombe Down. De Havilland had been asked to fit 65-gallon tanks in the fuselages allowing them to range far over the Bay of Biscay.

Two Mosquito FB. XVIIIs were allocated to 248 Squadron, based at Predannack, on 22 October, 1943. A fortnight previously a detachment of five crews from 618 Squadron arrived, along with thirty-four ground staff to service the aeroplanes. This was necessary since 248 was a Beaufighter Squadron. 618 Squadron, with much Mosquito experience, was an ideal choice for, here was a chance for crews denied their primary intent as described in Chapter 7, to fly on operations. These commenced on 24 October, 1943, when two Mk. XVIIIs, with Sq. Ldr. Rose in E:HX902 and Flg. Off. A. L. Bonnett in I:HX903, set off on a U-boat hunt over Biscay. On 4 November, when next they operated, they fired at a trawler, but tragedy followed. As Sqn. Ldr.

Operations: Coastal Strike

C. F. Rose, D.F.C., D.F.M., made his second attack the trawler's fire hit home. Smoke poured from the tail of his Mosquito. He jettisoned his hood and as he prepared to ditch, the aircraft hit the sea, disintegrated, and left no survivors. Flg. Off. A. L. Bonnett, R.C.A.F., flying I 'Item' on 7 November found a surfaced U-boat proceeding slowly to its base south of Brest on the cold, bleak morning. Its crew unexpectedly spotted the Mosquito approaching, firing hefty shells several of which struck aft of the conning tower producing black and yellow smoke as the submarine crash-dived.

After two such attacks the Germans were forced to provide a surface escort to ships and fighters for returning U-boats, but the attacks continued.

Notification reached 248 Squadron on 20 November that their Beaufighters were to be replaced by Mosquito VIs in mid-December. Conversion was via a T. III and two Mk. IIs, and training was necessary in the use of *Gee* and *VHF* radio. As conversion took place operations continued, using Beaufighters and the FB. XVIIIs. Squadron strength on 1 January, 1944, was established as 16+4 Mk. VIs for operation under 19 Group. Fighter reconnaissance was to be the primary role, secondly fighter support for strike squadrons. There were to be regular interceptor patrols over anti-shipping and U-boat operations in the Biscay area. These commenced on 20 February, 1944, three days after the squadron arrived from Portreath. Four or six Mosquitoes usually formed the force and ten patrols had been made by the close of the month.

Flt. Sgt. Doughty on a navigation exercise west of Ireland had a harrowing experience when, in LR346, he came face to face with an albatross. The giant bird attempted to swoop beneath his port wing as he lowered this to avoid collision. Although the bird collided with the wing leading edge, damaged the spar and No. 3 fuel tank, the Mosquito flew home to roost. Whilst this was happening another exciting moment came as the squadron's Mosquitoes first faced an enemy aircraft, a Ju 88 with which Q:MM430 and R:LR828 inconclusively exchanged shots.

On 10 March came the first major action. Four Mk. VIs escorting two XVIIIs found a German naval force covered by Ju 88s. Four were immediately engaged and one fell in flames. Another probable was scored by HJ828 drawing fire away from the XVIIIs. 'Y' chased a couple of 88s, claiming one. MM431 twice attacked but both times lost targets in cloud. Meanwhile the XVIIIs, L:MM425 and 'E' attacked the convoy, singling out a U-boat for special attention and damaging a destroyer. For good measure they, too, shot down a Ju 88. HP922 fired at a Junkers from 30 feet, before racing off to contact two Liberators which continued the convoy attack.

From 27 March the squadron's Mk. XVIIIs formed '248 Squadron Special Detachment' existent until January, 1945. Apart from gun-blast trouble Mk. XVIIIs performed well in service. Air Chief Marshal Sholto

Operations: Coastal Strike

Douglas, G.C.B., M.C., D.F.C., applied for 248s establishment to be raised to six, but the Air Staff turned down the idea, since rocket-armed Mosquitoes were already in being.

A tough engagement on 11 April was the highlight of that month. Six Mk. VIs and two Mk. XVIIIs took-off to attack a U-boat escorted by a Sperrbrecker, flak ship, two trawlers and eight Ju 88s. In the ensuing battle two Ju 88s were destroyed, the ships damaged but two Mosquitoes lost. Additionally one crashed on landing and another, lost soon after take-off, crashed into a hill. On 30 April came notification that 248 Squadron was to continue in its present role, and carry drop tanks but not bombs on the wing racks. Further, it was told to hold itself in readiness during the coming invasion to be called upon to attack land targets.

A Jaguar-class destroyer presented the most tempting target in May, but failing light prevented an assault and the eight aircraft despatched returned. Twenty-four operations were mounted in May, calling for 118 sorties and 400 hours' flying.

248 was one of the busiest Mosquito squadrons on D-Day flying sorties from 04.45 until 22.15 hours. Three tasks were allotted it, a) prevention of U-boats and ships from attacking allied craft, b) blockade of Biscay and Channel ports, and c) affording cover to air strike forces. Five operations were mounted, the third at mid-day by two Mk. XVIIIs MM424:H and NT225:O which found two freighters in the Gironde Estuary but held their fire hoping for more offensive targets. On the fifth and final operation ten Mosquitoes took off at 18.45 to cover seventeen anti-flak Beaufighters of 144 Squadron and fourteen R.P.-armed Beaufighters of 404 Squadron on a roving mission in the Biscay area. A U-boat was seen to crash-dive, and a few moments later the formation came across three destroyers steaming north line abreast at fifteen knots. 248 Squadron climbed to cover the Beaufighters, which set the middle ship afire then holed the rear one below the waterline leaving it sinking, and seriously damaged the leader with the R.P.s. Mosquito G:HR120 fired shots at a shadowing Ju 188 which F:LR339 then destroyed.

The nearest Mk. XVIIIs had come to sinking a U-boat was when MM425:L fired twelve 57 mm rounds at one on 7 June. The submarine crash-dived leaving a crew member to swim for home from a large oil patch, whilst overhead NT225 made dummy attacks because she suffered a stoppage in her heavy cannon. Two days later on their third attempt of the day Y:HR138 and I:HX903 found an enemy destroyer aground near Ile de Bas, north of the lighthouse. Seeing little point in attacking, they headed homwards. As 'Y' broke to port for landing the upper machine collided with it slicing off its tail. 'I' fell into the sea, while 'Y', with six feet chopped off the starboard wing and almost out of control, raced in to land with 15 degree flap and 160 knots on the clock.

June's activity continued unabated. On 10th four Mosquitoes attacked

Operations: Coastal Strike

U-boat U821, the crew from which jumped into the water leaving their submarine to be sunk by Liberator 'K' of 206 Squadron. NT225 scored two hits on the U-boat and five of the Liberator's nine rounds hit it, so in all fairness 248 Squadron could claim half a submarine at last. Three days later a Ju 188 fell to HJ828 and HP908 off Ushant, and at dusk that day the Command's prophecy of a ground-attack role for 248 came true when LR346 shot up Morlaix airfield. At 05.02 next day operations were resumed. There was no respite.

To widen further Mosquito usefulness, tests were undertaken with Mk. VI HR135, borrowed from 248 Squadron, assessing suitability for dropping wing-mounted depth charges or mines. Release and handling trials with the Mk. XI depth charge and A. VIII mine were undertaken at Boscombe Down in 1944. It was found that the Mosquito could be dived safely to an I.A.S. of 450 m.p.h. but mines needed to be dropped at no more than 250 m.p.h., otherwise they were likely to hang up. At a conference on 5 July, 1944, it was suggested that wooden noses be fitted to the mines, but this was abandoned and no mining undertaken by Coastal Command aircraft. First operational use of 25 lb. depth charges occured on 22 June, 1944, when Q:MM430 of 248 Squadron came across what appeared to be a periscope off St. Malo. Weapons were dropped upon it, but afterwards it was thought the target might be only a buoy. Next day a mixture of VIs and XVIIIs went after a convoy and a U-boat near the Ile de Groix, the leader attacking with depth charges and machine guns. Another operation to the same region the following day, resulted in four mine sweepers being attacked. The decision not to use bombs having been rescinded, 500-pounders were released on a T.T.A. of 1,000 tons and a tug boat, in the Gironde Estuary, by K:LR340 and C:LR378 of 248 Squadron on 25 June.

248 Squadron was no longer the only Mosquito strike squadron in Coastal Command for during June 235 Squadron, also at Portreath, received Mk. VIs. Sqn. Ldr. Barnes made the unit's first operational flight on 16 June. By the end of the month thirty-two sorties by the squadron's Mosquitoes had been completed. Beaufighter operations ceased on 22 June. On 27 June six of 235's aircraft joined with nine of 248 on the first of many similar missions. They were providing escort to Beaufighters of 144 and 404 Squadrons. No attacks were made upon ships found, for they were too awkwardly placed. The first successful strike by the two squadrons came on 29 June. Two dozen Mk. VIs and two Mk. XVIIIs attacked a convoy comprising a tanker, two escorts, two A.S.T.s and two trawlers. Ten Mosquitoes swept in with cannon and machine guns blazing, follow by two with 500 lb. M.C. bombs, and then Mk. XVIIIs. Considering that they had practised little the bombers did well, and the tanker was left blazing. Next day twenty-one Mosquitoes and nine rocket-firing Beaufighters shot up four naval vessels off Concarneau. During June 248 Squadron flew 274 sorties in the course of eighty-nine operations.

Operations: Coastal Strike

Twice in three months Flt. Lt. S. G. Nunn proved the remarkable staying powers of the Mosquito when, after its being badly shot up, he brought his aircraft home to make a belly landing. The first time was on 11 April, 1944, when, flying LR362:T of 248 Squadron, he fired at shipping and at its escorting Ju 88s, one of which put shots through his starboard propeller; this severed the hydraulic oil pipe supply and the aileron-trimming controls, and blew off the port flaps. Nunn dived to sea level and, escorted by a Mosquito of another squadron, cruised back to base 350 miles away on one engine. He could not lower his flaps or undercarriage, and made a belly landing. On 16 July flying 'T' of 248 Squadron, Flt. Lt. Nunn was leading a section of three Mosquitoes on an anti-shipping strike, again off St. Nazaire. Finding his cannon had a stoppage he dived to 200 yards to fire at a patrol vessel with machine-guns. There was a bang and a bump as flak from the ship hit his port engine. As he broke away he could see oil and coolant streaming from the engine, in which there was a red glow. He climbed as he flew over land fearing his fuel tank would catch fire, and warned his navigator to prepare to bale out, but the emergency exit door was jammed. Nunn headed out to sea and dived to sea level only to find himself between two ships which opened fire on him. On one engine he weaved his way out to sea. He now called up the formation, which he was able to join, and, flying on one engine, again slithered into base. There he found that the starboard mainplane and the rudder, elevator and trimmer were badly shot up. Following on site repairs LR347 served 248 Squadron until July, 1945.

July was less busy, as Allied forces advanced and the Germans were driven from the west of France. The highlight came on 21 July, when HR127 and LR346 of 248 Squadron were escorting Allied ships in an area where, on the previous evening, Dornier 217s had used Henschel 293 glider bombs. At 13.20 hours an enemy aircraft was seen to crash into the sea and the two Mosquitoes raced to the scene finding three Dorniers, two of which they promptly destroyed. Two relief Mosquitoes believed they engaged an He 177 which quickly entered cloud, after shooting down a pursuer.

Do 217s with glider bombs were again discovered on 9 August, when twelve Mosquitoes of 235 Squadron and two Mk. XVIIIs of 248 Squadron, on a Biscay patrol, finding four bombers, shot down two and damaged the others. Between them the two squadrons flew 339 sorties in August, mainly attacking shipping in the Gironde Estuary and during Biscay patrols. Thirty-five operations were mounted.

A brief concentrated period of operations came towards the end of August but by September enemy activity had dropped off rapidly, and it was decided to transfer the squadrons to Banff, Scotland, early in the month, thereby completing Operation *Outmatch*. Final sorties over Biscay came on 7 September, 1944, when six 248 aircraft searched in poor visibility for U-boat activity. At Banff the squadrons met 333

Operations: Coastal Strike

Norwegian and Beaufighter squadrons forming the Dallachy Wing. Between them they were to mount impressive assaults on shipping off Norway. Mosquitoes were to protect Beaufighters from enemy fighters, and to silence flak ships. 333 Squadron's skilled Norwegian pilots as outriders often finding or leading the way to pin-point targets situated in fiords which, by inexperienced eyes, were difficult to locate, particularly in frequently poor visibility.

Forty-four aircraft took off on the first armed reconnaissance of the Norwegian coast on 14 September. There were twenty-five Mosquitoes of 235 and 248 Squadrons, four Mk. XVIIIs of 248 Squadron, seven Beaufighters of 144 Squadron and twelve of 404 Squadron which led the formation, searching for ships between Egero and Stors Toreungen Light. The Mk. XVIIIs left the formation before it struck at four motor vessels sailing north, protected by two escort ships. Hits were claimed on all of them, and fires started.

A U-boat search came three days later. On 21 September the next large-scale strike was mounted, six Mosquitoes of 248 Squadron escorting twenty-one Beaufighters to the Kristiansund region. This was followed by a large *Rover* on 30 September by seventeen Mosquitoes of the three squadrons, and twelve Beaufighters.

Three large-scale strikes took place in October, 1944, and on the 26th Mosquitoes fired rocket projectiles for the first time, acquiring much increased potency. Effective when aimed with skill and precision, the aerial torpedo was a costly weapon, and even late in the war not without imperfections. Rockets—considerably cheaper and small enough to be carried in reasonable numbers—had much to commend them, and in the autumn of 1943 de Havilland had studied the possibilities of fitting rocket rails to the wings of the Mosquito. On 28 September the installation of four racks under each wing capable of carrying four 60 lb. rockets was approved. Flight tests made at Hatfield in mid-October and on 25–26 November at Boscombe Down, revealed an ideal dive angle of 20 degrees. Success of the installation was such that future Mosquito bombers and fighter-bombers were able to carry rockets, which produced no recoil stress to the airframe. Two types fired operationally were a 25 lb. solid armour-piercing type and, more usually, 60 lb. semi A.P. high-explosive-headed rockets, a salvo from which was equivalent to a broadside from a cruiser. On receipt of intelligence a handful of Mosquitoes could deliver, hundreds of miles away, equivalent fire power to a cruiser force, but with more tractability. They could repeat the dose a few hours later.

Rapid Mk. VI production permitted another squadron to replace Beaufighters by Mosquitoes, namely 143, which flew initial Mosquito sorties from Banff on 7 November, HR141 and PZ419 searching for enemy aircraft between Obrestad and Lindesnes. Armed with cannon, machine guns, rockets, bombs and the six-pounder gun Coastal Command Mosquitoes, expanded in number, participated in four major

Operations: Coastal Strike

operations during November. Eighteen Mosquitoes joined for the first time with six aircraft of Dallachy Wing on 13th, setting four ships on fire. On 21 November, 1944, Wg. Cdr. Sise, D.S.O., D.F.C., led the largest strike so far, against shipping at Aalesund. Thirty-two Mosquitoes, with an outrider from 333 Squadron and forty-two Beaufighters, given top cover by twelve Mustangs of Fighter Command, made the attack.

The closest Mosquitoes had yet come to sinking a U-boat was on 29 November, when hits were obtained by 20 mm cannon fire and a Mk. XVIII of 248 Squadron. The damaged U-boat crash-dived. An anti-shipping patrol by thirty-five Mosquitoes on 5 December was the first time a Mk. IV of 138 Wing attended to obtain a photographic record. Four motor vessels, two T.T.A.s and a tug, were set on fire for the loss of one aircraft. Strike Mosquitoes first came face-to-face with enemy fighters on operation on 7 December. Twenty-five from Banff with forty Beaufighters of the Dallachy Wing protected by 315 Squadron's Mustangs, were led to the Norwegian coast by Sqn. Ldr. Barnes of 235 Squadron. Near Gossen the Mustangs were engaged by about twenty Fw 190s and Bf 109s. In the ensuing battle 315 Squadron claimed four Bf 109s and two 190s collided. Two Mosquitoes, a Beaufighter and a Mustang failed to return.

German gunners firing from the shore and the sides of the cliffs failed to prevent Mosquitoes attacking a medium-sized merchant ship and its escort on 16 December. The vessels were hiding between rocks with steep cliffs on either side, at Kraakhellesund. A line-astern attack was first made because there was so little space for manoeuvre. Half an hour later a second wave attacked. Two Mosquitoes were lost. The second occasion when the enemy sent two formations of fighters to intercept a strike force, led by Sqn. Ldr. Norman Jackson-Smith, was on 26 December. A dozen Mosquitoes flying very low attacked two motor vessels in Leirvik harbour, leaving one sinking, another blazing. Of the twenty-four enemy fighters engaged one was shot down and one of 235 Squadron's Mosquitoes lost.

January 1945, a month of bad weather, brought less action. Flying through intense flak sixteen Mosquitoes attacked shipping with shells and rockets in Leirvik harbour on 15 January, 1945. Two large merchant ships exploded and an armed trawler was set on fire. They fought their way out through a formation of fighters. A battle with nine Fw 190s taking place a few hundred feet above the sea. One Mk. XVIII fired at the Fw 190s, which broke off their attack, then raced off to shoot at others firing at a lame Mosquito. Six Mosquitoes were lost.

Withdrawal from operations of 248 Squadron's Special Detachment of the Mk. XVIII came on 15 January, following an anti-shipping patrol between Karmoy and Marstein, when a Fw 190 was shot down. They were later used by 254 Squadron. The month's most important operation was directed at three ships in Flekkefiord on 11 January. Fourteen Mosquitoes accompanying eighteen Beaufighters led by 455

Operations: Coastal Strike

Squadron were intercepted off Lister. Two Bf 109s were shot down by Flt. Lt. N. Russell, D.F.C., and another Mosquito pilot claimed a Bf 109. Against fighters Mosquitoes were a ready match. Only Swedish ships were seen, before Fw 190s and more Bf 109s tackled the force. Four enemy fighters were then claimed as destroyed for the loss of a Beaufighter, a Mosquito and an A.S.R. Warwick.

235 Squadron made its first rocket strike in February, 1945. Bad weather restricted its operations to four during the month, two with R.P.s, two with bombs. Rockets were first fired against a 5,000-ton ship in Askevold Fiord. 11 February was one of those rare occasions when during a coastal search an enemy bomber was found, and Sqn. Ldr. Reid destroyed a Ju 188. Delayed-action bombs were soon after bounced down a 3,000 foot cliff into shipping concentrated in a narrow fiord off Midgulen along the Norwegian coast, then crews saw bombs bursting amongst ships. Black smoke began to rise, but the fiord was so narrow it was impossible to see which ships had been damaged. Not always did the Mosquitoes content themselves with coastal targets, for on 27 March Flt. Lt. R. G. Young, raked a freight train near Naerbo, after patrolling the Utsire-Naze area.

Mosquito squadrons were now given a variety of specific tasks, and no longer mixed with Beaufighters. A strike against seven self-propelled barges in the Kattegat on 7 March typifies one. 235 Squadron despatched two aircraft to shoot up flak ships with machine-guns and cannon. A dozen each from 235 and 248 with rockets followed and Wg. Cdr. Orrock led the force. 143 Squadron sent fourteen R.P. aircraft. Four outriders came from 333 Squadron. Top cover was given by a dozen Mustangs, and two Warwicks of 279 Squadron trailed behind to drop survival gear to ditched crews.

Wg. Cdr. Orrock reported 'there were eight barges going south and a large merchantman with an escorting flak ship going north. I decided to attack the eight, they seemed more important. They were obviously well laden.' There was no enemy interference, but when a similar force was despatched on 12 March eight Bf 109s gave battle and one was claimed as destroyed.

Most spectacular March strikes were those against Aalesund and Porsgrunn led by 235 and 248 Squadrons respectively. Six ships came under withering fire at Aalesund, one of the most heavily defended ports in Norway, on 17 March from thirty-one Mosquitoes using cannon and rocket fire. 'Two ships were lying in the inner harbour and four more were just outside as we came over the hills,' reported Flt. Lt. W. F. Clayton-Graham. 'There was quite a large amount of flak, some heavy.' Sqn. Ldr. Reid reported 'a concentrated attack. There was certainly plenty of flak meeting us, but everyone seemed to be scoring hits with rockets.' They scored 32 hits on one ship—14 below the waterline. Another on fire had received 37 hits, all but 6 below the waterline.

Wg. Cdr. Simmonds led the Porsgrunn attack of 30 March. Singled

Operations: Coastal Strike

out for special attention at 15.50 hours were the chemical shed and a large motor ship at Menstad Quay. Thirty-two rocket-firing Mosquitoes arrived. One flew so low that it struck overhead electric cables leaving the remainder to attack. Eight others kept watch for enemy fighters, and a photographic Mosquito covered the operation. The attack was at zero feet and one merchantman was hit by 28 rockets, another by 39 and a third by over 60. From this port the Germans mainly evacuated troops from Norway.

Flt. Lt. Royce Turner, who had taken part in many shipping strikes, reckoned this to have been about the best he had ever seen. 'As we went after the merchantman tied to the quay we could even see the plimsol marks on one. We scored hits with all the rockets aimed at one of the vessels, and as we came away three separate plumes of smoke were building up to a great height.' 'A gun position on the side of the hill was firing at our crews as they went in. There were about four guns. I swooped down and sprayed them with cannon and they did not bother later crews going in,' Flt. Lt. Clause added.

Apart from direct shipping attacks, searches were flown for U-boats creeping to German harbours. Because of the amount of traffic between Denmark and Norway the first of a large number of strong armed reconnaissance patrols over the Skagerrak and Kattegat were flown on 12 March. Forty-four Mosquitoes and twelve Mustangs taking part tangled with Bf 109s destroying one and claiming a probable. An impressive strike was delivered against a ship at Sandshavn on 23 March by 143 and 248 Squadrons, rockets ripping into the hull and tearing it apart.

April, 1945, brought a lessening of activity to many squadrons as the war in Europe sped to its end, but for Coastal strike squadrons there was no let-up, with Mosquitoes mounting five major strikes and busily scouring the sea around Denmark. The month opened with thirty-four Mosquitoes using 262 rockets to set fire to 4 out of 8 ships anchored in a heavily defended anchorage in Sande Fiord, scoring over 30 hits on a tanker in dry dock, subsequently seen to blow up taking the tanker with it. With the attackers went 4 other Mosquitoes expressly to deal with any interference and 333 Squadron's outriders helped formations to be maintained and ensured it found the correct fiord. As usual a Mosquito went along to film the proceedings.

On 9 April came a climax to the U-boat search when three—U804, U843 and U1065—were found and destroyed by thirty-seven Mosquitoes of 143, 235 and 248 Squadrons. 143 and 235 Squadrons attacked and sank a U-boat, Mosquito after Mosquito hurling its rockets into the submarine. 248 Squadron attacked the second, which blew up. An attack on the third by 143 and 235 Squadrons set it afire. The defenders put up sufficient fire to cause three Mosquitoes to land in Sweden, and a reconnaissance Mosquito spun into the sea.

Porsgrunn was again raided on 11 April, three ships being set ablaze

Operations: Coastal Strike

and one damaged. Wg. Cdr. A. H. Simmonds led the force through flak, to hit four vessels. Ten miles away *en route* for home they could see a dark grey pall of smoke rising to about 1,500 feet over the ships. Three times the formation was intercepted, each time escorting Mustangs driving away the foe. An interfering Bf 109 was destroyed, and two Mosquitoes lost from the thirty-five despatched.

U-boat U251 was sunk in the Kattegat on 19 April. Two days later Coastal's Mosquitoes had one of the highlights of their career, coming across eighteen Junkers torpedo bombers from Norway 150 miles off the Scottish coast and probably proceeding to attack shipping. Facing them were a dozen rocket-firing Mosquitoes of 235 Squadron bound for the Kattegat, 248 Squadron's ten rocket-firers and two others carrying depth charges. Fourteen more with rockets came from 143 Squadron, which supplied three fighters along with 333 Squadron's four. In the ensuing battle it mattered not which Mosquitoes attacked, all were equally capable. Four rocket-firers of 248 Squadron each destroyed a Ju 88. Four Ju 188s and an '88 fell to the other squadrons. Wg. Cdr. Foxley-Norris, force leader, reported that the Mosquitoes were returning to Banff when the enemy formation was sighted. The engagement was over so quickly that he never managed to get a bomber in his sights. Flt. Lt. J. R. Keohane said 'As soon as we saw the enemy markings, we went in to attack. They were right down on the water. I let one '88 have a burst, and its starboard engine caught fire. He tried to climb away, but just stalled and went flop into the sea.' Sqn. Ldr. A. H. Gunnis, D.F.C., recorded that the sea appeared full of blazing aircraft. 'Five times I got a Ju in my sights, and each time another Mosquito crew mixed in and shot it down before I could draw a bead.' For 143 Squadron this was a record day for operations, its maximum effort involving ninety-two hours of operational flying.

On 22 March 404 Squadron was told it would soon be flying Mosquitoes. Half the squadrons moved to Banff on 24 March to begin conversion, the remainder following on 3 April. Its first operation was flown on 22 April, when RF851:H on a shipping reconnaissance, using cannon and machine-gun fire, destroyed a Bv 138 flying-boat anchored off Kjevik. It immediately exploded, sending a smoke plume to 500 feet.

254 Squadron at North Coates armed with Beaufighter Xs concentrated attacks on shipping off the Low Countries. There were also U-boats here, especially midget submarines which attacked Channel shipping at night. To combat them five Mosquito XVIIIs were transfered from 248 Squadron to 254 Squadron in March. Extra ground crews, and the equipment for handling the new aircraft, arrived in March, 1945.

First action occurred on 12 April when 'A' fired its six-pounder at a submarine, supported by depth charges from a Wellington of 524 Squadron. Next day it was the turn of 'B' to fire and on the 16th 'D' attacked a schnorkel with unknown result. Two Mk. XVIIIs found five

Operations: Coastal Strike

Type 23 U-boats on 18 April, but before they had time to attack the submarines had dived and the two rounds fired were near misses. Eighteen Mk. XVIII sorties were flown in April, and many more by supporting Spitfire XXIs of 91 Squadron. During May a further eight were made and shots inconclusively fired at a Type 23 U-boat on 3 May.

As the area open to the enemy closed on land, so the region in which sea targets were found had a new look. Fearing a last stand in Norway it became important to scour the sea around Denmark. The campaign there increased in the early days of May preventing the escape of German forces. Twenty-seven Mosquitoes of 143, 235 and 248 Squadrons escorted by Mustangs of 19 and 65 Squadrons went on an anti-U-boat patrol on 2 May, six of 248 Squadron sinking U2359 with rockets and damaging another.

Next day the final shipping strike of the war took place in Kiel Bay. A force of seventy-two aircraft included eight of 404 Squadron, whose R.P. Mosquitoes were first taking part in a major shipping strike. On 4 May the Squadron sent seven Mosquitoes to the Kiel area, along with the forty-one from other squadrons, eighteen Mustangs and three A.S.R. Warwicks. 404 Squadron seriously damaged an R-boat. A seven-vessel convoy was attacked in the face of heavy flak, six ships being severely damaged. Wg. Cdr. Foxley-Norris, strike leader, reported, 'They met us with a box barrage of light and heavy flak. It was some of the heaviest we met, and the convoy was unusually well protected by escort vessels.' Flt. Lt. G. N. E. Yeats, D.F.C. and Bar, said, 'The gunners aboard an escort ship kept firing back at us, so we had to keep going at them. When I pulled out I could see the masthead coming straight at us and ducked automatically . . .' But into the nose of the Mosquito came part of the mast and when the aircraft was inspected back at base a German ensign was found embedded into it! Whilst this operation was going on Mk. XVIII Mosquitoes caused explosions in the stern of a vessel which quickly hoisted a white flag. East of Kiel they scored hits on a medium-sized merchantman.

Whereas VE-Day brought an end to hostilities for much of the Allied Air Forces Coastal Command continued operating. There was fear of enemy submarines unheeding or not knowing hostilities had ended. Thus, the Mosquitoes flew convoy escort patrols, a role for which they had been considered many years before, and searched for survivors from lost aircraft.

Operations ended on 21 May when 248 Squadron supplied two Mosquitoes and 143 Squadron two for an anti-U-boat patrol between Svinoy and Terningen. Eight E-boats were the only activity recorded.

Mosquitoes hindered the use of French west coast ports by surface ships and U-boats. They seriously interfered with shipping movement along the coast of Norway, employing speed, versatility and great hitting power. In the narrow fiords, where they attacked difficult targets from predictable angles, Mosquitoes faced fierce and costly opposition, in

Mosquitoes for the Royal Navy

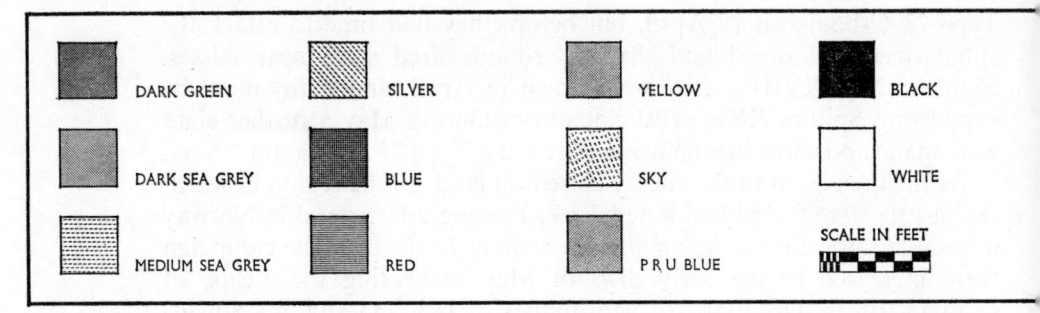

DARK GREEN	SILVER	YELLOW	BLACK
DARK SEA GREY	BLUE	SKY	WHITE
MEDIUM SEA GREY	RED	PRU BLUE	SCALE IN FEET

Mosquito TR. 33 fighter-bomber with ASV radar and wing folding. Upper surfaces extra dark sea grey, under surfaces Sky Type S. This was the standard colour scheme applied to Mk. 33s. Mk. VIs transferred to the Royal Navy had normal R.A.F. night-fighter camouflage colouring and, like their R.A.F. counter parts, naval Mosquitoes had fin stripes. The Royal Navy received PR. Mosquitoes, too, and these also had standard R.A.F. colour schemes like the T. Mk. 3s supplied for twin-conversion training.

The extra dark sea grey and Sky finish was applied to some Mk. VIs of Coastal Command in the last few months of the war, notably to the aircraft of 333 Squadron, whose squadron identity letters were then changed to Sky. Otherwise the strike Mosquitoes of R.A.F. Coastal Command had the usual night-fighter camouflage applied to Mk. VIs, with red letters, 'D-Day stripes', etc.

Mosquito FB. VI (Merlin 25) TE720 delivered to R.A.F. 3.8.45, passed to Royal Navy 27.8.45 and joined 762 Squadron at Dale. Later served at Halesworth and Ford, still for twin-conversion pilot training.

Mosquito TR. 33 (Merlin 25) TW256 delivered mid-1946 and passed to No. 771 Squadron at Lee-on-Solent where it was recorded in November, 1948.

Mosquito TR. 37 (Merlin 25) VT724, production version, used for trials.

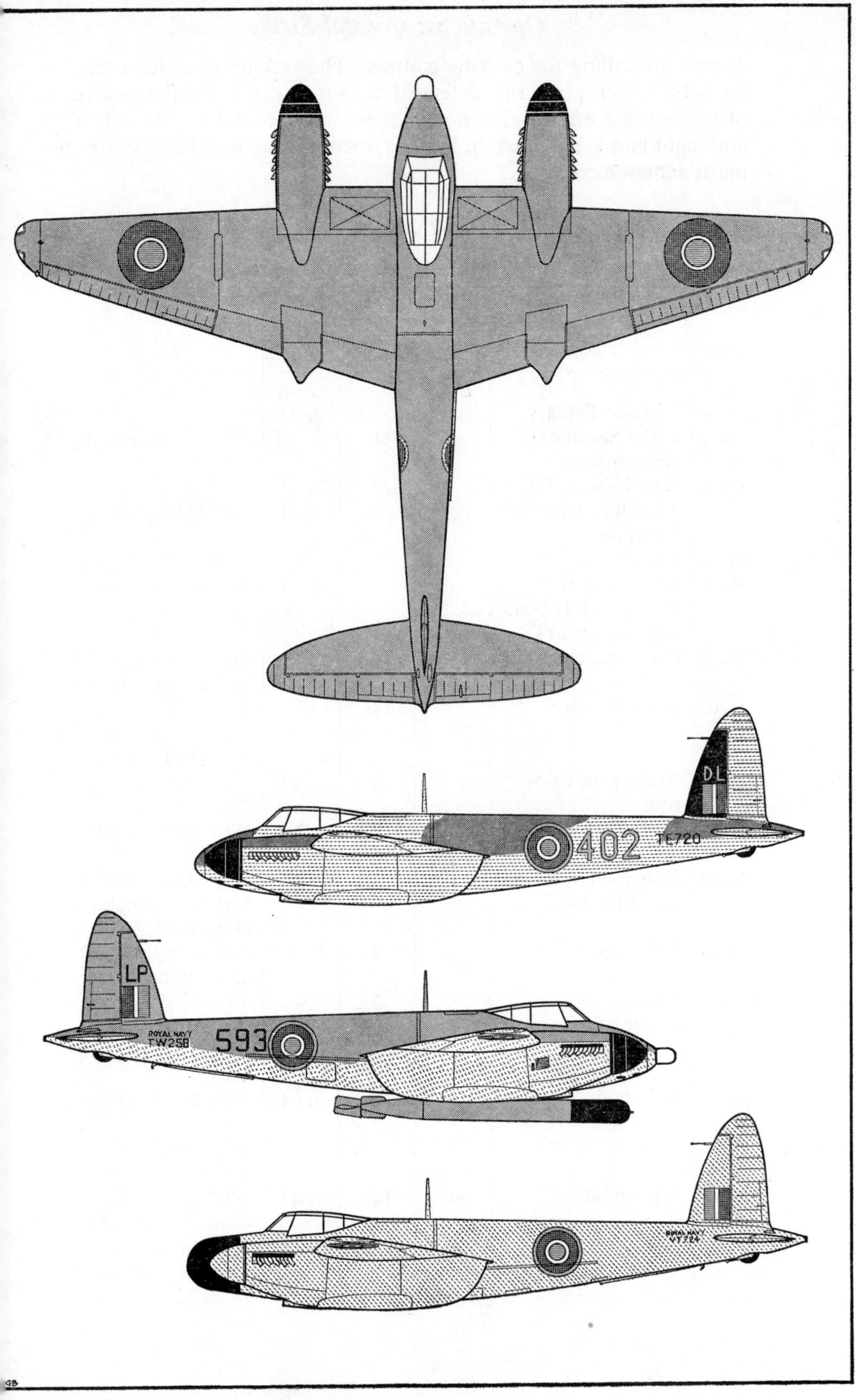
DL
402
TE720
LP
ROYAL NAVY
TW258
593
ROYAL NAVY
VT724
GB

Operations: Coastal Strike

operations calling for extreme courage. The sinking of at least ten U-boats by an aeroplane that defended Britain by night, was the spearhead of the bomber offensive, pestered enemy ground and air forces by day and night and kept watch on his movement throughout Europe, was no mean achievement.

Main Operations flown by Mosquitoes of Coastal Command
September, 1944—May, 1945

Date	Target	143 Sqn.	235 Sqn.	248 Sqn.	Notes
1944					
7. 9	Gironde Estuary			4, 2T	
14. 9	4 MV South of Kristiansund		11	14, 4T	With 19 Beaufighters
17. 9	Anti U-boat, Floro.		2	6, 2T	
	2 small trawlers nr. Ylterone		2	2, 2T	2 of 333 Sqn. led.
21. 9	Kristiansund			6	With 21 Beaufighters
28. 9	Kristiansund to Lister, anti-U-boat, attacked an MTB			4, 2T	Torpedo boat sunk
30. 9	Rover off Norway		10	6	Led by 333 Sqn., with 12 Beaufighters
15.10	Skaw to Naze		12	5	Two ships sunk. With 9 Beaufighters of 144 and 12 of 404 Sqns.
17.10	Norwegian coast rover		9	5, 4T	
18.10	Naze-Torfungen		8	3	With 24 Beaufighters No att.
21.10	Haugesund. 2 MVs and TTA fired		3	3	1 of 333 as outrider and 15 Beaufighters (144 and 404 Sqns.)
28.10	Kristiansund		12	8, 3T	2,000-ton ship. RP and bomb att.
4.11	Kinn		7	9	TTA fired
13.11	Rover off Norway		12	6	With 6 Dallachy Beaufighters. 4 ships fired
21.11	Aalesund	6	12	8, 6T	With 11 of 455 Sqn., 6 of 489, 12 of 404, 13 of 144, N of 333. Esc. by 12 Must.
25.11	Karmsund	11	14	9, 4T	With 42 Beaufighters from Dallachy. Only X/143 attacked an MV, RPs undershot

Operations: Coastal Strike

Date	Target	143 Sqn.	235 Sqn.	248 Sqn.	Notes
5.12	Nordgulen	12	8	10, 4T	Att. on 4 MV, 2TTA, tug fired
7.12	Aalesund	6	7	6, 4T	40 of Dallachy Wing, 2 of 333 Sqn. 12 Must. of 315 Sqn. Intercepted
12.12	Eldfiord 2 MV and Coaster att.	9	7	6	1 of 333 outrider
26.12	Leirvik	4	4	2, 2T	2 MV hit. Intercepted
1945					
9. 1	Karmoy-Marstein	7	6	5	With 2 of 333 and 12 Must. 8 ships att. in Leirvik Bay
11. 1	Flekkefiord	3RP, 3C	4	2, 2T	With 18 Beaufighters of 144, 404, 455. 3 ships att. Intercepted
15. 1	Karmoy-Marstein	6	4	4T	With 2 of 333. Fw 190s s/d 6 Mosquitoes
3. 2	Uyvaer-Bremanger	4	4RP, 2C	4RP	No action. 12 Must. esc.
8. 2	Rover, Lister-Kristiansund	6	6RP, 2C	6RP	No attacks
9. 2	Ylterone-Standlet	4	4RP, 2C	4	4 of 333; no attacks
12. 2	5,000-ton MV off Praesto	7	6RP, 1C	4RP, 1C	1 of 333; attacked ship
16. 2	Anti-ship patrol	1	3	3	4 of 333. 5 acft. used wing bombs on a ship
21. 2	Anti-ship recce Ylterone to Sandoy	7	—	4RP/HE	4 of 333. No attack
3. 3	Marstein-Rovaers-Olmen	7	5RP, 2C	5RP	2 of 333. 12 Must. No attack
7. 3	8-ship convoy attacked	14	12RP, 2C	12RP	4 of 333. 12 Must.
12. 3	Skaggerak-Kattegat	13	12, 2C	12RP	3 of 333. 12 Must. No attack, intercepted
17. 3	Aalesund. 6 ships att.	9RP	9RP, 2C	9RP, 2C	2 of 333. 12 Must.
20. 3	Skaggerak-Kattegat patrol	12	12RP, 2C	11RP	3 of 333. 12 Must. No attacks
23. 3	Tejenaes; ship at Sandshavn att.	15RP, 2C	—	11RP	Must. esc.
25. 3	Patrol Vadheim-Askvoll	9RP, 2HE	—	9RP, 1HE	1 of 333 with bombs. No attack
30. 3	Porsgrunn-Skein. MV and chemical shed att.	12RP	12RP	8RP	8 other Mosquitoes gave cover

285

Operations: Coastal Strike

Date	Target	143 Sqn.	235 Sqn.	248 Sqn.	Notes
1. 4	Sandefiord. 4/8 ships set on fire	12	11	11	6 Mosquitoes gave fighter cover
5. 4	Rover Kattegat	13	14	10	2 of 333 and Must. Several atts. on ships
9. 4	3 U-boats attacked	12RP	9RP	10	5 Mosquitoes gave fighter cover
11. 4	Porsgrunn; 3 ships on fire, 1 damaged	14	12	7	2 of 333 and Must. Intercepted
17. 4	Rover-Kattegat	9	2C	10	With Must. Abandoned
19. 4	U-boat Kattegat	4	9	5	With 2 of 333
21. 4	Skaggerak-Kattegat Engaged 18 enemy bombers	14	12RP	8RP, 2DC	7 of 143/333 esc.
24. 4	Kattegat-Skaggerak anti-U-boat. No action	10	8	9	2 of 333, 4 of 404 and 2 of 235 also, as esc.
2. 5	Anti-U-boat Kattegat	9	9	9	Must of 19 and 65 Sqns. 2 U-boats attacked
3. 5	Anti-ship Kiel Bay Last strike	12	9	9RP, 3C	1 of 143 outrider. 1 of 235 and 2 of 333 as fighter cover. 8 of 404 with RPs
4. 5	Kattegat anti-ship patrol	13	8	9	With 7 of 404 and 4 of 333. No action
21. 5	Last patrol anti U-boat. Svinoy-Terningen	2	—	2	8 E-boats sighted

Abbreviations:— MV: Motor vessel. MTB: Motor torpedo boat. RP: Using rocket projectiles. C: using cannon. HE: using bombs. Must: Mustang III. Esc: escorted. T: Tsetse Mosquito Mk. XVIII. DC: Depth charges. Att: attacked.

Chapter 19

Operations: Bomber Support

Under Air Commodore E. B. Addison No. 100 (Bomber Support) Group was formed on 8 November, 1943, to control the operation of radio counter measures from the air and ground, and to organize night-fighter protection for Bomber Command. Since 1940 de Havilland had considered the Mosquito an ideal long-range fighter, and in 1942 Sir Arthur Harris, C-in-C Bomber Command, had suggested to Sir Sholto Douglas that some Mosquito fighters be released for mixing in the bomber streams to engage enemy night-fighters whose presence was causing increasing embarrassment. To risk losing Mosquitoes and their radar over enemy territory was undesirable, yet to operate without radar was useless. Intruder Mosquitoes were already doing this, but relying upon visual identification of the foe was keeping successes down.

Fighter Command decided in June, 1943, that time was right for some experimental sorties over enemy territory with radar-equipped Beaufighter VIs of 141 Squadron, based at Wittering. These had Mk. IV A.I. and *Serrate*, a device allowing them to home on to enemy fighter transmissions from as much as a hundred miles away. It gave the enemy's bearing but not his range, so A.I. Mk. IV was used as the fighter closed to attack. *Gee* navigation equipment was installed and, after brief training, five crews set off to patrol the airfields at Deelen, Gilze Rijen and Eindhoven on 14 June, 1943, and bomber-support operations commenced. When they were halted on 7 September, 233 sorties had been flown—179 successfully—and thirteen enemy aircraft claimed as destroyed. A.I. Mk. IV was now being interfered with and at the end of the period the combat/sortie ratio had fallen from 1:11 to 1:35. Beaufighters had shown themselves neither fast enough nor sufficiently manoeuvrable, so Fighter Command suggested using the projected Mosquito XIV.

Before the *Serrate* trials were completed Mosquitoes of 605 Squadron had flown over the bomber stream, one reaching Berlin on 31 August and two on 3 September. Luftwaffe fighters which dropped lines of flares over the target to illuminate the bombers managed to evade the Mosquitoes.

Three Mosquito IVs arrived in November, 1942, on 1474 Flight (renamed 192 Squadron on 4 January, 1943). Apparatus was installed in them to detect enemy radio and radar transmissions up to over

Operations: Bomber Support

3,000 m/cycles. It was many months before their crews were fully trained and the moment for operations arrived. This happened on 11 June, 1943, when DZ410 took-off from Feltwell to patrol between Texel and Calais. DZ376 continued the task the following night. Fourteen such operations were flown in July over Western and North-West Germany. Patrols from Predannack were made by DZ375 over Biscay and the Western Approaches until a long-range fighter caught the Mosquito unawares as it spied upon enemy radar. Intercepted by a night-fighter near Bremen on 31 August, DZ376 arrived home with her port wing damaged. Five September sorties took the Mosquitoes over Germany, and although the squadron used mainly Wellingtons and Halifaxes, it managed a dozen more Mosquito missions before the end of the year, by which time it was at Foulsham and in 100 Group, and playing a vital role in the bomber offensive without dropping bombs.

Meanwhile it was decided to install a second radar transmitter, with an aerial beaming aft, in two Mosquito IIs of the Fighter Interception Unit, Ford (*see* Appendix 22). German fighters were known not to have rearward looking radar, and if the Mosquito had a rearward contact its speed would enable it to sweep round and attack the enemy unexpectedly from the rear. Operations commenced when the F.I.U. aircraft including DZ299 set off from Coltishall along with Mosquitoes fitted with A.I. Mk. IV, to cover a raid over Hanover. Subsequently F.I.U. Mosquitoes and some from 605 Squadron made numerous experimental operations, either flying alongside the bombers, or making for the night-fighter marshalling area. Flying under the code name *Mahmoud* they had limited success, which included a Bf 110 destroyed near Mannheim on 18 November. HJ702 was an F.I.U. machine fitted with rear-facing radar. HJ705 was the third, received on 8 October. Both HJ702 and DD715 were damaged by some of the ninety or so anti-personnel bombs dropped on Ford in September, 1943, one of those rare occasions when the enemy damaged Mosquitoes on the ground and, in this instance, held up experimental work. On 30 November F.I.U. received DZ659, the first of two machines in which A.I. Mk. X was experimentally installed. This, in the last year of the war, came into widespread use on British night-fighters. Well into 1944 F.I.U. carried on *Mahmouds* to such places as Kassel, Dusseldorf, Cologne and Frankfurt.

25 Squadron began night intruders against enemy airfields in May, 1943, using Mosquito IIs unassisted by radar and from 26–27 July was flying *Flowers*, bomber support operations within the same context. Six FB. VI Mosquitoes arrived on the squadron on 9 August and, with these, bomber support operations continued with varying intensity until the squadron was equipped with Mk. XVIIs in December, 1943, as operations by 100 Group commenced. Towards the latter part of the period 25 Squadron scored its only success when a Ju 88 fell to DD759 near Kassel. No. 456 Squadron also flew six *Mahmoud* patrols in the closing weeks of 1943 using its Mk. IIs.

Operations: Bomber Support

November, 1943, to April, 1944, marked the initial period of 100 Group's bomber support operations, in which Mosquitoes of the 2nd T.A.F. and later *Oboe*-equipped Mosquitoes of 8 Group assisted. At its commencement 169 and 239 Squadrons were at Ayr, Scotland, where crews of the planned Mosquito force were restless and eager to learn about their future employment in Bomber Command. Already *Serrate*-equipped Mosquito IIs of 141 Squadron, the first of which had been received from Colerne on 16 October, had been fitted with long-range tanks and commenced patrols on 3 November, when three aircraft patrolled the Bocholt area and DD725 chased a Bf 110 near the Hague. When next 141 operated, two patrols were despatched from Ford to Dijon on 10 November. Seventeen days later the squadron was ordered into 100 Group and to West Raynham, where it arrived on 4 December.

239 Squadron received its first Mosquito equipment at Ayr on 9 October, 1943, and its navigators underwent special training at Ouston. The squadron moved to West Raynham on 9 December, taking its two Oxfords and a Beaufighter, for not until 11th did two Mosquito IIs arrive from Colerne. These needed to be fitted with obsolescent radar for forthcoming operations deep into Germany. They were given major overhauls and fitted with fore-and-aft A.I., with common tubes for both, and a change-over switch, and special leading-edge aerials. 169 Squadron was the third *Serrate* unit. It had served as an army co-operation squadron flying Mustangs until October, 1943, when it reformed at Ayr prior to joining 100 Group at Little Snoring on 8 December, where, like 239 Squadron, it equipped with worn Mosquito IIs.

141 Squadron was the first of these to operate under 100 Group, despatching its first sorties on 17 December, when HJ659 claimed a Bf 110 as damaged fifty miles West of Berlin. Fog, frost and training delayed operations by the other squadrons, 169 not making its operational debut until 20 January, 1944, when the Commanding Officer, Wg. Cdr. E. J. Gracie, D.F.C., flew HJ707 on escort duty to the Hamburg area during a Berlin raid. He made three radar contacts with the enemy, met a fighter head-on, but scored no success. This came to the squadron on 30 January, when HJ711, having patrolled West of Berlin for twenty-five minutes, found a Bf 110 and destroyed it. 141 Squadron had opened the scoreboard for 100 Group two nights before when a Bf 109 fell to HJ941. Flt. Lt. Booth made 239's first patrol on 26 January, flying to the Freiburg area during a Berlin raid; two nights later a flaming Bf 110 fell to the squadron.

Three types of fighter patrols were being flown. Firstly, they were undertaken in the target area during or after an attack, the latter proving the more successful. Secondly, flights were despatched to enemy fighter assembly points and thirdly, escort given to the bomber stream from about forty miles away—some eight minutes flying time. Thus, the Mosquitoes might engage night-fighters at any time—and unexpectedly.

Operations: Bomber Support

By the end of April 220 sorties had been flown, and nine successful combats had arisen from 220 A.I. chases.

It rapidly became clear that the Mk. II aircraft were badly worn, a state reflected by low serviceability figures. In the early weeks of 100 Group's existence trouble was also encountered with the radar gear, but mostly it arose from the state of the engines, which were cutting on take-off and unreliable on operations because of their age. It was decided to reduce operational flying and install Merlin XXIIs, after which serviceability much improved. There were forty-one *Serrate* patrols in January; in April there were 175. Ten Mosquitoes gave escort to mine-laying aircraft over the Baltic and Swinemunde on 18 April, the best Mosquito effort yet.

515 Squadron formed on 1 October, 1942, to operate *Moonshine*, jamming equipment (fitted in Defiants) to disguise the numbers of aircraft approaching the enemy coast. From December, 1942, to July, 1943, a few Defiants flew from Tangmere, West Malling and Coltishall producing a *Mandrel* screen confusing enemy radar, but now that similar tasks had fallen to 100 Group the squadron was transferred from Fighter to Bomber Command. Mosquitoes came on to the squadron to replace its Beaufighters on 29 February, 1944, at Little Snoring. For training purposes it received four Mk. IIs pending the issue of Mk. VIs. With the improvement in operating techniques and serviceability 100 Group had decided to extend its sphere to include low-level intruding on enemy night-fighter bases, and this task was initially given to 515 Squadron, which began operations on 5 March using a Mk. VI borrowed from 605 Squadron. Wg. Cdr. E. F. F. Lambert, with Flt. Lt. E. W. M. Morgan as navigator, flew to Melun, then to Bretigny, where he destroyed an He 177 on its landing approach. The new phase in the offensive started auspiciously. Training alongside 605 Squadron to acquire the intruder technique continued during April, when eighty-eight sorties were flown. Some excitement at Little Snoring was occasioned on 18–19 April as a Mk. VI returning from Florennes had to be diverted to West Raynham to avoid anti-personnel bombs distributed by a raider.

Snooping upon radar signals by 192 Squadron's Mosquito IVs continued throughout 1944, fifty-five sorties being made in the first three months of the year, five to the Berlin area. When Bomber Command's heavies switched to attacking communication centres in France and the Low Countries 192's Mosquitoes accompanied them. DK327 was fired upon by another Mosquito over Northern France on 27–28 June, and, seriously damaged, crash landed at Friston. But 192's Mosquitoes had an enviable escape record. June, 1944, their busiest month so far, with thirty-six sorties flown, was beaten by the forty-two that July generated.

At the start of May about one hundred Mosquitoes were in the Group's squadrons. Claims rapidly mounted, Plt. Off. Miller and Plt. Off. F. C. Bone of 169 Squadron in DZ478 being extremely successful

Operations: Bomber Support

on 15th during the mining of the Kiel Canal by Mosquitoes of 8 Group. First, they claimed a Ju 88 at 10,000 feet, came down to 2,000 feet and shot down another and then at 10,000 feet again found a Bf 110, which they also claimed. Group sorties totalled 212 in May and eighteen enemy aircraft were claimed, making it the best month so far. A further attempt to improve tactics involved making a sweep ahead of the bomber force, but it proved none too fruitful. It was decided to equip Mk. VIs with *Serrate* and replace the worn Mk. IIs. 169 Squadron first took one on operations on 18 June, when Sqn. Ldr. Thorn flew NT113 over the Pas de Calais. Thirteen such sorties had been flown by 30 June and the first success was achieved on 28th when NT150 was responsible for the destruction of a Messerschmitt 110. All the *Serrate* squadrons were flying Mk. VIs by the end of July. NT121 proved to be a lucky aircraft for on 4 July a 110 fell to it during the Villeneuve raid, next night a Ju 88 and on 14th its guns seriously damaged a Ju 88 during the attack on Revigny. A Bf 110 was next to succumb, near Courtrai on 21 July. NT121's longevity was not exceptional for a Mosquito, although it continued in battle until seriously damaged in action on 24–25 March, 1945. Another noteworthy event in 169 Squadron's history was the honour of claiming 100 Group's one hundredth victim, a Bf 110 which fell to Wg. Cdr. N. B. R. Bromley, O.B.E., and NT113 near Courtrai on the night of 20–21 July, 1944.

Mounting bomber losses caused the transfer of two squadrons to the Group. 157 Squadron—first to equip with Mosquito fighters—arrived at Swannington on 7 May, a week after 85 Squadron moved in. The latter received Mosquito XIXs first. Work on this version, originally intended as a centimetric radar-equipped long-range fighter for Coastal Command, commenced in November, 1943. Powered by Merlin 25s it carried SCR 720 or A.I. VIII radar and on 21 April, 1944, was released for operational service, bringing to Bomber Command an advanced development of the Mosquito night-fighter, with long endurance and the latest radar with which to combat increasing successes by German night-fighters.

85 Squadron's XIXs were transferred to 157 Squadron when later in May XIXs fitted with A.I. Mk. X arrived on 85 and on 157 Squadron from 21st. Projecting a beam for over five miles the new radar scanned the area ahead of the fighter whereas the old Mk. IV despatched signals simultaneously in many directions and was much affected by land forms. A radarscope for the navigator now provided plots of the position of his aircraft relative to the foe. A.I. Mk. X aircraft had no backward looking radar so *Monica I* devised by the Bomber Support Development Unit was installed ahead of the pilot. Until it was fitted the aircraft were used in the low-level role, such patrols commencing on 5–6 June, 1944, when a dozen of 85 Squadron gave support and cover over Normandy and four of 157 Squadron watched over the airfields at Deelen, Soesterberg, Gilze Rijen and Eindhoven. They were carrying A.I. X

Operations: Bomber Support

over enemy territory for the first time and taking the Mosquito XIX into action initially too.

Operations were currently planned to have three phases. In No. 1 low-level intruders carrying two bombs patrolled airfields. Next, A.I. X aircraft flew low and then climbed ahead of the bomber stream to pick off the first enemy fighters, then continued patrolling until about 30 minutes after the raid. Finally, low-level intruders with bombs patrolled airfields awaiting landing night-fighters.

To MM630:E of 157 Squadron flown by Flt. Lt. J. G. Benson, D.F.C., went the first success with this version of the Mosquito, a Ju 188 destroyed near Compiegne on 12–13 June. MM671 accounted for a Ju 88 as it landed at Juvincourt on 14–15 June, the same night as 'J Johnny' opened the scoreboard for 85's Mk. XIXs by claiming a Ju 88 at Florennes. Two nights later 85 Squadron destroyed a Bf 110 at Soesterberg, by which time 157 had shot down another Ju 88 and noticed three V-1s in the Gravesend area. This was a pointer unbeknown, for the two squadrons, having such accurate radar equipment, were soon after transferred to the anti-V-1 campaign. Before this, and by the end of June, 176 intruder patrols had been despatched by the two squadrons, and 131 completed. Thirty-eight contacts were recorded, ten enemy aircraft destroyed and three damaged—all within sixty-two sorties flown between 11–12 and 16–17 June.

85 Squadron was borrowed from 100 Group to combat flying bombs on 25 June, Capt. Weisteen making first claim the following night to a V-1. On 27 June it was decided to put 85 and 157 Squadrons of 100 Group exclusively on to anti-diver patrols, and it was then that 157 Squadron commenced patrolling, destroying two flying bombs. The nose of MM630 stove in at the high speed at which it was flown, so all the noses of participating Mosquitoes were strengthened. Operations ceased on 8 July, when exhaust shrouds were changed for stub exhausts, the Merlin 25 engines adjusted to plus 24 lb. boost pressure and modifications made to allow the use of 150-octane fuel. Maximum speed now rose to about 360 m.p.h. at sea level. It was at this time that *Monica* was installed in the aircraft of the two squadrons under Flg. Off. Davies. A resumption of operations came on 16 July, two days after the squadrons had completely moved to West Malling. Busily engaged, 85 Squadron destroyed seventeen V-1s in July, thirteen in August. Patrols off the Dutch coast began on 3–4 August by 85 Squadron watching for suspected launchings from that direction, and continued until 29th when the squadron returned to Swannington to resume bomber support duties.

Meanwhile the *Serrate* squadrons also had been busy. 239 Squadron despatched seven aircraft on 11–12 June, among them DZ256 flown by Flt. Lt. D. Welfare, who shot down a Bf 110 near Paris, then flew through its wreckage. DZ256's tail unit was largely burnt away and the entire aircraft blistered, yet it satisfactorily flew 300 miles home. A Ju 88

Operations: Bomber Support

was claimed near Paris on 24–25 June by the same crew flying DD759, and for this they were awarded £5 National Savings Certificates since it was 239's twenty-fifth success. Flt. Lt. Welfare made his fifth claim of the month on 28 June when he engaged an Me 410 near Paris; his sixth was a Ju 88 found near Brussels. Once more he plunged through falling wreckage, which put his radar and gyro out of use. He then switched off his damaged electrics for the 200-mile journey home on one engine. A narrow escape came to Flg. Off. Bridges of 239 Squadron on 7–8 July, when, after shooting down a Bf 110 his aircraft DZ298 spun out of control for 8,000 feet near Charleroi. He nursed it home on one engine to force land at Woodbridge. At this period the three *Serrate* squadrons were dogged by misfortunes with 169 Squadron having many unserviceable aircraft, 141 with few crews and 239 Squadron shouldering much of the operational burden. Although they were now severely worn the Mk. IIs were giving surprisingly good service and whereas only 62·5 per cent of all sorties were completed in January, 1944, some 89·7 per cent were successfully flown (107 out of 117) by 239 Squadron in June.

After much enjoyable low-flying practice 23 Squadron resumed operations on 5–6 July as the second low-level intruding squadron in 100 Group. A Ju 88 which fell to PZ187 was the squadron's first success. The fortunes of 23 and 515 Squadrons now followed similar paths, leading to an interesting break-away from nightly operations on 4 August, when each despatched thirteen Mk. VIs led by Wg. Cdr. Murphy of 23 and Wg. Cdr. E. F. F. Lambert of 515 Squadron, to escort Lancasters making a day raid on Bordeaux. They left Little Snoring and flew to Plymouth, Coulre Point and thence to Bordeaux, from where the smoke was already curling to 11,000 feet. Homewards they flew over Vannes and Paimpol guarding the bombers, and avoided fog at base by landing at Winkleigh. After lunch next day thirty set off to escort Lancasters returning from Bordeaux. 515 Squadron alone supported a 5 Group Bordeaux raid on 11 August, when it escorted the bombers back to Quiberon Bay, repeating the task on the two following days.

On three other occasions Bomber Command's Mosquitoes flew daylight escort operations. Two Mk. VIs of 239 Squadron covered Fortress III HB772 of 214 Squadron, carrying out a jamming patrol off the Low Countries at mid-day on 11 September. On 14th and 16th of the same month 23 Squadron despatched four aircraft for similar purposes. Another occasion when 100 Group Mosquitoes operated on a day raid came on 4 October, 1944, when five of 23 Squadron and seven of 515 Squadron flew from Dallachy, escorting Lancasters to Bergen and part of the way home.

On 16–17 August, the first night of operations with Mk. VIs, 141 Squadron's HR213 was responsible for destruction of a Ju 88 and the damaging of a Bf 110 in the Ringkjobing area of Denmark. *Serrate* successes, however, fell away in the late summer of 1944. In August there were only nine from 331 sorties and in September only one for 240.

Operations: Bomber Support

Use of *SN 2* radar by the Germans, who were jamming A.I. Mk. IV, was responsible for the fall-off. Following the discovery in June of the frequency on which *SN 2* operated, *Serrate* Mk. IV was developed, although it was to be January, 1945, before it was ready for operational employment. Another development under way was *Perfectos*, a device which homed on to transmissions from the enemy 'identification friend foe' apparatus, and which came into service in November, 1944. Although *Serrate* patrols were continued by 141 Squadron the rewards were hardly fruitful. For 452 hours of operational flying—115 sorties—no enemy aircraft were claimed in November. With A.I. Mk. X the story was different and on the night of 4-5 November twenty Mosquito XIXs on patrol claimed six enemy fighters and damaged two.

When in September the Mk. X A.I. squadrons returned to bomber support duties the worn-out Mk. IIs were being withdrawn. 169 Squadron had flown its last Mk. II sorties on 17 July, 141 on 29 August, and the distinction of flying the last operational sortie in a Mk. II fell to Flg. Off. G. E. Johnson of 239 Squadron using DD789 on 26-27 October.

Three of the enemy were claimed in September by Mk. XIXs on low-level patrol, whereas non-radar equipped Mk. VIs flew 232 sorties, destroying five German aircraft during the same period. Because of this Mk. XIXs were transferred to high-level patrolling towards the end of the month. 141 equipped with *Serrate* Mk. VIs gave early support to the Arnhem operation, when six aircraft led by Wg. Cdr. C. V. Winn, D.F.C., strafed the airfield at Steenwijk shortly after dawn.

Night intruder operations with an experimental slant were carried out during the summer and autumn by Mk. VIs of the Fighter Interception Unit. After its return to Ford it was re-organized, becoming the Night Fighting Development Wing. Intruder patrols were resumed to assist 100 Group, and in the course of three days detachment to Swannington Flt. Lt. Hedgcoe of the Night Fighter Development Squadron destroyed three 110s and a Ju 88.

Bomber Support Development Unit formed at Foulsham in April, 1944, to handle the application of special radar to 100 Group's aircraft. Early in its existence it supervised the fitting of rearward looking *Monica IIIe* in a Mk. VI Mosquito and *Mk. VI Monica* in the Mk. XIX and XXX. *Monica IIIe* was also fitted in a forward-looking position in an attempt to produce an accurate ranging radar for use against flying bombs in July.

American A.I. Mk. XV 3 cm radar equipment subsequently known as *ASH* arrived at Foulsham in June, 1944. A wing position in a cylinder for its carriage by the Mosquito was first considered, then rejected when it was decided to fit it into the noses of the Mk. VIs of Nos. 23, 141 and 515 Squadrons in place of their machine-guns. Trials were undertaken in October, then 218 M.U. modified Mosquito VIs, equipping them with A.I. Mk. IX and *ASH* for the three squadrons. 85 and 157 Squad-

Operations: Bomber Support

rons were now operating with *Monica VI*, 169 Squadron received *Perfectos* in November, No. 85 Squadron was using *Perfectos II* soon after and No. 157's Mk. XIXs had *Serrate Mk. IV* installed.

Whilst the Wellingtons and Halifaxes of 192 Squadron searched for evidence that V-2s were being radio or radar controlled, investigation of radar by the Mosquito IVs continued, as evidenced by the following examples. DZ376 sought details of *FuG 200* transmissions on 5 October, and next night checked the density and characteristics of enemy A.I. on 90 Mc/s whilst DZ617 engaged in jostle jamming of V.H.F. signals. On 19 October DZ491 flew to Stuttgart to attempt to discover low-frequency *Würzburg* signals, as DZ292 detected a new A.I. band. DZ590 was sent to Berlin on 30 October to see whether *FuG 216* or *217* could be intercepted. Five of 192's Mosquitoes were out on 2–3 November, recording enemy radio transmission chatter, and on 4 December DZ491 flew to Karlsruhe in an attempt to discover whether coastal observation units were being employed on inland flak control. Each sortie with measured purpose was designed to gather special material, and with its speed and range the Mosquito was ideally suited to this task.

Night Rangers by 169 Squadron commenced on 29 October, and during November its 'A' Flight concentrated on *Serrate* patrols while 'B' Flight used *Serrate* and *Perfectos*, achieving first success with the latter in December. On 14 January, 1945, 'A' Flight collected five Mk. XIXs from Swannington and these it first operated on 21 January. Two of the squadron's Mk. VIs operated in daylight on 12 January, giving high-level and air-sea-rescue support during the Lancaster raid on Bergen. Near the target NS998 encountered five Fw 190s, attacked two of them and damaged one. Sqn. Ldr. J. A. Wright in NT176 was chased by the 190s.

An entirely new role for 100 Group's Mosquitoes and further evidence of the diversity of the Mosquito's service commenced on 27 November, when four from 141 Squadron were despatched on 'spoof' bombing raids dropping 40 lb. H.E.s on a train near Cloppenburg from 800 feet, the airfield at Quackbruck from 20,000 feet, lock gates near Lingen, and a train near Beilen. On the night of 28–29 November PZ234 positioned its bombs on to target indicators over Essen, to draw the attention of enemy fighters, and PZ235 attacked Bonn with the aid of *Gee*. Five bombing sorties were flown by Mk. VIs on 30 November. Not only did these attacks confuse the enemy as to the target for the heavies, they also provided the Mosquitoes of 8 Group with 'spoofs' for their raids.

Early in December it was the turn of enemy airfields and 239 Squadron dropped 250 lb. bombs on them. On 9 December three Mosquitoes made a high-level raid on Koblenz. HR213 took two 250 lb. bombs to Giessen on 15 December, three more carried 2 × 500 pounders and HR203 took a 500 lb. bomb and a 250 lb. yellow target indicator to mark for the small force which attacked from 20,000 feet through 10/10th cloud,

Mosquito Night-Fighters

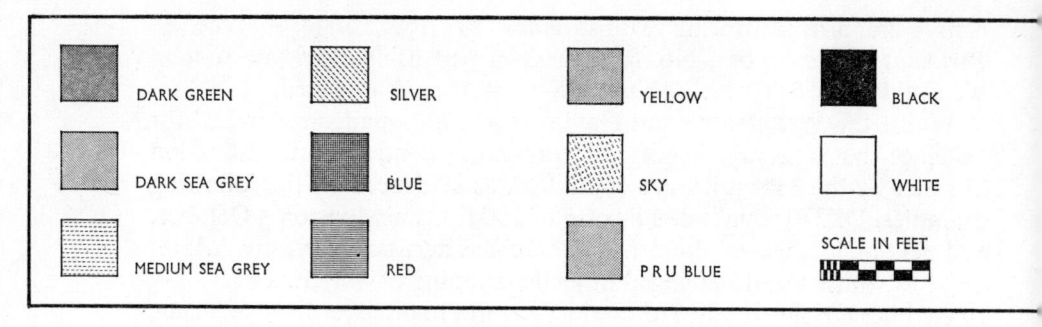

DARK GREEN	SILVER	YELLOW	BLACK
DARK SEA GREY	BLUE	SKY	WHITE
MEDIUM SEA GREY	RED	P R U BLUE	SCALE IN FEET

Mosquito NF. XII night-fighter showing standard camouflage pattern. Nose radomes on some machines were all black. Mosquito IIs and VIs used on bomber support duties with 100 Group generally had black under surfaces ending in a disruptive line along the fuselage sides at approximately the 60 degree tangent. Mk. XIXs and XXXs introduced to squadrons late in the war had the standard night-fighter finish. Some *Night Ranger* Mosquitoes of fighter squadrons had black under surfaces, and black was also featured on the aircraft of some night defence squadrons which assisted with bomber support. Limited A.E.A.F. stripes were applied to Mosquitoes generally operating at night beyond the shores of Britain.

Mosquito NF. XIII (Merlin 25) MM446 with early type of nose radome. To 27 MU 30.1.44, to 151 Squadron 22.2.44. Wg. Cdr. Goodman destroyed four He IIIs at Dijon in this aircraft 4.5.44 on *Day Ranger*. To 96 Squadron 11.8.44, anti-diver patrols off E. Coast. To 29 Squadron 14.12.44 for bomber support until 27.2.45. Broken up 28.5.45 at Cambridge. Camouflage as recorded March, 1945.

Mosquito NF. XVII (Merlin 21) HK286/G. After conversion reached 456 Squadron 29.1.44, to 51 OTU 4.1.45 until 1.7.45. SOC 21.6.47. Used by Wg. Cdr. Hampshire 27.2.44 to destroy two Ju 88s, shot down Ju 88 near Ford 24.3.44, another on 22.5.44 off Portsmouth and one more on 12–13.6.44. Recorded May, 1944.

Mosquito NF. XIX (Merlin 25) MM644 delivered 28.4.44. To 85 Squadron 20.5.44, 157 Squadron 8.12.44, 169 Squadron 14.1.45 and used for bomber support duties. To 9 MU 25.8.45 and sold for scrapping 27.10.48. Recorded April, 1945.

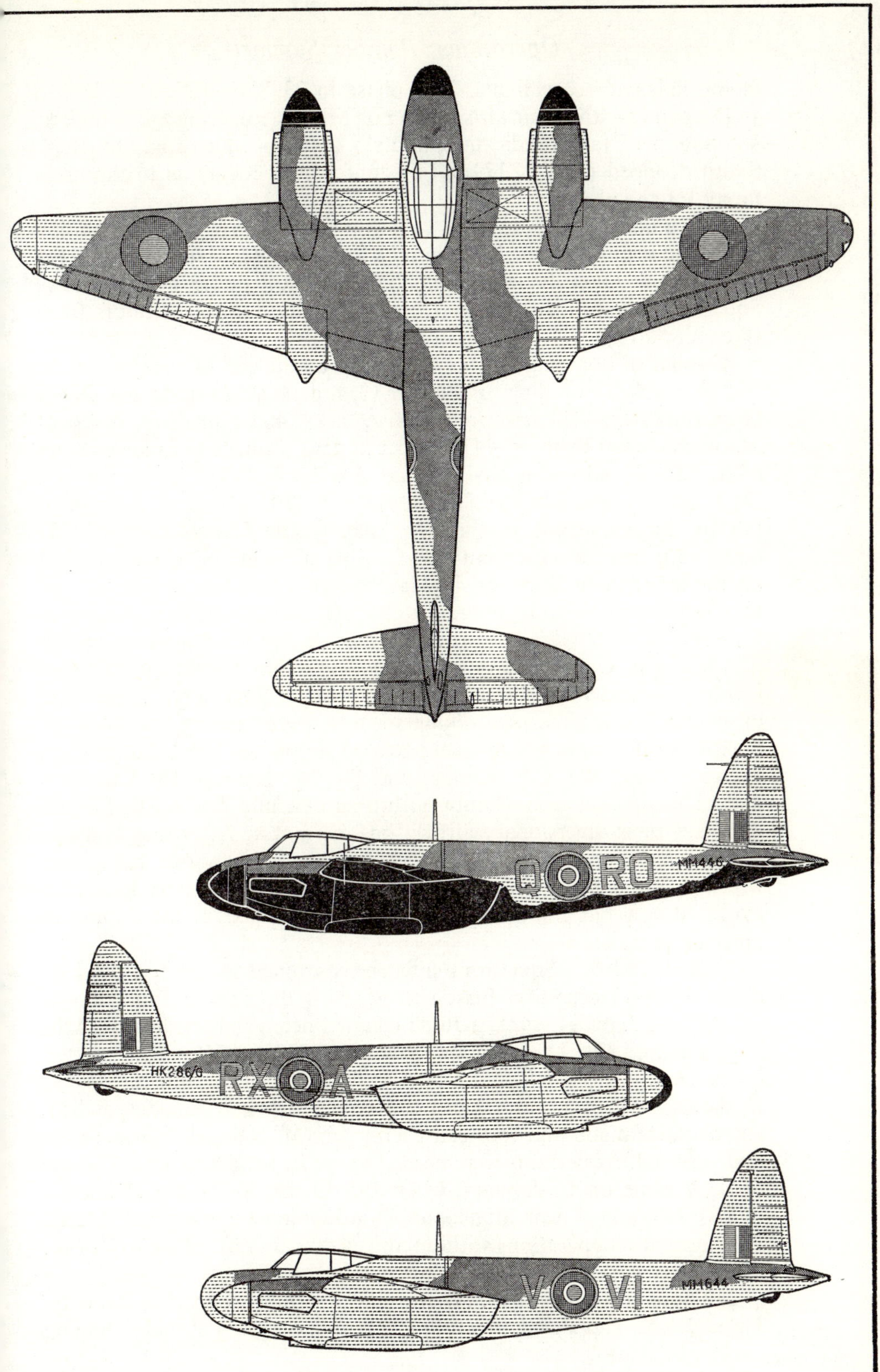

Q RO
MM446
HK286/G
RX A
V VI
MM644

Operations: Bomber Support

facing flak and searchlights. This phase in 239's operations ended on 17 December with a four-aircraft raid on Mannheim, using 250 pounders, 500 pounders, target indicators, and sky markers which Flg. Off. R. E. Smith dropped from PZ179, all to confuse the enemy as to the main target for the night.

A resumption of these raids came on 7-8 January, 1945, when 239 Squadron dropped 5,000 lb. of bombs on Stuttgart. Mannheim was the target for a 4,000 lb. load from 239 Squadron on 14/15th, and 169 Squadron joined in for the first time, dropping target indicators from three aircraft on which 239 aimed.

Absence of bombing attacks on Britain by the end of 1944 permitted complete release of three of Fighter Command's Mosquito NF. XXX squadrons from defensive to offensive tasks. 406 Squadron, based at Manston, began them on 5 December and 307 Squadron followed soon after. 151 Squadron meanwhile had *Monica* installed in its aircraft. On 1 January, 406 Squadron destroyed a Bf 110, first enemy aircraft to fall to the squadrons in the new role. Castle Camps received 307 Squadron, then *Monica* went into its aircraft, none of which was lost on the subsequent high and low-level patrols flown by the squadron, totalling seventy-nine in February and fifty-three in March. High and low-level bomber-support sorties were flown by 406 Squadron, which in April, 1945, claimed eleven enemy machines. No. 151 Squadron began low-level intruders and freelance sorties on 21 February. These were most effective against night-fighters which were landing and short of fuel. Returning from a high-level patrol in support of a Potsdam raid on 15 April, 1945, Wg. Cdr. Kimber and Flt. Sgt. Ryan of 151 Squadron had an engine cut near Hamburg, but successfully flew on to Linton-on-Ouse, their diversionary airfield on one motor. After they had run for 200 yards on the runway MT304 had an unlucky tyre burst, was wrecked and burnt. 456 Squadron began bomber support duties on 27-28 March but no enemy aircraft were engaged in the course of nineteen patrols.

23, 141 and 515 Squadrons intensively trained in the use of *ASH* radar during December. First success with the new equipment was gained on 1 January, 1945, a Ju 88 claimed near Ahlhorn by Sqn. Ldr. Tweedale and Flt. Lt. L. Cunningham in RS507. *ASH* patrols started on 21 December, 141 Squadron being first to operate with the device. By the end of the war more than 1,000 *ASH* sorties had been despatched, 100 contacts made and fourteen enemy aircraft claimed. It was found that very skilful operators were needed to use it. 141's first successful use of *ASH* came on 16 January, when Flt. Lt. D. H. Young claimed a Bf 110 on a patrol near Magdeburg. Throughout February and March 141 engaged on operations without any enemy aircraft falling to its Mk. VIs, simply because there were few to engage.

239 Squadron's first Mk. XXX sortie was flown on 21-22 January. Three XXXs supported the Frankfurt raid of 1-2 February, finding

Operations: Bomber Support

dense cloud and a rough ride which forced many attackers down to 500 feet. On this occasion they obtained their first kill using the Mk. XXX, a Bf 110 which fell to the C.O. in NT309. Seven of the squadron's aircraft were despatched on 13–14 February covering operations against Bohlen and Dresden; next night Chemnitz was their target area, but in neither case did they have any combats.

169 Squadron briefly participated in the special spoofs, dropping target indicators on Giessen and Heilbronn in mid-February. One of its Mk. XXXs gave high-level support for the Dresden raid. 169's role varied at this time, Mosquito XIXs being used for high-level patrols and intruders and Mk. VIs flying low-level rangers and intruders. An enemy jet fighter was probably encountered by an aircraft of 515 Squadron on 1–2 February, but unconfirmed, for this was one occasion when the Mosquito came off second best, and fell near Brussels.

To the end of the war 192 Squadron continued its highly specialised secret duties. An interesting addition to squadron strength on 20 November, 1944, was W4071, one of the original ten Mosquito P.R.U./Bomber Conversion aircraft, first operated by 192 on 5 January, 1945, in an attempt to jam A.I. radar around Hanover employing *Piperack*. Eight similar sorties were completed by W4071, the last being to Kiel on 9 April, the occasion of the final operational sortie by a Mosquito IV. Watching the launch of V-2s in case radio control was adopted had been amongst 4071's last tasks. PR. Mk. XVIs replaced the IVs in April after they had given outstandingly long and meritorious service of utmost importance. DZ376 was used by the squadron from 20 November, 1942, until 11 April, 1945, the day when DZ410 originally delivered 17 March, 1943, also retired to 44 M.U.

By March, 1945, the equipment of 100 Group Mosquito Squadrons varied. No. 23 had *ASH* in Mk. VIs which operated until 2–3 May, in common with the other 100 Group squadrons. 141 stood down from 15 March to 3 April to equip with Mk. XXXs, its last Mk. VI patrol having been flown on 8 March. 169 used mainly Mk. XIXs and 157, like 85, had received Mk. XXXs.

Flt. Lt. J. H. Leland flying NT369:A of 157 Squadron was held by searchlights over Berlin on 3 April. NT382 nearby observed a jet make four attacks upon his Mosquito scoring two hits upon the engine nacelles. To escape Leland spun his machine. On this occasion eight crews from the squadron were simulating a raid on Berlin, in support of 8 Group's Mosquitoes. Long-range tanks and the knowledge that safe landings were possible over much of the Continent now allowed deep penetrations of Europe and some very long endurance flights, like that of Sqn. Ldr. D. L. Hughes, D.F.C., of 239 Squadron, who shot down a Ju 188 in the course of a five-hour patrol in NT330 over the Nurnberg region. A more spectacular success was that recorded by the squadron on 10–11 April, when it destroyed an He 111 near Berlin, the bomber literally standing on its tail as it disintegrated. Low-level patrols now

Operations: Bomber Support

constituted 239 Squadron's main effort. When operations ceased it claimed fifty-five enemy aircraft as falling to its Mosquitoes.

Three of 141 Squadron's Mk. XXXs opened operations with this variant on 4 April, giving high-level support in the Leuna region. Three went to Hamburg on 8 April, after which the squadron concentrated on freelances for which the faster Mk. XXX with its A.I.X was very suitable. On 17 April bombs were dropped from the wing racks of a Mk. XXX for the first time, on the night-fighter control centre at Schleissheim. As often now the Mosquitoes operated from St. Dizier.

The idea of Master Bomber VHF controlled spoof attacks by 100 Group Mosquitoes was developed by 515 Squadron. Usually its aircraft dropped green T.I.s, leaving 141 and 169 Squadrons to bomb on them using anti-personnel bombs, 500 pounders with screamers attached and 250 lb. incendiary bombs. Thirteen 100 Group Mosquito bombing raids were controlled by a Master Bomber of 515 Squadron, the first occurring against Lubeck/Blankensee airfield on 13–14 April.

A further stage in the multiplicity of roles undertaken by 100 Group was reached on 18 April when, during a firebomb patrol of Munich's Neubiburg airfield, seven of 141 Squadron's aircraft each dropped two 100-gallon drop tanks filled with napalmgel whilst another Mosquito gave them high cover. Their attacks were low-level, one going in at 500 feet. Flensburg, Schleswig, Lubeck and Neubiburg again on 24 April all were targets for similar *Firebash* raids. The origin of Bomber Command's use of napalm lay in a conversation between an R.A.F. and a U.S.A.A.F. officer in a train as they were returning to their bases in Norfolk. Careless talk or not, it led to some spectacular raids on enemy military objectives.

Incendiary raids on enemy airfields were also begun by 23 Squadron in April, the load usually comprising eighty four-pound bombs or forty bombs and flares. On the final night, 2–3 May, Sqn. Ldr. Griffiths of 515 Squadron in RS513 controlled the *Firebash* on Hohn in which 23 and 141 Squadrons were engaged. 141 Squadron despatched thirteen Mosquitoes carrying napalm to attack Hohn and Flensburg. One pilot jettisoned his load on making a *Monica* contact, but this, the last of the war, proved fruitless.

For the final night's operations 169 Squadron divided its effort between Schleswig and Westerland. As '653' touched down at base one of its napalm tanks which had hung up fell off but, after moments of suspense, failed to ignite. There were searchlights and guns to face at the airfields, and two Mosquitoes failed to return. Control of another fire raid, on Jagel, had been in the hands of Lt. W. Barton of the S.A.A.F.

In seventeen months of operations Mosquitoes of 100 Group flew 7,884 sorties, claiming 249 enemy aircraft destroyed in flight and eighteen on the ground, this for the loss of sixty-nine Mosquitoes. Bombing and fighting under a cloak of great secrecy their exploits incredibly reflected the agility and versatility of the Mosquito.

. XV A.I. (High level) FTR E/A Claims Dst. P Dgd.	Sorties desp.	Mk. XV A.I. (Low level) FTR E/A Claims i) Air, ii) ground Dst. P Dgd. Dst. P Dgd.
— — — —	—	— — — — — — —
— — — —	—	— — — — — — —
— — — —	—	— — — — — — —
— — — —	—	— — — — — — —
— — — —	—	— — — — — — —
— — — —	—	— — — — — — —
— — — —	—	— — — — — — —
— — — —	—	— — — — — — —
— — — —	—	— — — — — — —
— — — —	—	— — — — — — —
— — — —	54	— 1 — — — — —
— 3 — 1	78	2 2 — — 1 — —
1 — — —	192	1 1 — 1 — — 2
— 1 — —	213	3 1 1 1 3 — 9
— — — —	226	1 4 — 1 5 1 —
1 4 — 1	763	7 9 1 3 9 1 11

ration *MAHMOUD*

gned to bring night-fighters into action *MAHMOUD* was not directly ected with Bomber Command operations. A.I. Mk. IV Mosquitoes flew y to one of twenty-two known night-fighter assembly points in the hope aking interceptions. Such operations began mid-August, 1943, sixteen es being completed that month. They arose from the finding, during *RATE* sorties, that single British night-fighters seemed an attractive t—until their superiority of equipment and radar came into play. use of identification difficulties *MAHMOUDS* were initially flown when bers were not operating, or well away from bomber streams. Mosquitoes them from Castle Camps, Ford, Coltishall, Hunsdon, Manston or West ing and these operations—like *FLOWERS*—were synchronized with der activity. *MAHMOUD* aircraft flew a steady course until making a act. When enemy range was not less than 5,000 ft. the Mosquito turned gh 360 degrees with throttles fully open to come up behind the foe. *HMOUD* was, from the outset, considered a temporary measure, super- d as later equipment and aircraft became available.

riginal attempts to entice the enemy to combat were not particularly essful. Attempts were then made to operate long-range Mosquito IIs bomber targets, but 100 Group began operations before any real con- ons could be drawn from relatively small-scale operations.

to face page 300

Chapter 20

Operations: Light Night Striking Force

Bomber Command was relieved of 2 Group and tactical bombing duties on 1 June, 1943. Its Mosquitoes went to 8 Bomber Group, already known as the Pathfinder Force, under the command of Air Commodore D. C. T. Bennett, C.B., C.B.E., D.S.O., where they joined 109 Squadron which had pioneered the use of *Oboe*.

105 Squadron, refurnished with Mk. IXs specially equipped with *Mk. II Oboe*, underwent a training course and became the second *Oboe* squadron. Scheduled for 139 Squadron was night high-level nuisance bombing. Crews were not pleased about this, since it meant losing their individuality after having made a name for themselves as low-level day attackers.

The type of operation to which 139 Squadron was committed bore similarity to the dawn attack with which the Mosquito bomber began its operational career. In February, 1943, Air Marshal d'Albiac conceived the idea of developing high and low-level Mosquito attacks using aircraft powered by Merlin 72s or 25s. For the high-level role the hope was that a Polish unit would form. At a working level, agreement was reached through Hereward de Havilland that the Company should produce the new version. When the re-organization of 2 Group took place plans for using Mosquitoes were revised and now 139 Squadron was to fly nuisance raids or spoofs—the squadron called them 'spooks'—to mislead the enemy as to Main Force targets.

Nuisance attacks had begun on 20–21 April with a raid on Berlin 'to celebrate Hitler's birthday', when 105 Squadron despatched nine Mosquitoes, and two were sent by 139 Squadron, of which one failed to return. The immediate success of the new tactics can be judged from a request sent by the A.R.P. authorities in Stettin, where the heavies were attacking, addressed to Berlin. From there the reply was that no help could come, for a raid was developing upon the Capital. Equally confused were the German night-fighters, one of which claimed that it was being attacked by a Mosquito fighter. That one of the attackers should be lost on the first operation was ironic, for the immunity of the Mosquito bomber to night interception was to become legendary. More so was it for 139 Squadron a shattering blow since the pilot of DZ386 was Wg. Cdr. Peter Shand, leader of so many daylight operations by

Operations: Light Night Striking Force

the squadron. He was shot down over the Dutch coast by a fighter.

Next came a three-Mosquito raid on Duisburg on 27 April, after a Main Force attack. By the end of May Berlin had been bombed five times, attracting 16/31[1] sorties despatched to targets including Düsseldorf, Kiel, Cologne and Mannheim. Berlin was easy to locate in fine weather by the Togel and Havel Lakes to the west of the city, but in bad conditions navigation equipment in the Mosquitoes was at that time too limited to provide accurate bombing. Flak over the city was inaccurate above 22,000 feet, but often intense. DZ597:L flown by Flt. Sgt. Walters over the city on 21 May was hit, and dived to 10,000 feet to escape further damage. The port engine gave trouble as the enemy coast was crossed, and the aircraft was again damaged. The remainder of Walters' journey was at 200 feet. In spite of its battered state the Mosquito brought him safely home.

105's final nuisance sortie to Berlin came on 21 May. For much of the remainder of the war the squadron was engaged on P.F.F. duties. On 22 June, whilst the main raid of the night fell on Mulheim, four of 139's aircraft bombed Cologne. At the end of the month they were being navigated with the aid of *Gee* fixes, used first during sorties to 26 June. Lübeck was the target for three Mosquitoes on 24 July, in the face of flak and searchlights. Leading them was Wg. Cdr. R. W. Reynolds, D.S.O. and Bar, D.F.C., in DZ601. Flares marking the route to Hamburg for the heavies were useful to the trio who, on return, stayed to watch the Hamburg raid and reported it to be 'very spectacular'. Half-a-dozen crews dropped incendiaries on to Hamburg next night to keep its fires burning. Three others dropped bundles of *Window* before bombing Essen, prior to the Main Force raid. On 26th and 28th the Mosquitoes again added to the torment of the Hamburgers, and on the next night they laid flares on the route, and dropped *Window* ahead of the heavies.

Enemy fighters were occasionally encountered, but the speed of the Mosquito and its high-flying ability were usually sufficient protection, as on 4–5 August, when Flt. Sgt. G. Cumming in DZ478:V found two fighters closing in. He kept them at 400 yards and they were forced to give up the chase. A week later in DZ476 he was intercepted, and although the enemy fired upon him he managed to get away. A spoof for the Nürnberg operation on 10–11th delivered against Mannheim was made by two of 139's aircraft, in the course of which they dropped T.I.s as well as bombs to mislead the enemy.

When the heavy Berlin raids began at the end of August, 1943, 139

---

[1] 16/31: 31 aircraft were despatched, of which 16 claimed or were known to have attacked the primary target. This standard method of indicating the effectiveness of sorties despatched appears in this and Chapter 23. The remaining sorties were not necessarily completely abortive, for some aircraft may have attacked secondary or other alternative targets. Losses might also reduce the number of known attacks on the primary target.

Operations: Light Night Striking Force

Squadron contributed nuisance attacks on the Capital. Eight Mosquitoes followed up the heavy raid of 23 August. Next night three Mosquitoes returned from Berlin with many flak holes, and on the 25th Grp. Capt. Slee flying DZ478 was coned for twenty-five minutes, and hit by flak. Further assistance to the four-engined bombers was given the aircraft on 3-4 September when 139 Squadron sent four Mosquitoes to drop white flares south-east of the city, in an attempt to draw the attention of enemy fighters to what would seem to be a turning point for the attack.

The rapid expansion of 139's Mosquito night bomber operations is indicated by the despatch of 145 sorties (117 successfully) in August. Hamburg was twice raided, Duisburg ten times, Cologne and Düsseldorf three times, Mannheim attacked once and eight raids delivered were on Berlin all for the loss of only two aircraft. Noticeable advantage was that the Mosquitoe's speed enabled even short breaks in the weather to be used for bombing. The Berlin round trip of 1,200 miles took a Mosquito about four hours.

Losses during autumn 1943 on nuisance attacks represented 1·75 per cent of the sorties flown. Set against this was the fact that not much material damage resulted from Mosquito raids, but clearly they confused defences and were valuable because the interruption they caused was out of all proportion to the expenditure of effort. B. Mk. IVs were attacking at around 27-28,000 feet at about 320 T.A.S. over the Ruhr, and were being damaged on about 50 per cent of their sorties, although, in most cases only slightly. Accurate fire from German 88 mm guns was now rising to 35,000 feet, thus an improvement in high-altitude performance was becoming an urgent need. No proof had yet come to hand that any Mosquito had fallen to the guns of a night-fighter.

One of the pilots flying with 139 Squadron was Plt. Off. Patient, who had spent some time in the Fitting Shop at Hatfield prior to joining the R.A.F. Whilst heading for Berlin the port engine of his Mosquito was put out of action by flak at 27,000 feet. He continued, dropped his bomb load, and then his machine was further damaged by gunfire. While descending the aircraft was shot up by two Fw 190s and the navigator was wounded. The fighters he shook off by diving to 200 feet. He successfully bellylanded the Mosquito at Manston, where a Typhoon returning from a night patrol cut the fuselage of the bomber in half while the crew remained aboard.

Throughout October 139 Squadron continued its nuisance raids, visiting Berlin, Cologne, Duisburg, Dortmund, Düsseldorf and Frankfurt. For the attack on Hanover on 3 October the squadron operated two Mk. IXs ML908:Y/909:Z for the first time. Each carried two 500 lb. bombs under its wings in addition to four in the bomb bay. This raid was a spoof for a Kassel raid and was the most successful so far. 'Success' in this instance meant that the Mosquitoes attracted many enemy fighters—and evaded them completely. Their immunity had an odd twist, however, for an enemy intruder damaged one of the squadron's other Mk. IXs at its base.

Five of 139's Mosquitoes constituted Bomber Command's only effort on 9 October, a foggy night when Berlin was raided and Flt. Sgt. Marshall in DZ388 had his helmet grazed by a bullet. He was coned by searchlights over Dortmund on 16th and his aircraft, DZ612:C, damaged by flak. Unknown to him when he landed, not all of his bombs had left the aircraft. Standing at its dispersal after the sortie the Mosquito shed a 500-pounder, causing no mean concern amongst the ground crew. This particular raid was of more than passing interest since it was the first occasion when 139 Squadron had mounted a miniature Main Force attack. One Mk. IV dropped T.I.s on radar and was followed by two backers-up releasing green T.I.s for the others to bomb on.

When Berlin was raided on 20 October, diverting attention from an attack on Leipzig, nine Mosquitoes dropped only red T.I.s and yellow drip flares, and did similarly on 22nd when Frankfurt attracted their load. So frequently had Cologne been the target that on 24th the squadron diary recorded a 'return to the old milk run', four Mosquitoes bombing the city on *Gee* and D.R. 'The only gunner, English or German to score a hit' recorded the squadron historian, 'was stationed at Aachen, and he hit Flg. Off. Denny's aircraft in the bomb door'.

Unlike their German counterparts attacking Britain, Mosquito crews were always given specific, and sometimes small, targets. Whereas an air-raid warning in Britain might herald the passing of a German fighter or bomber, at most a handful, a similar alert in Germany would be the prelude to either a vast saturation attack or a small Mosquito raid, designed perhaps, to bring loss of sleep, upset production at night or confuse the defenders as to the main target. By judicious planning later in the war it was possible to cause the air-raid sirens to wail on some nights throughout Germany. When sufficient aircraft became available Mosquitoes were organized to fly 'Siren Tours' bombing several targets on one sortie, as part of Operation *Ploughman*.

On average Mosquitoes remained about five times as long over enemy territory as did the German raiders of 1943/44. Over Britain a handful of Luftwaffe aircraft scattered in ones and twos over some 10,000 square miles were facing a loss rate of about 7 per cent. Mosquitoes had for six months in 1943 been flying virtually in formation towards recognized targets with a loss rate of about 1 per cent. Compared otherwise each missing Mosquito delivered close to the target 103,000 lb. of bombs, whereas each German aircraft destroyed over Britain had released 40,000 lb., often in open country.

On eight nights in November 139 Squadron visited Berlin, despatching sixty sorties there. A major task was now introduced to the squadron, that of flying ahead of the Main Force, dispensing *Window* prior to the arrival of the heavy bombers and confusing enemy radar. On 26 November three Mosquitoes four minutes ahead of the Main Force showered

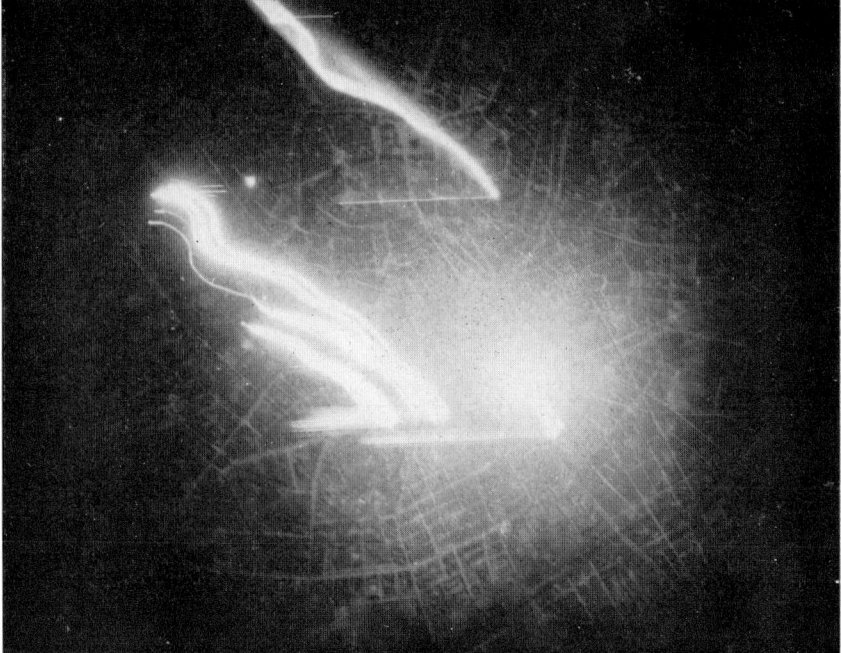

113. The trial-installation Mk. IV with bulged bomb bay, in November, 1943
114. Berlin being marked by P.F.F. Mosquitoes
115. Loading a 4,000 lb. bomb on PF432 for the Berlin raid of 21 March, 1945

116. Dicing tonight

117. Unarmed merchantmen of B.O.A.C.

118. B.O.A.C. Mosquito G-AGGD comes in at Leuchars

119. Mk. IX ML897 of 1409 Flight lands at Wyton after its 153rd sortie
120. Oboe-equipped Mk. IX LR504 after its 190th sortie
121. Mk. IX ML914 veteran of many operations

122.
A line-up of 109 Squadron Oboe Mosquitoes at Marham, 1943

123.
DZ319 : H of 109 Squadron after its 101st sortie

124.
A Mi. VI of 21 Squadron photographed from a 109 Squadron Mosquito during a *Noball* operation

125.
Loading target markers into a B. Mk. XX, KB162:J of 139 Squadron

126. Searchlight tracks and markers around Berlin on 11/12 August, 1944
127. ML963 'K-King' of 571 Squadron in April, 1945
128. MM156 Mk. XVI, bought with contributions from the Oporto British

129. 464 Squadron Mk. VIs at Hatfield, 2 June, 1944

130. A P.R.34 with swollen 'bomb bay' and 200-gallon wing tanks

131. NF. Mk. XIX, MM652

132. HK382, a Mk. XIII *Night-Ranger* aircraft of 29 Squadron

133. W4087, a Mk. II with Turbinlite airborne searchlight
134. A Mk. XVI converted to carry *H2X* for the U.S. 8th A.A.F.
135. A Mk. VI supplied, after refurbishing, to the Turkish Air Force

136. The second pre-production T.R.33 for the Royal Navy
137. A Mosquito T.T.39, PF606, before delivery to the Royal Navy
138. TW240 after conversion into the prototype T.R. Mk. 37

Operations: Light Night Striking Force

Window on to the approaches to Berlin opening a new phase in the Mosquito's versatility and invulnerability, for which reason it had been chosen for this vital role. In addition to the bundles of foil a bomb load was carried, which was dropped as the Mosquitoes returned across the target. *Early Window* sorties played an ever increasing part in the Mosquito night offensive almost to the end of the war, the wooden bombers heading the bomber streams in some of the war's most intensive raids, and sometimes finding themselves led by, or leading, Mosquitoes with specialized radar bombing aids.

Another type of attack was first tried on 9 November, when 109 Squadron marked Leverkusen by *Oboe* for nine 139 Squadron aircraft. I. G. Farben's chemical works was the target, over which moderate flak was encountered. On twenty-three nights in November 139 Squadron operated, making for example a three-wave attack on Berlin on 22nd, dropping flares and bombs in a spoof on Frankfurt on 18th, and despatching 149 sorties during the month, as well as carrying out trials of *G-H* radar.

Equipping existing squadrons with Mk. IX Mosquitoes, and a trickle of XVIs and Canadian-built Mk. XXs, released sufficient Mk. IVs for the formation of 627 Squadron, also 692 Squadron at Gravely on 1 January, 1944. Both were born under the aegis of the P.F.F. The former began operating 24–25 November, sending three aircraft—two unsuccessfully—to Berlin. The second commenced action on 1 February, despatching DZ547 and the customary four 500 lb. bombs. 692 Squadron's leader was Wg. Cdr. Lockhart, an officer who had shown great courage in the hands of the Gestapo and Spanish Fascists. Previously he had served with 627 Squadron.

From a nucleus of 139 Squadron the formation of 627 took place at Oakington on 12 November. By the end of that month it had flown eleven sorties, diversifying the Mosquito effort by attacking Berlin on 25 November, when 139 Squadron stood down, and Cologne on 29th whilst 139 Squadron bombed Dusseldorf.

It fell to Flt. Lt. G. Salter and W. Off. A. C. Pearson, D.F.M., to take a Canadian-built Mosquito, KB161:H, on its first operational sortie, on 2 December. This was one of twelve aircraft despatched by 139 Squadron for a three-wave assault on Berlin. Dogged by misfortune the machine encountered engine trouble and frozen controls, but its mission was completed. Four Mosquitoes flew ahead dropping *Window*, then bombs on to flares positioned by the second four. The remainder, and four Mk. IVs of 627 Squadron, bombed on to the fires which had been started. 627 Squadron suffered its first loss, DZ479:F.

Throughout December the two squadrons usually operated in concerted efforts occasionally attacking different targets. 627 concentrated on bombing, 139 dropped flares and T.Is. as well as making spoof and *Early Window* sorties. One of the largest Mosquito raids yet despatched was to I. G. Farben works at Leverkusen on 10 December, when five of

305

Operations: Light Night Striking Force

105 Squadron and three of 109 Squadron should have marked the target, using *Oboe*, for ten of 139 Squadron and four of 627 Squadron. Strong winds unfortunately ruined the operation. A second outstanding attack, the last of the year, came when seven of 139 Squadron and three of 627 raided Cologne. The intention was that Sqn. Ldr. Skene should lead the raid, dropping markers from DZ601 by the aid of *G-H*. Its unreliability resulted in a *Gee* and D.R. operation. Already 139 Squadron was taking over the task of marking for the Mosquito night bombing force, a story further detailed in Chapter 23.

During 1943 Mosquitoes of 139 Squadron flew 730 effective sorties at night. 627 Squadron completed 53. Cologne was attacked on 28 nights by 139 Squadron, Berlin 41 times, Duisburg 27, Düsseldorf 17, Dortmund 6. Forty-eight other operations had been mounted by 31 December. 627 Squadron bombed Berlin 6 times, Cologne 2, Düsseldorf 2, and made four other operations.

Consideration of the possibility of modifying the Mosquito IV to carry a 4,000 lb. bomb arose, in April, 1943, before Mosquito night raids began. This idea was then expressed to N. E. Rowe at the M.A.P. during a visit by R. E. Bishop. As a result the following letter arrived at Hatfield:

SB.42483

Ministry of Aircraft Production (DTD),
Millbank,
London, S.W.

29th April, 1943.

Dear Bishop,

You mentioned to me during your recent visit the possibility of carrying the 4,000 lb. bomb in the Mosquito, which could be done by comparatively straightforward changes to the main supporting structure and bomb doors.

It is now agreed that one aircraft should be modified on these lines forthwith for trials. Official confirmation will be sent to you at once,

Yours sincerely,
N. E. Rowe.

This decision was immediately seen to present a major change in policy. Operations by these new aircraft were considered unlikely during 1943. The Commander-in-Chief of Bomber Command, Sir Arthur T. Harris, was known to view the 4,000 lb. bomb as an anti-morale weapon and thus it was surmised that a new role was under consideration for the Mosquito. For destructive results Bomber Command preferred a load of $4 \times 1,000$ lb. bombs.

Two 4,000 lb. bombs had arrived at Hatfield by 12 June. One was the normal medium-cased variety, the other a new thin-cased bomb giving greater blast effect. A neat simple loading arrangement was designed, the bomb being hung from a single hook on a bridge of two spruce

Operations: Light Night Striking Force

beams fixed only to the front and rear spars of the wing. Slightly swollen was the fuselage belly. Bomb doors were redesigned, and a small fairing added aft.

On 19 June, 1943, the Director's Report noted that the first modified Mk. IV, DZ594, was almost ready to fly. Only seven weeks had elapsed since the conversion job had been authorized. Now the installation was able to accommodate three types of 4,000 lb. bomb—General Purpose, Medium Case and High Explosive. Tamblin's section in the Drawing Office were busy preparing drawings to make the Mk. IX, as converted for the 4,000 lb. bomb, adaptable to carry alternatively 4×500 lb. bombs. This was possible with a reduction of about 30 gallons of fuel.

Early in July DZ594 was first flown. It could not be tested carrying the bomb from Hatfield, for its all-up weight of about 21,500 lb. was more than could be safely taken off. Therefore in mid-July it was transferred to Boscombe Down.

Production of thirty sets of conversion parts was well advanced by 31 July but only three aircraft were to be converted at Hatfield. Whether the remaining twenty-seven would be Mk. IV or IX was undecided. Consideration was also given to such work being done on Canadian aircraft at Toronto.

Boscombe Down's initial trials were soon completed. The aircraft returned to Hatfield on 9 August due to a sticking bomb-loading winch. After modifications it returned to A.A.E.E. for bomb dropping trials. Final decisions on the 4,000 lb. bomb modification scheme were made on 9 October, 1943, whereby all Mk. IVs and IXs in service were to be modified and all Mk. IXs and XVIs in production would have 4,000 lb. provisioning. This was to be introduced to the lines at the end of October.

Before the end of the year take-off trials with DZ594 up to a weight of 25,200 lb. had been carried out, and stability considered reasonably satisfactory. At altitude with 4×500 lb. bombs the normal Mk. IV suffered fore-and-aft instability, so 3×500 lb. and 1×250 lb. bombs were generally carried, and the aircraft needed 'flying' all the time. At all altitudes the modified machine was divergently unstable but, as expected, it was much improved when the elevators had larger horn balances. Better stability was desirable before it could enter service. A limit clearly needed to be placed on the increasing all-up weight, and single-engine flight directional control was now unsatisfactory.

Whereas the initial work on the project had been so rapid, development by January, 1944, was disappointingly slow, partly due to bomb-winch trouble. During the month Boscombe Down decided to pass the modified Mk. IV—provisionally—with large elevator horn balance, rear camera removed and 60 lb. of ballast placed in the nose. 139 Squadron returned one of the earliest aircraft to de Havilland with a complaint of bad fore/aft stability above 15,000 feet with only a 1,500 lb. bomb load. Stability problems were clearly going to bedevil the 4,000 lb. bomb

Operations: Light Night Striking Force

aircraft. They were never fully cured on the Mk. IV. For this reason few Mk. IVs and IXs were modified.

Two modified aircraft reached 692 Squadron on 4 February, 1944, after a brief stay with 627 Squadron. Sqn. Ldr. Watts, D.F.C., with Flg. Off. Hassell, took off at 19.03 on 23 February in DZ647:B. Almost immediately Flt. Lt. V. S. Moore with Flg. Off. P. F. Dillon in DZ534:M followed. Two minutes previously Flt. Lt. McKeard in DZ637:C preceded them. Each aircraft was carrying one 4,000 lb. bomb, and these were the first Mosquitoes to drop one operationally. Backing them came Flg Off. Goodwin's DZ478:L loaded with 4×500 lb. delayed action bombs. Ahead flew six of 105 Squadron, three of which accurately positioned their red markers. Over the target, Dusseldorf, searchlights were active and there was moderate accurate flak. On to the markers Flg. Off. Hassell released a bomb at 20.45½ hours. Flg. Off. Dillon released his one minute later. Meanwhile Moore was handling the Mosquito in a cone of searchlights, and his aircraft was slightly damaged by flak soon after. Next night the three special aircraft went to Kiel where they bombed on T. Is set by 139 Squadron.

Throughout February 627 and 692 Squadrons worked closely together and despatched 124 sorties. DZ606:M took off from Oakington for Düsseldorf on 14 March, carrying a 4,000 lb. M.C. Twenty of these were dropped by this aircraft and DZ632 and '646 of the squadron during March. A variety of marking weapons also characterized the squadron's load during its 137 March sorties. A further 102 were flown in April up to the 13th, and twenty-two 'cookies' were delivered before the modified aircraft were exchanged for normal Mk. IVs at Upwood. The squadron then moved to Woodhall Spa beginning new duty as a marker squadron for 5 Group.

Not always, of course, were the Mosquito raids entirely successful, but those which can be judged failures stand out as exceptions. One such occurred when three modified aircraft of 692 Squadron went again to Dusseldorf on 29 February. They orbited the target six times before dropping their bombs. 105 Squadron had sent five aircraft to mark, yet only one was able to drop its T.Is. as detailed. In February 692 Squadron flew thirty-six to thirty-eight successful sorties and dropped a dozen 4,000 bombs.

From the outset the intention had been that the B. Mk. XVI Mosquito should carry a 4,000 lb. bomb. This, with its pressurised cabin and two-stage Merlin 72 engines, had an operational ceiling of 35,000 feet. Mk. XVIs arrived only just in time, for the enemy had developed special anti-Mosquito flak remarkably accurate to about 40,000 feet, or twice the height at which our heavy bombers usually went in. The new protective shield was particularly thick around Berlin, on the western approaches to the capital, and it often took the form of a carpet that moved up and down to catch approaching bombers at all heights. Sometimes predicted flak was mixed with the barrage.

Operations: Light Night Striking Force

Being able to fly high was a comfort in these conditions. All the Mosquito crews underwent training in decompression chambers to make sure that none was subject to aeroembolism, or 'bends'. The Mosquito cockpit lent itself conveniently to pressurisation, and in the Mk. XVI the pressure could be built up to 2 p.s.i., sufficient to reduce the effect of altitude by 5 to 6,000 feet. To have given a higher cabin pressure would have been dangerous, for a puncture by flak at altitude would have reduced pressure suddenly and injured the crew.

692 Squadron's first XVI operation took place on 5 March, when Flt. Lt. V. S. Moore and Flg. Off. P. F. Dillon attacked Duisburg from 27,000 feet. B. XVIs were now steadily introduced to the L.N.S.F., 692 being able to mount a six-aircraft spoof raid over Hanover on 22 March, marked by 139 Squadron. *Early Window* dropping, spoofs and all-Mosquito diversion raids of increasing strength were becoming nightly occurrences. Most important of all, though, was the beginning of the increase in the number of Mosquito bomber squadrons. Air Vice-Marshal D. C. T. Bennett expressed a desire for a fleet of 200 XVIs, but many of these were completed, instead, as P.R. aircraft.

Further expansion took place on 7 April, 1944, with the formation of 571 Squadron at Downham Market. It was intended to equip it with Mk. XVIs, but instead was decided to operate the unit at half-strength allowing more Mk. XVIs to reach 105 Squadron. Air crews and half of the ground crews of 571 Squadron were detached to Graveley for Mosquito XVI experience and from here Wg. Cdr. J. M. Birkin in ML942:D, and ML963:K flew the first attack by the squadron, to Osnabruck on 12–13 April. Five days later the squadron was instructed to move to Oakington by 24 April, where the detached party then arrived. 571 operated on nine nights from Graveley, using their own or borrowed aircraft and making twenty-nine sorties, eleven to Cologne, eight to Mannheim and four to Berlin. On the day the detachment joined the main party, 24 April, three Mosquitoes were despatched to Dusseldorf—two of them borrowed aircraft—and from this point onwards 571 began a sustained night offensive on German cities despatching in all 2,546 sorties. On 20 May the squadron was raised to a two-flight level.

Apart from frequent intrusions into the sleep of the Berliners, Mosquitoes were delivering almost nightly attacks on cities in western Germany, led by either *Oboe* or *H2S* Mosquitoes. Often straightforward, the operations were not entirely without moments of excitement, as on 6–7 April, when the twelve aircraft of 692 Squadron found *FIDO* equipment belching flames along the sides of Graveley's runway as they landed. An impressive spectacle. 692 dismissed it in their records with the laconic 'nothing much to it'. Of more account were 692's eight sorties to Berlin on 13–14 April, for this was the first time that Mosquitoes each carrying 2 × 50-gallon drop tanks and a 4,000 lb. bomb attacked the capital, or indeed such a distant target. *En route* they had to climb to 28,000 feet to cross towering storm clouds over much of Germany. T. Is.

Operations: Light Night Striking Force

were positioned three minutes late by 'Y' aircraft of 139 Squadron, and with their heavy loads the bombers circled in the face of 300 to 400 searchlights and moderate flak. Five were coned, two damaged. In this operation two of 571's crews took part and six of 627 Squadron, the latter making its last sorties as part of 8 Group's L.N.S.F.

Berlin was again the target on 18 April, when eight Mk. XVIs carried 600 gallons of fuel each. Two Mk. IVs each with 4×500 lb. bombs and a similar fuel load flew a longer route to mislead the enemy. Attack plans were becoming increasingly complex. On 27–28 April four Mk. IVs and four XVIs of 692 Squadron flew to Stuttgart, which the IVs bombed on markers laid by 139's 'Y' aircraft. The XVIs meanwhile carried on, to bomb Friedrichshafen through an area thick with fighter flares. Five XVIs of 571 Squadron also bombed Stuttgart. By the end of the month 692 Squadron had flown two hundred sorties, lost no aircraft and dropped nearly two hundred 4,000 lb. bombs.

The first two weeks of May, 1944, were mainly occupied by attacks on the giant I. G. Farben chemical works at Mannheim and Leverkusen. On 3–4 May, when Mannheim was the target, 139 Squadron did their customary marking, then nineteen Mosquitoes delivered 4,000 lb. bombs. Two aircraft that had dropped visual markers over the target gave VHF orders to the force. It was unexpectedly cloudy, and this made the task of the Master Bombers difficult. DZ640, held in searchlight for ten minutes, was holed by flak.

Enemy fighters appeared on 7–8th near Leverkusen. ML966:P was twice attacked, once by a single-engined fighter, but both enemy aircraft were eventually shaken off. When a mixture of bomb loads was taken to Berlin on 16–17 May by 571 and 692 Squadrons, Germany was again found to be covered by storm clouds, the tops of which rose to 28,000 feet. It was a question of trying to ride over them—or risk passage through them.

One of the most remarkable things about the Mosquito always was its versatility, the surprises it had in store. Towards the enemy on 12–13 May, 1944, it turned a new face. Earlier in the day Mosquitoes could be seen flying low over the East Anglian River Ouse. Three Mk. XVIs and ten Mk. IVs of 692 Squadron left in the early hours of 13 May to lay mines from a low level in the Kiel Canal. A stretch $3\frac{1}{2}$ miles long devoid of ground defences was chosen, between 54·08 N./09·21 E. and 54·11 N./09·26 E. A ragged formation flew at 10,000 feet to the north of Heligoland, then turned south-east. 139 Squadron fired green Verey lights marking the route. Run-in was flown at 8,000 feet, then the Mosquitoes dived for the last 14 miles at 240 I.A.S. knots. A two-wave attack developed, the first six aircraft laying mines from 300 feet. Target identification was aided by moonlight and the dawn breaking. Eleven crews reported successful sorties, leaving one to return with a load still aboard. Mines were laid from 50 feet by some aircraft, and all flew below 250 feet. Mosquitoes of 100 Group preceded them, shooting up ground

Operations: Light Night Striking Force

defences in the moonlight. Loads were dropped on timed runs from ground markers. As a result of the operation the Canal was completely closed for seven days, re-opened for reduced traffic for three days then closed for a further three. About a million metric tons of overseas cargo, and 350,000 tons of coastwise cargo, were held up. Twelve days after sixty ships were still waiting off one entrance and sixteen at the other end. Large ore cargoes from Sweden were delayed, also military supplies for Norway. Since the attack took place in the ice-free summer months it came when traffic was at its highest.

A second mining operation was flown by 692 Squadron on 1–2 June. Mk. III mines were laid by five Mosquito IVs in the Baltic, in the area 56·05 N./10·32 E. Drops were made from between 700 and 1,100 feet when in the far distance could be seen the lights of Aarhus. T. Is. of the marker aircraft failed to appear so both 8 Group Squadrons resorted to *Gee*/D.R. Contrails were a source of worry at this time of year at the height at which Mosquitoes operated; a further hazard was unexpected flak in regions other than the target area. Although flak belts needed to be avoided they were a hazard that had constantly to be faced, sometimes unexpectedly like that which claimed DZ608 over Osnabruck on the night of 10–11 June.

Whilst fifty-eight Mosquitoes undertook special D-Day marking operations 8 Group mounted an all-Mosquito raid on railway marshalling yards at Osnabruck working at full pitch to pass war material to France and Belgium for the expected invasion. 692 Squadron despatched eight B. IVs and two Mk. XVIs, and 571 sent nine Mk. XVIs. Take-off was about half an hour before midnight. Each bomber took a 'cookie', and the raid was led by 139 Squadron who positioned markers just before the final turn in for the bombing run. The cloud layer was 10- to 12,000 feet, but 139's green T.Is. could be seen glowing through. Bombing took place between 00.50 and 00.55 from heights varying from 24,000 to 28,000 feet. For the raid 139 Squadron had mustered six 'Y' and six ordinary aircraft, including six Mk. IVs, three Mk. XXs and three Mk. IXs.

After D-Day 8-Group squadrons began to turn their skilled attention to enemy oil targets, beginning the new phase on 12 June with a five aircraft raid on the Gelsenkirchen Scholven plant. It was a poor attack, so was repeated next night and on 15 June, but again *Oboe* marking failed. On 25th–26th the oil refinery at Homberg was the target, whither 692 sent its Mk. XVIs with four internally carried 500 lb. bombs and a 500-pounder on each wing. This proved to be a good attack, as was usually the case when *Oboe* was efficiently used.

An uncommon feature of an intended raid on Gottingen on 26–27 June by thirty-five Mosquitoes was that it was to be a *Parametta* attack delivered from 5,000 feet. Marking was, however, absent, for the *H2S* gear in 139 Squadron had failed, and the Master Bomber therefore faced the ensuing confusion, added to which bad weather forced them to be

Operations: Light Night Striking Force

diverted as they headed for home. Another unfortunate raid was that on Saarbrucken on 28–29 June. All six *H2S* aircraft of 139 Squadron had radar failures and the Master Bomber had to take over. By now flak was intense, and in order not to waste the effort entirely six Mosquitoes were directed to Metz marshalling yards. Next night 692 Squadron flew a Mk. IV, DZ611, for the last time, during an attack on Hombérg. From now on only Mk. XVIs were operated by the two squadrons.

Flt. Lt. Val Moore and his navigator, Pat Dillon, were nearing Berlin on 18 July. Searchlights illuminated them, as Dillon was settling at his bombsight. Almost immediately tracer whipped past. Moore called to his navigator to return immediately to his seat and strap himself in, to allow weaving action. Then he dived MM135 at almost 500 knots to shake off the fighter, before climbing steeply. As he did so a packet of *Window* landed on Dillon's lap, so violent was the motion. Once caught in the searchlights it was difficult to escape at high altitudes, and fighters also had an idea as to the positioning of the Mosquito. Ten times they fired at MM135 without hitting it. Next day engineers examined the machine. They found nothing wrong, although it had dived at far beyond the recommended speed limit.

608 Squadron began the war using Ansons, then Bothas and later Hudsons. It reformed at Downham Market on 1 August, 1944, armed with Mosquito XXs. KB242:B carrying four 500 lb. bombs and flown by Sqn. Ldr. Bolton was the first of its new aircraft to operate, the target being Wanne Eickel. On 23 August, '242 was attacked four times by a fighter, chased for fifteen minutes and damaged. It crash-landed at Woodbridge.

In the two months following D-Day 8 Group Mosquitoes flew about 14,000 hours, about two and a half times the pre-invasion rate. The latest version of *H2S* used by 139 Squadron was giving good definition over Berlin, and T.Is. could be placed within 1,500 yards of the aiming point. An average Mosquito raid using about twenty-five 4,000-pounders now stood a chance of doing some damage in the desired place. Over the period 15 July to 15 August Mosquitoes dropped 336 such bombs on Berlin; when 818 flying bombs landed in Greater London.

Bombing by the Mosquito force had often been outside the effective target area due to limitations of the Mk. IX bombsight. Early in 1944 R.A.E. were allotted a Mosquito, for their approval, using the Mk. XIV bombsight, which 8 Group in their keenness for results had installed and tested in eight days. After several months there was still no result from R.A.E. 8 Group went ahead and by August had twenty-eight of their seventy-four aircraft equipped. They overcame trouble arising from lack of suction at altitude by fitting two additional electrically driven pumps.

With a 4,000-pounder aboard and the Mosquito tanked up for a 1,200-mile sortie the worst moment for the crew was take-off. Fused or unfused a bomb of this size might go off on heavy impact, whereas the

Operations: Light Night Striking Force

alternative load of six 500-pounders was safe until fused from the cockpit. On take-off it was customary to keep the aircraft on the runway to within about two hundred yards of its end, even if it was 2,000 yards long, so as to have a better chance of take-off should one engine fail. Such a procedure in bad weather on a pitch-black night was not without its risks. An equally unenviable emergency was the need to return to base for any reason before the bomb had been dropped in the Wash or over enemy territory. MM143 flown by Flt. Lt. Galloway of 692 Squadron was taking off for Mannheim on 27 August when she swung and the resulting crash set her afire. The crew were lucky to scramble clear, and it was fortunate for others that the bomb burnt but *did not explode*. The following night there were further moments of suspense, as Flg. Off. B. D. McEwan returned from Essen with his 'cookie' still aboard MM141. He needed to make a very cautious landing.

After take-off the Mosquitoes sometimes climbed to their attacking height. Alternatively they flew one leg of the course to the enemy coast as low as 300 feet to fox the radar, and then steeply climbed. Even with a full load the Mosquito had sufficient climbing power to pass quickly through any icing layer, and the customary de-icing paste was therefore not applied to the leading edges and tail unit. With a heavy load it was advisable to avoid flying through cumulo-nimbus cloud formations with their bumpy up-currents. One unusual return, most satisfying, included a dive beginning at Over-Flakkee from 32,000 feet to 10,000 feet reached at Southwold. The 88-mile journey was completed in eleven minutes ... which was fast, even for a Mosquito.

August's raids were to the pattern of the period, and included attacks on synthetic oil refineries, the routine raids on Berlin, Dusseldorf, Hanover etc. and spoof attacks including one on Hamburg on 29th when ML959:G flown by Flt. Lt. Lockhart needed to dive 20,000 feet to shake off an enemy fighter. German jet fighters were now being reported at night. Near Magdeburg the crew of PF383 believed they saw one on 25–26 August. A second sighting was by Flt. Lt. Don who, in MM144, participated in the Nürnberg raid of 8 September. Mosquitoes damaged by flak and coming home on one engine were particularly attractive to the jets, which could be mastered if they were observed first and outmanoeuvred. Those who escaped from encounters reported a very fast foe.

In spite of the jets, massive flak barrages and searchlights, not to mention the increasing efficiency of German interception radar, Mosquitoes retained their almost negligible loss rate. To the end of September, 1944, only fifty-nine had failed to return from night raids from the total of 12,517 sorties flown, a loss rate of ·9 per cent. At the same time greater demands were being made on the Mosquito's high performance, as on 10 September, when the route 692 Squadron were called upon to fly to Berlin was the longest in the squadron's history, and then the attack turned out badly due to late T.I. release. One of the eleven from the squadron had to turn away and bomb Lübeck due to fuel feed trouble

Operations: Light Night Striking Force

with its long-range tanks. Three others were so short of fuel that on return they had to land at Woodbridge, the first airfield they came to. The following night, after bombing Berlin again, 692 Squadron recorded opposition as 'bitter'.

September 12 brought a new experience to 692 Squadron. Ahead of a daylight attack by heavies went four Mosquitoes, dropping *Window* and each delivering a 4,000-pounder on Wanne Eickel. They were not quite the first of the squadron to conduct a daylight raid, for MM149 and MM128 had half an hour earlier bombed Gelsenkirchen. It proved to be a busy day for the squadron, for two aircraft were sent to Frankfurt in the evening and seven by night to Berlin.

128 Squadron reformed at Wyton on 15 September. Previously it had led a somewhat mundane existence as a coastal reconnaissance and convoy protection squadron operating from West Africa, but now it was to take its place in 8 Group with the Light Night Striking Force. Its first two aircraft arrived from Upwood on 8 September, being two Mk. XXs, KB210 and KB353. Two nights later (and five before the squadron was officially formed!) '353 bombed Berlin. '210 was deprived of the pleasure by technical trouble. Next day they both went to the city. On 12th '210 bombed Frankfurt, but '353 overshot during night flying training, damaging her tailwheel and tail unit. Again '210 reached Berlin on 13th—and on the 15th, to celebrate the formation of her squadron. Her days were numbered, for on 16 September, '210 took off for Brunswick and failed to return. By the end of the month 128 Squadron had flown twenty-two sorties using only four Mk. XXs.

571 Squadron's one-thousandth sortie was flown on 3 October, when Flt. Lt. N. J. Griffiths took PF389, one of thirteen contributed to a raid, on Kassel. Only five of the squadron's Mosquitoes had so far failed to return, a loss rate of 0·5 per cent. For the second time Kiel Canal was mined, by four of 571's aircraft and others of 692 Squadron on 5 October. Defences there were stronger now, amounting to ninety-seven guns and twenty-five searchlights.

When MM116:M landed at Woodbridge it was with her pilot Sqn. Ldr. E. J. Greenleaf seriously injured and her navigator dead. For his immense courage the pilot was immediately awarded the D.S.O. Just after his mine was released an anti-aircraft shell burst alongside the cockpit, killing the navigator and injuring Greenleaf whose arm was put out of action. He had to fly the aircraft and navigate his way home literally single handed, without navigation aids and with his radio out of operation. For 400 miles he flew, to the emergency landing ground at Woodbridge. With 571 Squadron had flown six Mosquitoes of 692 Squadron, five carrying a 1,000 lb. mine and one a 1,500 lb. mine. They mined the Canal from heights varying from 150 to 250 feet, encountering light flak and searchlights. Six others from the Squadron were meanwhile engaged on an equally taxing task, following 4 hours 35 minutes flight plan in an attack on Berlin.

Operations: Light Night Striking Force

Jet fighters were again in action during the month, two crews reporting one in the Essen area on 3rd and near Osnabruck on 24th. On 28 and 29 October, they were seen near Cologne, but on neither occasion did they present any hazard to the Mosquitoes which, even in daylight, found it possible to out-manoeuvre them. With four squadrons comprising the L.N.S.F., in addition to the marker force often accompanying them, the number of effective sorties flown was rapidly increasing, the total for the four squadrons reaching 806 during October when eleven failed to return. B. XXVs had joined 128 Squadron, which despatched three to Hanover and two to Mannheim on 18 October. Previously this type had been used by 139 Squadron, and by 608 Squadron which sent a Mk. XXV KB441:Q flown by Flt. Lt. Bell to Berlin on 6 October. This was the first time the squadron operated the Mk. XXV which, for a while, nearly supplanted its Mk. XXs.

142 Squadron reformed at Gransden Lodge on 25 October, 1944. Its first two Mosquitoes arrived on the morning of 27 October, revealing themselves as Mk. XXVs KB430 and '460. Next night Sqn. Ldr. R. C. Don and Flt. Lt. A. W. D. James were alerted for a Berlin raid. Bad weather prevented it, so the first operation by the new squadron came on 29th, when both aircraft went to Cologne. Next night both reached Berlin and on 31st, Cologne.

November's outstanding event was the first major daylight raid by Mosquitoes of 8 Group. Three squadrons took part, despatching twenty-four aircraft, in addition to the eight marker Mosquitoes. Contact with the *Oboe* leader at the planned rendezvous failed so the formation flew on leaderless, the Mosquitoes paired astern with 692 leading. Unfortunately both heading aircraft now had equipment troubles, the lead passing to the next pair. 571 Squadron tagged on to the rear, and all the bombing was visually on the leader's release. Slight flak on the beam faced the formation on the run in, but there were no fighter attacks. Cover was given by Mustangs of Fighter Command. Thick cloud covered the target area, preventing a view of the ground. After the raid oily smoke curled through the white clouds as the undamaged Mosquitoes came home in pairs.

A second daylight attack came on 30 November, when the target was again the Gessellschaft Teerverwertung in Meiderich, a suburb of Duisburg. Eight Mosquitoes attacked, drawn from both 571 and 692 Squadrons, also seven from 608 and five from 128. Wg. Cdr. R. J. Gosnell led in ML942 of 692 Squadron. During the outward climb 692's formation became split up but succeeded in reforming, and it contacted the raid leader above the rendezvous point. The run in was good, bombs falling on to sky markers placed above 10/10th cumulus. Secondly came 571 Squadron and lastly a group from 128 Squadron which, as they left, saw a column of black smoke rising to 10,000 feet and claimed that the attack on the plant had been good. For the third day raid on Duisburg seven aircraft of 571, eight of 142 and eight of 608 Squadrons were

Operations: Light Night Striking Force

despatched. The *Oboe* leader flew too fast and the formation became scattered over the target. On 11 December 571, 608 and 692 Squadrons each sent off twelve aircraft for the final day raid on Meiderich.

Another day raid mounted on 11 December by two waves drawn from 128 Squadron bombed Hamborn, attacking in echelon starboard. The final 8 Group daylight operation of the year came on 22 December, when 4/6 of 692's aircraft bombed Seigburg, led by an *Oboe* Mosquito.

Customary night operations continued during December, main targets being Hanover, Nürnberg, Hamburg and Hagen. Three attacks were mounted against Berlin, bringing the total for 1944 to sixty-seven.

An extension of the specialized duties for which Mosquitoes became famous occurred when skilled attacks were delivered by four squadrons early in the morning of 1 January, 1945, on railway tunnels in western Germany. This was a further attempt to prevent vital supplies and troops from reaching Rundstedt's forces, already harassed at night by Tactical Air Force Mosquitoes. Delayed-action 4,000 lb. bombs were to be hurled into tunnels from an altitude of 200 feet. Seventeen Mosquitoes were prepared for the operation, but PF411 of 128 Squadron crashed on take-off, her crew being killed. The remaining five of the squadron attacked tunnels in the Koblenz area, with varying success. Six from 692 Squadron found tunnels difficult targets, but Sqn. Ldr. R. G. St. C. Wadsworth, D.F.C., in PF430, dropped his bomb from 100 feet into the mouth of a tunnel near Mayen. It exploded at the far end, from which smoke and dust bellowed forth. The hills around Kochem made Flt. Lt. T. H. Galloway's task difficult, but on his third attempt he dived MM 128, only to have his bomb overshoot into the nearby town. PF414 faired more disastrously, for, as Flt. Lt. G. D. T. Nairn brought her in to attack, she was shot down by gunfire from five light anti-aircraft guns. Flt. Lt. G. C. Crow managed to get his bomb into a tunnel near Scheven, Flt. Lt. Hill saw his land above the target and Flt. Lt. C. H. Burbidge claimed a hit.

Five crews set off from 571 Squadron, and of these Plt. Off. D. R. Tucker flying PF383 was probably the most successful for, after three dummy runs on three tunnels, he went back to the second one and bombed it. He flew across to assess the result, discovering that the tunnel had erupted and the nearby cliff had partly collapsed. Sqn. Ldr. Dawlish flying MM115 dropped his bomb on to the entrance of a tunnel. Such was the current serviceability that all three squadrons despatched their aircraft again for the following night's operations against Hanau and Hanover.

There now started the 1945 series of almost nightly raids on Berlin, attracting no less than 3,988 Mosquito sorties involving 3,766 attacks for the loss of fourteen aircraft, ·99 per cent of the attackers. It would, as can be seen, be wrong to assume that these 'small-by-other-standards' raids attracted no defence reaction. Indeed, the resistance during the

Operations: Light Night Striking Force

operations on sixty-one nights in 1945 was considerable, to the very end of the war. Although the Mosquitoes were frequently holed after their crews had faced barrages, searchlights in great numbers, and fighters, the aircraft as usual proved themselves able to take heavy punishment. Raids took place mostly around 9 p.m., but sometimes they were delivered later in order to achieve maximum effect upon morale. This further suffered when the raids comprised up to three waves, by dividing the efforts of the squadrons and routing the attackers by a variety of courses in order to disturb the largest number of people.

Some interesting sidelights on the effects of the Mosquito nuisance raids on Berlin were mentioned by the Directors of the Siemens Schukert and Halske factories during their post-war interrogation. The raids, they said, caused a constant drain on the production of both concerns, for their workers were sent to the shelters each time an attack developed. Those not on night shifts were kept awake, and frequently these were people who had been sent home early to avoid evening travel during the bombing of the city. Both firms operated a system whereby, after the raid, they notified their employees where the bombs had fallen to alleviate worry, and allowed those whose homes were affected to leave work on compassionate grounds. Siemens Schukert lost $1\frac{1}{2}$ million working hours in the last few months of the war, almost exclusively to Mosquito nuisance bombing. Halske reckoned their loss as 2 million hours.

On seven January nights attacks were made on Berlin, three of them two-wave raids. 367/409 sorties were despatched. One hundred and seventy-six aircraft despatched to eight targets made the heaviest attacks so far on 1 February, during which month there were raids on twelve nights and effective sorties totalled 813/851. These far from occupied the whole of the Light Night Striking Force (or Fast Night Striking Force as it became known)—for it frequently despatched small forces to drop *Window*, and delivered very sharp attacks on Wanne Eickel, Mannheim, Magdeburg, Hanover, Stuttgart, Frankfurt, Bremen and many other German cities. If the night was clear, then *H2S* Mosquitoes of 139 Squadron led the deeper penetrations and dropped ground markers; if it was cloudy the bombing was on skymarkers.

The force was by now much stronger, for 162 and 163 Squadrons had become fully operational. 163 Squadron reformed at Wyton on 25 January and received six Mk. XXVs next day. On the 28–29th four were sent on an *Early Window* operation to Mainz and the second raid again involved dropping *Window*, this time near Gelsenkirchen. Bombs were then directed on to Dortmund. Thereafter the squadron joined others for the attacks on Berlin, making its first contribution to one of two waves on 1–2 February.

One thousand six hundred and sixty-two sorties were flown by the L.N.S.F. during February, in addition to those of the *Oboe* squadrons. On thirteen nights Berlin was the target. Magdeburg was bombed five times. *LORAN* fixing was first used this month, 571 Squadron employing

Operations: Light Night Striking Force

it initially when six aircraft raided Magdeburg on 7 February. One of the six sent to Koblenz also used it. With the larger number of Mosquitoes available for night operations, an average now approaching two hundred, more elaborate raids were planned for the L.N.S.F.[1] 142 Squadron on 13–14 February, 1945, despatched two aircraft to drop *Window* over Bohlen and bomb Dresden, two for a spoof attack on Bonn and eight to Magdeburg. Night operations on 23 February took four to Berlin, four to Frankfurt and four to Darmstadt, there to deliver a mock marking attack. This squadron during the month dropped 156·075 tons of bombs and made 223 sorties, every one of which was effective; at this period it was flying raids from between 20,000 and 26,000 feet.

The largest Mosquito raid as yet on Berlin by ninety-one aircraft came on 1–2 February. 692 Squadron contributed eleven machines to the first wave and seven to the second. Of these MM224 was involved in a serious accident, when it overshot Rougham airfield on return and crashed into a civilian car. The following night 692 Squadron again found misfortune when one aircraft became bogged in the mud at Gravely immediately prior to take-off. But on the 8th the squadron noted —with some satisfaction that a 'less superior squadron had actually attacked a "dummy" instead of Berlin'—692 was again in fine form!

Mud also claimed D-Dog of 162 Squadron as she swung off the runway at Bourn on 1–2 February, leaving sixteen others to head for Berlin, During February 162 Squadron was to despatch 222 sorties and deliver 195·29 tons, and to raid Berlin on ten nights. To 10–11 February 3,015 tons of bombs had been dropped on Berlin by Mosquitoes.

For the Berliners the greatest Mosquito attacks were to come, for during March they raided the city on twenty-seven nights, making 1,222 sorties and losing seven of their number. On the squadrons the Berlin raids acquired the popular name for such regularity, 'the milk run'. Hits on the attackers were now less common and such losses as occurred were not necessarily due to damage sustained about the city area. The largest attacks so far came on 5th (76+9 'markers'), 9th (81+5), 11th (80+7), and 12th (72+8). Nineteen of the twenty aircraft 692 Squadron sent off on 21/22 March bombed Berlin, and this was the largest number of aircraft despatched by any of the squadrons. 162 and 571 Squadrons also each flew twenty sorties. 118/139 Mosquitoes claimed to have bombed the city during the night, on what proved to be the heaviest raid delivered by Mosquitoes on Berlin. That night's entry in the log book of a pilot of 142 Squadron read 'Big raid, over 100 aircraft. Did

---

[1] Operations on 14–15 February, 1945, typify those of this period. Chemnitz was the Main Force target. 329 heavies attacked it in Phase I, while 224 heavies and 8 Mosquitoes bombed Rossnitz. Feints were made by 12 and 19 Mosquitoes on Duisburg and Mainz respectively. 46 Mosquitoes bombed Berlin, 54 heavies laid mines and other aircraft flew diversionary sweeps. In Phase II Chemnitz was bombed by 388 heavies. Feints against Dessau, Frankfurt and Nuremberg attracted 14, 8 and 11 Mosquitoes respectively. 101 night-fighters gave support, with another 101 operating a jamming screen and dropping *Window*.

Operations: Light Night Striking Force

Dly. Telegraph X-word on way back.' This officer flew 30 operations on D-Dog, a Canadian B.Mk.XXV, including eight to Berlin.

Mosquito raids were now playing a very important part in the final destruction of the Third Reich, for Berlin was the jittery centre of a rapidly disintegrating régime. For this reason attacks were delivered on thirty-six consecutive nights beginning 20–21 February, 1,896 sorties being flown for the loss of eleven aircraft, a rate of ·58 per cent. Some Mosquitoes were returning with forty or more flak holes. In view of the large numbers of aircraft now involved in operations in a comparatively small space, it is little wonder that collisions from time to time occurred, but for the Mosquitoes, flying so high, these were fortunately rare. On 27–28th Flt. Lt. L. G. Smith believed he collided with another aircraft causing his Mosquito, MM202:V of 128 Squadron, to do a half spin, lose its starboard propeller and set its starboard engine on fire. He nursed the machine home, making a crash landing at Woodbridge.

The last Bomber Command Mosquito day raid was delivered on 6 March, when 128 Squadron sent five aircraft and 142, 162 and 571 Squadrons each despatched six; they set off around 15.00 hours for Wesel. Shortly before the attack the formation of Mosquitoes picked up a fighter escort. Over the target there was 10/10th cloud cover reaching to 15,000 feet. The attack was led by *Oboe* aircraft of 109 Squadron two of which were earlier seen to collide. In the absence of markers 571 Squadron jettisoned their bombs in the target area.

It was in March that 162 Squadron began operating as the second *H2S* Squadron, but this concerned only a few of its aircraft, leaving the remainder to operate as before with the main Mosquito force. Many of the sorties flown were similar in another way to those of 139 Squadron, for they were 'siren tours'. On 26 March 608 Squadron operated Mk. XVIs for the first time. By the beginning of April it was half-equipped with these and Mk. XXs, the latter being used on 9 April, when three raided Kiel and one Hamburg. On the same night the last Mk. XX sorties were flown, one each to Kiel and Hamburg by KB356 and KB 231. The L.N.S.F. flew 2,211 sorties during March.

An average of 203 Mosquitoes was available for operations in April and for 692 Squadron the month started with their greatest effort on any one night when 'for fifteen minutes aircraft roared down the runway till twenty were airborne, the last ten minutes late. Thin cloud, red T.I.s on the ground, contributed to a good attack. Red route markers laid for Berlin were rather confusing though' wrote the squadron historian. The squadron's twenty-one aircraft delivered 37 tons on Magdeburg on this occasion. The Berlin attacks continued on fifteen nights during April and totalled 831 sorties. The largest came on the nights of 3rd (83+9), 11th (85+17), 12th (77+16) and 15th (84+15). On 16 April a three-wave attack was delivered. The last attack took place on 20–21 April when 59+7 were despatched.

A tally at this point showed that Berlin had been the target for close

Operations: Light Night Striking Force

Mosquitoes despatched to Berlin on 21/22 March, 1945

Phase I attack	Phase II attack
139 Squadron (Marker squadron for the attack) KB204:A KB349:F KB217:H KB214:K KB399:M KB156:N KB192:S KB391:W	KB225:U KB148:L MM200:E RV313:V
128 Squadron PF443:A RV319:B PF461:C PF428:D RV297:F PF440:H PF405:J MM192:R MM223:S PF449:T PF457:U PF432:W PF413:Y PF458:Z	PF406:E RV307:K MM202:V MM204:X
142 Squadron KB436:A KB468:B KB435:C KB473:D KB450:E KB439:G KB487:J KB519:K KB432:M KB470:O KB457:R KB449:S KB466:U KB423:V	KB468:B
162 Squadron KB477:A KB407:R KB462:B KB415:C KB454:T KB465:F KB453:G KB497:J KB492:S KB458:U KB461:V KB509:Z KB483:D	KB509:Z KB462:B KB458:U KB465:F KB445:E KB407:R
163 Squadron KB488:D KB511:E KB526:F KB624:G KB518:J KB623:S KB502:U KB425:V KB427:X KB541:Y KB403:Z KB464:T	KB510:B KB624:G KB555:R KB502:U
571 Squadron MM169 ML963 MM179 RV323 MM145 RV305 PF387 PF389 PF433 PF438 MM119 RV315 PF394 PF383	ML963 MM148 PF387 PF438 PF394 RV315
608 Squadron KB236:A KB356:B KB493:C KB451:D KB438:F KB355:G KB347:H KB358:J KB411:M KB491:P KB346:R KB400:U KB405:Y KB298:X	KB236:A KB493:C KB411:M KB491:P
692 Squadron PF445:A MM133:D PF456:J PF397:I PF441:B PF388:C PF400:M PF455:P PF430:T MM172:V RV312:X PF392:R RV310:S ML970:O (FTR)	RV311:Q PF455:P PF430:T RV318:H PF448:F PF400:M

This operation was the largest ever carried out by Mosquitoes against Berlin. The first attackers took-off around 19.30 hrs. and attacked at about 21.30 hrs. The second wave left about 01.40 hrs., attacked at about 04.00 and landed back about 06.00. Twenty Mosquitoes are seen to have been despatched to the city twice in the one night, and 79/139 despatched were Canadian built machines. Only one Mosquito was lost.

Operations: Light Night Striking Force

on 3,900 sorties and that about 4,470 tons of bombs had fallen from the Mosquitoes of 8 Group on to the capital, between 1 January, 1945 and 21 April, 1945. One thousand four hundred and fifty-nine × 4,000 lb. bombs were dropped on the city. In addition 71/88 *Oboe* Mosquitoes bombed Berlin during April.

With the war drawing rapidly to its inescapable end 8 Group turned its Mosquitoes to attacks on airfields at Ingoldstadt, Wittstock, Schlessheim, Husum and Eggebeck. At Pasing, close to Munich, the important Reichsbann power station was situated. Twice it was bombed, on 24–25 and 25–26 April, attracting sixty-seven Mosquito sorties on the second occasion.

It was to Kiel that attention was now directed. This port was handling traffic with Scandinavia and there was the possibility that the enemy might attempt to fight to the bitter end in Norway. Kiel was attacked on five nights in April, then on 2–3 May it became the last German town to experience a Bomber Command Mosquito attack.

Eight squadrons were engaged and despatched sixty-three Mosquitoes for the first attack and fifty-three for the second. The first bombs fell shortly before 23.30 hours on the markers dropped by three aircraft of 139 Squadron and two of each *Oboe* squadron. The attack was delivered from 18–19,000 feet through 9/10th strato-cumulus. A large explosion of three to four seconds duration was seen by the crews. The second wave attacked about 00.20 hours from 16–18,000 feet through 4/10th cloud.

An analysis of the Mosquito night bombing effort divorced from the vitally important marking operations is almost impossible to detail, likewise one cannot confirm the effectiveness of each sortie flown. But in round figures the L.N.S.F. and 8 Group aircraft (including 105, 109 and 139 Squadrons) flew 26,255 sorties for a loss of 108 of their number. In addition to these, 88 Mosquitoes were written off as a result of battle damage. The dropping of about 10,000 'cookies' or 26,000 tons of bombs was surely an amazing achievement for a little unarmed wooden two-man aeroplane, which flew 68 per cent of its operations when heavy bombers were not operating.

Analysis of Night Sorties May to October, 1943

	105	ab.	109	ab.	139	ab.	ab/total sorties
5.43	19	3	88	21	12	1	25/119
6.43	—	—	89	19	45	3	22/134
7.43	13	5	65	35	111	13	53/189
8.43	30	5	43	19	145	28	52/218
9.43	48	13	43	17	127	17	47/218
10.43	64	7	69	30	106	9	46/239

Operations: Light Night Striking Force

Statistics Relating to Bomber Command Mosquito Operations
January–May, 1945

Mosquito Attacks on Berlin by Non-Oboe Mosquitoes

	Sorties	No. attacking	FTR	No. of 4,000 lb. bombs dropped	Tons dropped
January	409	367	1	132	437
February	851	813	3	309	955
March	1,687	1,596	7	546	1,850
April	953	919	3	472	1,228
Totals	3,900	3,695	14	1,459	4,470

Additionally there were 88 sorties by *Oboe* Mosquitoes in April, of which 71 attacked. On 1–2 February there were 176 Mosquito sorties, on 8 targets.

Sorties by Non-Oboe Mosquitoes other than on Berlin

	Despatched	Attacked	FTR	No. of 4,000 lb. bombs dropped	Tons H.E.	Tons Incend.
January	531	501	3	211	591	11
February	811	786	1	360	941	21
March	524	518	—	212	628	10
April	961	932	5	467	1,196	24
May	118	117	—	55	149	—
Totals	2,945	2,854	9	1,305	3,505	66

Sorties by Oboe Mosquitoes other than to Berlin

	Despatched	Attacked	FTR	No. of 4,000 lb. bombs dropped	Tons H.E.	Tons Incend.
January	124	111	1	52	140	5
February	252	229	—	118	280	17
March	194	175	2	27	124	23
April	270	223	—	56	196	20
May	24	21	—	14	25	2
Totals	864	759	3	267	765	67

Operations: Light Night Striking Force

**Mosquito losses on Night Bomber Operations
May, 1943 to May, 1945**

	IV	IX	XVI	XX	XXV	Total Sorties
May, 1943	1* 0	—	—	—	—	119
July, 1943	2 0	—	—	—	—	189
August, 1943	4 3	—	—	—	—	218
September, 1943	2 1	0 1	—	—	—	218
October, 1943	3 0	– –	—	—	—	242
November, 1943	2 2	2 1	—	—	—	368
December, 1943	2 0	1 2	—	—	—	347
January, 1944	3 2	0 1	0 1	—	—	525
February, 1944	3 0	0 1	0 1	—	—	490
March, 1944	1 0	– –	– –	—	—	951
April, 1944	1 1	– –	0 1	—	—	968
May, 1944	2 3	– –	0 2	0 1	—	1,343
June, 1944	4 1	1 0	1 5	1 0	—	1,487
July, 1944	1 1	2 1	5 0	– –	—	1,511
August, 1944	1 0	– –	2 5	4 1	—	1,628
September, 1944	1 0	– –	3 2	4 2	—	1,715
October, 1944	1 0	– –	8 3	2 1	0 2	1,863
November, 1944	2 0	– –	4 2	1 2	– –	1,859
December, 1944	– –	– –	2 4	1 1	0 2	1,777
January, 1945	– –	– –	3 11	0 1	2 1	1,299
February, 1945	1 0	– –	4 5	1 0	1 0	2,404
March, 1945	1 1	0 1	6 5	1 1	2 1	2,950
April, 1945	– –	0 1	1 0	2 0	3 1	2,324
May, 1945	—	—	—	—	—	141
Totals	38 15	6 9	39 47	17 10	8 7	26,936

Grand totals 108 plus 88

Figures in the first position in each column indicate number of aircraft that failed to return from operations.

Figures in the second position indicate number of aircraft written off as a result of battle damage and battle accidents.

* In addition to this loss at night 3 Mk. IVs were lost on day operations.

Operations: Light Night Striking Force

First Operations by Mosquito Bombers

	IV	IX	XVI	XX	XXV	VI
139	25. 6.42 (105 Sqn. acrft.)	3.10.43	10–11. 2.44	2.12.43	9.10.44	—
192	11. 6.43	—	2. 2.45 (PR)	—	—	—
128	—	—	23.10.44	10. 9.44	18.10.44	—
162	—	—	—	—	21.12.44	—
617	3. 5.44	5. 4.44	—	20. 7.44	—	18. 4.44
109	20.12.42	11. 6.44	1. 3.44	—	—	—
142	—	—	—	—	28.10.44	—
163	—	—	—	—	29. 1.45	—
105	31. 5.42	13. 7.43	2. 3.44	—	—	—
571	—	—	12. 4.44	—	—	—
608	—	—	26. 3.45	5. 8.44	6.10.44	—
627	24.11.43	—	21. 3.45	7. 7.44	11.11.44	—
692	1. 2.44	—	5. 3.44	—	—	—

Comparison of Effort and Effectiveness of Bomber Command Aircraft

Type	Total sorties flown	Total failed to return	Bomb load delivered (tons)	Losses: sorties ratio per cent
Stirling	18,440	606	27,821	3·81
Blenheim	12,214	443	3,028	3·62
Ventura	997	38	726	3·60
Wellington	47,409	1,332	41,823	2·80
Boston	1,609	40	952	2·48
Halifax	82,773	1,830	224,207	2·28
Lancaster	156,192	3,340	608,612	2·13
Mosquito	39,795	254	26,867	0·63

N.B. These figures include all sorties by the listed types operating under Bomber Command control.

Operations: Light Night Striking Force

Analysis of sorties flown by Mosquitoes of 8 Group from December, 1942 to May, 1945 (excluding 1409 Met. Flt.)

Date	(b) 105 Sqn.	109 Sqn.	139 Sqn.	(a) 627 Sqn.	692 Sqn.	571 Sqn.	608 Sqn.	142 Sqn.	128 Sqn.	162 Sqn.	163 Sqn.	Monthly total
1943												
May	19	88	12	—	—	—	—	—	—	—	—	119
June	—	89	44	—	—	—	—	—	—	—	—	133
July	13	65	111	—	—	—	—	—	—	—	—	189
Aug.	30	43	145	—	—	—	—	—	—	—	—	218
Sept.	48	43	127	—	—	—	—	—	—	—	—	218
Oct.	64	69	109	—	—	—	—	—	—	—	—	242
Nov.	113	96	149	10	—	—	—	—	—	—	—	368
Dec.	90	106	105	46	—	—	—	—	—	—	—	347
Total	377	599	802	56	—	—	—	—	—	—	—	1,834
1944												
Jan.	173	135	130	87	—	—	—	—	—	—	—	525
Feb.	138	114	114	91	33	—	—	—	—	—	—	490
Mar.	206	225	242	137	141	—	—	—	—	—	—	951
Apr.	154	188	227	170	181	48	—	—	—	—	—	968
May	285	305	256	112	250	135	—	—	—	—	—	1,343
June	380	367	212	83	229	216	—	—	—	—	—	1,487
July	412	366	204	80	225	224	—	—	—	—	—	1,511
Aug.	348	509	198	68	214	213	78	—	—	—	—	1,628
Sept.	397	426	227	95	249	148	151	—	22	—	—	1,715
Oct.	315	455	227	45	247	243	202	6	123	—	—	1,863
Nov.	302	306	210	53	225	239	223	88	213	—	—	1,859
Dec.	258	270	196	51	176	192	187	222	193	32	—	1,777
Total	3,368	3,666	2,443	1,072	2,170	1,658	841	316	551	32	—	16,117
Accum. Total	3,745	4,265	3,245	1,128	2,170	1,658	841	316	551	32	—	17,951
1945												
Jan.	126	129	182	33	146	144	109	155	136	135	4	1,299
Feb.	248	269	332	99	231	240	246	223	210	222	84	2,404
Mar.	328	343	301	125	274	275	222	283	271	274	254	2,950
Apr.	265	303	232	72	303	229	110	228	232	228	122	2,324
Apr. 'Manna'	16	16	—	—	—	—	—	—	—	—	—	32
May	12	12	14	—	24	—	15	16	16	16	16	141
May 'Manna'	64	74	—	—	—	—	—	—	—	—	—	138
Total	1,059	1,146	1,061	329	978	888	702	905	865	875	480	9,288
Grand Total	4,804	5,411	4,306	1,457	3,148	2,546	1,543	1,221	1,416	907	480	27,239

(a) Includes all operational sorties flown with 5 Group (first 473 were flown whilst with 8 Group).
(b) 109s earlier sortie totals were:—
    12.42: 23, 1.43: 50, 2.43: 19, 3.43: 18, 4.43: 39. Total 149.
All totals listed agree with Squadron Operational Record Book listings, and relate to the number of sorties despatched, not the number of aircraft that necessarily attacked, were PFF reserves, etc.

Two-Stage Merlin Mosquito Bombers

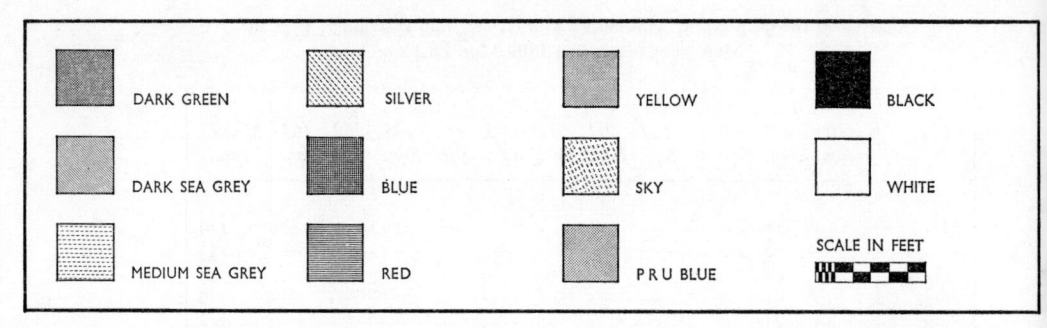

DARK GREEN	SILVER	YELLOW	BLACK
DARK SEA GREY	BLUE	SKY	WHITE
MEDIUM SEA GREY	RED	PRU BLUE	SCALE IN FEET

B. Mk. IX (Merlin 72) ML922 reached 105 Squadron 21.9.43. Night bomber camouflage, red serials, white *Oboe Leader* markings for day raids. As recorded at Bourn 16.12.44.

B. Mk. IX (Merlin 72) ML913 reached 105 Squadron 24.9.43. Flew very many sorties, shot down by flak near Scholven-Buer 5–6.7.44. Night bomber camouflage, as recorded at Bourn 25.3.44.

P. Mk. IX (Merlin 72) LR507 reached 109 Squadron 17.6.43, to 105 Squadron 5.7.43, operated almost continuously from 7.43 to 4.45. To 22 MU 21.9.45, SOC 15.5.46. Night bomber camouflage, as recorded 12.4.44.

B. Mk. XVI (Merlin 72) ML966 reached 692 Squadron 3.3.44, operated throughout 1944. Day bomber camouflage depicted as recorded 25.3.44.

B. Mk. XVI (Merlin 72) ML956 reached 109 Squadron 30.1.44 to 105 Squadron 5.10.44, to 27 MU 12.11.45, sold to General aircraft 3.6.47 then to Royal Navy. Day bomber camouflage, recorded Bourn 3.45.

B. Mk. 35 (Merlin 113) TA694 delivered 4.6.45 to 14 Squadron later, and used in Germany. Recorded 1949.

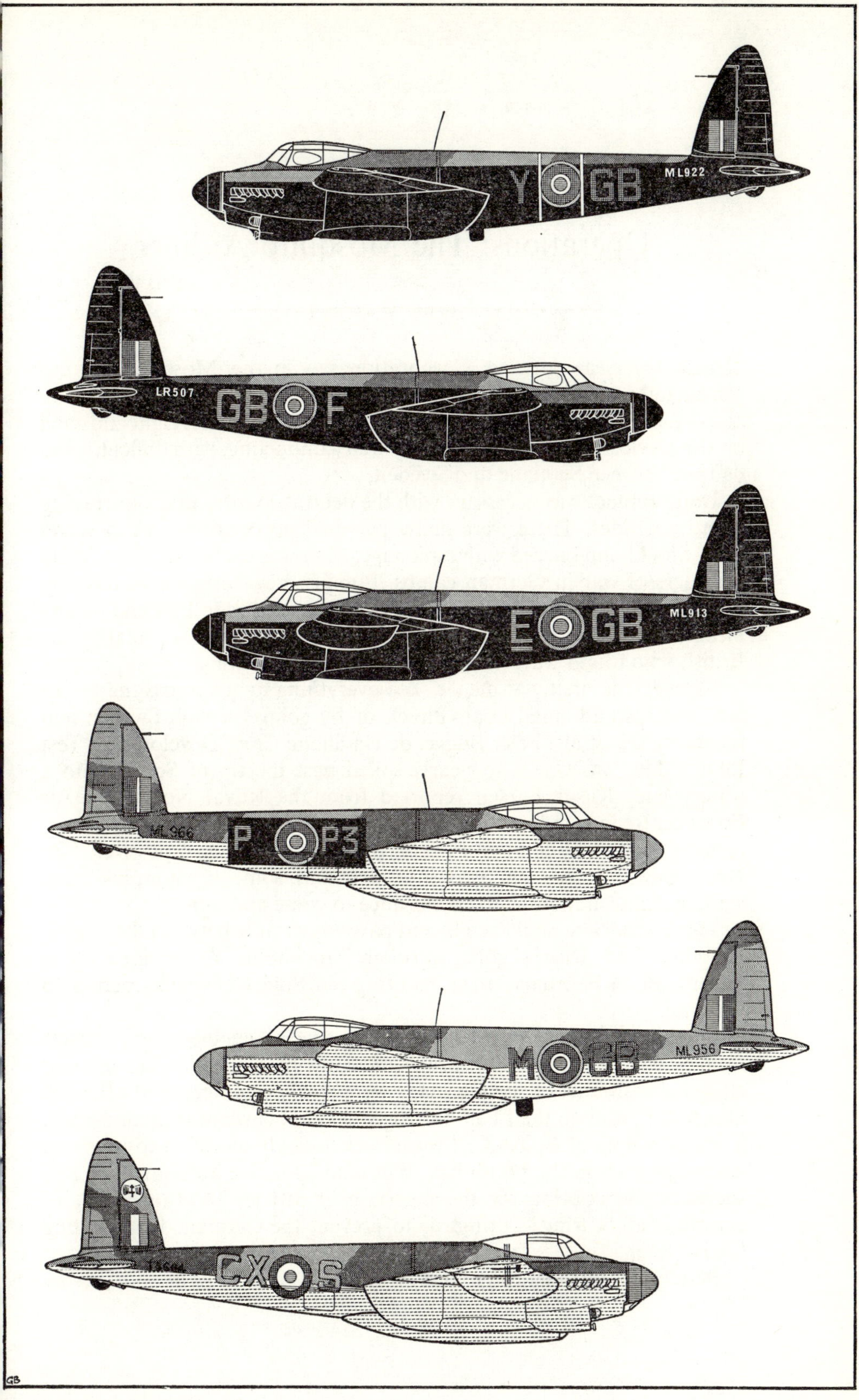

ML922
Y GB
LR507
GB F
E GB
ML913
ML966
P P3
M GB
ML956
CX S
GB

Chapter 21

Operations: The Mosquito Airliner

British Overseas Airways Corporation saw in the Mosquito an unarmed light merchantman that could outpace enemy fighters and fly above flak. This would be invaluable for carrying small freight and mail on the service which in 1941-42 they were conducting, with difficulty and danger, between Scotland and Sweden.

Daily contact was necessary with the neutral country that was trading with both sides. There were many personal negotiations, much secret mail, Stockholm buzzed with espionage. Mail was carried also for Allied prisoners of war in German camps. Bomber crews often came down in Sweden, or got there somehow, and tried to get back to England by any means. Even after American bombing developed it was still mainly the British who found ways of getting back into operations.

Norwegians: many airmen, risked everything to get across the North Sea in ships and small boats direct, or by going through Sweden and thence by sea or air. Peter Bugge, de Havilland Chief Development Test Pilot today, was one who nearly sailed past the tip of Scotland to a watery end. Riisar Larsen reported from the Royal Norwegian Air Force centre in London on 4 March, 1942, that President Roosevelt was helping them to obtain two Lockheed Lodestars to get, if possible, fifty Norwegians a week from Sweden to Scotland, some of whom had been waiting eighteen months for the chance to come and fight.

There was some mail, freight and passenger traffic between the United Kingdom and Russia going through Stockholm, exemplified by a request on 26 February, 1942, to bring ten Soviet engineers across to England.

Furthermore, Britain was obtaining special engineering products from Sweden, such as ball-bearings, machine-tool steel, fine springs and electrical resistances. The Germans were rival customers, and after the U.S.A.A.F. raid on their ball-bearing works at Schweinfurt in mid-1943, it was a couple of B.O.A.C. Mosquitoes, quickly modified to take one passenger each in the bomb bay, that made possible an immediate and successful negotiation for the purchase of Sweden's entire output of essential ball-bearings—primarily to prevent the Germans from getting hold of them.

Besides aircraft there were fast motorboats venturing eastward across

Operations: The Mosquito Airliner

the North Sea to make stealthy rendezvous with friendly ships near Scandinavian waters, and bringing valuable loads back.

B.O.A.C. pleaded for faster, higher-flying aircraft than the four Hudsons and two Lodestars which were being used in 1942. Their ceiling was not above 20,000 feet. They were obliged to avoid daylight, moonlight, Northern Lights, and clear weather—although cloud cover and general bad weather were disliked because they increased icing and navigational hazards. The light summer nights made the service practically inoperable from mid-May to early August. After an intensification of German activity in the Skagerrak, in the autumn of 1942, longer, more northerly routes had to be considered for these slow aircraft.

A.B. Aerotransport, the Swedish line, operated Dakotas to Scotland. The free Norwegians used mainly Lockheed 14s. Co-operation with the Swedes was a delicate matter for they wanted to avoid offending the Germans, yet they needed good communications with London as well as with Berlin. Anything might happen, and in April, 1942 the British Air Ministry were discussing with B.O.A.C. plans for evacuation if hostilities should break out in Sweden.

A.B.A. personnel were fairly friendly, though competitive with B.O.A.C. and their Dakota crews naturally had fewer navigation and identification problems. As the prospects of Allied victory improved co-operation became easier; this was noticeable late in 1942. But there was every appearance of correctness and the avoidance of favour. Any aircraft wandering from the narrow obligatory Swedish corridor was fired on, even if shots went conveniently wide, and a complaint was always lodged next morning with the British Ambassador. B.O.A.C. pilots often said their digression was due to faulty wind information from the Swedish Met. Office before departing from Bromma.

Identification and navigation were always difficult. Weather reports were inadequate, D/F bearings were unreliable at night, and there was no direct communication between Leuchars, Fifeshire and Bromma, Stockholm. B.O.A.C.crews kept almost complete radio silence, weighing the danger of a single short H/F transmission for identification to the Swedes (which could also alert the Germans) against the accuracy of Swedish gunfire. The Norwegians at Bergen and Stavanger were not allowed by the Germans to broadcast, but managed often to keep their carrier wave switched on to help R.A.F. and B.O.A.C. crews.

B.O.A.C. aircraft were flagged in on the Bromma apron, to park beside Ju52s of Deutsche Lufthansa, and crew who had been friendly at Croydon before the war would exchange a curt nod. The Germans had not to run the gauntlet of enemy fighter bases on their direct route into Germany.

Indeed, the Norwegians and the British in 1942 were having it tough. B.O.A.C. considered that a speed of at least 200 m.p.h., and this at 23,000 feet, was necessary to avoid icing and enemy action. Brighter prospects were offered in the Ventura and Hudson 3, or even the

Operations: The Mosquito Airliner

Beaufighter. But when the R.A.F. had to deliver some special mail to Sweden in deep secrecy on 6 August, 1942 it was the Mosquito (105 Squadron) that was picked for the job, and Flt. Lt. Parry and Plt. Off. Robinson in DK292

Could B.O.A.C. have Mosquitoes?

Suggestions of using the Whitley were regarded as 'retrograde', even if Merlin XXs could be made available. On 5 November, 1942, Air Ministry allotted an Albemarle for a month for suitability tests. But B.O.A.C. pressed for Mosquitoes—even one Mosquito—and sent Parker and C. B. Houlder to fly the type at Lyneham. At last a B Mk. IV, DZ411, was allotted, to be adapted by de Havilland as G-AGFV. Long-range tanks in the bomb bay would leave little space for payload, and pilots worried about the lack of de-icing. Separate ejector exhausts were fitted. The intention was, at first, to carry mail to Sweden, and bring back mail and cargo. Dakotas replaced B.O.A.C. Lockheeds, and, despite the visibility and route restrictions on them, were useful for handling up to ten or fifteen passengers.

G-AGFV was delivered on 15 December, and first operated to Stockholm on 4 February, 1943. The crew were Capt. Houlder and R/O Frape. Thereafter Houlder trained the pilots throughout, as well as doing a fair share of the operation. Two more round flights were made that month, eight in March, two in April. Training flights, also troubles with radio and fuel system, were behind this slow start with the Mosquito. On 12 April, it was confirmed that B.O.A.C. were to have six more; these were to be unarmed FB Mk. VI, HJ680, 681, 718, 720, 721 and 723, to be registered respectively G-AGGC to AGGH, and they were promptly delivered on 16, 17, 23, 24 April, and two on 2 May.

Clarkson wrote from Hatfield to B.O.A.C. Chief Project Engineer C. H. Jackson, 28 May, explaining differences in performance; the B. Mk. IV G-AGFV (authorized weight 21,000 lb.) had unshrouded multiple ejector exhausts which (compared with the standard saxophone on the six F.B. Mk. VIs) gave an increase in top speed (3,000 r.p.m., plus 9 lb./sq. in. boost) of 12 m.p.h., and in cruising speed (2,650 r.p.m., plus 4 lb./sq. in.) of 9 m.p.h. External wing tanks cost 5 m.p.h. while wing bombs with fairings had cost the R.A.F. 12 m.p.h. The faster Mk. IV was always a favourite with B.O.A.C.

The Germans intensified flak and fighter activity when Mosquitoes appeared on the route. As G-AGFV sped across the Skagerrak on the night of 22–23 April, 1943, an Fw 190 crept down on it, opened fire, severely damaged the wings, fuselage, hydraulic system, and shot off the escape hatch; out went the maps and papers. It was not possible to lower the undercarriage but the pilot, Capt. Gilbert Rae, made a belly landing at Berkaby near Stockholm. Mosquito operations abruptly halted until the Mk. VIs were ready, and it was 10 December before AGFV was repaired. There were only five round flights in May, but thirty were made in June.

Operations: The Mosquito Airliner

Practically every night in June and July there was flak above 20,000 feet, from ships in the Skagerrak, but the Mosquito sailed through, midnight sun and moon notwithstanding. Other B.O.A.C. aircraft were withdrawn at this period; occasional northerly Dakota flights were planned for late September or October. Their route changed nightly depending on weather, but generally tracks were between lat. 58 and 64, avoiding Trondheim, thence over Falun (main internment camp for British and Americans). A route by Wick and lat. 62 exposed aircraft to radar longer, while a route between Stavanger and Bergen (lat. 59–60) was the most dangerous for flak; a route as far north as lat. 65 was investigated.

But the Mosquito needed these not. Often a Mosquito would go northerly to spy out new German fighter fields or installations in Norway. Dakotas could be intercepted by fighters based or refuelled there. A field was spotted near Oslo, and another between Oslo and Stavanger. Sometimes B.O.A.C. flew north to search for crashed Lancasters, and for the secret lamp signal of their fugitive crews. They pinpointed several.

If engine or constant-speed unit necessitated feathering it was not unusual to maintain 10,000 feet on one engine. The 'cough' of the automatic supercharger change, descending through 14,000 feet, sometimes made a crew think that the other engine had gone.

Rae, with R/O Payne, was returning from Bromma on 24 June in G-AGGH when the port engine coolant temperature rose to 145°C. He was at 20,000 feet, and he feathered. He went on because the weather behind was bad. This time (flying blind) he got very low, and the aircraft was nearly lost.

It was soon after this episode that the ball-bearing deal called for the conversion of two Mosquitoes, in a matter of hours, each to carry a passenger in the bomb bay. The two Britishers got there an hour or two ahead of German negotiators. This successful mission led to a passenger being carried as regular practice. He reclined on a mattress, with reading lamp, wearing a flying suit and Mae West, flying boots and parachute harness, having temperature control and adjusting his oxygen-mask setting as instructed by the pilot through headphones. The flight took less than three hours, compared with up to nine hours for the slow aircraft.

Some notable passengers graced the bomb bay, such as Sir Malcolm Sargent and Sir Kenneth Clark, and although the crew were sometimes not told the name of a hush-hush traveller, Pelly (Flight Captain from late 1943) was certain one night that he recognized Marshal Timoshenko. The gentleman apparently spoke no English, but knew a few proper names, so it was decided that the intercom. message 'Betty Grable' would mean they were about to land; likewise 'Tolstoy' would mean they were going to bale out. All went well, but they hadn't arranged a code word for 'The confounded bomb doors are half open!', which indeed they were. However the notable passenger thought that it was quite normal, and took it very well.

Operations: The Mosquito Airliner

Rae and Payne in G-AGGC had another nasty time on 18 July. Just after leaving the Swedish coast on a clear night with full moon, 50 miles off Havstund, they spotted two contrails at 23,000 feet which appeared to belong to Fw190s. Rae turned, dived, weaved, almost down to sea level, then headed west. The usual rule when challenged while still in the Skagerrak was to turn back to Sweden—for fear of running out of fuel in a long chase towards Scotland. He continued homeward because his passenger was an important man whom the Germans had asked the Swedes to hand over, and because Capt. Steen with R/O Omholdt, in a second Mosquito, was following with a similar passenger. Rae was making 2,850 r.p.m., plus 15 lb., rad. 105° to 110°C., oil 85°, 80 lb; the A.S.I. was put out by high-speed stall, but he estimated his speed at 330–40 m.p.h. He jettisoned wing tanks. The enemy gave up the chase after about 30 minutes, presumably for lack of range. Steen, landing at Leuchars 32 minutes after Rae, reported no incident. They reckoned German Intelligence knew about the departure of these two men, and they were saved only by the Mosquito's speed. The elevator of Rae's aircraft had been deformed by the severe evasive manoeuvre; Hereward de Havilland got it to Hatfield, for the R.A.F. had experienced the same thing, and a modification substituting metal for fabric skin was introduced.

On 7 August, with their Lockheeds held off the northerly route, Dakotas in danger of interception, B.O.A.C. looked forward to getting a York for the bigger loads, or a Lancaster, and considered meanwhile using, for an occasional flight, two Liberator IIIs (with about fifteen passengers), then based at Lyneham. Neither Liberator nor Lancaster, however, could use Bromma on load until the runway (then 1,200 m.) was extended at least to 1,500 m. (It was 1,650 m. by January, 1944.) The Mosquito meanwhile remained the only B.O.A.C. aircraft.

Capt. Wilkins, heading the operation, and his radio officer N. H. Beaumont, left Leuchars in G-AGGF on 17 August, 1943. Soon after take-off he radioed that he was returning. No more was heard, but a gamekeeper on 8 September found wreckage in Glen Esk. The aircraft had hit a mountain at 2,500 feet on approach in bad weather, and had become B.O.A.C.'s first Mosquito loss. Capt. C. N. Pelly was brought home from B.O.A.C./R.A.F. Communications, North Africa, to replace Wilkins as Flight Captain.

A second tragedy occurred on 25 October when Capt. Martin Hamre, R/O Serre Haug (both ex-R.N.A.F.), and passenger Mr. Carl Rogers were killed. About 230 miles from Stockholm they radioed that the port engine of G-AGGG had failed. They continued westward and about a mile from Leuchars crashed, after a valiant attempt to reach home. In some 1,250 hours' flying this was the second serious accident.

A Dakota started operating via lat. 62 on 2/3 October, and a Liberator, G.AGFS, experimentally made Göteburg (Torslanda) early that month.

Operations: The Mosquito Airliner

But then A.B.A. lost two Dakotas and had to cease operations. Norwegians and B.O.A.C. were hard pressed. The British Ministry of Economic Warfare late in November had urgent need to fly 90 tons of machinery and material to Britain, of which 55 tons was bulky, calling for forty-five Dakota flights and one Liberator flight, while the other 35 tons could be moved by the Mosquitoes. A period of difficult weather set in which held up many flights and diverted Mosquitoes and Dakotas into Torslanda. The backlog of impatient passengers built up again. An entry records that the Christmas schedule had to be modified so as not to disturb the holiday timetable of Swedish personnel. In these conditions B.O.A.C. found the pathfinder technique a great help. The advance Mosquito would radio a single letter back indicating that the night was 'suitable' or 'unsuitable', meaning from both weather and tactical aspects, and the two code letters were changed nightly. G-AGFV was the favourite pathfinder; with 610 gallons, 6 hours' cruise endurance, and a better speed than the Mark VIs, it could get within a hundred miles of Bromma, and still turn back if things were bad. On the northerly route it could turn back from the Norwegian mountains; refuelling at Wick gave an extra hundred miles radius. The loss of payload eastbound on the one aircraft was outweighed by the gain in frequency.

When R.A.F. P.R.U. aircraft were out over Scandinavia they would report conditions to Leuchars, and save B.O.A.C. the trouble.

Flying the Skagerrak one night when cloud cover was reported good, with fog on near-by enemy airfields, a Norwegian, Piltingsrud, in a Lodestar left Leuchars at 21,000 feet, 20 minutes ahead of Pelly in a Mosquito. Interested in the effectiveness of Lodestar flamedampers, Pelly and his operator Jimmy Miller had arranged to keep a look-out at the estimated time for overtaking—and spotted them. They formated at 30 yards until noticed; then the other aircraft evaded violently. When Piltingsrud got into Bromma, Pelly said 'What made you windy? Didn't you recognize me at first?' The Norwegian replied 'I didn't see anything. I took the northern route—too many breaks in the cloud near the Skag.' (Nigel Pelly thinks).

With summer coming round again the decision was repeated to use only Mosquitoes over the four months of the light nights. Consideration of the newer high-altitude, high-performance Mosquitoes Marks IX and XVI, as replacements for lost and damaged aircraft, was dropped to avoid operating mixed aircraft and engines on a small establishment. Three more Mark VIs (Merlin XXIIIs) came to B.O.A.C. in the spring, HJ667 G-AGKO on 27 April, G-AGKR on 11 April, G-AGKP on 22 April. A Mark III trainer, HJ985 had been loaned to B.O.A.C. on 28 November, 1943, and was replaced by another LR524, on 21 February, 1944, which in turn was replaced by HJ898 in April, 1945.

A peak in Mosquito flights was reached in 1944. Between May and August 151 round trips were made. June was the busiest month with forty-two.

Operations: The Mosquito Airliner

On 4 July the undercarriage of G-AGFV was damaged at Stockholm due to A.S.I. trouble. Then came the third and fourth Mosquito losses. Gilbert Rae, whose services in about 150 round trips had earned him the O.B.E., was lost with his radio officer, D. T. Roberts, and their colleague Capt. B. W. B. Orton, flying as a passenger, on the night of 18-19 August, when G-AGKP, returning from Stockholm, came down in the sea mysteriously, in good weather, about 9 miles from Leuchars. The bodies of Orton and Roberts were recovered. Capt. White and Radio Officer J. C. Gaffeny were lost on 28-29 August when G-AGKR failed to arrive from Göteburg. The cause was never traced.

The changing situation, with the Allied drive across Western Europe, brought a review of the Scandinavian service and on 3 November A.O.M., London, advocated to D/A.D.C.(T) that Mosquitoes be withdrawn, as being, in winter months, wasteful from a crew point of view. They might be used again in the 1945 summer 'if this should be necessary'. Only 8 round trips were made in November by Mosquitoes, and they were suspended 30 November. One round trip was made in February, none in March. For a short while B.O.A.C. resumed the service 12 April, 1945. Air Ministry advised using the northern route as well as the Skagerrak, and to fly above the flak ships, not below the radar. As the Germans retreated so they looked to Denmark and Norway as places of refuge, bringing strong fighter and radar defences into those countries.

The Mosquito service ended 17 May, 1945. The terminal was moved from Leuchars to Croydon, and three Mosquitoes were handed over to the R.A.F. on 22 June. They were the Mark VIs G-AGKO, AGGE and AGGH. The first of the VIs, AGGC, somehow managed to stay in the merchant service until 9 January, 1946!

From 3 February, 1943, to 17 May, 1945, B.O.A.C. Mosquitoes made 520 round trips. They averaged about nine single trips a week for two and a quarter years. Several Mosquitoes and crew made three single trips in a night. Out of 1,040 flights four failed with the loss of those on board, eight crew and two passengers. In the two fiscal years April, 1943 to March, 1945, Mosquitoes made 783,680 service miles, Dakotas 517,752 miles, and Hudson, Lodestar, Liberator and York aircraft together made 67,648 miles. (See appendices.)

A book about the Mosquito is inappropriate for an account of the valuable contribution which the separate Norwegian unit made, but their mileage figures with Lodestars were substantial. They often operated the Bergen-latitude route, even through the summer of 1944, checking if necessary over the Swedish frontier whether conditions were 'suitable' for continuing to Stockholm, and having just enough fuel to return if they were not.

Operations: The Mosquito Airliner
Brief Notes on B.O.A.C. Mosquitoes

G-AGFV Mk. IV Unshrouded multiple exhausts (increased top speed by
DZ411   about 12 m.p.h.) Total weight 21,000 lb. 15.12.42. to
              B.O.A.C.
              23.4.43 Landed Barkaby (Sweden). Hydraulic trouble caused by enemy attack. Serviceable again 10.12.43.
              14.5.44. Wheel brakes burst as result of pulling up sharply, indirectly caused by excessive heating of port engine radiator.
              4.7.44. Stockholm. Airspeed indicator u/s. Capt. swung off runway, undercarriage collapsed.
              6.1.45. Returned to R.A.F. ($24\frac{1}{2}$ months with B.O.A.C.)

G-AGGC   VI 16.4.43. To B.O.A.C. Bramcote
HJ680   30.11.44. Withdrawn from service, retained Leuchars for emergency ops.
             14.3.45. To Croydon temporarily for C of A.
             9.1.46. Returned to R.A.F. (33 months with B.O.A.C.) (longest).
             23.3.46. To 22 M.U.
             15.6.50. Sold as scrap to J. Dale.

G-AGGD   VI 16–17.4.43. To B.O.A.C.
HJ681   3.1.44. Landed Sarenas, radio trouble. On landing, undercarriage seriously damaged—propellers bent, wings damaged, crew uninjured.
             Reduced to produce. ($8\frac{1}{2}$ months with B.O.A.C.)

G-AGGE   VI 23.4.43. To B.O.A.C.
HJ718   15.8.44. Propellers touched runway when landing Stockholm. Damaged.
             22.6.45. Returned to R.A.F. (26 months with B.O.A.C.)
             25.7.45. To Marshalls for major overhaul.
             28.6.46. To 15 M.U.
             3.9.47. Sold as scrap.

G-AGGF   VI 24.4.43. To B.O.A.C.
HJ720   17.8.43. Lost on service 11B452.
             8.9.43. Wreckage found at Invermaik, Glen Esk.
             20.8.43. Written off (4 months with B.O.A.C.). Capt. Wilkins and R/O Beaumont killed.

G-AGGG   VI 2.5.43. To B.O.A.C.
HJ721   25.10.43. Crashed near Leuchars returning from Stockholm on service 12B517 (6 months with B.O.A.C.) Capt. M. Hamre and R/O S. Haug killed.

G-AGGH   VI 2.5.43. To B.O.A.C.
HJ723   30.11.44. Withdrawn from service (crew shortage). Retained Leuchars.
             22.6.45. Returned to R.A.F. (26 months with B.O.A.C.)
             4.8.45. To 19 M.U.
             20.9.45. To 71 M.U. ROS then Cat. C.
             10.12.45. Became 5755M, thence to 41 Grp., then to CDT & R.E. Estletoke Camp from 19 M.U.
             8.2.46. SOC.

Operations: The Mosquito Airliner

G-AGKO VI 8.3.43. To 10 M.U.
HJ667 27.3.43. Cat. B flying accident.
  27.3.43. To 27 M.U., thence R.I.W.
  24.12.43. To 27 M.U.
  27.4.44. to B.O.A.C.
  14.6.44–30.11.44. Leuchars/Stockholm services.
  30.11.44. Withdrawn—crew shortage.
  22.6.45. Returned to R.A.F. (14 months with B.O.A.C.)
  24.7.45. To 19 M.U.
  14.12.45. Cat. E SOC.
G-AGKR VI 11.4.44. To B.O.A.C.
HJ792 28.8.44. Departed Gothenburg for Leuchars, failed to reach destination. Presume crashed in sea; no trace ever found. (4½ months with B.O.A.C.) Capt. J. H. White and R/O J. C. Gaffeny lost.
G-AGKP VI 22.4.44. To B.O.A.C.
LR296 19.8.44. Crashed in sea 9 miles from Leuchars (4 months with B.O.A.C.) Capt. G. Rae, R/O D. T. Roberts, Capt. B. W. B. Orton killed.
HJ898 III 22.4.45. To B.O.A.C.
  Swung off runway onto grass landing Leuchars, undergoing acceptance test. Port wing, undercarriage, airscrew badly damaged, starboard undercarriage badly bent.
  12.5.45. Returned to R.A.F. (3 weeks with B.O.A.C.)
HJ985 III 28.11.43. To Leuchars from R.A.F., loaned for crew training.
  26.1.44. Returned to R.A.F. (2 months with B.O.A.C.)
LR524 III 21.2.44. to B.O.A.C.
  4.12.44. Returned to R.A.F. due to suspension of Mosquito operations. (9½ months with B.O.A.C.)

B.O.A.C. Scotland—Sweden Mosquito Service
3 February, 1943–17 May, 1945
B.O.A.C. U.K.—Scandinavia Operations

Aircraft mileage	1.4.43–31.3.44		1.4.44–31.3.45	
	Service	Dead	Service	Dead
Mosquito	381,934	106,614	401,746	66,523
Dakota	161,131	34,637	356,621	49,704
Hudson	22,456	10,568	—	—
Lodestar	20,852	7,513	—	—
Liberator	21,967	6,025	—	339
York	—	—	2,373	1,074

Aircraft hours	Service	Dead	Service	Dead
Mosquito	1,510·04	480·21	1,492·20	283·04
Dakota	1,224·12	239·50	2,328·21	356·13

Engine hours	Service	Dead	Service	Dead
Mosquito (Merlin)	3,020·08	960·42	2,984·40	566·08
Dakota (Twin Wasp)	2,448·24	479·40	4,656·42	712·26

Operations: The Mosquito Airliner

Route data (all aircraft types)
1.4.43–31.3.44. Stage flights scheduled 767—completed 765 (99·7%).
　　　　　　　Load ton-miles 537,808. (87% of capacity ton-miles).
　　　　　　　Passengers 1,024.
　　　　　　　Total tons carried 643·57.
　　　　　　　Aircraft hours flown on route 3,155·59.
1.4.44–31.3.45. Stage flights scheduled 918. Completed 916. (99·8%).
　　　　　　　Load ton-miles 723,890. (91·7% of capacity ton-miles).
　　　　　　　Passengers 1,605.
　　　　　　　Total tons carried 896·79.
Aircraft hours flown on route 3,834·06.

Crews

Captains
Carroll, A. M.
Copeland, J. R. G.
Dykes, A.
Hamre, M. ex Royal Norwegian Air Force, lost, G-AGGG, 25/10/43
Houlder, C. B. Training Captain
Hunt, V. A. M.
Longden, C.
Orton, B. W. B., lost as passenger, G-AGKP, 19/8/44
Pelly, C. N. Flight Captain after Wilkins
Rae, G. lost, G-AGKP, 19/8/44
Rendall, A.
Steen, Capt. Niels
Tapley, R. H.
White, J. H. lost, G-AGKR, 28/8/44
Wilkins, L. A., Flight Captain lost, G-AGGF, 17/8/43

Radio Officers
Battye, G. A.
Beaumont, N. H., lost, G-AGGF, 17/8/43
Blackburn, R. J. G.
Burnett, J.
Dalglaish, I. A.
Frape, F.
Gaffeny, J. C. lost, G-AGKR, 29/8/44
Haug, S. ex Royal Norwegian Air Force, lost, G-AGGG, 25/10/43
Miller, J.
Omholdt
Parker
Payne, J.
Roberts, D. T. lost, G-AGKP, 19/8/44
Weir, J.

　　　By reason of postings this list cannot be quite complete.

Chapter 22

Operations: Intruders and Rangers

Intruding Mosquitoes, by night and day, accomplished some of the most amazing flights of the war. They swept across Germany, far out over Biscay, over France, Norway, Denmark and the Baltic, into Czechoslovakia and Austria on patrols of more than 1,000 miles, destroying aircraft on the ground or in the air and attacking all manner of targets. Speed, range, fire power, manoeuvrability and ability to survive punishment fitted the Mosquito for this work.

Early in the days of Mosquito fighter and bomber squadrons, indeed up to two months before they became operational, there was little interest in a Mosquito Intruder. Then the position rapidly altered and, throughout the succeeding years of greatest activity, Intruders did outstanding work.

On 1 April, 1942, Wg. Cdr. Paddy Crisham, C.O. of 23 Squadron using Boston Intruders, brought Sqn. Ldr. Sammy Hoare, his senior flight commander, to de Havilland to lunch to talk about Mosquito Intruder possibilities—and the chance of getting a few aircraft after P.R.U., night-fighter and bomber squadrons received their monthly quotas.

They were impressed with the performance and handling, the view, and armament of the Mosquito. Their Bostons had only four ·303 in. guns and barely three hours' fuel. Give them, they said, *all* the Mosquitoes (leaving maybe the odd one for P.R.U.) and they would write history across the troop concentrations, communications and airfields of Occupied Europe.

Further meetings took place in April, and there was then talk of having twenty-five Intruder Mosquitoes for trials, after two night-fighter squadrons had been equipped. The crews said that bombs were a secondary consideration, provided there was plenty of ammunition for the cannon. The N.F.II with A.I. radar removed was good enough, although they would have preferred better deceleration to avoid overshooting targets.

By 7 June 23 Squadron had one T. Mk. III at Ford, their South-Coast base. On 2 July the first operational machine arrived, DD670, the first Mk. II (Special Intruder) Mosquito. Sam Hoare and his navigator, Plt. Off. Cornes, were airborne at 02.23 on 6 July to commence Mosquito

Operations: Intruders and Rangers

Intruder operations and patrolled over Caen. Luck was out; all they encountered was flak around Le Havre.

Next night, again in DD670 (now S–Sugar), they were more successful. After patrolling Avord airfield they saw lights on an aircraft east of Chartres. They stalked it, a Do 217, then fired three bursts of cannon. It crashed in flames at Montdidier. Two nights later Salisbury-Hughes took his turn with the only operational Mosquito 23 Squadron yet possessed. He, too, claimed a Do 217 after patrolling three French airfields. After following it on two circuits he shot the bomber down near Etampes, and then finished off an He 111 circling Evreux. Hughes declared the Mosquito's manoeuvrability and hitting power much to his liking.

By 31 July twenty-eight sorties had been flown by the Mosquitoes, for the loss of only DD677 near Eindhoven. Four bombers had fallen to DD670, currently the top scoring Mosquito.

From July to Christmas, 23 Squadron worked hard with the fine new weapon, first from Ford, then from Manston (where the very poor airfield surface quickly led to two crashes) and soon after from Bradwell. Trade was good and there were many adventures and a few losses. On 23–24 August Sqn. Ldr. Starr in DD673:E set off to patrol Deelen in Holland. Ten miles off the Dutch coast his port engine failed and on landing at Manston he overshot—into a steam roller. Both crew members miraculously escaped.[1] Three aircraft failed to return from sorties over Holland on 8–9 September. In one, which ditched at about 02.00, the injured pilot sat in his seat afloat for six hours, only to die of shock and exposure after being taken aboard a rescue launch. A Do 217 of 1/KG 77 destroyed on circuit at Beauvais fell to the squadron, the night before the famed Czech fighter pilot, Flt. Lt. Kuttelwascher, tried his skill with a Mosquito around Orleans and Châteaudun.

Wg. Cdr. Sammy Hoare, now 23's squadron commander, had an exciting time on 13 September, shortly before he was given a D.S.O. to add to two D.F.C.s and taken off operations for a rest. He spotted an enemy machine near Twente with its tail light on, and suspected it was a decoy. Current German technique was to lead the Intruders over a gun concentration which opened up with a frightening display when the decoy's lights went off. Hoare kept wide while chasing his quarry which flew off in an unexpected direction. With the weather clamping on this dark night Hoare decided to close for the kill, but the Hun flew lower and lower then struck the ground. This was not the end; as Hoare crossed the coast at Hammstede, his starboard engine was put out by flak. After he had feathered, the port engine ran roughly and lost much power. Just to have something to go on with, he unfeathered the starboard engine, which ran for a short time and then finally packed up; but

[1] Eight months later DD673 was again flying after repair by Martin Hearn. After short service at 51 O.T.U. it was completely refurbished and passed to 141 Squadron on 24.3.44. On 9 June it was written off after battle damage.

Operations: Intruders and Rangers

before it did, the port one recovered somewhat. The battery was now too flat to feather the starboard propeller again. Then the radio failed. It was 'rather unpleasant' because it was very dark, with low cloud, and Hoare was losing height. Eventually he saw some pointer searchlights, which guided him to Hunsdon where he made a bellylanding having no time to lower the undercarriage. DD670 B–Beer was repairable.[1]

October and November, 1942, were dull months, for there was little night flying at training schools in France and the Low Countries. Bad weather also interfered with patrols, only forty sorties being possible in the period, during which trains and power houses were shot up because aircraft were rare.

A main requirement now was increased range. All the Intruder Mosquitoes were fitted with long-range tanks behind the cannon giving an extra $1\frac{1}{2}$ hours flying, and making the operation radius about 600 miles. On 7 December Hereward de Havilland was told by Fighter Command that 23 Squadron were to proceed to the Middle East after modification of their aircraft. It seemed that an end to intruding over Europe had abruptly come. But no.

In view of all that followed it is interesting to record that as early as 12 July, 1942, Basil Embry, Station Commander at Wittering, where the second Mosquito fighter squadron equipped, remarked, in a conversation at a dance, that one of the finest things that could be done to discourage the enemy would be to form an independent tactical force of Mosquito long-range intruders concentrating on such objectives as the German flying-training organization. He argued this officially and requested authority to convert eight of his night-fighters into Intruders.

On 10 December Air Marshal Leigh Mallory, C.-in-C. Fighter Command, visited Hatfield to report that it had been decided that Mosquitoes in his Command were to adopt more offensive roles, and that the company would shortly be asked to equip some night-fighters with long-range tanks enabling them to make daylight cloud-cover patrols. In December, and January, 1943, this work was done on six aircraft in each of five squadrons, Nos. 25, 85, 151, 157 and 264. Radar was removed and *Gee* navigation equipment fitted. Twenty-mm. ammunition capacity was increased from 175 to 255 rounds per gun. Thus, as 23 Squadron left, the Intruder campaign was actually to increase.

Intruding on enemy airfields and communications required special skill. To September, 1942, crews were provided by night-fighter Operational Training Units, but in October a squadron of 51 O.T.U. Cranfield, under Sammy Hoare, was transferred to Twinwoods to begin specialist training courses for seven crews a month. Following Fighter Command's decision to expand intruding 60 O.T.U. was reformed at High Ercall, from the nucleus of the Twinwoods unit in May, 1943.

[1] After a spell with Hunsdon's Station Flight DD670 served with 51 and 60 O.T.Us. It was struck off charge (obsolete) 31.1.46.

Operations: Intruders and Rangers

Gradually it expanded until it acquired fifty Mosquitoes, and responsibility for all Intruder training.

Another chapter in Mosquito operations began 15 December, 1942. Coastal Command aircraft were encountering stiff opposition from fighters over the Bay of Biscay, through which U-boats passed to French west coast ports. Beaufighters began patrols searching for the enemy in Autumn, 1942, but the Mosquito's extra speed and range made it more suitable for *Insteps*, as these patrols were called. The 59th and subsequent *Insteps* were flown by Mosquito Intruders, initially by 264 Squadron. To ease operations a detachment moved to Trebalzue, Cornwall, on 28 December, whence patrolling began 6 January. Apart from *Insteps* 264 maintained night defence patrols, and also had three crews at Bradwell Bay, Essex, from where on 18–19 January, they reopened the Intruder offensive.

Another form of Mosquito operation also was planned at this time. Instead of giving crews specific targets, a roving commission to a specified area was given, these flights being known as *Rangers*.[1] 264 Squadron were again first to operate along the new lines, when Wg. Cdr. Maxwell shot up Lorient Harbour on 4 February on a *Day Ranger* over NW. France. 25 and 151 Squadrons on 16–17 February were first to fly *Night Ranger* sorties. Two nights later a 264 Squadron Mosquito shot up a 2,000-ton ship, further extending *Ranger* possibilities of intrusion.

Insteps were now a daily feature of 264 Squadron life. Fifty-three had been flown by 31 March, yet only on one did action occur, when Flt. Lt. W. F. Gibb, Flg. Off. R. M. Muir and W. Off McKenzie found two Ju 88s about three miles away. They gave chase at 295 I.A.S., catching them in about five minutes. Gibb and Muir closed on to one 88, experiencing return fire which damaged a Mosquito's tail. Quickly the Ju 88 was despatched to a watery end. During this time McKenzie attacked the other Junkers, which had started to climb, his third burst sending it crashing into the sea.

March was a busy month for other squadrons, as *Night Rangers*

---

[1] The decision to undertake deep 'free-lance' penetrations known as *Rangers* was made in February, 1943. *Rangers* differed from *Intruders* in that only the latter were directed against definite enemy activity, known or anticipated. *Intruders* were held at 'readiness', whereas *Ranger* flights left at pre-determined times. *Intruders* were always in communication with home stations, whereas *Rangers* were not. The aim of *Ranger* operations was to force the enemy to maintain and disperse a fighter force in the West at high 'readiness' state, destroy his aircraft, disorganize training schedules and attack transport. Six Mosquitoes (initially 3) without A.I. were held by ten squadrons, (initially seven, Nos. 25, 151, 157, 264, 307, 410, 456) and Beaufighter squadrons were permitted each to have three aircraft for *Night Rangers* only. Only Mosquitoes made *Day Ranger* flights, and only when there was good cloud cover from well out to sea to the target area. *Night Rangers* could be flown in any weather, but against ground targets only on moonlight nights. *Rangers* started from forward bases, with 10 Group concentrating on West and Southern France, 11 Group the Eastern part of France, Belgium and Southern Germany and 12 Group Holland and Northern Germany.

Operations: Intruders and Rangers

commenced by Nos. 25, 157 and 456. In its closing days 151 Squadron began *Day Rangers*, and 410 Squadron on their first *Day Ranger* sent two Mosquitoes over NW. Germany. Trains and communications were the main targets, apart from airfields.

Intending rail passengers at Ijlist station south-west of Sneek, Holland, witnessed the end of a Mosquito. As Flg. Off. J. E. Leach and Flg. Off. R. M. Bull of 410 Squadron brought DZ743 to attack a steam locomotive they saw the station crowded, and each time they flew across very low the crowd waved to them, and no attack was possible. Once more they turned to make a pass, but they came too low, for both were killed as the Mosquito hit a tree.

From the Intruder angle, however, the great event of the month was the full-scale start of Mosquito operations by 605 Squadron, now committed to intruding. On their second sortie 605 Squadron destroyed a Do 217 over Holland. Next night another was claimed. 605 Squadron had long awaited Mosquitoes, and an enterprising member opened a book in November, 1942. The odds? 5:1 against delivery in February, 1943. A Mosquito first reached the squadron on 3 February.

Excitement was great also on 418 (City of Edmonton) R.C.A.F. Squadron, to whom a dual-control Mosquito was delivered on 18 February. Training and conversion was delayed by bad weather and it was 7 May before Plt. Off. Tony Croft and Earl Morton left on the first of 418's 3,000 Mosquito sorties, a patrol of the Melun–Bretigny area. All was quiet, until, near Nantes, they came by chance upon a Ju 88. A twelve-minute chase, two bursts from 200 yards, and the score board opened on 418's amazing record. Signs for their Mosquito operations were as good as those for 605, as both squadrons began months of friendly rivalry and outstanding successes.

Rangers and *Night Intruders* further increased in April, forty-eight sorties being made by 25, 151, 157 and 456 Squadrons. A typical combat report was filed by a 25 Squadron pilot who flew a *Night Ranger* to the Essen area, attacking a medium-sized lorry, its bright headlights blazing as it travelled for Hesepe. He attacked from the starboard quarter closing to 150 yards and firing for 2–3 seconds. Strikes were seen and a tongue of flame shot forth. Immediately afterwards a goods train travelling north was sighted and given a 5-second burst, the Mosquito closing from 700 to 150 yards. Strikes on the locomotive were followed by sparks, a vivid glow and clouds of steam. The train stopped. . . .

Whilst 418 and 605 Squadrons built up their Mosquito offensive the first FB. Mk. VIs were being tested at Hatfield, and on 11 May 418 Squadron were first to receive this version able to carry four 500 pounders and retain eight guns. Provisioning for wing drop tanks increased the Mosquito's operational status, bringing much of Europe into its radius of action.

In April–May seven enemy aircraft had fallen to Mosquitoes over enemy territory, and Mosquitoes had limped home on one engine from

Operations: Intruders and Rangers

Norway, Bordeaux and Essen. 456 Squadron flew their first *Day Ranger* on 6 May, 25 Squadron on 9 May. Next day 264 Squadron sent DD636 over France, where it inconclusively encountered two Fw 190s. The first successful battle with the vaunted German fighter came on 14 May when Flt. Lt. H. E. Tappin of 157 Squadron destroyed one on a *Night Ranger* over Evreux. Pilots flying Mk. IIs were now certain that Fw 190s without external tanks could catch them up at sea level without a dive, if the Mosquitoes had saxophone exhausts and +14 lb. boost.

The Polish 307 Squadron began *Night Rangers* 16 May, when DZ739 shot up a factory at Karlsruhe, then suffered an engine failure and flew home on one. Two nights later the squadron shot up the German flying-boat base at Concarneau on the Brest Peninsula, leaving a flying-boat burning. Another *Night Ranger* went to Bremen where a factory and goods yard were hit, and a train was stopped. Seven trains were attacked by another Mosquito, and four more locomotives, a signal box and three searchlights close to Bremen were shot up.

There now began a period of remarkable success by 418 and 605 Squadrons. Sqn. Ldr. C. C. Moran and Sgt. G. V. Rogers of 418 visited Avord on 27 June, destroying an He 111 and a Ju 88. At Bourges they bombed a radio mast, and shot up a train for good measure. 605 Squadron's June successes were obtained at Aalborg, Denmark, at St. Dizier in France and Venlo in Holland. Flg. Off. Smart, a clay-pigeon expert, was responsible for the Aalborg victory during a 1,000-mile flight. He recorded his success as an He 177, which crashed in flames after a three-second burst.

Over the Atlantic, too, the battle hotted up in June. *Insteps* were now flown by detachments of three aircraft each drawn from 25, 151, 157, 307, 410 and 456 Squadrons. June 11's was the most successful engagement to date. Three 25 Squadron Mosquitoes and two of 456 came across five Ju 88s. A heated engagement ensued and three Ju 88s retired damaged. Another, Ju 88 C6 360288:FB+HZ of V/KG 40, spiralled into the sea. Stung to action, the Luftwaffe despatched Fw 190s to patrol over Biscay on 13 June, and after tackling Ju 88s four Mosquitoes fought the single-engined fighters, claiming one. Only one Mosquito returned, but the loss was somewhat compensated for by 151 Squadron's destruction of an Fw 200 C4 0147:F8+CR of KG 40, the first to fall to Mosquitoes. Flg. Off. Boyle and Sgt. Freisner, and Plt. Off. Humphreys and Plt. Off. Lamb of 151 Squadron were responsible, about 250 miles west of Bordeaux. There was no return fire from the huge four-engined machine, the tail of which was shot off.

Four of 410's Mosquitoes destroyed a Bv 138 three-motor flying-boat on 17 June, and two days later 151 and 456 Squadrons came across a trawler being escorted by eight Ju 88s, one of which they destroyed and another they damaged. 307 and 410 Squadrons eliminated another Bv 138 flying-boat belonging to 1 (F) 129, 1075:KL+MA, on 19 June west of Bordeaux. Next day 264 Squadron damaged three Ju 88s, and

Operations: Intruders and Rangers

destroyed an 88 C6 360113:F8+AX of V/KG 40, south of Ushant.

This proved to be the height of Summer activity although the most spectacular operation came on the night of 20–21 June. Led by its C.O., Wg. Cdr. Allington, 264 sent four Mosquitoes to Lake Biscarosse, base of the enemy flying-boats recently encountered. After being out of sight of land for two hours, and in poor weather, the Mosquitoes found two circling Bv 138s, both of which they shot down. They then attacked machines floating on the lake, setting two on fire and damaging others. In the poor light they were uncertain of the true extent of their success, but a report obtained from a P.O.W. later confirmed destruction of two Bv 138s and two huge six-motor Bv 222 flying-boats belonging to 1(F) GR 129.[1] From an obscure position in the West Country 264 Squadron was now achieving as much as any Mosquito squadron. A unique honour conferred upon 264 on 25 June was their selection to give escort to H.M. King George VI. He was travelling in a York from the Middle East, but faster than planned for the Mosquitoes missed it by 30 minutes.

Three squadrons began using Mk. VIs for intruding in July, 1943, 418, 456 and 605. Next month three more, partly equipped, joined them. They flew *Rangers* and bomber support duties, which continued spasmodically and with various marks of Mosquito until May, 1945.

418 Squadron, during Summer, 1943, punctured twenty engines and damaged or fired, fourteen trains. Railway targets, marshalling yards and Seine barges—all were attacked. Sqn. Ldr. Moran was one who specialized in attacking trains. He strafed them then bombed them. But usually it was airfields that were raided. On 22 September Moran with Flt. Sgt. Rogers flew to Hopsten airfield, where lights were on and aircraft flying. All was rapidly stilled. Moran flew on to Achmer, where contact was soon made with an enemy aircraft. After a two-second burst it exploded, the crew of the Mosquito feeling a terrific jolt which hurled the shuddering aircraft about 1,000 feet upwards. At first Moran thought the Mosquito was on fire and ordered Rogers to bale out. In the confusion he did not understand. Control of the aircraft was meanwhile regained, and although it kept yawing and was very difficult to control they decided to try and make England. Off Holland Rogers obtained fixes and they headed for Manston. Sparks began to fly from the port engine, which soon after caught fire, and this could not be extinguished. When the wing caught fire Rogers baled out, while Moran radioed the position. As he tried to leave his harness caught in the seat as the machine fell out of control. He then climbed back into the seat to try and steady it to allow himself to bale out. Keeping his right hand on the stick as long as possible he slid out on his stomach and parachuted into the sea. He was picked up by a launch, but of Rogers, sadly, there was no trace.

418 extended their sphere of operations and began attacking airfields in Germany on 2–3 August. By the end of September nine enemy aircraft

[1] The Bv 138s were 310160 and 311019:NA+PS, the Bv 222s Nos. 0439 and 0005.

Operations: Intruders and Rangers

had been destroyed by them at night. Three of their crews, proceeded on special detachment to 617 Squadron in September. They began a week's training consisting mainly of night flying with 10° flap and throttled back to 170 m.p.h., in formation with Lancasters cruising at 150 feet. As eight Lancasters bombed the Dortmund Ems Canal on 15 September the Mosquitoes strafed guns and searchlights and held off enemy fighters.

It was now apparent that Mosquito sorties were giving very effective surveillance of enemy airfields, and offering a greater radius of rapid action. This was much apparent in the closing weeks of September when C/605 Squadron destroyed two Ju 88s off Denmark, 418 claimed two Me 410s near Stuttgart, T/605 destroyed two bombers at Achmer and an Fw 190 at Evreux. 605 were the top-scoring Mosquito squadron of September when they claimed ten destroyed, two damaged. At the end of the month, Sam Hoare took over command of 605, and on his first operation shot down a Do 217, at Dedelsdorf.

Airborne from Coleby Grange on 26 September, 1943, in DZ757 of 410 Squadron, Flt. Lt. M. A. Cybulski and Flg. Off. H. H. Ladbrook crossed the Dutch coast at 15,000 feet at 21.07 and patrolled for enemy aircraft. Finding none they recrossed the coast at 22.59. Six minutes later a radar contact was obtained and followed up. As they closed, the enemy, a Do 217, began to climb as the Mosquito dived towards its starboard side. Closing to 100 feet Cybulski gave it a three-second burst. Immediately the bomber exploded with a terrific flash, falling in flames. Burning petrol and oil flew back on to the Mosquito, scorching the fuselage from nose to tail, the port wing inboard of the engine, the bottom of the starboard wing, the port tailplane and the rudder fabric, which was torn away. Pieces of the Dornier struck the port oil cooler, resulting in loss of oil, and making it necessary to shut down the port engine. The pilot, blinded by the flash, was unable to read his instruments and the Mosquito went into a steep dive. Flames covered 757, which lost about 4,000 feet before the navigator obtained control. Fire was extinguished by the dive. After five minutes the pilot regained his vision. Course was set for base, but the port motor stopped; it was some minutes before the propeller could be feathered. Control of the aircraft was extremely difficult and it was necessary to keep both feet on the starboard rudder. After an extremely difficult flight Coleby was reached at 00.59.[1]

Instep patrols remained a daily feature. Usually there was nothing to report, but on 11 September a surge of activity began. In the course of two patrols 307 Squadron shot down a Bf 110 and damaged five more, then fought five Ju 88s off the Gironde Estuary. One (750333 of V/KG 40), was destroyed, and the others damaged. It was the turn of 456 Squadron led by Wg. Cdr. G. Panitz, D.F.C., on 21 September. HJ816 and '818 shared a Ju 88 C, one of eight which the Mosquitoes fought. Four days later 307 Squadron had another affair with KG 40, two of

[1] After repair at Hatfield '757 served 51 O.T.U. 28.6.44 to 19.7.45. It was SOC at 10 M.U. 21.5.46.

Operations: Intruders and Rangers

whose aircraft (360395:F8+FZ and 750412:F8+NY) were destroyed by DD724 and HX859, for the loss of one Mosquito.

Instep flights continued until the end of April, 1944, but never again did such severe fighting occur. Four Mosquitoes of 157 Squadron left at 11.55 on 20 November and searched for enemy aircraft north-west of Esparto Point. The intention was to catch Fw 200s and Bv 222s on their way to spot for U-boats. Instead, they came across a huge Ju 290 (159: CV+CK of 1/FAG 5) which they despatched with three engines blazing after chasing a group of eight Ju 88s four of which were damaged.

Acting out of character 418 Squadron sent Flg. Off. Charlie Scherf and navigator Flg. Off. E. A. Brown on an experimental daylight hedge-hopping flight to Biscarosse on 28 November. Approaching from the south-east to further the surprise they skimmed low across the water making one pass, avoiding gunfire, and damaging a Bv 222 and two Ar 196s. Although the risk of interception was greater, *Day Rangers* usually gave a better harvest and in the months ahead were to be a frequent feature of 418's operations—and their most remunerative, for 75 per cent of 418's successes were by day.

Only the successful flights of the Intruders hit the headlines. About 25 per cent of such flights were abortive in the last quarter of 1943, made when nobody really had any business being in the air at all, and when low cloud and haze covered all likely targets. There were times when crews saw the ground only on take-off and landing. On 4 December for example, Sqn. Ldr. D. MacDonald and Plt. Off. Wilson, along with Charlie Scherf and Al Brown, flew to airfields in the Leipzig area—1,100-mile journeys—without a pinpoint to guide them. The target area could be identified only by burning Leipzig, and there was no possibility of intruding.

A ban on ground strafing of rail traffic in France and the Low Countries came into force on 1 December, 1943, and remained for six months while these targets won the attention of Bomber Command. Airfields and aircraft therefore became main targets for *Intruders* and *Rangers*. On 10–11 January, Wg. Cdr. Hoare of 605 Squadron destroyed a Ju 188 near Chièvres, this being the squadron's one hundredth victory. 307 Squadron flying an *Outstep* patrol off Norway on 19 January, 1944, destroyed a floating Bv 138, damaged two more and destroyed a Ju 34 W.

Whereas *Night Intruders* and *Rangers* were made singly, Mosquitoes usually flew in pairs on *Day Rangers* to help follow-up attacks and afford protection from fighters. Low cloud, a nuisance at night, was essential by day, and the weather over N.W. France on 27 January, 1944, was perfect for *Ranger* flights, with a cloud base of 2,500 feet and unlimited visibility. Although it was raining at Ford, 418 Squadron's Wg. Cdr. MacDonald and his navigator Stan Wilson were airborne soon after lunch with Scherf and Brown. Scherf spotted an He 177 near Bourges, which MacDonald first fired at. As the bomber curled towards the ground Scherf set fire to its port engine. Almost at once Scherf spotted an

Operations: Intruders and Rangers

Fw 200 which MacDonald destroyed with an eight-second burst. Two minutes later MacDonald was finishing off another He 177. At the debriefing at Ford, MacDonald was heard to utter that, if the other pair also on a *Day Ranger*, Flt. Lt. J. T. Caine and Flg. Off. Boal and Flt. Lt. J. S. Johnson and Gibbons had not destroyed at least four enemy aircraft he would not talk to them. He talked, for they returned with a bag of one Ju 88 each, a pair of Ju 34s and a Ju 86 damaged. In thirty-two minutes and in eight minutes of combat the four Mosquitoes had destroyed seven of the enemy. This score 418 was to equal but never to surpass, and for days it was the talk of Ford.

157 Squadron flew some outstanding *Insteps* in February, 1944. On the 8–9th Flt. Lt. H. E. Tappin fired the centre section of a Bv 222, which crashed in the water at Biscarosse; on 12th the squadron fought no fewer than five Fw 200s 300 miles from England, one of which fell into the sea and another was damaged. A low-flying Ju 290 was destroyed on 19 February.

On 13 February a 418 Squadron Mosquito was almost rammed by an He 177 near Bordeaux. Officially on leave Bob Kipp and Pete Huletsky intruded on Juvincourt on 18–19 February as two Me 410s returning from a share in the Baby Blitz, ended in flames at opposing ends of the field. This was an intense month of night operations for both 418 and 605 as they shot up Luftwaffe bombers returning from raids on Britain to Ansbach, Brussels, Chièvres, Gilze Rijen, Handorf, and many other airfields scattered over N.W. Europe.

Most spectacular of all the sorties in February was, however, a *Day Ranger* on 26th by Scherf and Flg. Off. J. Finlayson with Flt. Lt. A. D. Cleveland and Flt. Sgt. Frank Day. First they shot up three aircraft dispersed at St. Yan, but the rest of the story is best told in Scherf's words: 'We headed south for twenty miles, then set course for Dole/Tavaux, having difficulty in finding the airfield as rivers appeared swollen, probably in flood. My observer, by excellent navigation, brought us to the town of Dole, and there we encountered a Bi-Heinkel, or He 111z, with its satellites. Owing to the low speed of the targets, I found it impossible in my climb to get into position to attack. I broke off and was delighted to see Flt. Lt. Cleveland knocking off the rear glider. I manoeuvred and came up behind the remaining glider and Heinkel. When the glider filled the ring sight I gave it a short burst. Pieces flew off, and I was very concerned in dodging debris as he went down out of control. I closed behind the Heinkel, opened fire and immediately saw strikes on two engines. I broke off, Cleveland carried on the attack. . . . When he had finished I came in and gave a burst of machine-gun fire as the Heinkel slipped to earth.

'As the Bi-Heinkel consists of two separate Heinkels, it is requested that a claim of one He 111 destroyed, each, be allowed to Flt. Lt. Scherf and Flt. Lt. Cleveland.'

The clumsy Heinkel glider tug and its two Gotha 242s was an Intruder's dream, but the two Canadians were not satisfied with just the glider

Operations: Intruders and Rangers

train, for they strafed another Gotha in a field near Dole. So impressed were the higher levels by the squadron's recent *Day Rangers* that they were told they could now fly them at their own discretion without the special permission hitherto required. 418 had claimed 22½ enemy aircraft destroyed in 22 months of operations; in the 23rd month their score alone was a further 24 for the loss of only one Mosquito! Not to be outdone 605 Squadron intruded on to Gardelegen on 5 March and bagged an He 177, an Fw 190 and two Me 410s.

Periodically some Mosquito squadrons carried out special duty flights. On one Sqn. Ldr. Lisson and Flg. Off. Franklin of 418 Squadron were sent to the Toulouse area and, failing, proceeded another 400 miles to Alghero. After a second attempt they made a supply drop on 6 March when heading home. While they were busy 1st Lt. J. F. Luma (U.S.A.A.F.) and Flg. Off. C. G. Finlayson were intruding on Pau-Pont Long airfield close to the Spanish border. They caught a long-nosed Fw 190D, the fastest and most formidable German aircraft at the time. As it was destroyed pieces flew into the starboard radiator leading to rapid engine overheating. Luma flew the Mosquito for 500 miles on one engine, and when he landed had completed 6 hours 22 minutes flying. 418 Squadron had now flown twenty-four *Day Rangers* and claimed fourteen enemy aircraft. Not until 9 March was a Mosquito lost. Such a nuisance were our Intruders over Brittany that a dense batch of searchlights wove their beams over the area, lighting the sky and the ground almost entirely in an attempt to prevent Mosquito operations.

Apart from *Intruders*, *Rangers* and *Insteps*, most squadrons dabbling in these also flew *Flowers*. Two of 418's crews destroyed seven aircraft on a *Day Ranger* to Luxeuil and Hagenau on 21 March. Another twelve aircraft were damaged on the ground, and for their exploit all the crews received D.F.C.s.

His tour of duty expired, Scherf left 418 Squadron for Intruder Control, but he often elected to spend leave at Ford with 418, and was soon off on a *Day Ranger*, once with Johnny Caine. There was no cloud cover and Caine quite rightly turned back; Scherf went on, all but cutting the grass from the French meadows as he hurried along. By the River Loire, Scherf came across a Fieseler Storch flying so slowly and so low that its crew were able to land it and make for nearby trees, before Scherf made a bonfire of it! Over Lyons he shot down two machines and, as he wended his way home, finished off two He 111s at St. Yan.

In the next six weeks 418 claimed thirty enemy aircraft destroyed in the air and thirty-eight on the ground. At Metz on 12–13 April an He 111 was destroyed and another and an He 177 were classed as probables; then came another great day for the Mosquitoes and crews of 418. R. A. Kipp and Flg. Off. Huletsky with Caine and Boal took off from Coltishall and flew to Kastrup. Over the Kattegat, they found four 'Mausi' Ju 52s minesweeping, all of which they put into the water. Three more enemy aircraft were destroyed on the airfield and two damaged. A pair of

Operations: Intruders and Rangers

Fw 190s from Grove whistled in to intercept, but the Mosquitoes' speed kept them well ahead. All four crew members were decorated.

Two days later their greatest success ever came to 418, for ten enemy aircraft were destroyed. Jasper-Martin and Harper-Rees flew to Luxeuil in the late afternoon, 35 miles from the Swiss border, where they wrought havoc amongst the Caudron and Ju 34 trainers and trainee pilots. Cleveland, Day and Kerr and Clerk flew around Central France dealing with parked aircraft at Toul. Near Thionville they subtracted a Ju 87 from Luftwaffe service.

Instep flights continued but there was little sign of the enemy, although on 11 April six Mosquitoes of 151 Squadron on air-sea rescue operations over the Channel sighted about fourteen Ju 88s at wave-top height. At once the 88s turned for home but 151 Squadron sped into the attack. Two fell to Playford in MM438, one to Harrison, another to Turner and to Wg. Cdr. Goodman's MM448, the one hundredth aircraft claimed by 151 Squadron. Other 88s withdrew damaged. Two Mosquitoes failed to return.

605 Squadron, whilst not equalling the scores of 418, because they flew only *Night Intruders*, accredited themselves remarkably well during the early months of 1944. In March they claimed seventeen aircraft for the loss of one, and on 15–16 March flew no fewer than twenty-two sorties. When Wg. Cdr. Hoare left on 17 April his score was thirty-three aircraft destroyed, two probables and thirty-five damaged on the ground. By D-Day another eleven were credited to 605 Squadron.

418's records seemed to be made only to be broken, for on 2–3 May Scherf was back again, and with Caine and their navigators hurried off to a stretch of water between Ribnitz and Putlitz, and attacked a group of He 115s and Do 18s. Next they strafed Barth airfield, destroying six bombers. Caine's port engine became infected with pieces of a Junkers so he turned for home and a 500-mile flight on one engine. Scherf continued eastwards destroying a Ju 86 and an He 111 at Griefswald. Passing Bad Doberan on the way home he added another Heinkel to the score. The listing at this time was ninety-six aircraft to 418 Squadron ... there was eagerness to catch the one hundredth. No. 97 was a Heinkel 111z glider tug at St. Yan. After destroying it MacFadyen was forced home by ground fire. Bob Kipp scored the next three successes the same night, all of them Fw 190s—to become top-scoring night-fighter pilot on a single night sortie without the aid of A.I. radar. He very nearly encountered disaster, however, because he flew into the wreckage of the third Fw 190 and only regained control a hundred feet up.

With the intensity of operations some Mosquito casualties were inevitable. In just over a week four of 418's were missing. On 16 May Cleveland and Frank Day accompanied by Scherf and Finlayson flying 'on just one last trip'. went to the Baltic. While strafing seaplanes Cleveland's aircraft was hit by flak. Instruments destroyed, starboard engine out of use, he turned MM421 for Sweden, but failed to make it by 3

Operations: Intruders and Rangers

miles. Rescue came in three hours, too late to save Frank Day. Cleveland eventually returned to 418, as their last wartime commander.

Scherf and Finlayson were meanwhile wrestling with their damaged Mosquito, which they brought home minus a port elevator. Over the German coast they frightened a flock of birds that the wily Hun had talked into collaboration. These rushed at the Mosquito, severely damaging its wings and fuselage. Nevertheless NS930–TH:U brought them safely home, with a bag of five air-to-airs, a score never exceeded on any trip by a Mosquito. Scherf's score now stood at $23\frac{1}{2}$, all but six obtained when visiting his old squadron during leave periods. The trip to the Baltic was, in fact, Scherf's last with 418. Tragedy, which must have so often been close to his great enterprise, was not to strike until July, 1949, when he died in a motor accident in Australia.

The outstanding period of *Day Rangers* ended 21 May, for after this date 418 flew mainly *Flowers*. The final sortie of May perhaps gave the accolade to the Mosquito, when Flt. Lt. C. J. Evans and Flg. Off. S. Humblestone flew to Pilsen airfield in Czechoslovakia, over 1,200 miles' journey, landing back with a mere fifteen gallons of fuel. A Ju 290 near Wurzburg in Bavaria, more flying-boats at Putlitz, an He 177 at Mont-de-Marsan, *Day Rangers* with Mustangs over Denmark, all were features of the final days of the most exciting period in the Mosquito Intruders' careers.

As Allied forces landed in Normandy on 5–6 June 605 Squadron had eighteen Mosquitoes airborne, shooting up troop positions around Caen. To 605 fell the first enemy aircraft shot down during the landings, an Me 410 caught by Flg. Off. Le Long and Sgt. McIlaren near Evreux. 418 Squadron harassed activity at twenty French airfields, assisting Bomber Command, and shot up trains, guns and searchlights in the Cherbourg Peninsula. Maximum effort was again called for the next night when 418 crippled five enemy aircraft, bridges on the Seine and Loire, and a train.

A skilled night-fighter with much experience on Beaufighters, Wg. Cdr. J. R. D. Braham, D.S.O., D.F.C., made his first *Day Intruder* flight in a Mosquito from Lasham on 28 February, 1944. It was not a success, but on 5 March he shot down an He 177 at Châteaudun. On 12 March near Beyeux his Mosquito was hit by flak. Usually he flew with 305 Squadron and whenever his duties at 2 Group permitted he was away to Lasham to fly with the squadrons there. Top-scoring night-fighter pilot, he managed an unofficial sortie to the Skagerrak on 12 May in a 305 Squadron aircraft and tangled with an Fw 190, which he eventually destroyed. During the engagement his fuel tanks were holed and while chasing the 190 his port propeller actually touched the ground. Over the North Sea he ran out of fuel, ditched near a convoy and was picked up. On 25 June Braham took LR373 of 21 Squadron, and, after refuelling at West Raynham, took off for Denmark. He was shot down near Ringkobing, coming down on a sandstrip and being taken prisoner.

Operations: Intruders and Rangers

When the V-weapons campaign opened it was not long before 418 and 605 Squadrons became fully involved, and the first V-1 shot down fell to 605 Squadron. Their places in the Intruder sphere were, to an extent, taken by 515 Squadron of 100 Group. Work commenced on 16 June when two aircraft shot up Leeuwarden and shot down two defending balloons. On 21 June Sqn. Ldr. P. W. Rabone, D.F.C., destroyed a Bf 110 as it took off from Eedde. PZ188 and PZ203 flew to the Jagel-Schleswig area on 30 June each shooting down an He 111. Additionally they returned home with a Ju 34 in the bag and an He 111 which was left burning. PZ188 and PZ184 came across a 3–4,000-ton ship off north-west Germany and shot it up on 4 July. Throughout July and August the squadron continued *Rangers* whenever they could spare the aircraft.

Permission to use A.I. VIII-equipped night-fighters over enemy territory was first given on 14 May to 29 Squadron, but only a few *Intruder* flights over France followed. On 4 July 406 Squadron first took two Mk. XIIs on a *Day Ranger* to south-west France, and subsequently there were many *Rangers* by this and other squadrons equipped with centimetric-radar aircraft.

418 and 605 Squadrons still appeared over many parts of Europe. On 28–29 August two of 605 Squadron flew a long *Ranger* to the Skagerrak, making four attacks on a 900-ton freighter off Norway. A new phase in operations was permitted by flying from Continental bases. Mosquitoes could now reach East Prussia, and Czechoslovakia more easily. Usually they continued operating in pairs, and for Czechoslovakian flights refuelled at St. Dizier and sometimes returned via Italy. Most sorties were at tree-top height, and a typical itinerary for a long *Day Ranger* would be: Leave Hunsdon 15.00, arrive St. Dizier 16.30. Leave next day at 05.35 arrive target area 06.40 and leave at 07.25. Arrive Italy 09.35, leave 15.00. Arrive Hunsdon 19.25. Two Mosquitoes operating from the Continent attacked an airfield near Konigsburg while a parade was in progress, destroying five aircraft and killing a number of troops.

A new interruption to 418 and 605's *Intruder* work came in September, 1944, for, with the firing of the first V-2s towards Britain, they were given the task of locating the release points. Even so, 418 Squadron managed no fewer than one hundred *Intruder* flights in September, and destroyed their one hundredth enemy aircraft in flight, a Ju 88 over Hailfingen. 418 Squadron began operating from St. Dizier on 30 September, at last bringing Austria into their sphere of operations, destroying three and damaging five aircraft at Erding and Eferding.

Flights to Austria, Czechoslovakia and East Prussia entailed tremendous mileages and were a great achievement. Use of an advanced base meant that all Germany and Occupied Europe was within range of Mosquitoes, from France, Britain or Italy. On 12 October, alone, two Mosquitoes of 418 Squadron destroyed nine other aircraft at Ceske Budojovice and Nemecky Brod, in Czechoslovakia, during a flight from Jesi, Italy. Three days later the most outstanding *Day Ranger* ever flown

Operations: Intruders and Rangers

came as a climax to 418's operations. Two Mosquitoes set off from Le Culot, Belgium, to north-west Germany. When they returned they had between them destroyed seven Ju 88s, a Bf 110, a Ju 87, a Bf 109, and damaged many other aircraft. For the last eleven months the *Rangers* and *Intruders* had worked at a tremendous tempo. Even so the loss rate was only 2·8 per cent, usually due to flak and perhaps flying accidents.

Outstanding operations were still being flown by 605 Squadron too. Flg. Off. Le Long destroyed five Do 24s in Jasmunder Bay and damaged a Bv 138. Soon after, he added another six aircraft to his score, a probable and five damaged. As he returned his Mosquito was damaged by flak over Schleswigland/Jagel, a fair price to pay. On 7 October two of 605 Squadron reached Vienna and, beyond its famous woods, left ten enemy aircraft destroyed and six damaged. In a week 605's score was sixteen destroyed, one probable, and eleven damaged.

With Germany and its stolen territory shrinking fast it was obvious that for 418 and 605 Squadrons the targets would soon decrease in number and area, and the decision was made that both could equally contribute to the war effort by joining the fighter-bomber squadrons of 2nd T.A.F., to which they were transferred in mid-November, and with whom they operated to the end of the war, playing, as ever, an outstanding part. These squadrons had flown the Mosquito across Europe at little cost, with outstanding courage, and in the best traditions of the Royal Air Force and the Royal Canadian Air Force. Between them they were responsible for, perhaps, the destruction of two hundred enemy aircraft, and they never used radar at any time.[1]

Intruders of 515 Squadron had continued with varying success by day and night. On 30 September, for example, PZ440 and NS993 left at midday for Holzkirchen airfield 20 miles south-east of Munich, where they were responsible for two wrecked Siebel 204s. Going home they attended to a group of Ju 86s at Neubiberg. Here NS993:S was hit by 20 mm. fire from the ground and, realizing that the damage was serious, the pilot headed for the Swiss border. Near Zurich/Konstanz four Swiss Air Force Moranes—scrambled to intercept—closed in on the crippled Mosquito. Passing Dubendorf airfield its starboard engine stopped, and since the port engine had already been closed there was little choice but to come down. Sqn. Ldr. Morley in the other aircraft headed home crashing *en route*.

Day Rangers were begun by 169 Squadron on 17 October. This and

---

[1] Successes / claims by 418 and 605 Squadrons (unadjusted by post-war research) as recorded in the squadron records almost entirely for their Mosquito operations are:—**418 Sqn.** flew about 3,000 effective sorties; destroyed 172½ enemy aircraft, had 9 probables and damaged 97; destroyed 83 V-1s, about 200 motor vehicles, 16 locos (23 probables and 36 damaged), sank 3 barges and several trawlers, lost 59 aircraft (92 crew members killed, 14 captured, 8 escaped and 27 missing), won 3 D.S.O.S, 9 bars to D.F.C.S, 43 D.F.C.S and 5 D.F.M.S, almost all when using Mosquitoes. **605 Sqn.** destroyed 96 enemy aircraft, claimed 7 probables and 78 damaged; destroyed 19 locos and damaged 306, and damaged 184 barges; 32 Mosquitoes failed to return.

Operations: Intruders and Rangers

others following, were to the Baltic coast. Soon after they largely ceased, because 100 Group's squadrons were equipping with new radar or new versions of the Mosquito, essentially night-fighters. There was resumption of free-lance ranging and intruding in the 418 and 605 tradition, but not until well into 1945. 169 Squadron introduced the NF. XXX to *day ranging* on 2 February, and on 12–13 April 406 Squadron despatched free-lance patrols to Prague. 'Come in and land' was the invitation flashed by a German airfield near Prague to an intruding Mosquito, when a controller mistook it for one of his own aircraft. The pilot replied by making three low level strafes across the airfield, scoring hits on dispersed He 177s. The Canadian pilot reckoned that, as he had circled, the Germans had mistaken his requirements. Ten days later MM727 of 406 Squadron shot up Rechlin, then dealt with a Ju 88 at Wittstock.

Only as the last few days of the war in Europe passed did the Intruder squadrons sit back, perhaps to admire the sheer beauty of the Mosquito's lines, to marvel that this little wooden aeroplane took them so far and in such safety, and perhaps to wonder to what peaceful purposes the skill of which it was born would now be applied.

Chapter 23

Operations: Pathfinders and Target Markers

A most important service performed by Mosquitoes was target marking for many important Bomber Command raids. The Mosquito suited requirements to a unique degree for, although it could accommodate special apparatus and just sufficient crew to work it, and sufficient fuel for the required range, it was scarcely slower, scarcely less manoeuvrable and scarcely larger than the best fighters. It could be flown fast through hotly defended areas, and very steadily on the dead-accurate track line the pathfinders had to make. Furthermore, its speed and high-altitude performance made it the only aircraft really suitable to carry *Oboe* (A.R.5513), an extremely accurate radar bombing aid. *Oboe*'s range was limited only by the altitude at which the aircraft carrying it flew, so the high-flying Mosquitoes were ideal machines to use it.

H2S radar was also fitted to Mosquitoes for pathfinder operations, and 627 Squadron developed special visual marking techniques. Between them the marker squadrons played a major part in Bomber Command's operations; indeed, they formed its spearhead.

(i) The Oboe Squadrons

On 21 July, 1942, DK300 reached 109 Squadron at Stradishall, there to have special radar installed. After moving to Wyton in August the squadron received eight more Mk. IVs which were used for *Oboe* trials. Grp. Capt. D. C. T. Bennett, C.B., C.B.E, D.S.O., had overall command to develop a technique which, it was hoped, would make possible accurate night bombing and attacks through cloud cover. There was great enthusiasm for the aircraft filled with hush-hush gadgetry, the installation of which was largely worked out by Wg. Cdr. MacMullin and Sqn. Ldr. H. E. Bufton.

Oboe was devised in 1941. In the system two ground stations about 100 miles apart transmitted pulses at different rates to the Mosquito. The 'cat' station measured its range accurately by radar and directed its track, whilst the 'mouse' was responsible for calculating and signalling the moment of weapon release. Pulses received by equipment in the aircraft extracted, then presented, relevant data to pilot and navigator. These signals travelled at a tangent to the earth, thus limiting the range

Operations: Pathfinders and Target Markers

of *Oboe* operations to the height at which the receiving aircraft could be flown.

Range increasing repeater equipment was developed by Flg. Off. Steed at Marham for carrying in a Mosquito, but this was not widely used until late 1944. Only one Mosquito could initially be directed at a time, about six per hour, four minutes being allowed for each target run up. Target markers burnt for about six minutes. Mk. III *Oboe*, allowing operation of four aircraft on one wavelength and using varying pulse recurrence, came into use in April, 1944. Early *Oboe* gave good results to a range of about 275 miles at 28,000 feet, but its success depended largely upon manoeuvrability, range and stability. Since only the Mosquito could fulfil the requirements it shouldered great responsibility in bomber operations. This it did in almost complete immunity from enemy interference, even in the hazardous run-up to the target.

Oboe was first operationally used on 20 December, 1942. Six Mosquitoes left Wyton to bomb the coking plant at Lutterade, Holland. Each, *Oboe* controlled, carried three 500 M.C. bombs and was briefed to attack from 26,000 feet. Away first, at 17.55, was DK331:D which attacked at 19.27. Her bombs fell 200 yards from the aiming point. DK332 attacked at 20.57 and her bombs landed 500 yards from the target. DZ317 bombed wide, following a bad run-in. The sixth aircraft DZ319:H went to Duisburg and, like DK321, aborted and joined the Main Force attack.

By 30 December 109 Squadron had despatched fifteen more sorties, attacking from 27–28,000 feet. Photographs taken by Spitfires revealed craters around the targets, but none could be attributed to *Oboe* attacks. After an early operation the enemy obligingly reported results providing a useful check on accuracy. A crew admitted a $1\frac{1}{2}$ second time lag, releasing 3×500 lb. bombs through dense unbroken cloud. The German report stated that bombs had fallen in a cemetery—and maps indicated one 700 yards beyond the target.

A Mosquito introduced a further item to improve the efficiency of night bomber operations on 31 December. Sqn. Ldr. Bufton took DK331 to Dusseldorf ahead of eight 83 Squadron Lancasters and, guided by *Oboe*, dropped sky markers over the aiming point on to which the heavies bombed.

109 Squadron operated on fifteen nights in January, 1943. Evidence pointed to an effective attack on Duisburg on the 8th, led by two Mosquitoes. Aachen was four times raided, Essen was bombed and Düsseldorf was the target for five sorties. It was here that, on 27 January, DK333:F, DZ356:J and DK318:B were the first Mosquitoes to drop ground target indicators, 250 lb. marker bombs which scattered clusters of coloured flares that burnt on or near the target. Four Lancasters of 83 Squadron then dropped 40×3-minute markers, the first one minute after the Mosquito flares ignited. Main Force then delivered its attack.

On 2 February DZ319 and DZ356 went to Cologne, dropped red

Operations: Pathfinders and Target Markers

T.I.s, and were followed by eleven Lancasters as *H2S* equipped backers-up. But still the results were inconclusive. Another fifteen sorties were flown by 109 Squadron that month, including DK333's as leader of a St. Nazaire raid.

Sensing that the recent attacks were accurate, Bomber Command decided to launch a large-scale raid led by the *Oboe* force. Four hundred and forty-two heavies were despatched to the Krupp works in Essen on 5 March, to a target that had many times been listed but rarely identified and hit. Eight *Oboe* Mosquitoes set out, to be backed up by eight of 83's Lancasters. In three Mosquitoes *Oboe* failed over Germany, but the remainder DZ436, '425, '433, '435 and '430—placed red ground T.I.s on the huge factory over a period of 39 minutes, through thick ground haze. Subsequent intelligence proved this to have been the most effective raid Bomber Command had yet delivered. *Oboe*, and the fast high-flying Mosquito, were necessary items with which to attack the Ruhr.

From 11 March *Oboe* Mosquitoes were only to be used for large-scale attacks, lest any of these valuable items were lost. By 20 April, 109 Squadron had made only six more raids. 8/10[1] attacked Essen on 12 March, again achieving excellent results, and accurately placing ground markers. Duisburg was the target on 26th, when DK318 crashed in the North Sea, the first *Oboe* Mosquito missing. Lorient was bombed on 2 April and next night 7/10 Mosquitoes again marked Essen.

Essen on 1 and 25 May, Dortmund on 4 May, Duisburg on 13th, and Bochum; all featured in the target lists for 109. Essen was sky marked on 27th, 11/12 Mosquitoes taking part. During this operation DZ432 flown by Flt. Sgt. C. K. Chrysler was shot down in flames. To Hereward de Havilland Sir Arthur Harris, C-in-C Bomber Command, now confided that he considered the Mosquito an 'indispensable' pathfinder.

The most noteworthy June sorties took place on 11th. 11/12 of 109 Squadron included, for the first time, B. Mk. IX LR497:Z flown by Sqn. Ldr. F. A. Green, on the Düsseldorf raid. Eleven bombed the city, the unlucky twelfth being '497 whose *Oboe* failed to function. Improved performance allowed the Mk. IX to attack from 30,000 feet at 320 m.p.h. increasing its safety and simultaneously extending *Oboe's* range. Eight B. IXs were already on 109's strength, the first having arrived on 21 April. A further three briefly served before proceeding to 1409 Flight, with whom the Mk. IX first flew operationally. Against Krefeld on 21 June B. IXs first delivered a successful raid, when LR496, '499 ,'500 and '503 attacked the city. From the end of the month mixed forces of IVs and IXs—flying at different levels—were usual items.

A major policy change came into force on 1 June, 2 Group's squadrons leaving Bomber Command except for its two Mosquito Squadrons,

---

[1] Eight attacking out of ten despatched were DZ319H, 425K, 427, 429L, 430, 433O, 435, 436T. When *Oboe* failed Mosquitoes found and attacked/bombed their target by *Gee*, or dead reckoning, or brought target markers home. 'Copes' were successful *Oboe* sorties.

Operations: Pathfinders and Target Markers

105 and 139. Both remained at Marham but now came into 8 Group, the pathfinder group which controlled the activities of 109 Squadron. 105 was now trained to become a second *Oboe* squadron and 139 adopted a nuisance bombing role as related in Chapter 20. On 5 July 105 Squadron was assigned five *Oboe* Mk. IXs, transferred from 109 Squadron.

Four major operations were flown by 109 Squadron in July, before the offensive against Hamburg which lay beyond *Oboe* range. During July a third *Oboe* station opened, and eighteen Mosquitoes could be controlled for one attack. 105 took its place alongside 109 Squadron on 9 July when Sqn. Ldr. Bill Blessing flew DK333 and Flg. Off. W. E. G. Humphrey DZ485 to Gelsenkirchen. Four nights later 105 began using Mk. IXs (LR506 and 507:F) when attempting to mark Cologne.

105's first major operation was the Essen raid of 25 July, when 7/12 of 109 and four of 105 Squadron marked for 599/705 heavies which again wrought frightful havoc at Krupps. On 30 July seven Mk. IXs comprised the entire marker force for the Remscheid raid, and thirty-two Mk. IX sorties had been flown by the end of the month.

Now that the heavies were attacking beyond *Oboe* range a new task was allotted to the Mosquito squadrons, destruction of specialized plants in western Germany. For this the Mosquito was ideally suited. These operations commenced in September with an attack on Emden. In October several small targets were selected for attack. They included the power stations at Knapsack and Brauweiler near Cologne, the August Thyssen foundry at Hamborn and Bochumer Verein, in addition to large factories such as the Rheinmetall-Borsig design offices and works in Dusseldorf, I.G. Farbenindustrie (Leverkusen) and the 'Fried Krupp Works', in 105's wording, in Rheinhausen.

For these raids Mosquito IXs carried 6 × 500 lb. bombs, including one on each wing rack which was soon frequently to carry extra fuel tanks. Under *Oboe* guidance 80 per cent of the load was being placed within 100 yards of the aiming point from 28–30,000 feet. During one raid lock gates at Dusseldorf received a direct hit from 29,000 feet. In daylight this would have been a creditable achievement; at night it was incredible. With their Mk. IXs the squadrons were able to fly higher than before, with no less accuracy but with less risk of flak damage to the aircraft. Searchlights were a dangerous enemy, however, for once coned a Mosquito attracted concentrated flak. So successful were these precision raids that Lord Trenchard paid a lunchtime call at Marham on 3 December, to offer his and the Air Staff's congratulations.

Hereward de Havilland, reporting on these operations, recorded that the two squadrons had flown over 750 night sorties for the loss of only three aircraft (·4 per cent) 'It will', he wrote, 'shortly be possible to attack any target within about 400 miles of the English coast with 4,000 lb. bombs by day and night, in any but the worst kind of weather, with considerable accuracy and low losses; if the repeater system is successful,

Operations: Pathfinders and Target Markers

targets 700 miles away will be within range, with a smaller bomb load. One feels, in view of the bombing accuracy now obtainable from high and low levels, the great confidence all crews have in the Mosquito, and its relative immunity from attack—that it is not being pushed anything like hard enough as an offensive weapon.'

A new demand was answered on 16 December, when 105 Squadron first marked a V-1 site near Abbeville. Many similar sorties were to follow, by day and night, until August, 1944. Great precision was called for in these sorties, since the targets were small and in the territory of allies. But for the first few weeks of 1944 night-fighter airfields in the Low Countries attracted most attention from the *Oboe* Mosquitoes along with factories in western Germany.

Squadron delivery of the pressurized Mosquito B. XVI began on 19 December, when ML928 and '930 reached 109 Squadron. 139 Squadron operated XVIs first, and 109 despatched its first Mk. XVI sortie, to Deelen, on 1–2 March. Next night 105's first Mk. XVI sortie followed, Wg. Cdr. H. J. Cundall attempted to mark Meulan-les-Mureaux in ML938:D, but his *Oboe* equipment failed at the target. 109 marked instead.

Oboe Mosquitoes commenced daylight operations on 1 March, two from 109 Squadron marking a V-1 site near Conches. Eight such operations were flown in March. Both squadrons concentrated on western Germany, airfields and rail targets being marked for the heavies. On 1 March, for example, 105 Squadron supported 600 bombers raiding Stuttgart with high altitude attacks on airfields in Holland. As Sqn. Ldr. R. J. Channer, D.F.C., was about to take-off on 5 March, *Oboe* gear in ML920 exploded, leaving Flt. Lt. Kenneth Wolstenholme to mark Duisburg in ML913 'E', whose starboard undercarriage and flap suffered flak damage. Trappes, Le Mans, La Chapelle, Laon, Amiens, Lens, Villeneuve, Acheres and other railway centres were marked by handfuls of Mosquitoes drawn from the two squadrons. Led by Mosquitoes Bomber Command neutralized the railway network of France and the Low Countries.

Raids continued during May. Gradually the emphasis switched to flying-bomb sites and the offensive against enemy oil acquired more importance. 109 Squadron's DZ319 on 8 May made its one hundredth sortie, and marked Bruneval V-1 site from 15,000 feet. Precision attacks by Mosquitoes on Gelsenkirchen and Sterkrade came in May and June.[1] Ever increasing demands now came to the Mosquitoes, particularly on the night of 5–6 June[2] when Nos. 105 and 109 each despatched twenty-five aircraft to mark ten gun emplacements on the Normandy coast for heavy bombers. Next night they marked rail targets close to Caen—at

---

[1] Mosquito raids were made on oil targets:— Gelsenkirchen 12–13, 14–15, 15–16, 17–18 June; 25–26 Homberg and again on 30 June.

[2] Targets marked by *Oboe* were gun batteries at Maisy, Mont Fleury, Moulgate, Longues, St. Pierre du Mont, Ouistreham, Crisbecq, Merville, St. Martin de Verreville and La Peronelle.

Operations: Pathfinders and Target Markers

St. Lo, Lisieux, Coutances etc. Then the target area widened, and attacks were made on railway installations deeper into France. Between 1 and 10 June, 105 Squadron flew 138 sorties and had only one early return.

The V-1 attack on Britain made it even more urgent to knock out French launching sites. More important was the destruction of supply and storage depots. Such targets called for skilled marking on day raids. Both squadrons swung into action by day, on 19 June, first against Watten, on 21st against Oisemont and St. Martin, and next day Sqn. Ldr. Bishop led 4 Group's Halifaxes to Siracourt. Each squadron was now composed of three flights, meaning that each unit could attack widely separated targets, operating in pairs or threes. As the hectic days of June passed it was obvious that a record number of sorties would be flown.

Another important task handed to 105 and 109 Squadrons was the marking for large forces of British bombers making daylight raids in support of troops in Normandy. The first such raid took place during the evening of 30 June, each squadron sending five Mosquitoes to lead 250 heavies to Villers Bocage. Six dropped their markers with great care for the Lancasters. Two Spitfires attacked ML982, Wolstenholme's Mosquito, ten miles north-east of the target, forcing him to turn away for safety.

Late evening on 7 July each *Oboe* squadron despatched ten Mosquitoes to lead 467 heavies in attacking troop concentrations by Caen. Amongst those taking part was Sqn. Ldr. Bill Blessing (R.A.A.F.), D.S.O., D.F.C. 'A' Flight Commander of 105 Squadron, veteran of many 105 Squadron raids, who was to mark a primary aiming point. Running in he was intercepted by an enemy fighter ML964:J was seriously damaged and Blessing ordered his navigator, Plt. Off. D. T. Burke, to bale out and called Biggin Hill on V.H.F. reporting the situation. Soon after the Mosquito went into a spin, then broke up. 105 Squadron lost an outstanding officer, who had joined them in October, 1942, and won a D.S.O. for his part in the Jena day raid.

Both squadrons again led a similar attack, Operation *Goodwood*, on 18 July. 15/25 Mosquitoes led waves of bombers, 1,996 in all, attacking five aiming points near Caen. Mixing fast and slow bombers needed excellent timing. Following the success of these massive daylight raids Mosquitoes were next positioned as *Oboe* leaders to smaller formations of heavy bombers. Ten formations of Halifaxes bombing a V-weapons target in the Forêt de Nieppe on 28 July were each led by two Mosquitoes, but the differences in speed made these operations tricky. Many were nevertheless flown in August.

23/28 *Oboe* Mosquitoes led a battle area raid on 30 July, and again for Operation *Totalize* on 7 August when five battle zones were bombed by 642 heavies. Great responsibility fell to the *Oboe* markers, for tactical attacks were made very close to Allied troops and absolute accuracy was accordingly demanded and achieved. Again, on 14 August, a battlefield

Operations: Pathfinders and Target Markers

was the target for 776 heavies, following twenty-two of 105 and twenty of 109 Squadrons marking for Operation *Tractable*. Marking was conducted from over 30,000 feet.

Although both *Oboe* squadrons operated in daylight many times, their raids had been on French targets. Therefore a daylight raid, leading 202 Halifaxes to Homberg, was an innovation. There was intense flak, fortunately inaccurately fired at the fourteen Mosquitoes at 31,000 feet. Such operations were the exception, however, for V-weapons sites, depots and oil targets in Germany attracted the major proportion of the effort. On 25 August a site of eight heavy guns at Brest was chosen for neutralization. For 105 Squadron it was the most successful operation so far. 12/13 attacked, supported by 109 Squadron, from 30,000 feet.

A specialized operation by 109 Squadron on 9 August called for markers to be placed from 35,000 feet on the Dortmund Ems Canal, 2/4 Mosquitoes managing accurate drops. On 15 August four of 109 raided Volkel airfield in Holland, in daylight, positioning their markers right in the centre of the runway! Airfield attacks continued during September as a form of bomber support, but the main objectives lay close at hand, German forces holding out at Le Havre, Boulogne and Calais. Upon these Bomber Command rained vast tonnages during a series of attacks lasting many hours made by wave upon wave of Lancasters and Halifaxes—invariably with Mosquitoes at the helm. On 10 September, for example, coastal defence guns at Le Havre were bombed during the morning by 50 heavies, marking for which was done by 17/23 of 105 and most of the eighteen sent by 109 Squadron, which sent four more to mark for an afternoon raid. On 17 September, for four hours, Boulogne was bombarded by 765 heavies aided by small groups of Mosquitoes leading each wave.

On the same day new targets came into 105's sights, heavy guns at Flushing and Westkapelle on Walcheren Island prohibiting entry to the Scheldt. Still the offensive against German oil and airfields continued unabated, with the Mosquitoes operating nightly. Fifty marker sorties by the two squadrons were made in two attacks on 28 September delivered against guns at Cap Gris Nez, upon which the heavies placed 3,600 tons of bombs. Next day the garrison surrendered.

Two hundred and fifty-nine Lancasters breached the Westkapelle sea wall on 3 October, led by five Mosquitoes. Flooding of the island then made it almost untenable, but the heavy gun positions were held. Another heavy raid, Mosquito led, came on 28 October, softening up the positions for the final landing. Twenty of 105 and 109 took off at 07.00 on 14 October to mark for 998/1,065 heavies making a day raid on Duisburg, followed by a night raid on the city by 1,007/1,063 on which both were marking. In each attack a greater weight of bombs was dropped than in any other raids on industrial cities. The 9,000 tons delivered was greater than the total bomb load dropped by the Luftwaffe in all its raids on Britain in September, 1940. Essen suffered an attack

Operations: Pathfinders and Target Markers

by 995/1,055 heavies on 23 October, and again *Oboe* Mosquitoes led the raid. Two days later a heavy raid on the city was Mosquito led.

On 29 November 8 Group despatched three squadrons of 'non-musical' Mosquitoes to attack an oil target at Duisburg in daylight, the first raid of a series (see page 315), led by six of 105 and 109 Squadrons. Sqn. Ldr. J. B. Burt of 109 Squadron flying ML992:U was *Oboe* Leader, upon whose markers others bombed. Next day another eight markers headed a repeat raid, and Mosquitoes of the two squadrons led similar attacks in December.

Following the Duisburg day raid of 30 November 109 Squadron again marked the city the following night. But already the winter weather was beginning to reduce the number of sorties.[1] Attacks were mainly delivered against large targets. An exception was the Urft Dam, marked three times by 105 Squadron leading small Lancaster formations. December's most memorable raid took place on 23rd, when 109 Squadron led three formations of Lancasters to bomb Cologne in daylight. Attacking from 17,000 feet the Mosquitoes were lower than usual. 109 Squadron also crewed three marker Lancasters borrowed from 35 Squadron, and 105 supplied two Mosquitoes to support two Lancasters flown by 105 Squadron crews—all *Oboe* equipped. Near the target the leading Lancasters entered heavy flak and one, captained by Sqn. Ldr. R. A. M. Palmer, was set afire, although it completed its bombing run before falling to earth. To mark his courage Palmer was awarded the V.C. Attention then switched to ML998:E flown by Flt. Lt. Carpenter, D.F.C., on to which four enemy fighters closed then shot it down in flames. Both 105's Lancasters were damaged and the resistance for once prevented successful *Oboe* attacks. It was a noteworthy operation for this was the only time when *Oboe* was really successfully intercepted, by night or day.

As the bomb line moved into Germany and the area for attack lessened, *Oboe* aircraft found themselves leading all-Mosquito night raids, or continuing the marking of oil targets. Cleve and Goch heavily bombed on 7 February by 338/550 heavies were marked by 105 Squadron Mosquitoes. So successful was it that only 1/13 needed to mark each target, supported by two of 109 at Cleve and three at Goch. Wesel also suffered by day on 17 February when 4/12 of 109 and 3/12 of 105 Squadron led the formation.

Returning from Erfurt, Sqn. Ldr. Ackroyd and Flg. Off. Casey found themselves short of fuel and baled out. Casey landed 4 miles from Brussels, and Ackroyd set himself down on a Brussels hotel roof.

There were some noteworthy operations in March. Flt. Lt. McGreal of 109 Squadron in PF446 was chased by two Fw 190s near Cologne on 2 March, but they could not close to nearer than 1,000 yards. 105 alone flew thirty-two sorties bombing Wesel for part of the time and leading

---

[1] In September 105 Sqn. flew 397 sorties, in November 302 and in December 258. Comparative figures for 109 Sqn. were 426, 306, and 270.

Operations: Pathfinders and Target Markers

'non-musical' Mosquitoes in their afternoon raid of 6th. 105 Squadron continued the attack throughout the following night assisted by twenty-five of 109 Squadron, whose crews all attacked and won a barrel of beer from the C.O. as a result. At 10.00 on 11 March crews of 105 and 109 were briefed for a massive day raid on Essen, 28/36 Mosquitoes being despatched. Next day 21/36 marked for the biggest ever raid on Germany, the bombing of Dortmund by over 1,000 aircraft. Mosquitoes laid sky and ground markers and dropped 500 lb. bombs. At Bielefeld there was a vital viaduct carrying main rail routes towards the battle front. This it was decided must be destroyed, and Lancasters of 617 Squadron one carrying a 22,000 lb. bomb, set off on 14 March to destroy it. Ahead flew RV304 and MM191 of 105 Squadron with PF408 and RV308 of 109. Only Flg. Off. G. W. Edwards in MM191 was able to mark the viaduct for 617's successful attack. Four other Mosquitoes attempting to mark the Arnsberg viaduct all had technical failures.

105 Squadron led a large Lancaster force attacking Wesel prior to the crossing of the Rhine on 23 March. Next day 109's crews were awakened at 06.30, briefed at 07.00 and despatched in threes to Gladbeck, Sterkrade, Bottrop and Dortmund. 105 was also busily engaged.

There were constant trials to improve the efficiency of *Oboe* and blind bombing. One Mosquito had installed an auto pilot which received radio signals from the ground, an experiment conducted from Farnborough. For many months experimental flights to such distant targets as Gotha, Weimar and Jena had been made by *Oboe* Mosquitoes carrying new forms of radar. They permitted strikes deeper into Germany, using transmissions from Continental stations, but not until April, 1945, were these really ready for use.

On both 105 and 109 Squadrons there was always the wish that the prime city of Germany was within their range. New radar brought this possibility and on 8 April, 105 Squadron sent LR513 and LR507 to Berlin, along with ML927, RV308 and RV318 of 109 Squadron. Next night LR513 marked for the third Mosquito raid of the evening on Berlin. LR507 and RV322 dropped markers for the first and ML936 for the third wave of seventy-seven Mosquitoes attacking on 10 April. By 21 April, when bombing of Berlin ceased, 105 Squadron had managed twenty-six copes on forty-six Berlin sorties, and 109 Squadron had despatched forty-six sorties there. It was an *Oboe* Mosquito flown by Flg. Off. A. E. Austin that was the last to bomb Berlin.

No worthwhile targets lay beyond *Oboe* range now. Rail yards at Leipzig were marked and Munich frequently visited, also Potsdam, Wismar and Oranienburg. Even though the end of the war was at last in sight there was no respite, targets around Kiel and Wilhelmshaven being heavily raided.

It is impossible to measure the contribution the *Oboe* Mosquitoes of a mere two squadrons made to the winning of the war, but certain it is that a handful of crews contributed greatly and out of proportion to

Operations: Pathfinders and Target Markers

their numbers. They made possible accurate and effective bombing and, as had been forecast, with amazingly low losses.

(ii) 139 Squadron, G–H and H2S

Oboe Mosquitoes operated with the heavies against special targets, whereas 139 Squadron's aircraft with *G-H* and, later, *H2S* installed in the noses of their fuselages, usually led the L.N.S.F. They made 'siren tours' and operated entirely by night. Thus Cologne, Hanover, Magdeburg, Hamburg, Frankfurt and especially Berlin figured largely as targets.

Apart from pioneering the use of Canadian-built Mosquito bombers, 139 Squadron introduced *G-H*, a radar device in which the navigational aid *Gee* was combined with the *H* system. The latter worked like *Oboe* in reverse, position plotting being done in the aircraft. Initially it went into Mk. IXs two of which, ML908:Y, with ML909:Z (Flg. Off. T. M. Mitchell/Flg. Off. F. E. Hay), introduced it unsuccessfully over Dortmund on 3–4 October, 1943. Next night it again failed, also on 6–7th when Hamborn was the target. Dortmund was raided by one to two Mosquitoes on 10–11th, and *G-H* was first successfully employed. Thereafter it was periodically used, but before it was fully worked out a simpler vista opened.

The ranges of *Oboe* and *G-H* were limited. Transmissions with *H2S* were made entirely from and to the aircraft, therefore it had much to commend it to Mosquito operations; although it did not have the high accuracy of other aids, it had no range limitations. Trials began with a Mosquito IV, then 139 Squadron commenced operational tests in January, 1944, with *H2S* fitted in DZ476:U. Berlin was its target on 1 February, when *H2S* was first used for marking. 139 Squadron was now operating a mixture of Mosquito IVs, IXs, XVIs[1] and XXs, and into one of the latter, KB329:L, *H2S* was also now fitted. Using *G-H*, *H2S*, the new Mk. XVIs and 4,000 pounders (first dropped by the XVIs ML941:N and '942:F on München Gladbach on 2–3 March), 139 Squadron had a wide range of equipment to use.

Next came the installation of *H2S* in the Mk. XVI, operations with which commenced in March. Markings for other Mosquitoes, spoof raids, direct support to Main Force attacks, all were 139's role. It concentrated attacks upon Germany. A typical night's duty of this period was 30–31 March, 1944, when two 'Y' aircraft (*H2S* machines) and three others attacked Nuremburg, and one flew on to Schweinfurt. Two 'Y' Mosquitoes went to Kassel supported by 4/5 others. There were now four 'Y' machines available for operations. Gradually the number increased until 139 flew mixed Mosquito versions, about half of which had *H2S*.

[1] Flt. Lt. E. A. Holdaway, D.F.C., was on 10.2.44 the first pilot to fly a Mk. XVI bomber on operations, when he took ML940:O to Berlin.

Operations: Pathfinders and Target Markers

Flying high, and relatively immune to enemy interference, unpleasant incidents over enemy territory were fewer than in many other bomber squadrons. Flg. Off. Cassells had an unnerving moment on 6–7 May, when flak hit the windscreen of KB162 and left by the side blister during a Ludwigshaven raid. KB161:H crashed on return from Mannheim on 10–11 May after a flare caught fire near base. DZ423:K flown by Flg. Off. T. Dickinson was returning from Düsseldorf on 28 May, when he was attacked by an enemy fighter 50 miles off Southwold at 22,000 feet. Two attacks were made, and the Mosquito almost half rolled to escape. It crash landed at Wittering.

Whilst the Normandy landings took place 139 Squadron had six 'Y' and six other Mosquitoes marking Osnabruck for the L.N.S.F. Later in the month it despatched Mosquitoes to join *Oboe* aircraft attacking oil refineries at Scholven-Buer.

For the rest of 1944 the mixture for 139 Squadron remained as before, increasing numbers of its Mosquitoes marking and bombing German cities, oil refineries and communications targets. In the Berlin raids, however, it primarily made its mark. Usually these followed a similar pattern, searchlights, flak and fighter activity varying according to weather conditions and operations intensity. Sometimes Mosquitoes were the only aircraft operating. On 27–28 October a two-wave attack on Berlin took place. Flying high, 139's Mosquitoes encountered cirrus at 28,000. In KB354 flew the Master Bomber crew, Wg. Cdr. Voyce and Sqn. Ldr. Gallienne. As they returned they saw a jet fighter near Neinburg, but it did not attempt to intercept.

'G-George' took part in the first official 'Siren Tour' on 1 January, when bombs fell on Cologne and Mannheim. Such operations were designed to cause the greatest general annoyance to the enemy, and often the flights were long lone penetrations of enemy air space. Being a lone operator had advantages; but there were times when, lacking the shield of *Window*, Mosquitoes were undeniably at a disadvantage. This was one—KB482 landed back with seventeen flak holes acquired over Mannheim.

162 Squadron formed at Bourn on 18 December under Wg. Cdr. J. D. Bolton, D.F.C., and, equipped with Canadian Mosquitoes, specialized in 'Siren Tours'. It began operating on 21 December by raiding Cologne and Berlin. During February it started to work up with *H2S* and commenced marker raids on Berlin on 3 March, sending KB191 and KB214. Bombing and marking of 'the Big City' occupied the squadron most nights, but it also took part in major attacks on Chemnitz, Munich and Kiel.

Almost nightly 139 Squadron bombed Berlin in 1945, and led each of thirty-three consecutive night attacks which Mosquitoes made from the start of February. During March, 159 sorties were despatched to Berlin and the thirtieth attack was celebrated in no uncertain manner. Six markers led other bombers in the first wave, three the second. There were

Operations: Pathfinders and Target Markers

fifteen raids on Berlin in April, and on all but two the squadron operated, despatching eighty-three sorties.

Although losses were few there were occasions when very heavy flak was encountered, as on 23 March when one Mosquito was coned by six searchlights and heavily engaged. Two others, KB367 and KB390, never returned that night.

It was not, of course, only Berlin to which 139 Squadron attended. Six aircraft marked Dessau on 8–9 April and Munich, a very distant target, was raided on 16–17. On 12–13 April, 4/5 Mosquitoes went to Berlin, the only ones that night. Incendiaries were carried on this operation, in 4,000 lb. bomb casings. There were only five bursts of flak, but one piece of shrapnel from the last burst to rise above the city hit PF465, then pierced Flt. Lt. Wilson's flying boot.

There were six Berlin raids on 20–21 April, the last night of its ordeal. Violent storms, huge cumulus clouds and lightning set a dramatic scene as the squadron left Upwood. Soon after, Flt. Lt. Coke-Kerr landed KB124 at Woodbridge without any instruments. Although the enemy had almost been vanquished, the weather remained a factor to consider when attacking distant targets.

The regularity and intensity of 139's operations bore testimony to the excellence of the Mosquito. Its work was closely associated with other L.N.S.F. squadrons, and its operations are also dealt with in Chapter 20.

(iii) 617 Squadron

Most famous bomber squadron, No. 617, 'the Dam Busters', found the Mosquito had special value to them. They received their first on 30 August, 1943, when the squadron moved to Coningsby. It was, however, 4 April, 1944, when Grp. Capt. G. L. Cheshire with Flg. Off. P. Kelly as navigator made first use of a Mosquito, ML976:N marking an aircraft factory at Toulouse for 617 Squadron. Attacking from between 800 and 1,000 feet he needed only two red spot fires to mark the target successfully.

He flew the same aeroplane when leading the destructive raid on the German signals depot at St. Cyr on 10 April. A dive from 5,000 to 1,000 feet was used to place spot fires on a corner of the building. That day 617 had received four Mosquitoes, two of them FB. Mk. VI judged to be most suitable for dive precision attacks. Cheshire used one first, NS993:N, raiding Juvisy marshalling yards, but his spot fires hung up. Flt. Lt. G. A. Fawke, who made many sorties with Mosquitoes then marked from 800 feet, so the other aircraft were not needed. Three out of four sent to La Chapelle on 20 April made accurate drops.

There followed, on 24 April, Grp. Capt. Leonard Cheshire's courageous leadership of the Munich raid when he dived from 12,000 to 3,000 feet in NS993. His aircraft was then coned by searchlights as he flew repeatedly over the city at no more than 700 feet. ML976:L flown by

Operations: Pathfinders and Target Markers

Sqn. Ldr. Shannon was also coned, after diving into the attack from 15,000 to 4,000 feet. The markers hung up on ML975, but four spots fell to the city centre from NS992. At one time Cheshire was blinded by searchlights and almost lost control. For 12 minutes he was under fire as he left the target area. For this attack, and showing repeated courage on operations, he was awarded the Victoria Cross.

No. 617's only operation in May was flown on the 3rd, when Mailly-le-Camp was the target. Cheshire and Shannon each made successful dive drops, and two other aircraft on loan from 627 Squadron re-marked the NW. area of the target later in the attack.

To prevent the Germans bringing supplies to France following the invasion, two Mosquitoes laid markers prior to 617's dropping of *Tallboys* on the Saumur tunnel. Cheshire placed markers in a cutting at the tunnel mouth. DZ421 then dropped flares to light the area. Le Havre was twice attacked in mid-June, NS993 carrying two 500 lb. M.C. bombs as well as two spots. They were placed with care on a flak post in the area. An important target was the weapons site at Watten, again marked by Cheshire in NS993 with two red spots on 19 June. On 24 June it was the turn of Flt. Lt. Fawke in DZ415 to point out Wizernes, after diving from 17,000 to 6,000 feet.

A further attempt to improve marking technique led to 617 Squadron acquiring a Mustang III first used by Cheshire on 25 June to drop two spot fires at Siracourt, the V-weapons depot. Shannon flew in NT205:L and released smoke markers and two red T.Is., so successfully that Fawke in the third machine was not needed. On 6 July the Mustang and Flt. Lt. Fawke in NT205 led a raid on Mimoyecques from 500 feet. NT202 and the Mustang flown by Wg. Cdr. J. B. Tait were responsible for the markings of Wizernes on 17 July.

Using the Mustang and Mk. 6 Mosquitoes it was possible to mark small targets for precision attacks by the skilled Squadron. A further use for NT202 came during August when she undertook photographic reconnaissance trips to La Pallice and Brest prior to attacks by the Squadron. NT205 spied out the land at La Pallice, Brest and Ijmuiden aided by 627 Squadron which largely took over 617's Mosquito work.

(iv) 627 Squadron

It was 617 Squadron that smashed the Mohne Dam using a larger version of *Highball*, with which the Mosquitoes of 618 Squadron might have sunk the *Tirpitz*. Eleven months after their exploit 'The Dam Busters' were joined at Woodhall Spa by 627 Squadron, which they dubbed the 'Model Aeroplane Club'. When 627 began operating from Oakington in 1943 617 Squadron was using a new technique to attack pin-point targets in France. A crew identified the target from very low-level runs, then marked it. Results achieved were excellent, but Lancasters were vulnerable. A high-level decision was made, transferring 627

Operations: Pathfinders and Target Markers

Squadron to 5 Group for training as the Group marker squadron, and to work closely with 617 Squadron.

Under Wg. Cdr. R. P. Elliott, D.S.O., D.F.C., dive marking practice began at Wainfleet, and concentration lay on placing a marker within 50 yards of any given target. At first the Mosquito IVs dived from 18,000 to 12,000 feet to release the markers, but the aircraft's speed in these dives became excessive and accuracy was insufficient. Lower and lower the flights were made until finally, accuracy was found best by diving from 3,000 to 1,000 feet.

On 20 April twelve Mosquitoes went to La Chapelle marshalling yards, Paris, with experts from 617 Squadron to mark the target with phosphorus bombs and act as backers-up. Twelve attacked Munich on 24 April. Four *early window* and five marking sorties were made on 26 April against Schweinfurt. Sqn. Ldr. Nelles dived DZ477:K from 5,000 to 400 feet to mark the target. On 28 April markers were placed by Mosquitoes operating from Lossiemouth on the roof of a target factory at Kjeller. Flg. Off. J. A. Saint Smith in DZ516:O delivered his from 100 feet. Next night the explosive works at St. Meddard en-Jalles were attacked, and by the start of May 627 Squadron were ready to mark the whole of 5 Group.

Attacks on railway targets in France were now 5 Group's prime employment. It was important that the surrounding civilian population should be as little affected as possible. Here 627's technique was of great importance. Each end of the installation under attack would be marked, and the heavies placed their loads between. An attack on the Usine Lictard, Tours, showed just how effective these operations could be. Several days before, this small factory had been the subject of a high-altitude day raid by the Americans and remained untouched. Mosquitoes marked the target with ease, the only hold-up occurring when the first markers fell *through* the glass roof of the factory and were barely visible from outside!

Two nights later ten Mosquitoes set out for the 21st Panzer Division depot at Mailly-le-Camp near Epinay. Fighters rose in strength, and the heavy bombers had to fight their way through. Others, arriving early orbited, and the losses—amongst the heavies—were high. Three Mk. IVs (DZ516:O, DZ518:F and DZ477:K) dropped markers at the three points of the L-shaped barracks, and six attacked anti-aircraft guns. The target was pulverized.

A ball-bearing factory at Annecy, the Gnôme Rhône Gennevillers works, Bourg Leopold camp—all were marked with care. On 24th the Philips works at Eindhoven was saved only by bad weather. A pilot searching for it suddenly had a searchlight shine directly in his face. So startled was he that he pulled backwards on the control column of the Mosquito, roared up into cloud and found the searchlight above his head! Commenting, later he said 'I half-rolled out . . . and came home'.

Wg. Cdr. Goodman recalling an attack on a heavy gun at St. Martin

Operations: Pathfinders and Target Markers

de Varreville on the Cherbourg Peninsula on 28 May recorded 'At first we could not identify the target in the absence of light flak, the gun itself being well camouflaged. We went in five minutes ahead of zero, but time ran out and the heavy bombers were forced to orbit the target area. Finally, one of the Mosquito pilots noticed a number of tracks leading to a large empty space in a field. He reasoned that if a number of tracks led to a blank space it couldn't be so blank. So the spot was marked and one hundred Lancasters destroyed the target.'

On the night of 5/6 June four Mosquitoes were sent to each of two gun sites, to La Pernelle where the bombing was so scattered that re-marking was required, and St. Pierre du Mont in support of the landings. A month later eight Mosquitoes marked the V-1 depot at St. Leu d'Esserent using Mk. XXs for the first time. These began to be used for daylight reconnaissance flights from 17 July. Prior to the large-scale raid, Caumont-Villers Bocage was photographed on 29th for Command. Pictures were taken from 1,000 feet, for the cloud base was at 2,000 and it was raining.

Day operations had commenced on 29 June, four Mosquitoes marking a weapons site at Beauvoir. As they flew home at low level a V-1 was launched from a ramp below them, and crashed. Its blast destroyed one Mosquito, then a higher one crashed following engine failure. The pilot of the latter was injured and the navigator escaped only to fall down a well. After an eventful evasion exercise he was captured—a few weeks later, by British troops.

August's operations included many reconnaissance and wind finding flights, gathering material and photographs to guide 617 Squadron's special raids with *Tallboy* bombs. Brest, La Pallice and Lorient figured in the sorties and KB215:H, most frequently used, filmed 617's attack on Ijmuiden with heavy bombs on 24 August.

Nine Mosquitoes, in three waves, set out for München Gladbach on 19 September. After an exemplary attack the markers turned for home. What happened aboard KB267, carrying the Master Bomber[1] will never be known, but in the town of Steenbergen, in the Dutch Province of Brabant, townsfolk heard a low flying aircraft whose engines were running rough. Amongst those who ran out was M. de Bruyn, night watchman of a nearby factory, who, on looking up, saw cockpit lights on in the aircraft. Suddenly its engines cut. There was a jet of flame and, in a slight arc ending in a vertical dive, a Mosquito crashed 100 yards from the factory.

Sometime later, near the crash, local people found documents which had come from the machine. These they hid from the Germans who had

---

[1] A Master Bomber, using VHF, controlled many large-scale attacks directing Main Force crews and backers-up supplying more markers and commenting upon the progress and requirements during the attack. KB267 reached the U.K. 30.5.44. After fitting out at 13 MU it passed to 139 Sqn. on 8.7.44 and 627 Sqn. on 18.7.44 where it became AZ-E.

Operations: Pathfinders and Target Markers

ordered the burial of Sqn. Ldr. J. B. Warwick, D.F.C., and an unknown airman. Not until after the war was a name added to the grave of the other crew member. Applied was the heading 'W/C G. P. Gibson, V.C., D.S.O., D.F.C., pilot, Royal Air Force, 19th September 1944. Age 26.'

This loss brought stunning sorrow to Bomber Command, but the offensive ploughed relentlessly on. A series of marker operations on the Dortmund–Ems Canal followed. Rail yards at Kaiserslautern were hit for the loss of KB366:O to light flak. Sqn. Ldr. R. G. Churcher and Flt. Lt. W. W. Yeadell marked the sea wall at Flushing on 7 October, whilst a raid on the Kembs Barrage near Bâle was filmed by KB215. Flt. Lt. B. D. Hanafin flying the ubiquitous DZ414 photographed a raid on the Sorpe Dam on 15 October. Two days later Westkapelle's sea wall was marked, and on 28th Flg. Off. F. A. Saunders took KB345 to the submarine pens at Bergen to discover wind speed for 617's attack. Next, the guns of Walcheren were the aiming point. In November 627 marked the Mittelland Canal, Trondheim U-boat pens and the Dortmund Ems Canal for 5 Group and the Dam Busters. Mk. XXVs came into use, Flg. Off. F. J. Nash making the first sortie in KB416.

New employment commenced on 29 November, when five Mk. IVs flew to the River Weser, only to find the weather conditions impossible for mining, even at 100 feet. 5/7 Mosquitoes flew to the Elbe on 29 December and, in excellent conditions, dropped mines in the river: two attacking from 120 feet. Bigger fish than ever used the reaches of that river had already been illuminated by 627 when, on 13 December, the Cruisers *Emden* and *Koln* were targets in Oslo Fiord.

At Woodhall Spa on 31 December there was great excitement. 627 was to have its own show and mount a roof top raid on Oslo, where eighteen months before Mosquitoes had delivered one of their most famous attacks. Two groups of six[1] Mosquitoes assembled and each, carrying 4×500 lb. or 2×1,000 lb. bombs, found the target area cloudless although visibility was spoilt by smoke haze. Again the target was the Gestapo H.Q. Leading was Wg. Cdr. G. W. Curry, who began his diving assault from 7,000 feet. The remainder followed, scoring hits on the building. By the time the second wave led by Flt. Lt. P. F. Mallender arrived there was so much smoke that their attack was impossible.

January witnessed deep penetrations of Germany, to Politz, Brux and Munich seeking oil refineries and communications centres. Dresden, industrial centre, signals centre, railway centre and troop control point, was marked for the crushing raid of 13 February. Such was Bomber Command's supremacy now that 627 was able to mark the Dortmund Ems Canal in daylight for a raid on what had been such a difficult and costly target.

Filming, wind finding, reconnaissance flights, all continued on behalf

[1] Aircraft in the first wave were P:KB416 (leading), U:DZ461, G:DZ611, N:DZ530, H:DZ606, J:KB345. Second wave comprised D:DZ633, O:DZ643, C:DZ641, X:DZ637, T:KB122 and K:KB362.

Operations: Pathfinders and Target Markers

of 5 Group and, when the Arnsberg Viaduct was bombed on 15 March, KB433 was there to film the operation. The Weser was mined on 21 March, by six Mosquitoes, while 8 Group made its biggest ever raid on Berlin. Twice more the Elbe was mined, and again the Weser; distant Leipzig and Pilsen fell under the glow of the squadron's markers too.

Unassisted by mechanical aids, 627 Squadron did for 5 Group what the *Oboe* Mosquitoes achieved for much of Bomber Command. Whereas they, using two-stage Merlin Mosquitoes, relied upon height and speed for safety, 627 utilised the Mosquito's manoeuvrability and high speed at low levels as they courageously roared over their targets at roof top height. Again, the Mosquito's versatility was to the fore.

Finale

In the briefing rooms of 1, 5 and 8 Groups, there was a buzz of excitement on 25 April. Rumour, ever ready to expand the trivial into the mighty, abounded. German propaganda exhibited a similar failing, claiming the existence of an impregnable 'National Redoubt' in the Harz Mountains, a region to which the German armies could fall back and last indefinitely. Proving the futility of this belief attacks were to be made this day upon the Eagle's Nest, Hitler's Chalet and the S.S. Barracks at Berchtesgaden where, it was believed, the Nazi leaders might be hiding.

Three hundred and fifty-nine Lancasters roared along the runways of eastern England and, in loose formations, headed south with their crews emotionally charged as never before. At Bourn and Little Staughton seventeen crews were briefed as *Oboe* leaders for the raid, but as they ran in for the attack their disappointment was bitter—*Oboe* failed in every aircraft! They could but watch the activities of their bigger brothers in immunity, as the flak reached out for the Lancasters. At the same time other Mosquitoes were over Wangerooge marking and observing an attack on heavy guns, and leaving when the dust reached 12,000 feet.

Although one more attack was to follow on 2–3 May in which Mosquitoes were the only participants, an entirely new and satisfying task had already been given to the *Oboe* squadrons when, on 24 April, they 'bombed' eight P.O.W. camps with news and medical supplies using Monroe bombs, and marked these small targets for three Lancasters also carrying medical supplies. Even as Berchtesgaden was raided, leaflets fluttered to the prison camps telling of the progress of the war, and giving instructions to the inmates.

One more series of missions needed to be flown—and quickly. Flooding, the retreat and trouble on the railways combined to make the food situation in Holland desperate. The only way in which adequate supplies could reach the Dutch was by air, and on 29 April four aiming points for Lancasters were marked each by four Mosquitoes of 105 and 109 Squadrons, on the Hague racecourse, Waalhaven, Valkenburg and

Operations: Pathfinders and Target Markers

Ypenburg. Next day the sorties were repeated, so dire was the people's plight, and again on the first five days of May *Operation Manna* continued.

As the jubilation at the end of the war rang throughout Europe on VE-Day the snarl and crackle of twin Merlins still swept across bomber airfields in eastern England. At Bourn six Mosquitoes trundled around the 'peri-track', passing many others upon whom the covers of peace had already been tied. Staughton presented a similar picture as, at midday, the little wooden aeroplanes roared away on what were to be their last operational sorties. At the Hague, Ypenburg and Rotterdam they found their aiming points without the worry of flak. The Lancasters delivered their food parcels, and the only clamour came from Dutchmen some of whom, years before, had doubtless bravely waved the Mosquitoes on their way. Markers bounced, and their brilliance shone, perhaps more brightly than ever before. The rush remained, but not the worry.

At 15.02 LR513 in the hands of Flg. Off. C. K. Ouin crossed the Bedford Road, engines crackling, to a perfect touchdown at Bourn. Eighteen minutes later Flt. Lt. D. P. Dalcom was taxiing LR508 to its dispersal point at Little Staughton. The mixture had been right, the job was done.

APPENDICES

APPENDIX 1

Full photograph captions

Between pages 32 and 33

1. The wing of the prototype, W4050, in the Salisbury Hall hangar on 16 June, 1940. An engine nacelle mock-up is in the foreground. At this time there were about thirty senior designers, aerodynamicists and stressmen engaged, and nearly a hundred people working in the prototype shop.

2. The Mosquito mock-up at the stage it had reached when the German armies were entering Paris, 16 June, 1940. The pilot's view over the engine was an important consideration, and the value of a blister side window for a view aft was being tried. With the fall of France the Mosquito project was officially stopped, but persuasion from Hatfield got it reinstated on 12 July—on a low priority.

3. The prototype assembled at Salisbury Hall, prior to dismantling and transfer to Hatfield.

4. E0234 taken out for engine runs on 19 November, 1940.

5. Painted yellow to be easily spotted by our anti-aircraft gunners, and covered in tarpaulins against the eyes of German bomber pilots snooping around Hatfield in daylight, W4050 is being prepared on 19 November, 1940, for fuel flows and engine runs. Hereward de Havilland discusses matters with the company chairman, Alan S. Butler.

6. Outside the flight shed on 19 November the chairman and the managing director of The de Havilland Aircraft Co., Ltd., Alan S. Butler and Francis T. Hearle, check progress with the prototype six days before its first flight.

7. Covered in canvas, the wing (with engines) and the fuselage of W4050 were transported on 3 November, 1940, by 'Queen Mary' tractor-trailer from Salisbury Hall to a protected flight shed at Hatfield, there to be re-assembled. The aircraft flew twenty-two days later.

8. The second landing by W4050, at Hatfield on 29 November, 1940. John Walker, engine installation designer, was the observer on the first and second flights. The scene looks unfamiliar because the factory building had been camouflaged to resemble terrace houses. Behind the Mosquito is seen the skeleton of the '94 shop', bombed out on 3 October. The power-house chimney is unavoidably conspicuous.

Full photograph captions

9. With 'Flamingo-type slots' on the inboard of the engine nacelles, aft, also beneath the stub wings, endeavours were made to rectify the slipstream without having to lengthen the nacelles. However, longer nacelles proved necessary, and involved complications with the flap gear. This picture shows wool tufts for tests on 10 January, 1941.

10. The engine nacelle in partially extended form, at the stage of development reached by 12 February, 1941. Long nacelles were standardized from the eleventh aircraft.

11. EO234 (later W4050) stands in the hangar mouth at Hatfield on 21 November, 1940.

12. In discussion just after the first flight on 25 November, 1940, Geoffrey de Havilland is with Frank Hearle, Charles Walker and Alan Butler. Behind is M. Herrod-Hempsall, aerodynamicist, also (in belted coat) C. T. Wilkins, then senior designer. To the left of Wilkins is Rex King, experimental shop foreman; further left is Fred Plumb, his superintendent, while on the extreme left is the Ministry resident technical officer, R. W. Fitch.

13. In July, 1941, W4050 flew with a mock-up turret, as depicted, for drag tests.

14. The men—or most of them—who, under the technical leadership of Captain de Havilland and C. C. Walker, were mainly responsible for the design and development of the Mosquito. From left to right: R. E. Bishop, chief designer; F. W. Plumb, experimental shop superintendent: W. A. Tamblin, senior designer; P. F. Bryan, chief draughtsman; R. M. Clarkson, assistant chief engineer, head of aerodynamics; D. King; J. K. Crowe; G. W. Drury; C. T. Wilkins, assistant chief designer; R. Hutchinson; C. F. Willis; Rex King; R. H. Harper, chief structural engineer; D. R. Newman; F. T. Watts; R. M. Hare; M. Herrod-Hempsall; F. J. Hamilton; J. P. Smith; E. H. King; R. J. Nixon; A. G. Peters; G. C. I. Gardiner. Unavailable at the time were C. G. Long, chief development and purchasing engineer, subcontracting manager; Geoffrey de Havilland, chief test pilot; J. E. Walker, engine installation designer; C. C. Jackson, electrics; A. W. Fawcett, and a few others—where does one stop?

15. H.M. King George VI, visiting the Hatfield works on Thursday, 15 August, 1940, during the Battle of Britain. Three forces of German bombers were reported north of Hatfield the following afternoon. Walking with His Majesty is A. S. Butler, behind whom Captain de Havilland can be seen, followed by Sir Arthur Tedder, later Lord Tedder. Lee Murray, head turned, is seen behind the King.

16. Hatfield workpeople had their first appreciation of the manoeuvrability of the fighter prototype, W4052, when Geoffrey de Havilland gave them a show on the evening of 5 September, 1941. He is seen talking afterwards with his father and mother. Behind is A. S. Butler, in Home

Full photograph captions

Guard uniform, with C. C. Walker. Overseas people often remark upon British ways of expressing enthusiasm.

17. The Rt. Hon. Winston Churchill, Prime Minister, after walking through the Hatfield works, 19 April, 1943, departs for London, seen off by Capt. de Havilland. He had watched demonstrations by the Gloster E28/40 (Whittle jet engine), the Mosquito and the Hawker Tempest. Behind de Havilland is seen Sir Stafford Cripps. On the left is Group Capt. David, the M.A.P. Overseer at Hatfield.

18. W4050, depicted with Merlin 61s, was flown thus between 14.7.42 and October, 1942. John de Havilland can be seen near the nose of the machine, on the left. In MS gear at 17,800 lb. all-up weight the Merlin 61s gave 4050 a top speed of 414 m.p.h. Between 8.10.42 and 1.7.43 W4050 had Merlin 77s and made twenty-three flights.

19. W4051 reached A.A.E.E. 28.6.41 and P.R.U. 13.7.41. It went to Hatfield for adjustments, returning 10.8.41. On 20.9.42 it joined 521 Squadron. Ten days later it arrived at Benson for 540 Squadron. On 31.8.43 it passed to 8 O.T.U. W4051 crashed on 19.7.44, was returned to Hatfield for repairs but SOC instead on 22.6.45.

20. W4052, the prototype fighter, photographed during A.A.E.E. trials in October, 1941. Overall finish was matt black, and A.I. Mk. IV radar aerials can be seen.

21. W4052 after being skilfully, if embarrassingly, put down at Panshanger by Geoffrey de Havilland on 19 April, 1942. The small amount of damage (resulting from a belly landing) to a Mosquito is readily evident.

22. W4065 was assigned to Duxford for operational trials on 7.11.41. On 27.2.42 it was delivered to 105 Squadron, becoming GB:N. From its twelfth operational sortie, on 19.8.42 it failed to return.

23. W4068 and two other Mosquito 1 (P.R.U./Bomber Conversion) aircraft at Hatfield. Their finish is dark green and dark earth, with duck-egg green undersurfaces. These colours were but briefly worn for, by July, 1942, they had changed to grey and green with grey undersides to tone with the sea on low-level missions.

Between pages 48 and 49

24. MP469, photographed on 16.9.42 at Hatfield, fitted with long-span wings and four machine guns after hurried conversion from pressure-cabin bomber prototype into high-altitude fighter.

25. MP469 later in the year, now fitted with nose radome and high-altitude fighter finish, overall deep sky.

26. Unlike other Mosquito fighters, the Mk. XV was fitted with a V-shaped windscreen. A production Mk. XV, DZ385, is shown here.

Full photograph captions

27. Felled ash and other lumber was brought to the High Wycombe works of Dancer and Hearne to be cut and stored for seasoning—expecting a long war. This company made wing spars and skins, fins, etc. and brought valuable experience in timber and skills in joinery to the building of the wooden aircraft.

28. The carapace fuselage shell had two thin skins of birch plywood stabilized by a thick cemented interlayer of balsa wood. This was located within spruce framework, including frames around bomb doors and hatches that were cut out afterwards. Steel bands tightened by turnbuckles held skins compressed while the cement set.

29. There were seven fuselage bulkheads, first fitted into slots in the jig, later attached through to spruce rings locally replacing the balsa interlayer. For the attachment of equipment to the fuselage plastic plugs were inserted; each had a glued plywood flange to spread the load. Fuselage halves were scarfed together with longitudinal V notches, ply inserts above and below, and an overlapping inside plywood strip.

30. The one-piece wing comprised two box spars, of laminated and plywood construction, with double plywood skin. Ribs divided the wing into tank bays. The upper skins were separated by spruce stringers. The lower skins largely comprised tank doors which, like the fuselage, were of plywood and balsa sandwich construction.

31. Wings for the Second Aircraft Group at Leavesden, showing the one-piece construction with large unobstructed bays for fuel tanks. Tank doors, of irregular quadrilateral shape and curvature, quite easily made on jigs, are seen standing vertically, waiting to become stress-bearing sections of the lower wing skin.

32. Women undertook fairly heavy work, for instance assisting the operator of a 10-ton drop hammer used for deep-drawn sheet-metal components. Women comprised about one-third of the labour force in the Mosquito main factories, Hatfield and Leavesden. Late in 1942 of the 18,546 people employed by de Havilland in all divisions (apart from 5,715 staff) in Britain, 13,088 were men and 5,458 were women. Approximately this proportion was maintained with the bigger payrolls of 1943–44.

33. Avoiding practically all of the precision machining necessary in manufacturing the usual hydraulic compression leg, the Mosquito undercarriage was made with a case of two folded halves of sheet metal containing a series of rubber blocks in compression.

34. Ernest Grinham's track system, with U-shaped floor layout, as it was on the fighter assembly line at Hatfield in August, 1943. Pits, at first wide but narrowed as the result of experience, enabled undercarriage retraction tests to be done without delaying other operations.

35. At Hatfield on 2 February, 1944, the array for flight test is headed up by a PR. Mk. XVI. Sixteen weeks before the invasion of Normandy,

Full photograph captions

Hatfield was turning out 25 Mosquitoes a week (15 FB. Mk. VI, 7 PR. Mk. XVI, 2 B. Mk. XVI and one FB. Mk. XVIII), while the Second Aircraft Group, Leavesden, was producing 15 a week. (13 NF. Mk. XIII, and 2 dual-control T. Mk. III). Standard Motor Company had delivered 18 FB. Mk. VIs in January. Percival Aircraft and Airspeed were yet to contribute. Canadian output was 21 B. Mk. XX in February, and rising fast. Australia's first delivery was on 4 March, 1944.

36. Production of Mk. IIs at Leavesden in June, 1943. All these aircraft were converted to NF. XVIIs at Cambridge. HK290 on the left destroyed two enemy aircraft and probably a third in the hands of 456 Squadron. HK286/G in the centre of the picture was a top scorer of the future for, with 456 Squadron as RX-A, she claimed eight enemy aircraft. Mingled on the lines are a few T. Mk. IIIs.

37. The Mosquito final assembly shed at the Standard Motor Co. works near Coventry in January, 1944. All Mk. VIs, production ranges from HP928 to HP969. '928 (nearest) had a brief career, going to 613 Squadron 20.2.44 and crashing on 17.3.44. Most of the others were used overseas.

38. The salvage hangar of the Mosquito Repair Organization at Hatfield on 18.5.44. ML957, a B. XVI in night-bomber colours HS-D of 109 Squadron, was seriously damaged on 9.4.44 after an operational sortie and the fuselage was used for spares. HK404:ZJ of 96 Squadron had an accident on 6.5.44, and beyond lies the fuselage of SB-Q of 464 Squadron.

39. This photograph of 30 September, 1943, exemplifies the invaluable work of the Mosquito Repair Organization, headed at Hatfield by A. J. Brant. The aircraft, damaged in a belly landing, had to be cut open by a rescue team to get the crew out. The R.A.F. returned the Mosquito to de Havilland as Category E (reduce to produce), but the M.R.O. decided to repair it, re-categorized it B, and soon had it in operation again. Damaged wings were often sawn off, and a new mainplane extension was butt-jointed on—the simplest and quickest of major repairs. One in four of the Mosquitoes going onto squadron strength in 1941 to 1945 came from a repair depot.

40. to 42. The cockpit layouts of these three Mosquitoes differ mainly in detail. No. 40 shows that of a P.R.U./Bomber Conversion aircraft with spectacle control column. No. 41, taken 24.2.42, shows the interior of W4052, the fighter prototype. The third picture (42) is of the cockpit of the Mk. 37, and was taken in November, 1946.

Between pages 112 and 113

43. The manufacturing simplicity of the undercarriage compression leg with its column of rubber blocks is evident from the modest layout and equipment of Cundall's assembly shop.

44. Drilling and riveting of panels to assemble the engine nacelle and undercarriage doors, at Briggs Motor Bodies.

Full photograph captions

45. The thousandth Mosquito fuselage left the works of Wrighton Aircraft, Ltd., Walthamstow, on 8 July, 1944. The risk of building major components in quantity in the heavily bombed areas had to be weighed against the advantages of employing an efficient organization where it was. The Wrighton Brothers, with the team, typified the energy, practicality and spirit of the Mosquito subcontractors, organized by C. G. Long.

46. The factory of Perfecta Motors was in a part of the Midlands industrial region which suffered severely from enemy action, and practical problems of ensuring complete black-out after dark interfered with natural lighting in daytime. Here large quantities of canopies and windscreens were assembled.

47. Hoopers, the famous coachbuilders, adapted their works at Park Royal, North-West London, to produce Mosquito wings, tailplanes and wing tanks. Vertical jigs were used for wing and tailplane assembly. The factory was in a heavily bombed urban area, but maintained a high output, employing a large number of women workers.

48. Of a number of village groups that were organized to do sub assemblies required in quantity the smallest was that which Mr. and Mrs. Bertram Hale, retired from a busy life in the Far East, operated on electrical parts in a garden shed beside their shelter at 'Fung-shui', Old Welwyn, Hertfordshire. The last bomb of the war fell half a mile from here, at Easter, 1945.

49. The first Australian-built Mosquito, the F.B. Mk. 40, A52-1, which was flown on 23 July, 1943, by Gibson Lee. This picture was taken later, and the aircraft was delivered to the R.A.A.F. on 4 March, 1944. The first flight was just about a year after the start of the project in Sydney.

50. Pat Fillingham, who introduced Hatfield flight-test methods for the Mosquito, first to the Canadians and then to the Australians, seen here at Sydney beside the British-built T. Mk. III and an Australian F.B. Mk. 40.

51. A52-1005, one of the fourteen British-built trainers which were supplied to the R.A.A.F., sets the pace when flying with an F.B. Mk. 40, A52-1, the first Australian-built Mosquito. From the F.B. Mk. 40 de Havilland evolved the Australian trainer, T. Mk. 43, with Packard Merlin 33s. Of this variant, twenty-two conversions, A52-1050 to 1071, were made.

52. The assembly line at Bankstown, Sydney, as it was in mid-1944, with the ill-fated A52-12 (destroyed in June) in the fore-ground.

53. A52-327 was the last of twenty-eight P.R. Mk. 41 aircraft which de Havilland evolved with Packard Merlin 69 engines, from the last batch of Mk. 40s. This version did much good work after the war with Japan ended, particularly on air-survey operations in Australia.

Full photograph captions

54. Ralph Spradbrow, then the chief test pilot of the Canadian de Havilland Company, shakes hands with Ralph Bell, Canadian Director of Aircraft Production, after flying the first Canadian-built Mosquito, at Downsview on 24 September, 1942, just a year from the start on the project there. In charge of production, Harry Povey (wearing spectacles) is behind Spradbrow, and on the extreme left is Pepe Burrell, observer on the flight.

55. 43-34943, an F-8 supplied from Canada to the U.S.A., being towed into a hangar at Toronto.

56. Russ Bannock and a flight test observer stand by one of the Mosquitoes assembled after the war for use by Nationalist China.

57. LR503 'F̄ Freddie' of 105 Squadron completed 213 sorties. It was flown then to Canada by Maurice Briggs (extreme right) and John Baker. The picture shows from left to right Jack Lightcap (War Finance Committee), the navigator, Wg. Cdr. Cooper of the repair depot at Winnipeg, Mr. Tanner and Briggs.

58. KA102 was the third of the three Canadian-built Mk. 21s, fighter-bombers with two-stage Packard Merlins. Many of the Canadian Mosquitoes had Sky Type S coloured spinners and rear fuselage bands common to R.A.F. fighters, but KA102 appears to have one wrongly positioned.

59. Ill-fated KA970 bears the scar caused by the explosion during her delivery to Britain on 17 April, 1945.

60. Mk. IV DK296 was delivered to 105 Squadron 21.6.42. It became GB-K, was damaged in action, became GB-G and flew fifteen successful sorties (two with 139 Squadron). On 24.8.43 it went to 10 M.U. where it was earmarked for passing to the Russians. Although it reached Errol in September, 1943, it was 20.4.44 when it finally left and 31.8.44 before the Russians accepted it.

61. Pilots of the Air Transport Auxiliary, under Miss Pauline Gower (extreme left), who were stationed at Hatfield early in the war; the picture was taken on 10 January, 1940. These ladies (including at least one grand-mother, Mrs. A. S. Butler), delivered Mosquitoes and aircraft of all operational types up to the heaviest four-engined bombers.

62. Aircrew of 604 (County of Middlesex) Squadron, Royal Air Force, in 1944, flying the N.F. Mk. XIII, with the Universal Radar Nose. This picture typifies the young men who fought to keep the invaders out of Britain in the war of 1939–45. One may conjecture as to what meaning it is likely to have to the inhabitants of Britain two or three generations later.

63. Flg. Off. Victor Ricketts, D.F.C., at Hatfield, describing to Mosquito workers the trip over Paris in W4060 on 4 March, 1942, with his

Full photograph captions

observer Flg. Off. Boris Lukhmanoff, when in bad weather they obtained low-level photographs of destruction accomplished by the R.A.F. at the Renault works at Billancourt.

64. One of 150 low-level obliques of the Renault works, Billancourt, secured by Ricketts from W4060 on 4 March, 1942.

65. Wg. Cdr. Merifield, the persistent photographic reconnaissance pilot and his observer, stand by a Mosquito used by them for a post-war record flight.

66. LR432, a P.R. IX, arrived on 540 Squadron 4.9.43. It went to 544 Squadron 8.10.43. On 15.9.43 it opened 544 Squadron's day operations with Mk. IXs and two days later Merifield flew it to Poznan. He tried again next day but collided with a flock of birds and turned back. On 21 October Douglas-Hamilton flew it to Lyons and covered part of the Pyrenees. It made a 5 hour 20 minute flight to Besançon and Nancy on 29 December. Its sorties took it to many parts of Europe but rarely was it intercepted. One occasion was on 6 October, 1944, when a fighter chased it over Utrecht and Lingen. Whilst Merifield flew MM355 to Holland on 21.8.44 Wg. Cdr. Steventon took LR432 on a similar low-level mission to Ijmuiden RDF station. Flt. Lt. J. C. Webb and Plt. Off. C. D. Smith flew LR432 on its forty-third and last sortie, to Hemmingstadt and Heligoland, on 29 November, 1944. LR432 passed to 8 O.T.U. on 22.1.45 and was SOC 11.9.45. The photograph shows LR432 coded L1, in the markings of 544 Squadron.

Between pages 144 and 145

67. To left of centre can be seen *Scharnhorst* and in the centre at the foot of the picture the German carrier *Graf Zeppelin*. Taken from W4060 over Gdynia on 2.6.42.

68. Another fine photograph by Merifield, this time an oblique of a Limoges factory after an attack by Lancasters of 617 Squadron. It was taken from 1,000 feet on 10.3.44 from LR422.

69. Night-photographs like this played a major part in defeating the Wehrmacht after D-Day. Road transport can here be seen moving through a French village-hoping for immunity, doubtless.

70. An RDF station at Bergen-Am-Zee, Holland, photographed almost from ground level by Wg. Cdr. J. R. H. Merifield, D.S.O., D.F.C., and Flt. Lt. W. N. Whalley, D.F.C., D.F.M., of 540 Squadron flying MM355 on 21.8.44. They left Benson at 14.00 hrs., flew to Southwold and crossed to Holland fast and at sea level. This superb photograph was obtained from the forward facing camera and a fourteen-inch lens.

71. Geoffrey Raoul de Havilland, the eldest son, who conducted flight development of the Mosquito through all its stages. He was responsible

Full photograph captions

also for the test flying of the Albatross, Flamingo, Moth Minor, Vampire and the Hornet which was evolved from the Mosquito. He lost his life on 27 September, 1946, three years after the death of his younger brother, when flying a tail-less experimental aircraft the D.H. 108.

72. La Pallice photographed from W4055 on the first Mosquito sortie, 17.9.41, before cameras ceased to function.

73. The Philips works at Eindhoven was attacked by 2 Group on 6.12.42, very effectively as can be seen. After the raid a 139 Squadron Mosquito crew flying DZ314 evaded flak and fighters to secure photographs from 800 feet of the burning factory.

74. DZ353 and DZ367 of 105 Squadron during a press visit in December, 1942.

75. DD712, a Mosquito II (Special) intruder without A.I., photographed 16.9.42. It had joined 23 Squadron 8.9.42, and made its first intruder operation, in the hands of Wg. Cdr. Hoare, to Beauvais and Creil on 24–25.9.42. It completed three more operations before failing to return from its sixth sortie, on 29.11.42.

76. On 23–24.8.42 Sqn. Ldr. Starr set off to patrol Deelen in YP-E:DD673. His port engine failed ten miles from the Dutch Coast. He turned back, and overshot Manston on landing and collided with a steamroller. The crew escaped, but the aircraft was categorized B. Martin Hearn repaired it and despatched it to 51 O.T.U. on 17.5.43. It then went to 60 O.T.U., in whose hands it again crashed (on 6.6.43) and then went to Hatfield for repair. On 24.3.44 it joined 141 Squadron and on 6.6.44 as TW-T patrolled over Paris in the early hours of the invasion. On 9.6.44 it was written-off as a result of battle damage.

77. DZ238:YP-H reached 23 Squadron 9.12.42. After 138 hours flying it returned to Hatfield from Malta on 5.5.43. After overhaul it became YP-Z, and on 17.8.43 was the last Mk. II to operate with 23 Squadron. It wears the grey-green-black finish also worn by intruders of 418 and 605 Squadrons and some *Ranger* aircraft.

78. DD750, a standard Mk. II night-fighter, was delivered to the R.A.F. 7.9.42 and after fitting out reached 151 Squadron on 6.10.42 with whom it served until transfer to 264 Squadron on 29.7.43. It had latterly been used for *Day Rangers*. On 14.8.43 157 Squadron received it; they operated it on several successful *Insteps*. 410 Squadron also used it, and following overhaul it passed on 25.4.44 to 239 Squadron. It was damaged beyond repair on operations 28.6.44.

79. The four-gun Bristol turret as fitted to the two Mosquitoes for trial purposes in the early phase of exploiting the fighter possibilities of the Mosquito design.

Full photograph captions

80. HP850 was a Mk. VI delivered to 157 Squadron for *Ranger* operations on 14.7.43, and is depicted here at Hunsdon. It became SB-Q of 464 Squadron on 6.9.43, and failed to return from Woippy on 9.10.43.

81. HK428 was a Mosquito XIII fitted with A.I. Mk. VIII in a 'thimble' nose. Delivered 14.11.43 it joined 29 Squadron 28.1.44. As RO-K it is believed to have shot down a Ju 88 on 17.6.44. On 22.10.44 it was damaged, then sent to Hatfield for repair. It later served the Central Gunnery School.

82. MM466, probably the top-scoring Mosquito fighter. A Mk. XIII it served with 488 and 409 Squadrons making its last sortie on 2.5.45. (See Appendix 10)

83. There were numerous occasions when Mosquitoes were badly scorched when they flew through the wreckage of aircraft they had set ablaze. The charred remains of this Mk. XIII of 264 Squadron were safely brought to an airfield in Normandy in August, 1944.

84. Mosquito IVs of 105 Squadron at Marham taxi out in December, 1942. GB-A:DZ360 flew three sorties with 139 Squadron and was shot down 22.12.42 flying with 105 Squadron, after encountering flak at Dunkirk. DZ353:GB-E flew eleven sorties with 105 Squadron, had two months with 139 Squadron on night operations, and reached 627 Squadron 24.11.43. On 8.6.44 it failed to return from raiding the marshalling yards at Rennes. DZ379:GB-H operated with 105 Squadron until 25.7.43, when it passed to 139 Squadron and failed to return from Berlin 17.8.43. DZ367:GB-J failed to return after eight sorties. DZ378:GB-K after two sorties was damaged in an accident on 20.12.42. As 3905M it was positioned at 13 O.T.U. DK338:GB-P was used for multiple-exhaust tests at Marham.

85. Sqn. Ldr. Ralston, D.S.O., D.F.M., and Flt. Lt. S. Clayton, D.S.O., D.F.C., D.F.M., standing by a 105 Squadron Mosquito in December, 1942.

86. Wg. Cdr. Wooldridge (closest to the Mosquito 'Popeye') and a group of 105 Squadron crew. Sqn. Ldr. Blessing is pointing, Flg. Off. Burke is behind Wooldridge.

87. Back from Berlin, the Mosquito crews on 31 January, 1943. From left to right they are Flg. Off. R. C. Hayes, Flg. Off. R. C. Morris, Flt. Sgt. F. I. D. McGeehan, Sgt. J. Massey, Flg. Off. J. T. Wickham, Flt. Lt. J. Gordon, Sqn. Ldr. R. W. Reynolds, Sgt. R. C. Fletcher, Plt. Off. E. B. Sismore and Plt. Off. E. D. Makin.

88. Bombs were bursting on the railway workshops at Trier on 1 April, 1943, as a 105 Squadron crew secured this picture. Over the target flies another Mosquito.

Between pages 208 and 209

89. Trier under attack on 1 April, 1943.

Full photograph captions

90. 139 Squadron Mosquitoes at Marham. Wg. Cdr. Shand is in the foreground in white jacket. He stands before XD-G:DZ421, which flew with the squadron from 31.12.42 until it passed to 627 Squadron 21.4.44. It crashed at Wistow, Yorks., on 25.7.44 when used by 1655 M.T.U. DZ373:XD-B, next in the line, came down in the Scheldt Estuary 12.3.43, caught by flak after bombing Liège. XD-T:DZ423 and XD-K: DZ428 come next in the line.

91. The briefing room at Marham in December, 1942. Wg. Cdr. Peter Shand is addressing the crews of 139 Squadron.

92. DD744 wearing all-silver finish was delivered to the Middle East on 22.1.43. A Mk. II, it was then modified into a reconnaissance aircraft for 60 Squadron, S.A.A.F., which it joined on 8.8.43. It was SOC in November, 1943.

93. Lt. Archie Lockhart-Ross with NS520 in which he and Capt. Pienaar returned to base after being shot-up by an Me 262 near Munich on 15.8.44.

94. 'Lovely Lady' the most famous PR. IX of 60 Squadron, S.A.A.F. undergoing maintenance at San Severo.

95. HJ743 an early production Mk. VI photographed at Hatfield 28.5.43, the day before delivery. 418 Squadron received it on 3.6.43, 25 Squadron on 9.8.43, 487 Squadron on 21.9.43 and 21 Squadron 29.9.43. Further details of its career are obscure; it was struck off charge 21.6.47.

96. Amiens prison during the attack to breach the wall, the aiming point of which is indicated by an arrow.

97. The day following the attack DZ414 photographed the gaol, its first picture being this. Damage to the buildings and the wall may be clearly seen. (See Appendix 11)

98. A typical FB. Mk. VI being turned round after a sortie from Lasham in the winter of 1943/44. LR366 was delivered to 613 Squadron on 10.1.44 and suffered battle damage on 5.2.44 when as SY-L it was one of seven 613 Squadron aircraft which attacked the V-weapon site at Motteville. 107 Squadron took it on charge on 27.7.44, and as OM-L it was shot down by flak on 17.9.44 when attacking barracks at Arnhem.

99. A 613 Squadron Mk. VI makes its getaway after adding its bombs to the Central Registry building in the Hague on 11.4.44.

100. Mosquitoes bombing Gestapo offices in the Aarhus University building on 31.10.44.

101. NS787, a PR. Mk. XVI of 684 Squadron, wears the special aluminium finish applied to many Mosquitoes in the Far East in the closing months of the war.

Full photograph captions

102. VL619, a PR. 34 which served after the war, first in 13 Squadron in the Middle East and then with 58 Squadron in England. 200-gallon tanks are seen under the wings. Extra fuel tanks fitted in the 'swollen bomb bay' boosted the range of the P.R. 34 to 3,000 miles, and more, in still air at 20,000 feet.

103. One of the photographs of the summit of Mount Everest taken from a Mosquito by Sqn. Ldr. Andrews and Flt. Lt. Irvine on 16 June, 1945.

104. An FB. Mk. VI of 143 Squadron, PZ446, being equipped with rockets and made ready for a strike off Norway in December, 1944. It was fitted out for Coastal Command duties at the Preparation Pool, Bircham Newton, in October, 1944. After many sorties it was passed to B.A.F.O. and to 4 Squadron. After overhaul in 1946–47 it was handed to the R.N.Z.A.F., on 3.5.48.

105. 143 Squadron Mosquitoes attack a ship at Sandshavn, Norway, on 23.3.45. U/143 recorded its own hits from 300 feet.

106. A close-up of RS625:NE-A of 143 Squadron reveals its nose-mounted camera, tiered rockets to compensate for fitting of wing drop tanks, paddle-bladed airscrews and the five-port exhaust stacks common to Mosquitoes late in the war. Delivered 31.12.44, RS625 served progressively with 143 Squadron, 248 Squadron, 143 Squadron, 4 Squadron, with 204 A.F.S. as FMO-V from 2.2.49 until 4.6.52, and was sold as scrap 19.11.53.

107. V/235, a Mk. VI, strafing a steamer on 19.9.44 at 61° 24′ N/05° 00′ E. Shots were also being fired from Y/235, which secured the photograph.

108. An attack on shipping at Nordgalen by 235 Squadron on 5.12.44. Machine-gun fire can be seen peppering the sea around two ships. The steepness of the fiord walls calls for dive attacks. E/235 secured the picture at 1,000 feet as she dived to attack.

109. Two U-boats under attack at 57° 58′ N/11° 15′ E on 9.4.45 by 143 Squadron, recorded by the camera of N/143.

110. HJ732/G, the prototype Mk. XVIII 'Tsetse', with the nose-mounted 57 mm. gun, at Hatfield in the summer of 1943.

111. 0/618 during *Highball* trials at Loch Striven. The spherical weapon, back-spun, has just bounced on the water and skips in decreasing bounces towards the target ship.

112. NT585, a Mk. 30 of 264 Squadron, photographed on 10.12.45 flying from Church Fenton. Delivered 4.4.45, it went to 307 Squadron and to 125 Squadron 26.4.45, but never flew on operations. When 125 Squadron became 264 Squadron it retained the VA coding for several months. NT585 later served 151 Squadron.

Full photograph captions

Between pages 304 and 305

113. DZ594/G, photographed 29.11.43, was the Mk. IV 4,000 lb. bomb conversion aircraft. After trials it was prepared at 44 M.U. for squadron use, and 627 Squadron received it on 15.2.45. As AZ-X it mined the Elbe on 22 and 27 March, and the Weser on 30 March. A month later it was damaged in an accident and was struck off charge 28.6.45.

114. This photograph shows target indicators bursting on the ground to mark Berlin for a Mosquito raid, led by 139 Squadron. The two light streaks uppermost are searchlight beams.

115. Loading a 4,000-lb. 'cookie' into a Mosquito B. XVI of 128 Squadron at Wyton on 21 March, 1945, prior to the heaviest night raid on Berlin, by two waves of Mosquitoes.

116. Scrubbed?—No, dicing tonight. The signal for the night's entertainment is duly displayed in the mess at Little Snoring, a base for 100 Group Mosquitoes.

117. Capt. C. N. Pelly (second from the right) with colleagues at Leuchars, Fifeshire, terminal of the B.O.A.C. war-time service to Stockholm. Nigel Pelly was flight captain from September, 1943, to the end of the operation. From 3 February, 1943, to 17 May, 1945, Mosquitoes made 520 return trips, averaging nine single trips a week for $2\frac{1}{4}$ years. Four Mosquitoes were lost, with eight crew and two passengers.

118. Mosquito VI G-AGGD comes in to land at Leuchars in July, 1943. It served for $8\frac{1}{2}$ months on the Sweden run.

119. ML897:D, a B. Mk. IX of 1409 Met. Flight in night-bomber camouflage and black and white 'D-Day stripes' landing at Wyton in November, 1944, after its 153rd sortie with Bomber Command. From July, 1943, until November, 1945, ML897 operated in a met-recce role, latterly with Transport Command.

120. During 674 hours of flying LR504 flew 190 sorties and was then photographed, at Little Staughton, on 1.3.45, in the hands of 'A' Flight, 109 Squadron. It flew in all 200 sorties. LR504 began operating on 21.6.43 and thereafter was frequently employed. It participated in route marking to Berlin on 31.8.43 and with a more usual load of 6×500 lb. bombs made precision attacks on Western Germany. On 13.1.44 it marked Essen, then followed attacks on communications targets. On 13/14.3.44 before being officially assigned to new owners, 105 Squadron at Bourn, it was operating against Le Mans; Dieppe, Orly, Caen and Trappes were also lit by its T.I.s. On 5/6.6.44 it marked St. Martin de Verreville. LR504 led a day raid on Domleger on 28.6.44 after marking Wizernes, a V-weapon target to which it returned on 5/6.7.44. Flying-bomb sites were now frequently its targets, but it found time to mark Homberg on 11/12.7.44, Wanne Eickel on 25/26.7.44 and the Normandy battle

Full photograph captions

area on 30.7.44. It was a leader for 286 Lancasters bombing Le Havre on 5.9.44, and subsequently marked for the massive daylight raids on Boulogne and Calais. When used by 105 Squadron it was coded GB:H.

Almost immediately after its return to 109 Squadron it went into action, as HS-K, as reserve marker at Cleve on 7.2.45. Subsequent marking operations took it to Pforzheim 23.2.45, Kamen 24.2.45, Cologne 2.3.45, Wesel 6.3.45, Essen 11.3.45, Munich 8.4.45, and again on 12.4.45 and Wangerooge for the final attack of the war. It made two *Oboe* sorties to Berlin, on 9 and 10 April. Worn out by its persistence it was struck off charge on 14 September, 1945. In all respects it was an outstanding aeroplane.

121. ML914, a B. IX purchased with money donated from Nigeria, served 105 Squadron as GB-K, GB-C and GB-N from 24.9.43 until 16.4.45. It participated in precision attacks in the winter of 1943/44, commencing operations on 10.12.43 with an intended raid on Leverkusen, from which it had to make an early return due to control troubles. Its targets included Essen on 26.3.44, Aulnoye 10/11.4.44, and on 3/4.5.44 it was 105's only marker for the devastating Mailly raid. On 5/6.6.44 it marked guns at St. Pierre and later in June V-1 sites. After taking a leading part in the day raids on Le Havre it switched to several marking operations against the Scholven-Buer synthetic oil refinery. On 18 July it was again over Normandy, marking Cagny in Operation *Goodwood*, the break-out of Allied troops from Caen. ML914 took part in the first Mosquito day raid on the Ruhr in 1944, namely Homberg on 27.8.44 and made sorties against Calais in September. It operated through the winter of 1944-45. On 2.3.45 it took part in the massive onslaught on Cologne, and it participated in the mass attack on Wesel on 6.3.45. It made in all 148 sorties. 627 Squadron received ML914 on 16.4.45, but never flew it on operations. After a brief spell with 1317 Flight (21.6.45 to 21.7.45) it returned to 627 Squadron, becoming part of 109 Squadron on 1.10.45, with whom it spent some months. It was SOC after a glittering career on 27.1.47.

122. Mosquito B. IVs of 109 Squadron at Marham in 1943. DZ425:HS-K is nearest with DZ429:HS-L, DK331:HS-D, HS-C and HS-W visible beyond. All of these aircraft contributed considerably to the success of Bomber Command attacks in 1943.

123. DZ319:HS-H, an *Oboe*—fitted Mk. IV of 109 Squadron, pictured in 1944 after its 101st sortie. The nose transparency has been over-painted, and therein is placed the radar gear. Delivered to 109 Squadron 26.9.43, it took part in the first *Oboe* operation, but a fault caused diversion to Duisburg. Two nights later it bombed Rheinhaussen using *Oboe*. Thereafter it was repeatedly in the fray taking part in the second Essen raid on 12.3.43 and that of 1.5.43. It operated frequently until damaged when raiding Leverkusen 10.5.44. After repair at Hatfield it went to the P.N.T.U. on 4.1.45, where it remained until 10.8.45. It was SOC 31.1.46.

124. LR356:YH-Y of 21 Squadron making a high-level attack on a V-weapon

Full photograph captions

target. Such raids were, somewhat confusingly, known as 'Highballs'. Above, a 109 Squadron Mosquito was flying, having dropped markers for this experimental raid. Bombs dropped from the Mk. VI can be seen bursting.

125. Bombing-up KB162:XD-J 'New Glasgow' of 139 Squadron, at Upwood. The nose shape, reminiscent in contour of the T. Mk. III, carries *H2S*. KB162, one of the first Canadian-built B. XXs to operate, was used by 139 Squadron from 23.11.43. On its seventy-ninth operational sortie 11/12.10.44 it was attacked by a Ju 88 near Berlin, but successfully evaded its fire. On 13.10.44 it took off on its eightieth mission, but crashed a few moments later and was destroyed, at Warboys, Hunts.

126. Berlin under attack on 11/12.8.44. Markers have been laid by 139 Squadron on the left of the picture. From top to bottom searchlight tracks can be seen, also in the bottom right corner and across the centre.

127. The beautiful Mosquito outline is well portrayed in this view of B. XVI ML963:8K-K of 571 Squadron photographed on 30.9.44. ML963 was delivered to 109 Squadron 9.3.44, 692 Squadron 24.3.44 and to 571 Squadron 19.4.44. It was damaged in battle when operating against Brunsbuttelkoog on 12.5.44 and returned to Hatfield for major repair 9.7.44. 571 Squadron again took it on charge 23.10.44. It failed to return from Berlin on 10/11.4.45 when making the squadron's 2,368th sortie.

128. Subscribed for by the 'Oporto British', MM156 a B.XVI was delivered to 109 Squadron 13.7.44 and to 571 Squadron 13.9.44. It was despatched to Hatfield for repairs 4.10.44 and to 44 M.U. 1.12.45. General Aircraft received it for conversion to T.T. 39 standard and it was delivered to the Royal Navy 13.8.47.

129. 464 Squadron's Mosquito VIs visited Hatfield on 2 June, 1944, arriving from Gravesend.

130. The swollen bomb bay of the P.R. Mk. 34 and its large wing tanks can clearly be seen in this photograph.

131. MM652, a Mk. XIX delivered 2.5.44. It reached 157 Squadron 27.5.44 and as RS-S was used against V-1s, and was damaged in action on 9.8.44. On 29.8.44 it returned to use, flying many bomber-support sorties before being passed to 169 Squadron on 18.4.45. After overhaul it was sold to Sweden.

132. Mk. XIII HK382 served 96 Squadron from 30.10.43 until damaged in a flying accident 24.11.43. After repair it became RO-T of 29 Squadron from 20.9.44. Black undersurfaces denote its main employment on *Night Rangers*. It was destroyed in a crash on 16.3.45.

133. W4087 was a Mk. II in the nose of which was installed a Turbinlite, an airborne searchlight, as well as A.I. Mk. V radar. In the hands of 1422

Full photograph captions

Flight the machine flew many hours, and was experimentally used by several squadrons too. In 1944 it was employed upon experimental radio and radar work in connection with N.P.L. programmes.

134. Six PR. XVIs had *H2X* installed in their noses for the U.S. 8th A.A.F The Americans carried out experiments with them and considered their possible used as P.F.F. aircraft or formation leadships, although their high-speed made them difficult to deploy. The first machine is seen here, photographed at Hatfield 11.2.44. *H2X* was the American equivalent of the British *H2S* radar.

135. Turkey took delivery of a considerable number of Mk. VI Mosquitoes, after they had been refurbished by Fairey Aviation at Stockport, near Manchester. This all-silver machine, 6652, is unusual because it has four-bladed propellers.

136. TS449 was one of two specially navalized Mosquitoes built at Leavesden and delivered in 1946. Note the four-bladed propellers, RATOG and arrester hook. Wing folding was manual. LR359, the first navalized Mosquito, appeared in 1943. Guns were deleted and an arrester hook added. It was a modification of the Mk. VI and began flight trials in November, 1943. It was based at Renfrew for trials off the Scottish coast in August, 1944, and written-off in an accident at Arbroath on 9.11.44. On 29 March, 1944, it made the first deck landing by a Mosquito —indeed, by any multi-engined aircraft—by a touch down on H.M.S. *Indefatigable*.

137. A Mosquito T.T.39, PF606, converted from a Mk. XVI bomber, for Royal Navy use, and photographed in September, 1948.

138. TW240 served as prototype for the Sea Mosquito T.R.37 This photograph taken in November, 1946, shows an experimental radome for A.S.V. Mk. XIII radar, and four-bladed airscrews common to navalized Mosquitoes.

APPENDIX 2

Abbreviations, etc.

A.A.E.E.	Aeroplane & Armament Experimental Establishment, Boscombe Down
A.I. Mk. IV, etc.	Airborne Interception radar Mk. IV, etc.
A.F.D.U.	Air Fighting Development Unit
A.S.W.D.U.	Air/Sea Warfare Development Unit
a.u.w.	All-up weight
Bf/Me	Bayerische Flugzeugwerke/Messerschmitt. German records list the commonly called 'Me 109' and 'Me 110' as Bf 109 and Bf 110. Both titles are given in this book since in some instances quotations from combat reports of the period are included using Me, etc.
C.C.P.P.	Coastal Command Preparation Pool, Bircham Newton.
cv	Converted/conversion
dev.	Development
E/A	Enemy aircraft
FTR	Failed to return from operations
/G	Suffix letter to aircraft airframe number indicating aircraft carrying special gear to be guarded
M.A.C.	Mediterranean Air Command
Mod.	Modified
M.U.	Maintenance Unit
Noball	Attack against V-weapons' site
Outstep	Patrol seeking enemy air activity off Norway
R.A.E.	Royal Aircraft Establishment, Farnborough
R & D	Research and Development
S.A.G.	Second Aircraft Group, Leavesden
SOC	Struck off Air Ministry charge
S. of T.T.	School of Technical Training
T.I.	Trial installation, or target indicator/marker
Trop.	Tropicalized
T.F.U.	Telecommunications Flying Unit, Defford
w.o.	Written off
'Y' aircraft	Mosquito bomber fitted with *H2S* radar

APPENDIX 3

Mosquito Genealogy

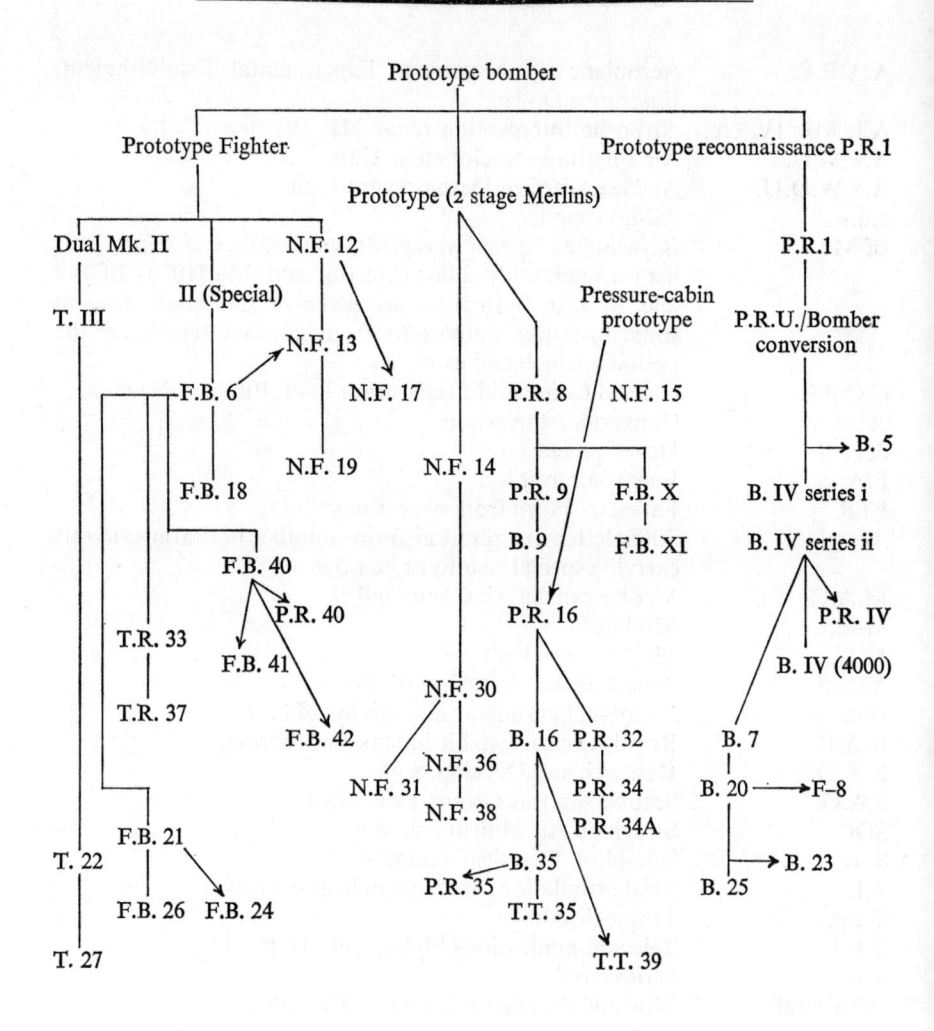

```
                                    Prototype bomber
         ┌──────────────────────────────────┼──────────────────────────────────┐
   Prototype Fighter                                                 Prototype reconnaissance P.R.1
                              Prototype (2 stage Merlins)
 Dual Mk. II        N.F. 12                                                    P.R.1
                                                  Pressure-cabin
   II (Special)                                     prototype          P.R.U./Bomber
  T. III                                                                 conversion
                     N.F. 13
                                                                                    → B. 5
         F.B. 6      N.F. 17      P.R. 8    N.F. 15
                                                                          B. IV series i
         F.B. 18     N.F. 19     N.F. 14    P.R. 9   F.B. X
                                             B. 9    F.B. XI             B. IV series ii
              F.B. 40                                                            → P.R. IV
                 P.R. 40                   P.R. 16                         B. IV (4000)
   T.R. 33       F.B. 41
                                   N.F. 30
   T.R. 37          F.B. 42      N.F. 31  N.F. 36   B. 16  P.R. 32   B. 7
                                         N.F. 38           P.R. 34   B. 20 ──→ F-8
         F.B. 21                                           P.R. 34A
   T. 22                                     B. 35                        → B. 23
         F.B. 26  F.B. 24                  P.R. 35                  B. 25
                                                    T.T. 35
   T. 27                                                   T.T. 39
```

APPENDIX 4

Summary of Mosquito Variants

Note. Weights quoted are, wherever possible, listed against aircraft on which they were measured. Frequent weight variations existed between individual aircraft, and the many modifications to equipment naturally altered weights. Therefore, these notes can only give general weight indication. They are based upon records in de Havilland archives, and in the reports issued following trials at the Aeroplane & Armament Experimental Establishment, Boscombe Down. Two fuel totals, where given, such as 547/410, indicate maximum fuel load including any in drop tanks/fuel load with useful operational load.

Mk. 1 (Merlin 21) Photo-reconnaissance aircraft based on W4050. First flight 10.6.41. *Fuel:* 690 gals. max., oil 44 gals. *Weights:* (lb.) tare 13,009; useful operational load 17,940, a.u.w. 18,050. Two were long-range aircraft, with operational weight of 19,310 (W4060). Two others similarly modified were tropicalized. All P.R. 1s had short nacelles. Nine built, in addition to the prototype. Had three vertical cameras and one oblique. Span 54 feet 2 inches. (W4050's wing span was 52 feet 6 inches.)

P.R./Bomber Conversion Type (Merlin 21) Bomber conversion of P.R. 1 airframe. Two thousand lb. bomb load. Re-designated B. IV Series i, prior to squadron service. All had short nacelles, to speed production. Max. fuel load 690 gallons. First flew September, 1941. Nine produced.

F. Mk. II (Merlin 21, 22, 23) Day and night long-range fighter, and intruder. First flew 15.5.41 from 19.25 to 19.40 hours. *Fuel:* 547 gals./410 gals. with op. load. Four 20 mm. Hispano cannon and four ·303 inch Browning guns. *Weights:* (lb.) tare 13,356; usual loaded weight 17,700, full operational loaded weight 18,649 (20,048 with Turbinlite). Tare weight with turret (W4073) 13,812, including turret and guns but no ammunition. All IIs had long nacelles. Twenty-five (Special) Intruders for 23 Sqn. had A.I. removed and additional tankage. Many were refurbished and re-engined for operations with 100 Group 1943/44. Fitted with A. I., IV and V. A few converted for P.R. duties. Length 41 feet 2 inches (42 feet 11 inches with A.I. IV.)

T. Mk. III (Merlin 21, 23, 25) Dual-control trainer, used during the war and more extensively later. First flew January, 1942. *Fuel:* max. 716 gals. (with 2 × 50 gal. drop tanks); 453 gals. normal with operational load. *Weights:* (lb.) tare 13,739; usual loaded 17,188, fully loaded 19,554; fully loaded and with 2 × 100 gal. drop tanks 20,319. Unarmed.

B. Mk. IV Series ii (Merlin 21, few had Merlin 23) Day and night bomber.

Summary of Mosquito Variants

First production aircraft flew March, 1942. *Fuel:* max. load 539 gals. Loaded weight 21,794 lb. (22,380 with full load). Fifty gal. wing drop tanks could be fitted. Two thousand lb. bomb load. Length overall 40 feet 9½ inches.

4,000 lb. bomb conversion: (8 Group recommendations): Maximum permitted all-up take-off weight 22,570 (20,500 maximum landing weight), with max. fuel capacity of 497 gals. Target ceiling 25,000 feet. Still-air range 1,430 miles max. op. radius 535 miles.

Highball conversions (DZ547/G data): Weights little altered, but speed at S.L. reduced by 18 m.p.h. and by 22 m.p.h. at 12,000 feet.

Highball trials aircraft (DK290): In original B. IV state with 4 × 500 lb. G.P. short-tailed bombs, a.u.w. 20,670 lb. With bomb bay modified but without drive, but with mock stores each of about 1,100 lb., take-off weight was 18,760 lb., cruise speed 260 m.p.h. Envisaged take-off weight 21,000 lb. with full load and fuel.

P.R. Mk. IV (Merlin 21, 23) Day and night photo-reconnaissance aircraft, conversion of B. IV series ii. First flew April, 1942. Loaded weight 18,000 lb. *Fuel:* 660 gals. with 50 gal. drop tanks/539 gals.

B.V. (Merlin 21) Projected bomber with 2 × 50 gal. drop tanks or 2 × 500 lb. wing-mounted bombs. Many projected variants. W4057, prototype for trial installations, was the prototype for Bomber/P.R. Conversion and B. IV, *Weights:* (lb.) 12,881 empty; 18,994 loaded (for W4057, known as B. Mk. V).

F.B. VI (Merlin 21, 22, 23, 25) Day and night fighter-bomber/intruder/long range fighter. First flew 1.6.42, production aircraft first flew February, 1943. Four 20 mm. cannon and four ·303 inch Browning guns in the nose, 2 × 250 lb. bombs in belly bay and 2 × 250 lb. bombs on wing racks (500 lb. later on Series ii aircraft), or 4 × 60 lb. rocket projectiles beneath each wing, or one mine or depth charge beneath each wing, and combinations of drop tanks and offensive weapons.

Fuel loads (A & A.E.E. listing):

Outer fuel tanks (2 tanks)	48 gal.
Outer fuel tanks (2 tanks)	68 gal.
Inboard tanks (2)	131 gal.
Inboard tanks (2)	156 gal.
Centre section	50 gal.
Total normal fuel load	453 gal.
Overload fuselage tanks	66½ gal.

A 50 or 100 gal. drop tank could be carried beneath each wing.

Weights (A.A.E.E. listing) (lb.):

	INTRUDER		LONG-RANGE FIGHTER	
	Normal	Overloaded	Normal	Overloaded
Tare	13,727	13,792	13,777	13,842
Service load	3,192	3,497	2,170	2,385
Fuel	2,962	3,222	3,701	3,701
Oil (32½ gal.)	293	293	293	293
Flying weight	20,114	20,804	19,941	20,221

Summary of Mosquito Variants

	ESCORT FIGHTER	
	Normal	*Overloaded*
Tare	13,797	13,862
Service load	2,696	2,911
Fuel	3,701	3,701
Oil	293	293
Flying weight	20,487	20,767

N.B. Added to these weights must be respective weights of operational loads, in lieu of normal service load, etc. With 2×250 lb. bombs the prototype HJ662 weighed 20,835 lb. in operational state. The L.R. fighter carried no bombs when weighed, escort fighter had 2×250 lb. bombs and extra belly tank. With S.C.I. smoke canisters, a.u.w. was 20,950 lb. With drop tanks HJ679 had a.u.w. of 21.020 lb. (42 gal. per tank) and 22,764 lb. when with 2×100 gal. tanks. A.A.E.E. tests with rocket-equipped F.B. VI (Merlin 23) revealed:

	M.S. gear		F.S. gear, at
	At sea level	*At 13,900 feet*	*20,700 feet*
Max. speed	278 m.p.h.	327 m.p.h.	329 m.p.h.
Max speed (rails fitted only)	286 m.p.h.	336 m.p.h.	337 m.p.h.
Max speed (No. R.P. gear)	296 m.p.h.	351 m.p.h.	358 m.p.h.

Gee, A.I. Mk. IV, V and XV variously fitted to Mk. VI.

B. Mk. VII (Merlin 31) Canadian bomber based on B. Mk. IV. First flown 24.9.42. Weights, performance similar to B. IVii. Used in North America only.

P.R. Mk. VIII (Merlin 61) Photo-reconnaissance version, conversions on line of Mk. IV, based on P.R. IV but with two-stage Merlins. First flew 20.10.42. *Fuel:* 860 gal. inc. 2×50 gal. drop tanks/539 gal. *Weights:* (DK324) (lb.) tare 14,252, a.u.w. 21,395.

P.R. Mk. IX (Merlin 72/73, 76/77) Photo-reconnaissance version for home and overseas service. First flew April, 1943. *Fuel:* 860 gal. inc. 2×50 gal. drop tanks/539. Fuel loads varied considerably, and could include 2×100 gal. and 2×200 gal. drop tanks. *Weights:* (LR405, recorded 19.4.43) (lb.) tare 14,380, a.u.w. 21,894. With *Rebecca, Boozer,* and 2×50 gal. tanks, a.u.w. 22,007; with 2×200 gal. drop tanks (1,255 gal. load) a P.R. IX tipped the scales on 18.7.44 tare 14,348 and a.u.w. 25,160. Trial installation of Merlin 67 (R.M.10 S.M.) in MM229, Dec. 1943. Overall length 41 feet 6 inches.

B. Mk. IX (Merlin 72/73, 76/77). Bomber with two-stage Merlins based on P.R. IX. First flew 24.3.43. *Fuel:* 697 gal., 497 with useful load. *Weights:* (lb.) Recorded on LR495: 14,976 tare, 22,823 all-up, incl. 2×50 gal. drop tanks, 24,753 with 2×100 gal. drop tanks. With 4,000 lb. bomb bay mod.: tare 14,644, 23,745 a.u.w. with 503 gal. and 4,000 lb. bomb. Max. permissible a.u.w. with additional radio etc., 24,865 lb. With *H2S Mk. VI* (4,000 lb. conversion) a.u.w. 24,441 or 26,001 carrying 2×100 gal. drop tanks—this lay beyond permissible take-off weight. LR495 with four fuselage and two wing

Summary of Mosquito Variants

500 lb. bombs had a.u.w. of 22,835 lb. A.u.w. of standard Mk. IX with *Oboe* in nose and 4,000 lb. bomb 23,590 lb., 24,386 with 100 gal. drop tanks. Only a few Mk. IXs were modified to carry 4,000 lb. bomb.

N.F. Mk. X (Merlin 61) Night-fighter with two-stage Merlins. Not proceeded with, although ordered in quantity. Length overall 41 feet 2 inches.

F.B. Mk. XI (Merlin 61) Fighter-bomber identical to Mk. VI with two-stage Merlins. Not proceeded with.

N.F. Mk. XII (Merlin 21, 23) Night-fighter conversion of Mk. II fitted with A.I.Mk. VIII in 'thimble nose'. First flew August, 1942. *Fuel:* 547 gal., 403 with useful load. *Weights:* (lb.) DD715 tare 14,640, a.u.w. 18,441. Typical production aircraft a.u.w. 18,441, HK242 weighed 14,575 tare. Overloaded (with 2 × 50 gal. tanks and additional equipment) it weighed 20,297 lb. Four 20 mm. cannon only. Length overall 40 feet 5 inches.

N.F. Mk. XIII (Merlin 21, 23, 25) Night-fighter with wing similar to that of Mk. VI ('Basic'). A.I. Mk. VIII in 'thimble' or Universal ('bull') Nose. Provision for drop tanks, or rarely used bomb load. First flew August, 1943. *Fuel:* 716 gal.; 453 gal. with useful load. *Weights:* (lb.). For HK363: tare 13,948, usual load 19,390, a.u.w. with 50-gal. tanks, etc., 20,278. Official trials comparing the N.F. XII and N.F. 30 were flown at Boscombe Down, revealing: Still-air range at recommended cruising speed 1,260 ml. at 20,000 feet; still-air range continuous cruising at F.S. 1,000 miles; maximum speed F.S. gear 394 m.p.h. at 13,800 feet and 379 m.p.h. at 6,000 feet on M.S. gear; max. speed S.L. 350 m.p.h. Initial climb 1,870 f.p.m., operational ceiling 28,800 feet, continuous recommended cruising speed 220 m.p.h. at 20,000 feet, 296 m.p.h. at S.L., 354 m.p.h. on F.S. gear at 18,500 feet. Carried 4 × 20 mm. cannon.

N.F. Mk. XIV (Merlin 67, 72) Two-stage Merlin Mosquito based on Mk. XIII. Not proceeded with, superseded by N.F. XIX and N.F. 30.

N.F. Mk. XV (Merlin 61, 73, 77) High-altitude fighter with two-stage Merlin engines, and A.I. Mk. VIII. First flew 9.42 (see Chapter 5). *Fuel:* 335 gal. in op. state. *Weights:* (lb.) Recorded on MP469 original state: tare 14,430, 21,602 operational loaded, 22,485 with pressure cabin fitted. MP469 as N.F. XV with 4 × ·303 inch guns in under-belly pack 13,768 tare, 16,661 normal, 17,395 a.u.w. Recorded on DZ385: tare 13,651, loaded 16,658, a.u.w. 17,391. Tare (4-bladed airscrew), 13,746 lb. Wing span N.F. XV production type 59 feet.

P.R. Mk. XVI (Merlin 72/73, 76/77) Photo-reconnaissance aircraft with two-stage Merlins and pressure cabin. First flew July, 1943. *Fuel:* 860 gal., 539 gal. with useful load. *Weights:* (lb.) Standard a.u.w. 21,916, 23,630 with w × 100 gal. drop-tanks. With *Rebecca* and *Boozer* a.u.w. 22,244. With *H2X* 16,361 tare, 23,180 max permis. NS729 fitted with deck landing gear weighed 16,944, 22,207 a.u.w. (539 gal. and 2 × 50 gal. tanks), a.u.w. 23,630 with 2 × 100 gal.

Summary of Mosquito Variants

tanks, 21,425 without L.R. tanks. MM328 fitted with four-blade propellers for Welkin and 2×200 gal. tanks had a.u.w. 22,982 (21,426 without L.R. tanks).

B. Mk. XVI (Merlin 72/73, 76/77 Bomber with two-stage Merlins and pressure cabin. First flew 1.1.44. *Fuel:* 860 gal. max., 539 gal. with useful load (4×500 lb. bombs and 2×100 gal. drop tanks); 597 gal. max. fuel load or 497 gal. plus 4,000 lb. bomb and 2×50 gal. tanks. *Weights:* (lb.) a.u.w. with small bomb bay 23,646 (24,152 carrying 2×100 gal. tanks), a.u.w. with 6×500 lb. bombs on Avro carrier 24,030. With 4,000 lb. bomb: 14,901 tare, a.u.w. 25,412; a.u.w. with 2×100 gal. tanks 25,917. With *H2S* a.u.w. 24,064 when carrying 4×500 lb. bombs and *Boozer* with *Oboe* and 4,000 lb. bomb; a.u.w. 25,501 when also fitted with *Album Leaf, Fishpond, Boozer* or *Monica*. 23,888 lb. when fitted with *Oboe, Gee,* VHF and 4,000 lb. bomb bay only. No. 8 Group recommended for the 4,000 lb. bomb Mk. IX/XVI, on 20.1.44, maximum take-off weight as 25,200 lb. and landing weight of 20,500 lb., fuel capacity being 597 gal., operating from an 1,800-yard runway. Initial operating ceiling was set by this order at 28,500 feet, rising to 29,500 feet at target. Still-air range was set at 1,470 miles, max. op. range as 1,100 miles and op. radius at 550 miles. *Speeds:* (8 Group listing) 329 m.p.h. at S.L., max. after leaving target 333 m.p.h. S.L., max. 408 m.p.h. F.S. gear at 28,500 feet before target, 419 m.p.h. after, all with unshrouded exhaust machines (standard). Length overall 40 feet 6 inches. Declared obsolete 14.4.49.

N.F. Mk. XVII (Merlin 21, 23) Night-fighter fitted with SCR720/729 or A.I. Mk. X radar. First flew March, 1943. *Fuel:* 547 gal. max. load with 50 gal. tanks, 403 gal. with useful load. *Weight:* (lb.) HK195/G: tare 13,526, a.u.w. 19,220 (20,393 with 2×50 gal. tanks); HK324/G with A.I. X and backward-looking radar and perspex tail cone: 13,546 tare, 18,720 loaded; 4×20 mm. cannon fitted, with 500 rounds.

F.B. Mk. XVIII (Merlin 25) Ground attack and anti-shipping fighter-bomber. First flew 8.6.43. *Fuel:* 668 gal. max., 403 gal. with useful load. Long-range fuselage tank for 65 gal. *Weights:* (lb.) HJ732: 18,160 lb. (as Mk. VI) loaded and 21,304 a.u.w. as Mk. XVIII. HX902: tare 14,756, loaded 21,257, a.u.w. (2×100 gal. tanks) 23,274, a.u.w. (with 8 R.Ps. 60 lb.) 22,255 lb. Molins 6-pounder 57 mm. gun weighed 1,580 lb.; 25 rounds carried. 4×·303 inch guns retained.

N.F. Mk. XIX (Merlin 25) Night-fighter with A.I. VIII or X/SCR720 and 729 in 'thimble' or Universal Nose, pased on N.F. XIII. First flew April, 1944. *Fuel:* 716 gal. max. with drop tanks, 453 gal. with useful load. *Weights:* (lb.) MM624: tare 14,471, loaded 20,420, with 2×100 gal. tanks 21,728, max. permis. 21,750 lb. With paddle-blade propellers: 14,598 tare, 20,547 loaded, 22,620 fitted with 2×100 gal. tanks. Official trials revealed top speed of 378 m.p.h. at 13,200 feet with op. load. Typical variations in Mosquito weights are evidenced by these tare weights for XIXs: TA404 15,291, TA435 15,377 and TA447 15,550. Declared obsolete 13.5.48.

B. Mk. XX (Packard Merlin 31, 33) Canadian version of B. IVii with similar weights, loads etc. *Fuel:* 860 max. gal., 539 with useful load. Loaded weight 21,980 lb. Declared obsolete 29.8.46.

Summary of Mosquito Variants

F.B. Mk. 21 (Packard Merlin 31) Canadian version of F.B. VI superseded by F.B. 26.

T. Mk. 22 (Packard Merlin 33) Canadian version of T. Mk. III, developed from F.B. 21.

B. Mk. 23 (Packard Merlin 69) Projected Canadian bomber comparable to B. Mk. IX.

F.B. Mk. 24 (Packard Merlin 301) High-altitude fighter-bomber based on F.B. 21, not proceeded with.

B. Mk. 25 (Packard Merlin 225) Revision of B. XX with improved single-stage Merlins; 2,000 lb. bomb load.

F.B. Mk. 26 (Packard Merlin 225) Canadian version of F.B. VI fighter-bomber, revision of F.B. 21. Weights and loads similar to F.B. VI. a.u.w. 21,473 lb.

T. Mk. 27 (Packard Merlin 225) Dual-control trainer developed from T. 22. with improved engines.

Mk. 28 Allocated to Canada, not taken up.

T. Mk. 29 (Packard Merlin 225) Dual-control trainer developed from F.B. 26. Declared obsolete 8.5.47. All were conversions from F.B. 26s.

N.F. Mk. 30 (Merlin 72 or 76) Development of N.F. XIX with two-stage Merlins. A.I. Mk. X. First flew March, 1944. *Fuel:* 716 gal., 453 with useful load. *Weights:* (lb.) MM686: tare 15,241, loaded 21,715, loaded (with 2×50 gal. tanks) 22,510, loaded (2×100 gal. tanks) 23,275. Loaded, with N_2O installation 21,913; loaded with *Perfectos* and 2×100 gal. lb. tanks 23,650; loaded when tropicalized and carrying 2×250 lb. bombs 22,731. Loaded, desert equipment and 2×100 gal. tanks 23,496 lb. First Mk. 30 trial installations aircraft was HK364 (SCR720 radar) which weighed: tare 15,156, normal loaded 21,105, with 2×50 gal. tanks 22,413 lb. Performance tests cf. Mk. XIII were conducted with a Mk. 30 at A.A.E.E. revealing: still-air range at recommended cruising speed 1,180 miles at 30,000 feet; still-air range continuous cruising F.S. gear 1,010 miles, max. speed F.S. gear 424 m.p.h. at 26,500 feet, 400 m.p.h. at 13,500 feet, 338 m.p.h. at S.L. Initial climb 2,250 f.p.m., operational ceiling 35,000 feet; cruising speeds 220 m.p.h. at 30,000 feet, 288 m.p.h. at S.L., 380 m.p.h. in F.S. at 30,500 feet. Length overall 41 feet 4 inches.

N.F. Mk. 31 (Packard Merlin 69) N.F. 30 with American Merlins. Project only.

P.R. Mk. 32 (Merlin 113/114) High-altitude photo-reconnaissance aircraft with two-stage Merlins, pressure cabin and specially lightened and extended wing tips (span 59 feet 2 inches). Based on P.R. XVI. First flew August, 1944. *Fuel loads:* Outer tanks 116 gal., inner tanks 287 gal., centre tanks 136 gal., fuselage auxiliary tank 121 gal.—total 660 gal. *Weights:* (lb.) Recorded on NS586:

Summary of Mosquito Variants

	2 × 50 gal. tanks	2 × 100 gal. tanks
Tare	14,281	14,281
Service load	1,227	1,272
Fuel	5,476	6,196
Oil	373	373
Flying weight	21,357	22,122

NS587: tare 14,247, a.u.w. 21,323 (with 2 × 50 gal. tanks), 22,122 (with 2 × 100 gal. tanks). Max. loaded 20,528, max. overload 21,323, max. landing weight 20,500 lb. Length 40 feet 6 inches.

T.F./T.R. 33 (Merlin 25) Torpedo-reconnaissance fighter/fighter-bomber for carrier operations. On 10.6.43 it was agreed to be just possible to meet Spec. S. 11/43 with the Mosquito, although it would need a larger fuselage to accommodate a third crew member. Max. speed would be 350 m.p.h., cruising 260 m.p.h. Span would need to be about 60 feet, and performance suffer. On 14.7.43, a carrier fighter (A.I. VIII) was considered, and a naval escort fighter. Ultimately a deck-landing version of the Mk. VI emerged as the T.R. 33, to Spec. N. 15/44 with LR387 as prototype. The T.R.33 had upward folding wings, arrester hook, four-blade propellers, oleo-pneumatic landing gear in place of rubber-in-compression, American ASH radar, JATO gear, four 20 mm. cannon and provision for under-belly 2,000 lb. Mk. XV or XVII torpedo or bomb or mine. 2 × 500 lb. bombs could alternatively be carried in rear bomb bay, and wing loads were as for the Mk. VI. Length 42 feet 3 inches. LR359 with A-frame deck-landing hook, reinforced fuselage, increased power and four-bladed propellers but with fixed wings was the first Mosquito to deck-land, doing so on H.M.S. *Indefatigable* on 25.3.44.

P.R. Mk. 34 (Merlin 114) and **P.R. 34A** (Merlin 114A). Very long-range reconnaissance two-stage Merlin aircraft. *Fuel:* 1,269 gal. incl. 2 × 200 gal. drop tanks. 1,192 gal. in belly tank. *Weights:* 25,000 a.u.w. carrying 1,226 gal. Speed reduced to 250 I.A.S. when carrying 200 gal. tanks. Max. permis. 25,500 lb. All armour and tank bullet proofing removed to give 3,000 feet ceiling increase. Swollen belly cut max. speed by 6 m.p.h. T.A.S.

B. Mk. 35 (Merlin 113A/114A) Ultimate Mosquito bomber. *Fuel:* 597 gal. including 2 × 50 gal. drop tanks. *Weights:* 25,200 a.u.w. with 597 gal. and 4,000 lb. bomb. Almost exclusively served postwar, in BAFO, 2nd T.A.F. and Nos. 109 and 139 Squadrons until replaced by the Canberra. Conversions to target tug (T.T. 35) and special conversions to P.R. Mk. 35 for special reconnaissance.

N.F. Mk. 36 (Merlin 113/114, 113A/114A) Night-fighter with the later type Merlins and later A.I. radar, equipping post-war squadrons until 1953.

T.F. Mk. 37 (Merlin 25) Torpedo-fighter/bomber with ASV. Mk. XIII otherwise similar to Mk. 33. First flew 1946. Max. speed 345 m.p.h. at S.L., 383 m.p.h. at 20,000. Normal still-air range (405 internal gal.) 1,100 miles.

N.F. Mk. 38 (Merlin 114A) Mk. 36 with later engines, some fitted with

Summary of Mosquito Variants

A.I. Mk. IX. Not used by front line R.A.F. squadrons, most sold to Yugoslavia. *Weights:* (lb.) tare 18,229, loaded 21,400. Wing loading 47 lb./sq. ft. Max. speed 404 m.p.h. at 30,000 feet. Op. ceiling 36,000 feet. Four 20 mm cannon. Overall length 41 feet 5½ inches.

T.T. Mk. 39 (Merlin 72/73) Target tug conversion of Mk. XVI for Royal Navy to Specification Q.19/45 to replace Miles Monitor. Lengthened fuselage, to 43 feet 4 inches overall. *Weights:* (lb.) tare 15,980, 21,500 normal loaded, max. overload 23,000 lb. Max towing speed with 32 feet span target 279 m.p.h., with 16 feet span target 292 m.p.h., with M3 sleeve target 283 m.p.h., with M4 sleeve 300 m.p.h. Take-off to 50 feet (still air max. normal wt.) 767 yards. Home and overseas service, conversion work by General Aircraft Ltd.

F.B. Mk. 40 (Packard Merlin 31, 33) Fighter-bomber built in Australia similar to British F.B. VI. Conversions to P.R. 40.

P.R. Mk. 41 (Packard Merlin 69) Two-stage Merlin reconnaissance aircraft built in Australia bearing relationship to Mk. IX and 40.

F.B. Mk. 42 (Packard Merlin 69) Two-stage Merlin fighter-bomber, prototype only.

T. Mk. 43 (Packard Merlin 33) Trainer similar to T. Mk. III.

APPENDIX 5

Mosquito Operational Performance and Loads

Versions fitted with Merlin 21, 22, 23, 31

CLIMB: Recommended climb 2,650 R.P.M.+4 lb. boost at 170 I.A.S. Bombers reached 20,000 in about 22 minutes over 75 miles on 40 gallons of fuel.
Maximum rate of climb on 2,850 revs. +9 lb. boost had no effect on range, but put strain on the engines.
CEILING: Operational ceiling of the fully loaded bomber was about 27,000 feet, homeward journey ceiling about 30,000.
CRUISE: Recommended cruise speed was 220 I.A.S. to about 25,000 feet outward and 210 I.A.S. at 30,000 homeward. Air miles per gallon were about 3·1 outwards and 3·4 homewards. A.M.P.G. at 5,000 feet: 2·8, at 10,000 2·9, at 15,000 3·1.

Approximate I.A.S. for maximum continuous cruise:

Altitude ft.	I.A.S. m.p.h.	T.A.S. m.p.h.	A.M.P.G.
S.L.	285	283	2·1
10,000	280	322	2·4
20,000	252	341	2·7
25,000	225	329	3·0

RANGE: At economical cruising:

Altitude ft.	Bomber/P.R.: 540-gallon load miles	Fighter 410-gallon load miles*
S.L.	1,040	780
5,000	1,110	830
20,000	1,220	900
25,000	1,210	890

*The above figures are for clean condition, without wing drop tanks etc. Mk. VI with 100-gallon drop tanks (650-gallon fuel load) had range of 1,325 miles at S.L.

At high speed cruise:

Altitude ft.	Bomber/P.R.: 540-gallon load 7 lb. boost miles	14 lb. boost miles	Fighter 410-gallon load 7 lb. boost miles	14 lb. boost miles
S.L.	805	700	610	530
5,000	860	750	645	560
20,000	990	860	735	640
25,000	—	1,110	—	820

Mosquito Operational Performance and Loads

These figures take into account using 20 gallons fuel for warm-up and taxiing, M gallons for climb to 5,000 feet over 12 miles, or 40 gallons to 20,000 feet over 75 miles or 52 gallons for climb to 25,000 feet over 100 miles.

Approximate stalling speed clean, m.p.h.:

	20,000 lb.	17,000 lb.
Flaps up	130 I.A.S.	120 I.A.S.
Flaps down	112 I.A.S.	103 I.A.S.

Maximum level speed: Accurate top speed tests showed a variance of about 8 m.p.h.

Approximate maximum speeds (clean state), m.p.h. were:

17,000 lb. weight M.S. gear—10,000 feet—2,650 R.P.M. plus 7 lb. boost
280 I.A.S.

17,000 lb. weight F.S. gear—20,000 feet—2,650 R.P.M. plus 7 lb. boost
252 I.A.S.

Multiple ejector exhausts increased maximum level speed by about 12 m.p.h. T.A.S. Under-wing bomb reduced speed by about 15 m.p.h. T.A.S. 19,000 lb. weight, Merlin 25s, multiple exhausts gave the T.R. Mk. 33 a top speed of 389 T.A.S. at 14,500 feet.

19,000 lb. weight, Merlin 25s, manifold exhausts gave the F.B. Mk. VI a top speed of 378 T.A.S., at 13,000 feet.

Versions fitted with Merlin 72/73, 76/77. (Mks. VIII, IX, XVI)

CLIMB: Reached 30,000 feet in about 50 minutes over 180 miles using 180 gallons.

CEILING: Operational ceiling of the loaded bomber versions was about 31,000 feet outwards, 36,000 homewards.

CRUISING: Outward recommended speed was 220 m.p.h. I.A.S. and 210 I.A.S.

Economical cruise speeds (clean condition) were:

Altitude ft.	A.M.P.G.	T.A.S.	FUEL CONSUMPTION gallons per hour
OUT:			
20,000	3	295	98
25,000	2·8	320	114
30,000 plus	2·6	350	135
HOME:			
20,000	3·2	280	88
25,000	3·1	305	98
30,000 plus	2·9	335	115

Maximum continuous cruising (clean condition):

Altitude ft.	I.A.S.	T.A.S.	A.M.P.G.
S.L.	287	285	2·1
10,000	276	316	2·3
20,000	260	349	2·4
30,000	236	378	2·5

Mosquito Operational Performance and Loads

RANGE: At economical cruising speeds:

Altitude ft.	Bomber (540 gallons fuel) miles	P.R. (760 gallons) miles
S.L.	1,080	1,540
10,000	1,120	1,620
25,000	1,070	1,580
30,000	1,000	1,500

At high speed cruise:

Altitude ft.	Bomber (540 gallons fuel) miles	P.R. (760 gallons) miles
S.L.	820	1,170
10,000	880	1,260
25,000	890	1,300
30,000	895	1,320

Approximate stalling speeds clean, m.p.h.:

	22,000 lb.	18,500 lb.
Flaps up	137 I.A.S.	125 I.A.S.
Flaps down	118 I.A.S.	107 I.A.S.

Maximum level speed:

19,000 lb. weight M.S. gear—10,000 feet—2,650 R.P.M. plus 7 lb. boost
278 I.A.S.
19,000 lb. weight F.S. gear—25,000 feet—2,650 R.P.M. plus 7 lb. boost
242 I.A.S.

Pressurisation: The Westland pressure valve began to build up pressure at 15,000 feet, to 2 lb./sq. in. at 30,000 feet and above. This reduced pressure in the cabin at 35,000 feet to the level at 10,000 feet.

APPENDIX 6

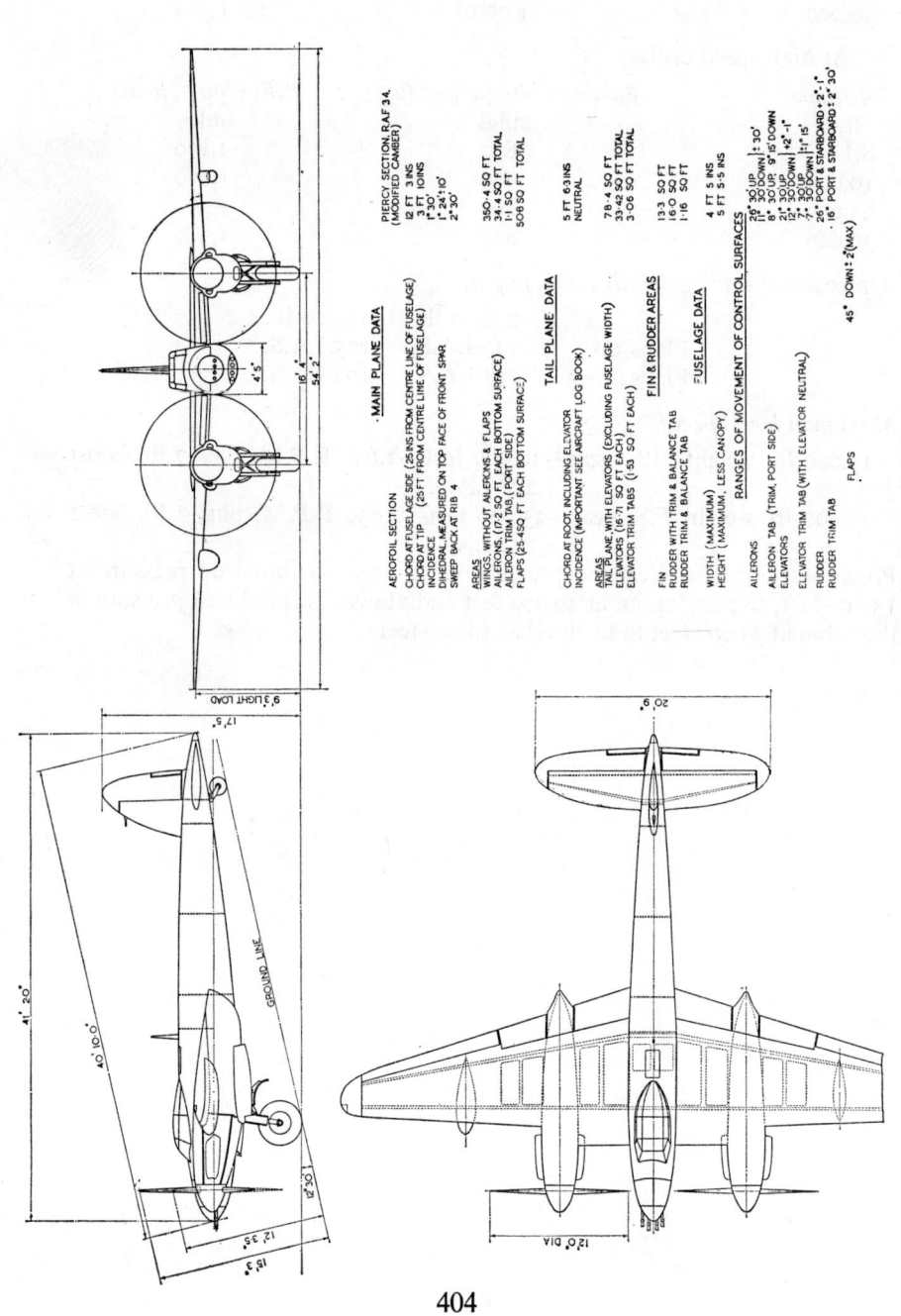

MAIN PLANE DATA

AEROFOIL SECTION — PIERCY SECTION, RAF 34. (MODIFIED CAMBER)
CHORD AT FUSELAGE SIDE (25 INS FROM CENTRE LINE OF FUSELAGE) — 12 FT 3 INS
CHORD AT TIP (25 FT FROM CENTRE LINE OF FUSELAGE) — 3 FT 10 INS
INCIDENCE — 1° 30'
DIHEDRAL, MEASURED ON TOP FACE OF FRONT SPAR — 1° 24' 10"
SWEEP BACK AT RIB 4 — 2° 30'

AREAS
WINGS, WITHOUT AILERONS & FLAPS — 350·4 SQ FT
AILERONS, (17·2 SQ FT EACH BOTTOM SURFACE) — 34·4 SQ FT TOTAL
AILERON TRIM TAB, (PORT SIDE) — 1·1 SQ FT
FLAPS (25·4 SQ FT EACH BOTTOM SURFACE) — 50·8 SQ FT TOTAL

TAIL PLANE DATA

CHORD AT ROOT, INCLUDING ELEVATOR — 5 FT 6·3 INS
INCIDENCE (IMPORTANT SEE AIRCRAFT LOG BOOK) — NEUTRAL

AREAS
TAIL PLANE, WITH ELEVATORS (EXCLUDING FUSELAGE WIDTH) — 78·4 SQ FT
ELEVATORS (16·71 SQ FT EACH) — 33·42 SQ FT TOTAL
ELEVATOR TRIM TABS (1·53 SQ FT EACH) — 3·06 SQ FT TOTAL

FIN & RUDDER AREAS

FIN — 13·3 SQ FT
RUDDER WITH TRIM & BALANCE TAB — 16·0 SQ FT
RUDDER TRIM & BALANCE TAB — 1·16 SQ FT

FUSELAGE DATA

WIDTH (MAXIMUM) — 4 FT 5 INS
HEIGHT (MAXIMUM, LESS CANOPY) — 5 FT 5·5 INS

RANGES OF MOVEMENT OF CONTROL SURFACES

AILERONS — 26° 30' UP ± 30'
                    11° 30' DOWN ± 30'
AILERON TAB (TRIM, PORT SIDE) — 8° 30' UP, 9° 15' DOWN
ELEVATORS — 21° 30' UP +2° -1°
                    12° 30' DOWN ± 15'
ELEVATOR TRIM TAB (WITH ELEVATOR NEUTRAL) — 7° 30' UP
                    7° 30' DOWN ± 15'
RUDDER — 26° PORT & STARBOARD ± 2° · 1'
RUDDER TRIM TAB — 16° PORT & STARBOARD ± 2° 30'
FLAPS — 45° DOWN ± 2° (MAX)

GENERAL ARRANGEMENT MK VI

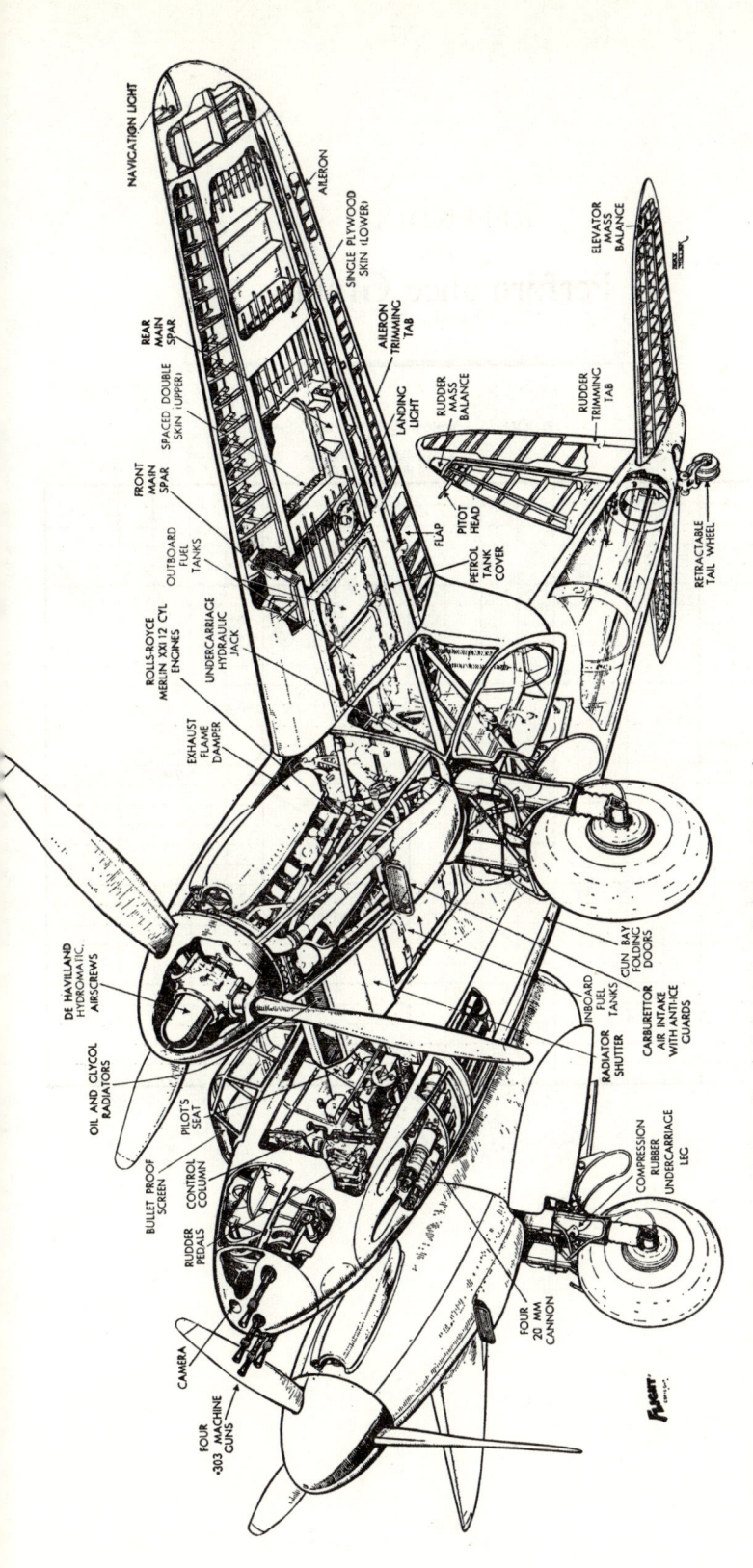

A cut-away drawing of a standard F. Mk. II Mosquito fighter prepared for *Flight International* by Mr. M. A. Millar in 1943.

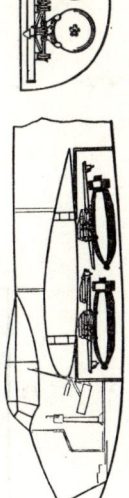

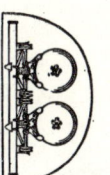

Side and front elevations of the bomb bay of the B. Mk. IV, showing four 500 lb. bombs being carried.

APPENDIX 7
Performance Graphs

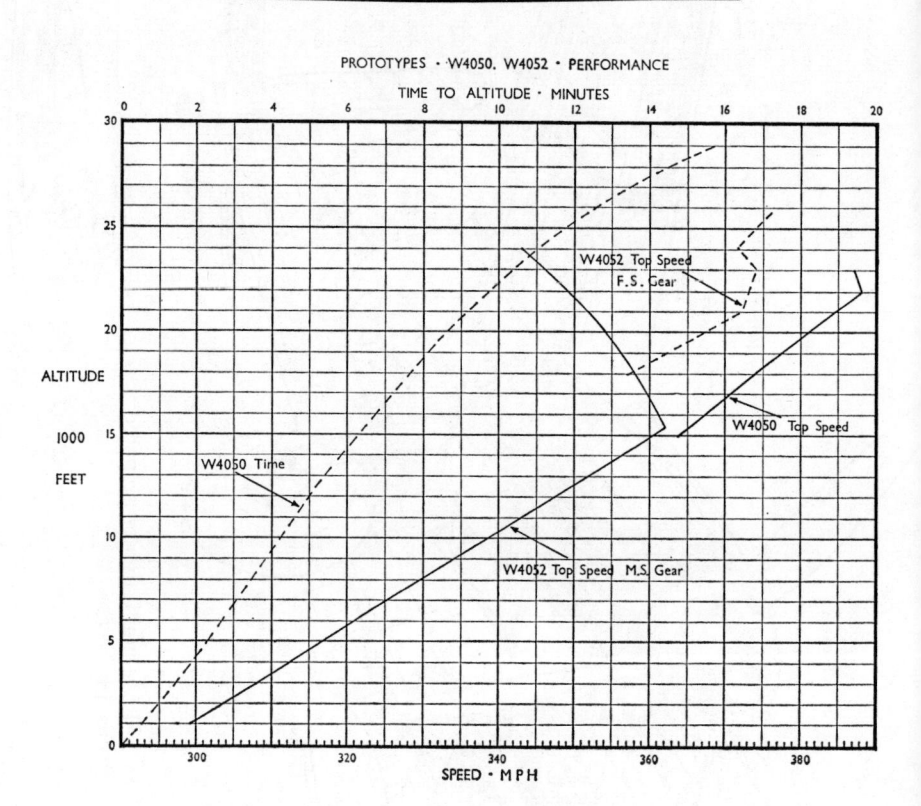

Performance Graphs

B.MK.IV - DK290 - PERFORMANCE

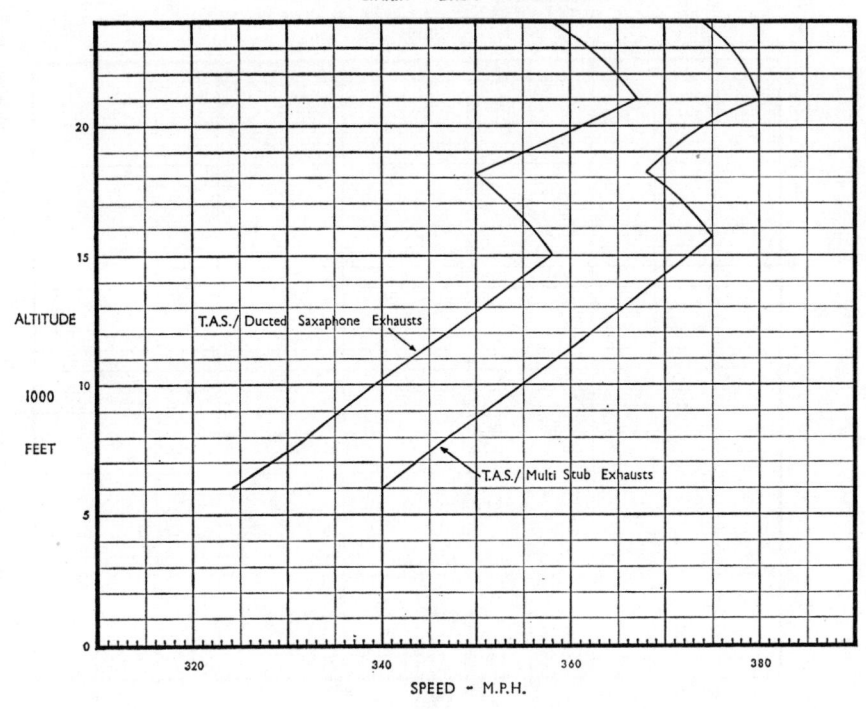

MAXIMUM SPEEDS

COMPARISON OF MAX. LEVEL SPEEDS MOSQUITO B.IV AND FW 190 A-3

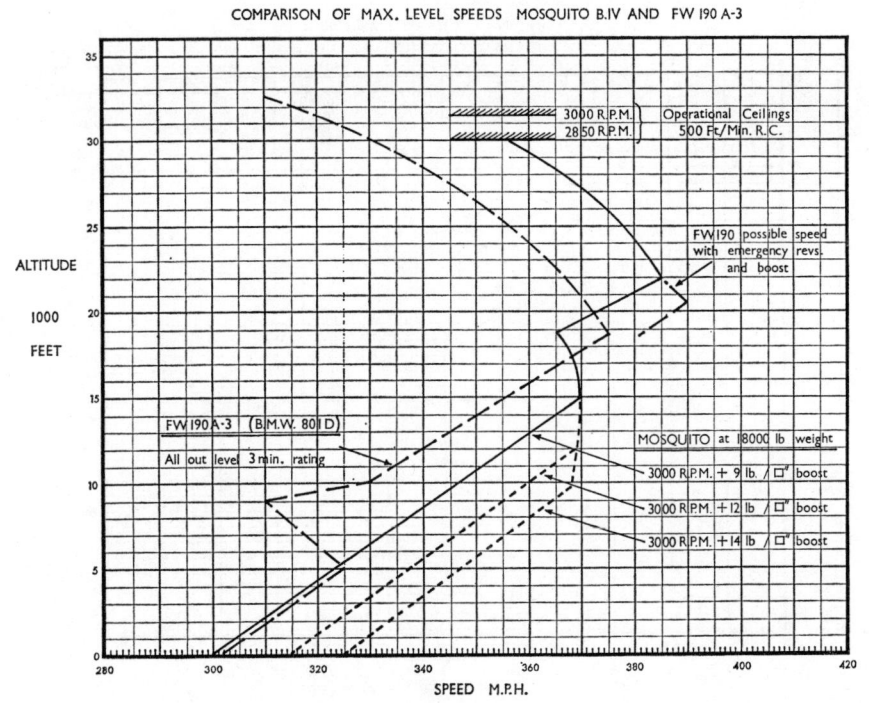

Performance Graphs

FB MK VI · HJ679, HX802 · PERFORMANCE WITH DROP TANKS. HJ679 - 20 000 LBS. MEAN WEIGHT

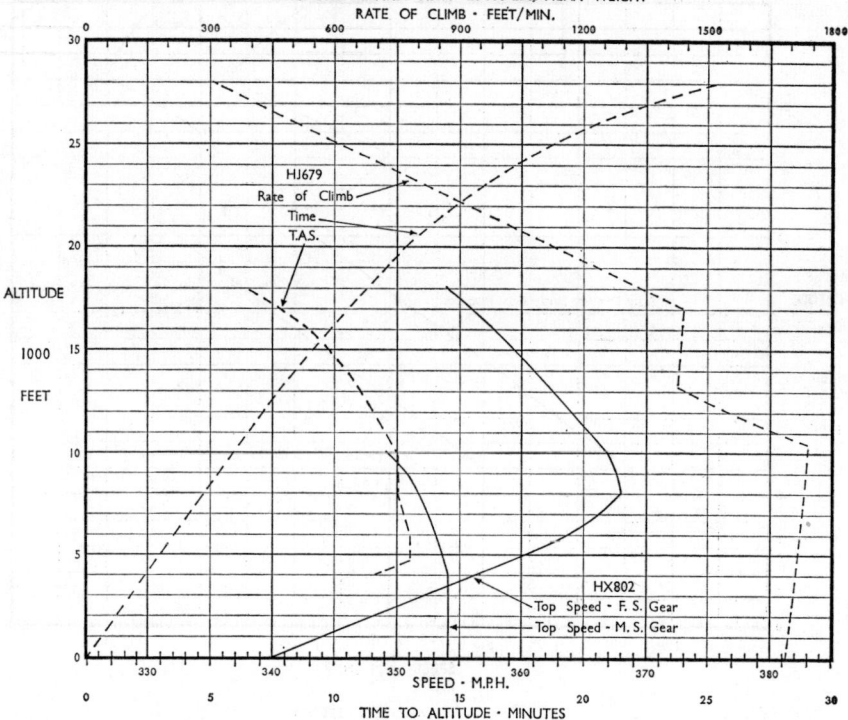

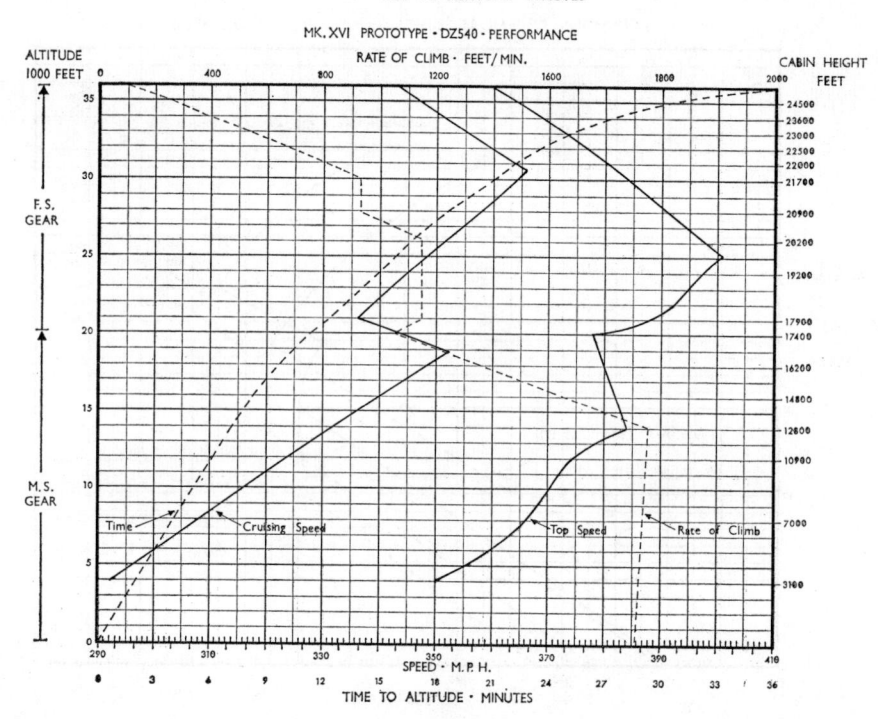

MK. XVI PROTOTYPE · DZ540 · PERFORMANCE

APPENDIX 8

General Statement of Weights, Armament, Bomb Load and Performance of Bomber, Fighter-Bomber and Torpedo-Reconnaissance Variants

N.B. The weights, armament, bomb load and maximum dive speed quoted in this note are those at which the Mosquito was limited for peace-time flying in the R.A.F. Under war-time conditions these aircraft flew at very much higher all-up weights and hence were able to carry a much larger quantity of fuel and greater bomb load. It was considered, that these high weights were not sufficiently safe for peace-time operations.

Type	Bomber	Fighter Bomber	Torpedo Reconnaissance
Altitude	High	Low	Low
Machine Mk. No.	35	6	33
Power unit Merlin engine	114	25	25
Maximum all-up weight lb.	22,000	21,700	21,000
Span	54 ft. 2 in.	54 ft. 2 in.	54 ft. 2 in.
Gross wing area, sq. ft.	450	450	450
Crew	2	2	2
Bomb load, total lb.	1,500	1,500	2,000 or 2,000 lb. torpedo
Underwing equipment	2 × 500 lb. bomb	2 × 500 lb. bomb	2 × 1,000 lb. bomb
20 mm. guns	Nil	2	4
Ammunition for 20 mm. guns	Nil	300 rounds	600 rounds
0·303 in. Browning guns	Nil	4	Nil
Ammunition for 0·303 in. guns	Nil	2,000 rounds	Nil
Fuel total gal.	539	453	405
Oil total gal.	31	31	31
Dive speed m.p.h. A.S.I.	400	400	400
Span loading lb./sq. ft.	7·5	7·4	7·1
Wing loading lb./sq. ft.	49	48	47
Still-air range at recommended cruising speed miles	1,600	1,120	930
Altitude for above ft.	25,000	Sea level	Sea level
Speed for above m.p.h. T.A.S.	300	250	240
Still-air range at maximum continuous cruising speed miles	1,250	960	825

General Statement of Weights

Type	Bomber	Fighter Bomber	Torpedo Reconnaissance
Altitude for above ft.	37,000	Sea level	Sea level
Speed for above m.p.h. T.A.S.	375	296	300
Patrol duration hours	—	4·8	4.2
Altitude for above ft.	—	Sea level	Sea level
Maximum speed in F.S. gear m.p.h.	425	378	387
Altitude for above ft.	30,500	13,200	13,500
Maximum speed in M.S. gear m.p.h.	410	361	372
Altitude for above ft.	17,500	5,500	5,500
Maximum speed at sea level m.p.h.	326	336	344
Initial rate of climb at sea level at full climb power ft./min.	2,500	1,870	1,870
Operational ceiling at 3,000 r.p.m. (500 ft./in. rate of climb) ft.	37,000	26,000	26,000
Take-off distance to 50 ft. screen from rest in still-air at maximum weight yd.	980	900	600

General Statement of Weights

MOSQUITO 6, 33 AND 35

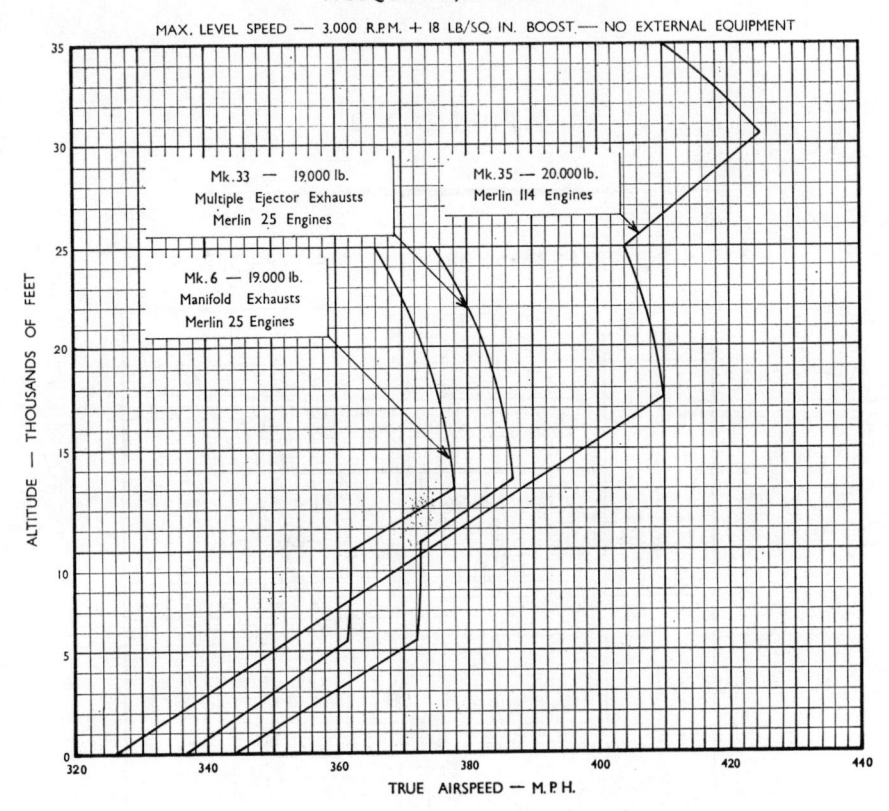

MAX. LEVEL SPEED — 3,000 R.P.M. + 18 LB/SQ. IN. BOOST.— NO EXTERNAL EQUIPMENT

Mk.33 — 19,000 lb.
Multiple Ejector Exhausts
Merlin 25 Engines

Mk.35 — 20,000 lb.
Merlin 114 Engines

Mk.6 — 19,000 lb.
Manifold Exhausts
Merlin 25 Engines

ALTITUDE — THOUSANDS OF FEET

TRUE AIRSPEED — M.P.H.

APPENDIX 9

ROLLS-ROYCE MERLIN PROGRESSIVE IMPROVEMENT

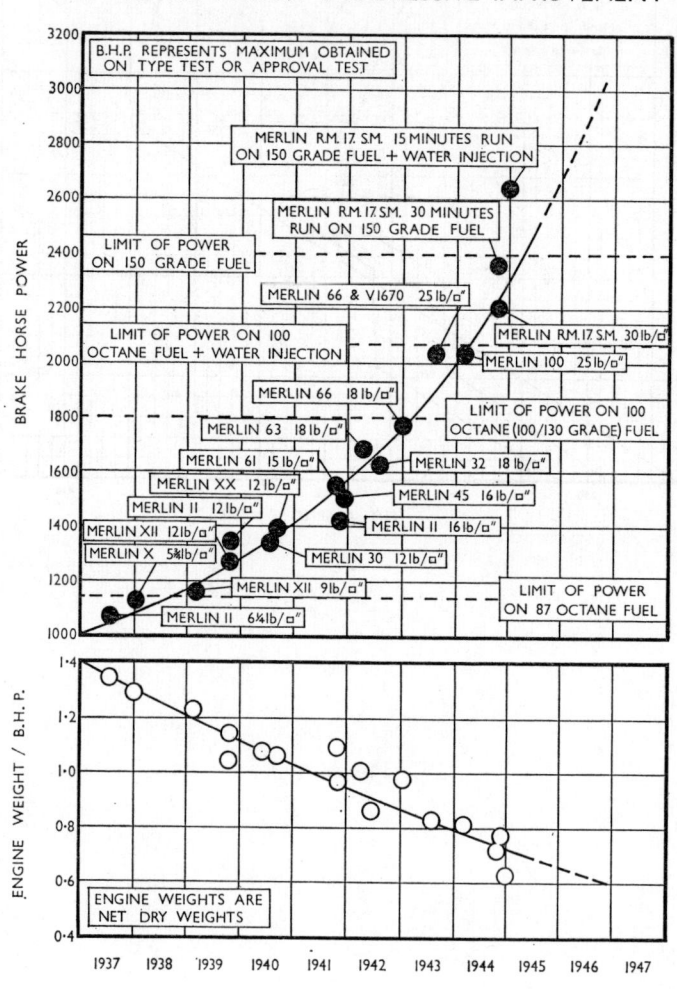

APPENDIX 10

Notable Mosquitoes

The yardstick measuring meritorious service by individual Mosquitoes is much controlled by role. Success favouring one aeroplane often depended upon chance, although crew skill must naturally not be overlooked. Extensive search through relevant records has revealed the following Mosquitoes to have had outstanding careers.

Fighters

Claims to the destruction of enemy aircraft appearing in this book, have, wherever possible, been checked against Official British and German records. Such cross-checking of claims and losses is extremely difficult, but on the available evidence Mosquito NF. XIII MM466 would seem to be top scorer. Delivered to 10 M.U. 11.2.44 it reached 488 Squadron 25.2.44 becoming ME-R. Its first patrol, from 02.55 to 04.00 hours, came on 1.3.44. Many followed, but not until 29/30.7.44 did Flt. Lt. G. E. Jameson, D.F.C., with Flg. Off. A. N. Crookes, D.F.C., as navigator bring success to '466. About thirty bombers set out to raid Allied lines around Caen, among them Ju 88 A4 B3+BH of 1/K.G.54 which set forth from Juvincourt. It became a victim of '466 and crashed 4 miles from St. Lo. Jameson chased another Ju 88 blazing into cloud and destroyed another Ju 88 south of Lisieux. A Do 217 he sent spinning into the Channel—and all within 20 minutes. On 3 August two Ju 88s were shot down, near St. Lo and east of Avranches. Flt. Lt. P. F. Hall destroyed another, on 15 August, 25 miles south of Caen. MM466 passed to 409 Squadron on 9.11.44. Its first sortie in new hands was made on 18.11.44 by Flg. Off. R. E. Britten and Flt. Lt. L. E. Fownes, who on 27 December destroyed two Ju 88 Gs. Flg. Off. M. G. Kent claimed another over the Scheldt Estuary on 23 January, and a fourth on 3 February. MM466's eleventh victory came on 21.3.45 when Britten and Fownes claimed a Bf 110 over the Ruhr. Its last patrol came late on 2.5.45. 84 Group Support Unit received '466, 6.7.45 and it was SOC 10.9.45.

Second highest scorer was almost certainly Mk. XVII HK286/G delivered to 27 M.U. 24.7.43, passed to Marshalls for conversion 12.9.43 and eventually delivered to 456 Squadron 29.1.44 where it became RX-A. Its defensive patrols started 20.2.44 flown, as was to become usual, by Wg. Cdr. K. M. Hampshire, D.S.O. On its fourth patrol it claimed a Ju 88 which fell in the sea off Beer, Devonshire. Hampshire then shot down a Ju 88 near Isle Brewer, which also had been detailed to attack Bristol. Scrambles and patrols followed until 24.3.44, when Wg. Cdr. Hampshire intercepted Ju 88 3E+AP of 6/K.G.6 which, with about five others, had left Le Culot to attack

Notable Mosquitoes

Government buildings in London. Instead its shattered remains landed at Walberton, near Arundel railway station. Eight sorties later, on 28/29.4.44, a probable Do 217 was claimed and next night HK286 engaged two Me 410s without success. Patrolling off Portsmouth, Hampshire claimed a Ju 88 on 22.5.44. Over Normandy's beaches he took HK286 on 5/6.6.44 and next night destroyed an He 177 stalking shipping off Cherbourg. Hampshire's final success with 286 came on 12/13.6.44, a Ju 88 shot down in Caen Bay. HK286 left when the squadron received Mk. 30s, reaching 51 O.T.U. 4.1.45. On 1.7.45 it went to Hatfield for an overhaul never completed.

MM671 delivered 13.5.44 to 218 M.U. and another NF. XIX, TA404, were next highest scorers each credited with 6 E/A. MM671 arrived on 157 Squadron 27.5.44, became RS-C, made its first sortie on 14/15.6.44 and claimed a Ju 88 at Juvincourt. Next night another, at Creil, fell to its guns. It claimed a V-1 on 3/4.8.44, a Ju 88 on 4.12.44, a Bf 110 (which crashed in Limburg) and a Ju 88 15 miles south-west of Giessen on 6.12.44, and on 24.12.44 chased a Ju 88 from Bonn to within three miles of Cologne before destroying it. TA404:RS-K arrived on 157 Squadron on 17.9.44. It first operated 4/5.10.44 on a low-level intruder to Grove. Its recorded scores were a Ju 88 north of Mannheim on 19.10.44, Bf 110 on 6.11.44, Ju 88 22.12.44 and another 10 miles N.E. of Koblenz next night and two Bf 110s the following night. It failed to return from the Kamen area 3/4.3.45, by which time it was 'RS-M'.

Mosquito II W4097 was credited with five E/A, likewise Mk. XIII MM517. W4097 claimed two victories with 151 Squadron, two Bf 110s on 7/8.7.44 in the hands of Sqn. Ldr. J. S. Booth, D.F.C., of 239 Squadron and a Ju 88 near Stettin on 29/30.8.44. Four of MM517's scores came to it when in the hands of 604 Squadron (a Ju 88 north of Le Havre on 3.7.44, Ju 188 east of Falaise on 7/8.8.44 and a Ju 88 near Conde soon after, and another '88 near the Dutch–German border in January, 1945). After transfer to 409 Squadron on 19.4.45 it destroyed a Ju 290 on 24.4.45.

Mosquitoes claiming four E/A were MM589 (409 Sqn.), HK255 (25 Sqn.), HK323 and MM499. Twenty-one laid claims to three.

Photographic reconnaissance

The measure of noteworthiness used here is the number of sorties flown. Not always were these successful, due to cloud, etc. Ignoring such misfortunes LR422 of 540 Squadron with sixty-nine sorties credited seems to have been the most successful machine, closely rivalled by LR423 of 544 Squadron (67 sorties) and LR417 of 544 Squadron (64 sorties). LR423 began operations 9.11.43 flying to the Franco–Spanish border. LR417's first foray was to airfields around Berlin on 4.10.43, in the hands of Merifield and Walley. All experienced dangerous moments, exemplified when LR417 was intercepted over Utrecht on 6.10.43. Its last flight was to Heligoland on 29.11.44. LR422 on its first operation 16.8.43 to Mannheim, Friedrichshafen and Nürnberg faced flak at all three targets. LR415 operated from July, 1943, to February, 1945, made fifty-one sorties and LR426 flew fifty, both with 540 Squadron.

Entering service later Mk. XVIs flew fewer sorties each. MM358 of 540 Squadron made thirty-five between 6.7.44 (when it flew to Arras at 25,000 feet in the hands of Flt. Lt. W. K. Watson, dived to 2,000 feet to photograph the

Notable Mosquitoes

V-1 site at St. Pol Saintcourt, and returned home at ground level), and 25.4.45 when it secured photo coverage of Czech railways, and probably flew the most sorties of any P.R. Mk. XVI.

Day Bombers, May, 1942–31.5.44

Using the number of sorties flown as the fairest measure of success, DZ408: GB-F with 26 and two abortives credited would seem to be the most 'successful' of the day bombers. It served 105 Squadron until written-off after battle damage as 'GB-H' on 21.1.44, by which time it had flown 26 night sorties. DK338:GB-O, between 12.9.42 and 1.5.43, flew 23 sorties and aborted on two others. DK302:GB-D between 11.7.42 and 13.5.43 made 21 successful and 9 abortive sorties with 105 Squadron and '4 plus 1' with 139 Squadron. DZ467:GB-P flew 16 successful and 3 abortive sorties, DK337:GB-N 16 and 3 abortive and DZ464:XD-C 15 and one abortive before being shot down attacking Orleans engine sheds 21.5.43.

Night bombers

Many Mosquito night bombers flew more than 50 sorties with the L.N.S.F., but those with the outstanding records were found in the three P.F.F. marking squadrons. LR503:GB-F̄ credited with 213 sorties was foremost. 109 Squadron received it 28.5.43, passed it to 105 Squadron 10.3.44 and almost at the end of the war it went to Canada for a victory tour. Flt. Lt. Maurice Briggs, D.S.O., D.F.C., D.F.M., and Flg. Off. John Baker, D.F.C. and Bar, were selected to make the tour with what may well have been the most successful bomber of all time, and one whose exploits were certainly without equal. The end of the Mosquito and its crew, veterans of 107 sorties, was cruel and sudden, as the aircraft unexpectedly plunged into Calgary Airport on 10 May, 1945.

LR504 was delivered to 109 Squadron 31.5.43 and to 105 14.3.44. From 6.12.44 to 5.2.45 it was out of action due to battle damage, then resumed operations with 109 Squadron making 200 sorties in all. LR500, 503 and 504 all made their first sorties against Krefeld on 21.6.43 and operated almost continuously to the end of the war. Other outstanding 105 Squadron aircraft were LR507 (148 sorties), ML914 (148 sorties), ML922:Y (111 sorties), LR508:G (96 sorties).

Long serving Mk. IVs were mainly found with 109 Squadron. They included DK331:HS-D which crash landed at the end of its 100th sortie, on 18.4.44, and DZ319:HS-H which flew its 100th sortie on 8.5.44 and made 102 in all. Two other long service Mk. IVs were DZ433 (68 successful sorties) and DZ439. 139 Squadron had fewer long service machines because it frequently changed equipment. DZ601:B later 'Z' made 57 night attacks, DZ478:V 49 and DZ482:P 44.

Fighter-Bombers

Due to the nature of their operations—often low-level in daylight and against strongly defended targets—sortie totals for these aircraft were not as high as for bombers. Apart from heavy losses during Operation *Clarion*, however, the Mk. VIs operated with low losses after 1944. Records for some

Notable Mosquitoes

squadrons are incomplete and therefore comparison by sorties is difficult, although it is known that LR385 flew 104 sorties with 21 and 487 Squadrons. Outstanding Mk. VIs by length of service were:— *235 Squadron* HP982 (25.6.44 to 2.2.45), HP989 (30.6.44 to 14.1.45), HR129 (18.6.44 to 18.5.45). *333 Squadron* HP858 (KK:K later KK:O 17.9.43 to 1945), HP862 (KK:Q later KK:K 2.10.43 to 1945), HP910 (KK:L 28.2.44 to 10.4.45), HR262 (KK:N 28.6.44 to 11.3.45 FTR). *418 Squadron* HR195 (4.7.44 to 10.4.45), HX803 (16.7.43 to 10.5.44), HX819 (25.7.43 to 12.12.44). *464 Squadron* HX919 (9.9.43 to 14.9.44), HX920 (9.9.43 to 22.2.45 FTR).

APPENDIX 11

Versatile Mosquito

Mosquito DZ414 'O for Orange' won top marks for versatility, being familiar with every kind of Mosquito operation from nought to 25,000 feet. She covered, almost faultlessly, over 20,000 miles often in the hands of Sq. Ldr. C. E. S. Patterson, D.S.O., D.F.C., carried bombs, and usually a movie camera operated by Flg. Off. Lee Howard. Patterson chose her from one of three good Mosquitoes at Hatfield in December, 1942. Her first trip, early in 1943, took her to film the results of a raid on Lorient. Soon after she crossed the Alps, filming a raid on Turin for 50 minutes. Later she sped across Nuremberg at 7,000 feet during a raid on the city. Her first daylight trip was to Lorient, where she went at dawn. On Hitler's birthday she bombed Berlin by moonlight and was hit by flak, one fragment of which nearly severed an elevator wire. In a low-level attack on the power station at Thionville she flew for 800 miles over enemy territory to bomb and film workshops there. Although enemy fighters often pursued her they were never near enough to catch her. Orange lost herself in cloud during the Jena raid. Instead of wasting the effort made she bombed Weimer station and blew up a couple of goods trains. A spell of night bombing followed and, being one of the fastest Mosquitoes, she easily eluded an enemy fighter over Berlin. Although she encountered heavy flak over Wilhelmshaven at 8,000 ft. she amazingly remained untouched.

She left for Denain 20 minutes after a force of forty Bostons had taken off from the same airfield—and returned 10 minutes before them with her pictures. After filming an attack on a Breton power station she flew off to record a similar one elsewhere. Here a 20 mm. cannon shell burst on the fuselage side and would have injured the navigator had he not gone forward to operate the camera. It destroyed much of the canopy and the radio, but Orange flew on well and her pictures were fine.

Then followed a spell of photographing raids on V-1 sites about which she recorded much vitally useful material. She was over the beaches during the invasion and filmed the first fighters landing in Normandy, rushing the pictures home without landing in France. During the V-1 onslaught she twice followed two across the Channel with her camera firing. On all but two of her operations into late 1944 Patterson was her pilot.

APPENDIX 12

Salient Data Relating to German Aircraft Engaged by Mosquitoes

Type	Role	Engine(s)	Max. speed with normal op. load m.p.h./ft.	Service ceiling (feet)	Normal op. loaded wt. (lb.)
Do 217 E4	Bomber	2 × BMW 801C	305 at 18,000	21,500	33,000
Do 217 M1	Bomber	2 × DB 603A	348 at 18,700	24,000	36,817
He 111 H	Bomber	2 × Jumo 211F	255 at 16,000	26,000	25,000
He 177 A3	Bomber	2 × DB 610A/B-1	305 at 20,000	21,000	60,000
Ju 88 A4/14	Bomber	2 × Jumo 211J-1	295 at 14,000	24,200	30,000
Ju 88 S1	Bomber	2 × BMW 801D	370 at 26,000	35,000	24,000
Ju 188 E	Bomber	2 × BMW 801G-2	325 at 20,000	33,000	32,000
Me 410 A1	Bomber	2 × DB 603	395 at 22,000	34,000	23,500
Fw 190 A3	Fighter	1 × BMW 801Dg	385 at 19,000	36,000	9,750
Fw 190 D9	Fighter	1 × Jumo 213	435 at 37,000	39,000	8,160
He 219 A	Nt. Fighter	2 × DB 603E	385 at 21,000	30,500	28,000
Ju 88 C6	Nt. Fighter	2 × Jumo 211J	347 at 20,000	33,200	27,500
Me 109G	Fighter	1 × DB 605A	400 at 22,000	38,500	7,500
Me 110 G	Nt. Fighter	2 × DB 605B	360 at 20,000	34,800	19,500
Me 163 B	Rocket fighter	1 × HWK 109A-2	596 at 30,000	52,500	9,500
Me 262 A	Jet fighter	2 × Jumo 109-004B	540 at 20,000	37,500	14,100
Bv 222	L.R. flying boat	6 × BMW-Bramo 323R	210 at 6,500	26,000	102,000
Fw 200 C	Recco-bomber	4 × BMW-Bramo 323R	220 at 15,500	20,500	46,300
He 111 Z	Glider tug	5 × Jumo 211	275 at 19,500	?	63,000
Ju 290 A2	Recco-bomber	4 × BMW 801D	280 at 18,000	20,000	100,000

The above figures are included to afford comparison between German types and the Mosquito. Details refer to the usual operational configuration, but it must be remembered that many other factors would affect maximum speed and performance, and that individual aircraft performance was always variable. Trials at R.A.E. with a captured Fw 190 A3 in 1942, for instance, revealed its top speed as 380 m.p.h.

APPENDIX 13

Disposition, Losses and Disposals of Fighter and Reconnaissance Mosquitoes

Mark	Missing from operations	To Med. Air Cmd.	To S.E.A.C.	Sold to Belgium	Sold to Czecho-slovakia	Sold to France	Sold to R.N.Z.A.F.	To S.A.A.F.	To Australia	Sold to Sweden
PR. I	5	2	—	—	—	—	—	—	—	—
F. II	136	16	4	—	—	—	—	2	1	—
PR. IV	8	1	—	—	—	—	—	—	—	—
FB. VI	314	158	356	2	19	57	74	—	46	—
PR. VIII	1	—	—	—	—	—	—	—	—	—
PR. IX	14	3	37	—	—	—	—	—	—	—
NF. XII	10	18	—	—	—	—	—	—	—	—
NF. XIII	26	45	—	—	—	—	—	—	—	—
PR. XVI	6	63	55	—	—	29	—	14	29	—
NF. XVII	11	—	—	—	—	—	—	—	—	—
NF. XIX	12	—	151	—	—	—	—	—	—	46
NF. 30	25	21	1	24	—	23	—	—	—	—

Other foreign sales of the FB. VI were:—
    To Norway   22, and 4 T. Mk. III
    Dominican Republic   6
    Israel   about 60 and 5 PR. XVI
    Turkey   96 Mk. VI and 10 T. Mk. III
    Yugoslavia   46, and 60 NF. 38s, also 5 T. Mk. III

APPENDIX 14

Increase in Mosquito Distribution 1942–1943 in R.A.F. Units (Excluding MUs)

Unit	7.6.42	31.7.42	30.9.42	31.11.42	31.1.43	31.3.43	31.5.43
i) Recco. acft.							
1 P.R.U.	11	11	14	—	—	—	—
540 A Flt.	—	—	—	10	15	25	24
540 B Flt.	—	—	—	6	7	7	—
1477 Flt.	—	—	—	—	—	—	5
521 Sqn./ 1409 Flt.	2	1	2	5	8	9	9
8 O.T.U.	—	—	—	—	7	9	10
ii) Bombers							
105 Sqn.	9	16	19	20	23	20	19
109 Sqn.	—	—	8	8	20	23	31
139 Sqn.	—	—	3	5	10	16	14
618 Sqn.	—	—	—	—	—	—	12
1474 Flt./ 192 Sqn.	—	—	—	3	3	5	5
1655 C.U.	—	—	—	10	11	14	13
iii) B.O.A.C.	—	—	—	—	—	—	7
iv) Fighters							
25 Sqn.	—	—	—	17	19	22	28
29 Sqn.	—	—	—	—	—	—	4
85 Sqn.	—	17	20	19	19	27	20
151 Sqn.	19	21	18	20	22	26	24
157 Sqn.	19	21	23	27	22	27	24
256 Sqn.	—	—	—	—	—	—	19
264 Sqn.	16	23	23	25	20	24	27
307 Sqn.	—	—	—	—	21	22	23
410 Sqn.	—	—	—	6	20	25	26
456 Sqn.	—	—	—	3	16	19	26
51 O.T.U.	—	—	—	—	—	4	—
60 O.T.U.	—	—	—	—	—	—	10
v) Intruder/MAC							
23 Sqn.	1	13	15	23	18	18	18
418 Sqn.	—	—	—	—	—	3	6
605 Sqn.	—	—	—	—	—	18	18
TOTALS	77	123	145	207	281	363	422

Material within this survey has been taken from de Havilland records.

APPENDIX 15

Mosquito Build-up, 1942–1944

Item	30.11.42	30.6.43	31.12.43	29.2.44	1.10.44
Total U.K. production	415	1,010	1,685	1,959	3,804
Total in op. units	204	490[1]	685[2]	805[2]	1,355[3]
Total overseas	—[5]	—[5]	75[2]	100[2]	250[2]
Acft. totally destroyed	23	80	171	223	425
Missing on ops. (U.K.)	39	122	217	249	489
Damaged (Cat. A.C.)	63	272	468	590	—[5]
Damaged (Cat. B.)	21	92	161	188	—[5]
E/A claimed	20	85	192½	266	—[5]
E/A probables	9	28	40	52	—[5]
E/A damaged	9	45	84	110	—[5]
Total flying hours[4]	23,258	79,695	155,546	187,146	440,203

[1] includes aircraft overseas
[2] approximate totals
[3] includes aircraft in training, operational, overseas units
[4] excludes test flying by makers
[5] figures not available

APPENDIX 16

Confirmed Sinkings of U-boats by Mosquitoes

Date	U-boat	Squadron(s) involved
9.4.45	U.843	235
9.4.45	U.804	143, 235, 248
9.4.45	U.1065	143, 235, 248
19.4.45	U.251	143, 235, 248, 333
2.5.45	U.2359	143, 235, 248, 404, 333
3.5.45	U.2524	236, 254
4.5.45	U.236	236, 254
4.5.45	U.2503	236, 254
4.5.45	U.2338	236, 254
4.5.45	U.393	236, 254

APPENDIX 17

Survey of Mosquito Production and Serial Number Batches

Listed here are Mosquito production batches by inclusive serial numbers. Hatfield production was covered by Contract 555/C.23(a), Leavesden by Contract 1576/SAS/C.23(a), Percival Luton Con. No. 3047, Standard Motors Canley by Contract 1680 and Airspeed by Contract 3527. Following notes on engines fitted are inclusive dates for the batch and pertinent notes on special aircraft.

50 aircraft (D.H. Nos. 98001–98050) to Contract B.69990/40

Placed 1 March, 1940. Amended to 21 aircraft in June, 1941. Remainder produced to B.1355522 16 November, 1940. (Fighter prototype, 26 fighters plus dual-control aircraft with A.I. Mk. IV.) These Merlin 21 aircraft were produced as follows:—
W4050 prototype (flown as EO234 on first two flights)
W4051 P.R. Mk. I prototype
W4052 F. Mk. II prototype
W4053 Turret Fighter, T.III prototype
W4054–56 P.R. Mk. I
W4057 Bomber prototype, B. Mk. V
W4058–59 P.R. Mk. I
W4060–61 P.R. Mk. I (Long-range)
W4062–63 P.R. Mk. I (Long-range/Tropicalized)
W4064–72 P.R.U./Bomber Conversion Type, or B. Mk. IV srs.i
W4073 second turret fighter prototype, modified to T. Mk. III.
W4074 first production Mk. II (Single control)
W4075, 77, 79, 81 Production Mk. III (or II dual control)
W4076, 78, 80, 82, 83, 84–99 Mk. II delivered 27.12.41–1.3.42.

150 ordered 9.2.41. (Hatfield)

DD600–644, DD659–691, DD712–759, DD777–800, F. Mk. II (Merlin 21/22.) 25.2.42–15.10.42.
DD670–691 and DD712–714 F. Mk. II (Special) Intruder for 23 Sqn. DD715 prototype Mk. XII, DD759 to Mk. XII. DD613 A.I. clearance R.A.E. DD735 tailplane incidence, stick load tests D.H., DD664 6.9.42 to Australia, became A52-1001. DD668 A.I. Mk. V trials F.I.U. 7.42. DD723 Lancaster 'power-unit' trials, and chin radiators.

50 ordered 21.4.41. (Hatfield)

DK284–303, DK308–333, DK336–339.
B. IV Series ii (Merlin 21/23), 11.4.42–13.9.42.

423

Survey of Mosquito Production

DK324 P.R. VIII to B. IX, DK287 to Canada on 'Oregon' 9.9.42.

400 (Hatfield)

DZ228–272, DZ286–310.

70 F. Mk. II (Merlin 21/22), 12.10.42–30.12.42.

DZ302 cv to NF. XII by Marshalls. DZ294 stability investigation D.H., R.A.E.

DZ311–320 then DZ340–338, DZ404–442, DZ458–497, DZ515–559, DZ575–618, DZ630–652.

250 B. Mk. IV Series ii (Merlin 21/23) except four P.R. Mk. VIII (DZ342, DZ364, DZ404, DZ424) and four NF. XV (DZ366, DZ385, DZ409, DZ417).

a) 27 cv to P.R. Mk. IV, 5.12.42–21.3.43:—
DZ411, 419, 431, 438, 459, 466, 473, 480, 487, 494, 517, 523, 527, 532, 538, 544, 549, 553, 557, 576, 580, 584, 588, 592, 596, 600, 604.

b) 20 to carry 4,000 lb. bomb, delivered 1944 included:—
Modified by D.H.: DZ594 (prototype), DZ534.
Modified by Vickers-Armstrong (Weybridge) and Marshalls (Cambridge) DZ599, 606, 608, 611, 630, 631, 632, 633, 634, 636, 637, 638, 639, 640, 641, 642, 643, 644, 646, 650.
Modified by Vickers and D.H. DZ647.

c) Cv into XVI pressure cabin prototype (Merlin 73), DZ540, to A. & A.E.E. 5.11.43, SOC 28.7.45.

d) Modified by Vickers-Armstrong (Weybridge) and D.H. for *Highball* for 618 Sqn.:—
DZ471 prototype; DZ531/G first mod. by V.A.; DZ520, 524/G, 529, 530/G, 533, 534/G, 535/G, 537/G, 539, 541, 542, 543, 546, 547, 552, 554, 555, 556, 559, 575, 577, 578, 579, 581, 583.
Mod. by Vickers-Armstrong (Weybridge), Airspeed and Marshalls with Merlin 24s, arrester hook, new windscreen, armour plating, motive power for stores spinning August–September, 1944, for 618 Sqn. *Highball* operations:— DZ520, 524, 529, 531, 537, 539, 541, 542, 543, 546, 552, 554, 555, 556, 559, 575, 577, 578, 579, 581, 582, 583, 585, 586, 618, 639, 648, 651, 652.

e) B.O.A.C.: DZ411 (PR. IV) 10.1.43–6.1.45.

f) To M.A.C.: DZ382.

g) First DZ434 became HJ662; replaced by a second.

h) Research aircraft: DZ350 (R.A.E.), DZ345 and DZ412 (T.F.U.).
DZ541 for R. & D. at Vickers to 4.48.
DZ653–661, DZ680–727, DZ739–761.

80 Mk. II (Merlin 21/22) 31.12.42–22.3.43. 5 to M.A.C., DZ659 (Merlin 21) SCR720, SCR729, at F.I.U. 4.43–5.45 (called 'Eleanora') universal radar nose T.I. DZ700 to R.N.

150 (Hatfield)

HJ642–661, HJ699–715

37 Mk. II (Merlin 21) 12.3.43–20.5.43.

HJ659 bomb gear T.I.s.

HJ662–682, HJ716–743, HJ755–792, HJ808–833.

113 FB. Mk. VI Series i (Merlin 23/25).

HJ662/G Mk. VI prototype began as DZ434, made several flights as this. Built by S.A.G. Delivered A.A.E.E. 13.6.42, W.O. 30.7.42.

Survey of Mosquito Production

HJ666 A.F.D.U. trials 1943. HJ672 to 60 Sqn., S.A.A.F.
HJ679 radar research at A.A.E.E., Hatfield, A.S.W.D.U.
HJ732 (Merlin 23) Mk. XVIII prototype. A.A.E.E. 12.6.43.
10 M.U. 24.7.45, SOC 31.5.46.

450 (Leavesden)

a) HJ851–899, HJ958–999.
   91 Mk. III (Merlin 21) 14.5.42–22.12.43.
b) HJ911–944.
   34 Mk. II (Merlin 21/22) 9.9.42–9.1.43.
c) HJ945–946, HK 107–141, HK159–204, HK222–235.
   97 Mk. II to Marshalls (Cambridge) 8.1.43–6.6.43 for NF. XII cv (A.I. Mk. VIII). HK186 initially to Southern Aircraft, later Marshalls for cv. Delivered as NF.XII 10.2.43–14.7.43.
   HK195 to Marshalls 19.5.43, cv to Mk. XVII (SCR 720 radar) later at Defford, Ford, 54 O.T.U.
   HK236–265, HK278–327, HK344–362.
   99 Mk. II (Merlin 21, 23), to Marshalls for cv to NF. XVII (SCR 720). To Marshalls as Mk. II from 6.6.43.
   Delivered as Mk. XVII 4.9.43–26.2.44.
d) HK363–382, HK396–437, HK453–481, HK499–536. (HK535/536 renumbered SM700/701, serials duplicated Lancasters.)
   129 N.F. XIII (Merlin 23) 15.9.43–2.2.44. Some universal nose, others thimble nose. A.I. VIII.
   HK364 to XIX prototype.
   HK369 trials of A.I. VIII/c.f. Mosquito XII/XIII at F.I.U. 12.43, and universal nose.
   HK472 to R.N.

500 (Standard Motors)

HP848–888, HP904–942, HP967–989, HR113–162, HR175–220, HR236–262, HR279–312, HR331–375, HR387–415, HR432–465, HR485–527, HR539–580, HR603–649.
F.B. VI (Merlin 23, 25). 16.6.43–17.12.44. Many used overseas.

130 (Hatfield)

HX802–835, HX849–869, HX896–922, HX937–984. F.B. VI (Merlin 21, 23, 25) 14.7.43–11.11.43, except HX902–904 Mk. XVIII (Merlin 25) 5.9.43, two on 6.10.43.
HX809 A.A.E.E. performance tests 12.43.
HX918 A.A.E.E. R.P. tests 11.43.
HX849–850 not delivered.
HX902/G A.A.E.E. 5.9.43, 248 Sqn. 23.10.43.

200 (Hatfield)

a) LR248–276, LR289–313, LR327–340, LR343–389, LR402–404.
   118 F.B. Mk. VI (Merlin 25 mainly, 21, 23) 17.10.43–8.1.44.
   LR387 folding wings, hooked, cv August, 1945, to Mk. 33. To R.N. 1946.
b) LR405–446, LR459–474, LR478–481.
   60 P.R. IX (Merlin 72) 20.5.43–21.10.43.

Survey of Mosquito Production

    LR410 Rolls-Royce 6.43–5.46.
    LR418 R.A.E. 7.43–4.44.
c) LR475, 476, 477, 495–513.
    22 B. Mk. IX (Merlin 72) 17.4.43–10.11.43.
    LR495 first trials aircraft to A.A.E.E. 17.4.43. Destroyed on take-off overload trials 29.1.44.
    LR503 Victory tour of Canada 1945.

59 (Leavesden)

LR516–541, LR553–585.
59 T. Mk. III (Merlin 21) 30.12.43–9.10.44.

300 (Hatfield)

a) ML896–920, ML921–924.
    29 B. IX (Merlin 72) 10.7.43–20.10.43.
    ML914 trials with 4,000 lb. bomb and 6-store Avro carrier.
b) ML925–942, ML956–999, MM112–156, MM169–179, MM181–205, MM219–226. (MM170 undelivered.)
    151 B. XVI (Merlin 72/73) 24.11.43–4.12.44.
    cv by General Aircraft to TT. 39 for R.N.:— ML935, 956, 974, 980, 995, MM112, 117, 156, 142, 177, 192.
    ML926 radar bombsight, *Oboe* repeater, *H2S* trials at Defford.
    ML932 rogue aircraft tests.
    ML937 A.A.E.E. load and perf. tests.
    ML994 200 gallon drop tanks, Mk. VIII mine drops, dev. aircraft.
    MM133 to Netherlands.
    MM175 *H2S* trials, Defford.
c) MM227–236.
    10 P.R. IX (Merlin 72) 29.8.43–19.11.43.
    MM229 Merlin 67 (R.M. 10 S.M.) installation trials; used by 105, 627 and 109 Sqns. SOC 15.1.46.
    MM230 exhaust shroud tests D.H. and F.I.U. T.I. aircraft.
    MM235 Hamilton paddle blade prop. tests A.A.E.E. 12.43.
d) MM237, 238, 241.
    3 B. IX (Merlin 72) 11.43.
e) MM239–240, MM243–257.
    17 P.R. IX (Merlin 72) 7.10.43–3.11.43.
f) MM258, MM271–314, MM327–371, MM384–397.
    104 P.R. XVI (Merlin 72/73) 24.11.43–4.5.44.
    MM258 T.I. aircraft.
    20 to U.S.A.A.F. in U.K.:— MM308 (*H2X*) 'Mickey' in nose, 310, 337, 338, 340, 342, 344, 345, 346, 364, 367, 368, 370 Trop., 371 Trop., 384 Trop., 385 Trop., 386 Trop., 388, 391, 393.
    MM273, 293, 309, 342, 346, 361, 364, 368 to R.N.
    MM328 T.I.s for PR. 32.
    MM363 assymetric loads, drop tank tests A.A.E.E.
g) MM398–423, MM426–431.
    32 F.B. VI (Merlin 25) 6.1.44–9.3.44.
h) MM424–425. Mk. XVIII (Merlin 25) 12.4 and 2.2.44 to 248 Sqn.

Survey of Mosquito Production

300 (Leavesden)

MM436–479, MM491–534, MM547–590, MM615–623.
126 N.F. XIII (Merlin 25) 23.1.44–12.5.44.
MM439, 462 to R.N.
MM624–656, MM669–685.
50 N.F. XIX (Merlin 25) 17.4.44–27.5.44.
MM626, 630, 635, 636, 638, 642, 644, 651, 656, 670, 675, 682, 685 to Sweden.
MM631 to Belgium.
MM686–710, MM726–769, MM783–822.
109 N.F. XXX (Merlin 72) 28.4.44–1.9.44.
MM764, 765, 769, 821 to U.S.A.A.F. N.W. Africa; later these and 4 more to R.A.F., M.A.C.

MP469 (Hatfield)

Pressure cabin prototype bomber, modified to N.F. XV.

45 (Leavesden)

MT456–500 (MT480 undelivered).
N.F. Mk. XXX (Merlin 76) 30.8.44–29.9.44.
MT462, 464, 465, 478, 479 to 416 Sqn. U.S.A.A.F., in N.W. Africa.
MT477 to R.N.
MT466 Merlin 113 (R.M. 16 S.M.) fitted 10.44 for N.F. 36 programme.

50 (Leavesden)

MV521–570.
N.F. Mk. XXX (Merlin 76).
MV569 to MAC (255 Sqn. 1945).

500 ordered 17 March, 1943 (Hatfield)

a) NS496–538, 551–596, 619–660, 673–712, 725–758, 772–816.
  250 P.R. XVI (Merlin 72/73 or 76/77), except NS586, 587, 588, 589 P.R. 32s. (Merlin 76/77); 24.3.44–31.12.44. (P.R. 32 13.9.44–18.10.44.)
  NS508, 509, 512, 513, 514, 515, 516, 518, 519, 533, 534, 535, 537, 538 (*H2X*), 551, 552, 553, 554, 555, 556, 557, 558, 559, 568, 569, 570, 581, 582, 583, 584 (*H2X*), 590, 591, 592, 593–596, 619, 620, 625, 630, 635, 638, 650, 651, 676, 686, 707, 709, 711, 712, 725, 730, 740, 742, 743, 744, 745, 748, 752–754, 756–758, 772–775, 782–783, 785, 792, 793, 794, 796, 804, 811, 812, to U.S.A.A.F. in U.K. NS538 *H2X* prototype for U.S.A.A.F.
  11 to R.N.
  Civil cv. NS811 G-AIRU, NS812 G-AIRT, NS735 G-AOCK, NS639 G-AOCI.
  NS561 special grey-green/grey finish mid-1944.
  NS729 arrester gear T.I. 9.44.
b) NS819–859, NS873–914, NS926–965, NS977–999, NT112–156, NT169–207, NT219–238.
  250 F.B. VI (all Merlin 25) 25.1.44–19.5.44 except NT220, NT224, NT225, F.B. XVIIIs.
  NT220 nose like B. IV, torpedo rack, R.P.s, used at R.A.E.
  NT181 *Monica IIIE* trials, radar ranging trials F.I.U. 1.7.44.
  NT224 to 'E' 248 Sqn. 2.6.44, FTR 7.12.44.

Survey of Mosquito Production

NT225 to 248 Sqn. 5.6.44, Martin Hearn 6.11.44, D.H. 22.2.45, 248 Sqn. 17.4.45, Bircham Newton 22.3.45, 254 Sqn. 16.5.45, 9 M.U. 20.8.45. SOC (RTP) 20.11.46.
(NS496/816 originally ordered as P.R. IX and NS819/NT238 as F.B. X bringing totals then on order to 625 and 304 respectively, and 1,930 Mosquitoes to contract 555.)

300 ordered March, 1943 (Leavesden)

NT241–283, NT295–336, NT349–393, NT415–458, NT471–513, NT526–568, NT582–621.
N.F. Mk. XXX (Merlin 76) 4.11.44–2.5.45.

245 (Percival)

a) PF379–415, PF428–469, PF481–511, PF515–526, PF538–579, PF592–619.
195 B. XVI (Merlin 72/73) 15.5.44–5.12.45.
cv to TT. 39: 439, 445, 449, 452, 481–483, 489 (2nd proto), 560, 562, 569, 576, 599, 606, 609; others earmarked only.
PF620–635, PF647–680.
50 built as P.R. 34 (Merlin) 6.9.45–26.7.46.
PF652, 656, 662, 669, 670, 673, 678, 679, 680 cv to P.R. 34A.

250 (Hatfield)

PZ161–203, PZ217–259, PZ273–316, PZ330–358, PZ371–419, PZ435–476.
241 F.B. VI (Merlin 25) 16.5.44–19.6.45 and 9 Mk. XVIII (PZ251, 252, 300, 301, 346, 467–470: Merlin 25).
Mainly to Coastal Command and 100 Group, none overseas.
To R.N.Z.A.F.: PZ196, 200 (to ZK-BCY), 237, 254, 297, 310, 313, 330, 403, 413, 444 (ZK-BCW), 447, 474 (ZK-BCV).
PZ281 Vickers 16.4.45–12.3.48 *Highball* development.
PZ202 RP/drop tank tests A.A.E.E. 1945.
PZ251 C.C.P.P. 13.7.44, 248 Sqn. 21.10.44, FTR 21.10.44.
PZ252 C.C.P.P. 13.7.44, Z/248 Sqn. 25.8.44, 254 Sqn. 16.3.45, 9 M.U. 23.8.45, SOC 25.11.46.
PZ300 C.C.P.P. 14.8.44, 248 Sqn. 25.8.44, 254 Sqn. 21.3.45, RIW 4.6.45 Cat. E 5.12.45.
PZ301 C.C.P.P. 18.8.44, T/248 Sqn. 25.8.44, 254 Sqn. 16.3.45, 9 M.U. 3.9.45, SOC 31.12.46.
PZ346 C.C.P.P. 5.9.44, Z/248 Sqn. 28.10.44. FTR 7.12.44.
PZ405 C.R.D. Hucknall 20.10.44–20.8.45.
PZ467 C.R.D. D.H. 5.10.44, 27 M.U. 12.12.44, Pershore 9.3.45, to U.S.A. 9.4.45 arrived 27.4.45, Dorval; reached Patuxent (U.S.N.) 30.4.45.
PZ468 3504 S.U. 2.11.44, D.H. 1.11.44, 248 Sqn. 12.12.44, 254 Sqn. 21.3.45, 9 M.U. 20.8.45, SOC 25.11.46.
PZ469 27 M.U. 4.12.44, A.A.E.E., 27 M.U. 8.8.45, 57 M.U. 17.10.45, Lanzer & Co. as scrap 7.8.47.
PZ470 27 M.U. 14.1.45, SOC 23.7.46.

300 (Standard Motors)

RF580–625, RF639–681, RF695–736, RF749–793, RF818–859, RF873–915, RF928–966.

Survey of Mosquito Production

F.B. VI (Merlin 25) 16.12.44–5.6.45.
To R.N.Z.A.F.: RF597 (ZK-BCU), RF595, 709, 719, 753, 837, 849, 856, 857, 885, 903, 908, 910, 935.

200 (Hatfield)

a) RF969–999, RG113–158, RG171–175.
   82 P.R. XVI (Merlin 76/77) 12.12.44–3.5.45.
   RF979, 982, 996, RG113, 145, 146, 156, 157 to U.S.A.A.F. in U.K.
   RG171–173 hooked to R.N.
b) RG176–215, RG228–269, RG283–318.
   118 P.R. 34 (Merlin 113/114) 6.1.45–10.1.46.
   RG176–178, 181, 189, 194, 195, 198, 201, 202, 205, 207, 231, 233, 236, 238, 240, 252, 259, 262, 265, 268, 300, 302, 314 modified to P.R. 34A.
   RG176 A.A.E.E. trials.

365 (Leavesden)

a) RK929–954.
   27 Mk. XXX (Merlin 76) 27.4.45–12.6.45—nine saw wartime squadron service.
   RK945/G, Vickers, Wisley 30.8.45–28.12.45.
b) RK955–960, RK972–999, RL113–158, RL173–215, RL229–268.
   163 Mk. 36 (Merlin 113) 16.5.45–23.3.47.
c) RL269–273, RL288–329, RL345–390 cancelled.
d) RL248 N.F. 38 prototype.

50 (Leavesden)

RR270–319, T. Mk. III 12.10.44–25.7.45.

109 (Hatfield)

RS501–535, 548–580, 593–633.
F.B. VI (Merlin 25) 13.10.44–22.1.45, many to Coastal Command.

300 (Airspeed)

a) RS637–680, 693–698.
   50 F.B. VI (Merlin 25) 8.4.45–7.6.46.
   RS646 (NZ2375), RS670 (NZ2357), RS693 (NZ2348) to R.N.Z.A.F.
   RS657 hooked, R.N. service 14.8.47–31.3.49.
b) RS699–723, 25 B. Mk. 35 (Merlin 113/114) 28.2.46–11.4.47.
   Postwar bomber service with 109 and 139 Sqns.; cv to T.T.35:— RS701, 702, 704, 706–710, 712, 713, 715, 717, 719, 722.
   RS700 first P.R. 35.
c) RS724–725, 739–779, 795–836, 849–893, 913–948, 960–999.
   RT105–123, cancelled.

60 (Hatfield)

a) RV295–326, 340–363, 56 B. XVI (Merlin 76/77) 11.12.44–21.3.45 except RV348–350 B. 35 (Merlin 113A) 29.3.45–3.4.45, cv to T.T.35. RV295, 296, 303, 308 to R.N.
b) RV364–367 4 B. Mk. 35 (Merlin 113A/114) 29.3.45–9.4.45, last three cv to T.T.35.

Survey of Mosquito Production

500 (Hatfield)

a) SZ958–999, TA113–122.
52 F.B. VI (Merlin 25) 4.1.45–23.3.45.
SZ994 to R.N.Z.A.F.

b) TA123–156, 169–198, 215–249, 263–308, 323–357.
180 N.F. XIX (Merlin 25) 27.9.44–19.9.45.
117 overseas, mainly SEAC 22, 89, 176 Sqns. for 255, 600 in M.A.C.
Many to M.U.s only. TA343 to Flt. Refuelling 21.2.49 as G-ALGU.
TA229 became G-ALGV.

c) TA369–388. 20 F.B. VI 9.3.45–10.4.45.
To R.N.Z.A.F.: TA373, 383, 385 (NZ2376).

d) TA389–413, 425–449. 50 N.F. XIX (Merlin 25) 20.6.44–25.11.44.

e) TA469–508, TA523–560, 575–603, 107 F.B. VI 12.4.45–15.12.45.
TA488 ASWDU camera tests, TA501 A.A.E.E. for R.P. flight path trials.
To R.N.Z.A.F.: TA491 (NZ2370). 577 (NZ2371), 578 (NZ2360), 597,

f) TA614–616, 3 P.R. XVI delivered 11.3 and 14.3.45. TA614 became G-AOCN.

g) TA617–618, TA633–670, 685–724.
80 B.35 (Merlin 113/114) 29.3.45–16.7.45.
cv to T.T.35: TA633, 634, 637, 639, 641, 642, 647, 649, 651, 660–662, 664, 669, 685, 688, 699, 703, 705, 710, 711, 718, 719, 720, 722, 724. TA650 to P.R. 35.

320 (Standard Motors)

TE587–628, 640–669, 683–707, 708–725, 738–780, 793–830, 848–889, 905–932.
266 F.B. VI. 27.5.45–21.12.45.
To R.N.Z.A.F.: TE739, 746, 751, 752, 755, 757, 758, 765, 766, 830, 856, 861–864, 874–876, 878, 880, 881, 883, 885, 887, 888, 905, 910–913, 925, 927, 740.
To R.N. 701, 704, 708, 710, 711, 717, 719–725, 741, 742, 813, 823, 826, 829.

70 (Hatfield)

TH976–999, TJ113–158. Ordered 24.5.44 as B. XVI, built as B.35 (Merlin 113/114) 11.7.45–9.11.45.
TH977, 978, 980, 981, 987, 989, 990, 991, 992, 996, 998, TJ113, 114, 116, 119, 120, 122, 123, 125–127, 128, 131, 135, 136, 138, 140, 147–149, 153–157. cv to T.T.35.
TJ124, TH985, TJ145, TH989 cv to P.R. 35.

94 ordered 3.7.44 to Contract 4221 (Hatfield)

TK591–635, TK648–679, TK691–707. Ordered as N.F. 30. Only TK591–635 and TK648–656 completed, as B.35 (Merlin 113/114) 11.10.45–12.5.46.
TK591–594, 596, 599, 603–610, 612–613, 616 cv to T.T.35.
TK615, 632, 650 cv to P.R. 35.

Cancellations included 300 (Leavesden) with Merlin 113

TN466–497, TN510–530, TN542–590, TN608–640, TN652–674, TN690–736, TN750–789, TN802–838, TN850–864.

2 Special prototypes (Leavesden)

TS444, 449. (Merlin 25) prototype Mk. 33s for service trials. Delivered 1946.

Survey of Mosquito Production

50 (Leavesden)
TV954–984, TW101–109, T.3. 20.7.45–31.5.46.

100 ordered January, 1945 (Leavesden)
TW227–257, 277–295. T.R. 33 (Merlin 25) 46–47. First 13 had fixed wings. TW240 T.R. 37 prototype. TW228 trials of *Card* (*Highball* development). TW230/G for *Highball* trials. 50 cancelled.

50 (Leavesden)
VA871–876 (to R.A.F.); VA877–881 (to R.N.); VA882–894, 923–928 (to R.A.F.) T. 3 6.6.46–1.47. VA929–948 cancelled.

13 on contract 4204 (Hatfield)
VL613–625. P.R. 34 (Merlin 113/114) 16.1.46.
VL621 and 623 fitted with *H2S*, 621 to Met. Res. Flt. 1.2.52–29.5.55.
VL625 P.R. 34A.

70 (Airspeed)
VL727–732 only built, F.B. VI July, 1946.
Used in U.K. and B.A.F.O.

25 (Airspeed)
VP178–202 B. 35 (Merlin 113A/114A) 14.3.47–3.10.47
VP178, 181, 191, 197 cv to T.T. 35. VP183 P.R. 35.

14 (Hatfield)
VP342–355 only, built as T. 3 1944–6.6.47.

40 (Hatfield)
VR330–349 T. 3 only built 11.46–4.48.

15 (Airspeed)
VR792–806 B. Mk. 35 (Merlin 113/114) 10.7.47–13.2.48.
VR802 to T.T., VR793 banner T.T. 35 prototype, trials at Cambridge.

44 (Hatfield)
VT581–596, VT604–631. T. 3 6.6.47–10.48.
VT595, 611, 615, 616, 619, 623, 627–631 to R.N.

50 (Hatfield: Chester)
VT651–683, VT691–707, N.F. 38 (Merlin 113/114) 2.48–1.50.
21 to Yugoslavia.

14 (Chester)
VT724–737 T.R. 37 (Merlin 25).

51 (Chester)
VX860–879, VX886–916 (Merlin 113/114) N.F. 38 2.50–12.50.
33 to Yugoslavia. Others cancelled.
VX916 completed November, 1950, was the 7,781st and last Mosquito built.

APPENDIX 18

Dispersal of Work People from Hatfield 1941

Site	Work	25.1.41	22.3.41	30.8.41
Old Rubber Works, St. Albans	Wood details	120	130	126
Salvation Army Printing Works, St. Albans	Pipe fitting	44	48	48
Andrew Buchanan & Son, Welwyn Garden City	Sheet Metal	78	200	225
Unity Heating, Welwyn G.C.	Coppersmith	143	211	203
Alfa Strings Ltd., Welwyn G.C.	Fitting shop	120	159	177
Naphos Distributors, Welwyn G.C.	Tank shop	29	—	5
Keystone Mills, Boreham Wood	Capstan Operators	16	248	489
F. Coleman & Son, Welwyn G.C.	Technical School	40	58	42
Paul Walser & Co., Luton	Press, foundry	—	98	96
Amalgamated Studios	Paint shop/D.O. and accounts	—	—	260
Verulam Motor Co., St. Albans	Machine shop tools	—	186	135
Thorpe & Co., St. Albans	Machine Shop	—	66	—
Nicholson & Co., St. Albans	Electrics, canteen	—	—	36
L.P.T.B., Hatfield	Stores	—	—	8
Totals		590	1,396	1,850

COMPARISON OF LABOUR FORCE DISPERSED AND AT HATFIELD

	Dispersed	Hatfield
30.8.41	1,850	3,426
21.3.42	2,066	4,157
5.9.42	1,552	5,200
7.11.42	1,595	5,749

EMPLOYEES OF DE HAVILLAND ON 27.6.42
(Two main aircraft factories only)

	Staff	Labour Force Male	Female	Total
At Hatfield	2,300	3,658	1,299	7,257
Second Acft. Grp.	733	1,119	286	2,138

APPENDIX 19

Makers of the Mosquito

The policy of subcontracting in time of war has been explained. More than four hundred subcontractors, spread throughout Britain, co-operated with the designers and production engineers of the Mosquito. Many are listed here. It is true to say that, in the main, it was these concerns which manufactured the 5,570 Mosquitoes that were produced in Britain to the end of the war.

Addis Brush	Hertford	*Machined components*
Aero & Auto	Wood Green, London	*Hydraulic and fuel pipes*
Aeroflex	Camberley	*Fuel pipe fittings and hose*
Aero Pipe & Glass	Willesden, London	*Hydraulic and fuel pipes*
Airwork & General Trng.	Hounslow	*Tanks*
Alford & Alder	Walworth, London	*Tail wheel units and flap jacks*
Alltools	Brentford	*Fuel galleries and bomb racks*
Almarco	Wellingborough	*Radar fittings*
Alta Car & Marine	Kingston	*Machined components*
Aluminium Plant & Vessel	Putney, London	*Fuel and oil tanks*
Anglo Cutting	Romford	*Machined details*
Artel Manufacturing	Slough	*Machined components*
Apex Production	Harrow	*Machined components*
Ashmore Bros.	Birmingham	*Exhaust manifolds*
Associated Brass	Willesden, London	*Control gears*
Auto Dairies	Wembley	*Cowling and sub assemblies*
Automotive Products	Leamington Spa	*Hydraulic units*
Aviation Development	Welwyn Garden City	*Control details*
Walter Baker	High Wycombe	*Wood components*
Beautility	Walthamstow	*Wing components, fuselage bulkheads*
Bell & Frome	Edgware	*Cowling panels, gun chutes, oil tanks*
Bell Punch	Uxbridge	*Control columns*
Joshua Bigwood	Wolverhampton	*Exhaust shrouds*
Birch Bros.	High Wycombe	*Tank doors*
Birmingham Guild	Birmingham	*Engine mounts, flaps*

433

Makers of the Mosquito

J. I. Blackburn	Godalming	Compression legs
J. R. Bramah	Sheffield	Fuel tanks
Bratt Colbran	Wembley	Gun fairings, radiator fittings
Botco	Wolverhampton	Machined parts
Briggs Motor Bodies	Dagenham	Engine cowlings and manifolds
British Emulsified	Islington	Fuel and oil pipes
Broadway Engineering	Southall	Machined parts
Brylock Equipment	Letchworth	Machined components
British Insulated Cables	London	Cables
British Manufacturing & Research	Grantham	Machined details
British Panels	Coventry	Cowlings
Burgess Products	London and Gateshead	Electrical assemblies
Cam Tools	High Wycombe	Wing fittings
Campbell Engineering	Ravenesbourne	Machined components
Carbodies	Coventry	Canopies
Castle Bros.	High Wycombe	Fins, fuselage details
Caxton Engineering	Thames Ditton	Machined fittings
C.B.A. Radiators	Acton, London	Fuel and oil tanks
Challis Bros.	Berkhampstead	Machined components
Chater Lea	Letchworth	Machined components
Chillington Tool	Wolverhampton	Manifolds, pressings
Chiltern Aircraft	Hungerford	Metal fittings
Church Bros.	Chertsey	Metal fittings
Clavis Tools	Balham, London	Wing fittings
Claude Butler	Clapham	Machined components
Collins Engineering	Kingston	Machined fittings
Co-operative Society	Enfield	Wings
Cornercroft	Coventry	Exhausts shrouds, gun chutes
Courtney Pope	Stamford Hill, London	Canopies
Cundall Folding	Luton	Compression legs, tail-wheel units
Contemporary Woodwork	Park Royal, London	Wood components
P. B. Cow	Wembley, London	Fuel tanks, moulded blocks
Cox	Watford	Plastic radome assemblies
Cutrock	Finchley, London	Machined components
Dancer & Hearne	High Wycombe	Wing skins, spars, fittings, assemblies
Decca Records	Merton	Hard chrome plating
de Havilland Engine Div.	Edgware	Hydraulic jacks
Delaney Galley	Cricklewood	Oil tanks and coolers
Durion	North Acton	Hard chrome plating

Makers of the Mosquito

Electro-Hydraulics	Warrington	Jacks
Electrolux	Luton	Fuselage and wing fittings
Elliott Bros.	Newbury	Tailplanes, wood components
E.M.C. Engineering	Tottenham, London	Metal fittings
Expert Heat Treatment	Mitcham	Hardening
E.R.A.	Dunstable	Machined fittings
Education Supplies Association	Stevenage	Wings
Essex Aero	Gravesend	Fuel tanks
Fireproof Tanks	Portsmouth	Fuel tanks
Flexo Plywood Industries	Chingford	Ply panels, bulkheads
Franco-British	Hendon	Engine mounts, fire proof bulkheads
General Account	Feltham	Sheet metal work
Gear Products	Tottenham	Pressings
E. Gomme	High Wycombe	Fuselage shells
Grice & Young	Dunstable	Aileron differential gear
Gurney Nutting	London	Metal fittings
Gyral Gears	Wandsworth	Control Assemblies
Harris Lebus	Tottenham	Fuselages and wings
Harley Landing Lamps	St. Neots	Fuselage joint fittings
H & B Engineering	Finchley	Machined components
S. Haskins	Larkswood	Engine mounts, rudder pedals
Hearle Whitley	Letchworth	Machined components
J. B. Heath	High Wycombe	Wing spars and leading edges
Heller & Sons	Edmonton	Machined components
Henderson Safety Tanks	Elstree	Fuel tanks, oil tanks
Herbert & Son	Tottenham	Machined components
Hertfordshire Rubber	Letchworth	Rubber components
Henschell & Son	Byfleet	Ammunition boxes, exhaust shrouds
Heston Aircraft	Heston	Cowlings, fins
E. Hines	Northwich	Aileron differential units
Hobbies, Ltd.	Dereham	Metal fittings
Hobart Manufacturing	New Southgate, London	Machined fittings
Hoopers, Coach Builders	Park Royal, London	Wings, tailplanes, drop tanks
Hymatic Engineering	Redditch	Control valves
Igranic Electric	Bedford	Pressing
Inglis Aircraft	Canonbury, London	Sheet metal work

Makers of the Mosquito

G. K. Jensen	Willesden Green, London	*Machined components*
Jigs (Leavesden)	Letchworth	*Machined components*
Johnson & Holland	High Wycombe	*Wing details*
Jowett Cars	Bradford	*Machined fittings*
K & C Ironfounders	Letchworth	*Metal fittings*
Kenilworth Mfg.	West Drayton	*Metal fittings*
Kigass	Leamington	*Machined fittings*
J. F. Kenmure	Feltham	*Metal fittings*
King Aircraft	Hayes	*Metal fittings*
Kingsbury Engineering	Hendon	*Machined fittings*
Kingston Jigs	Kingston	*Machined fittings*
Lancefield Coachworks	West London	*Ammunition boxes, pipes, manifolds*
Lane Motor Panel	Burnt Oak	*Cowlings, cannon chutes*
Walter Lawrence	Sawbridgeworth	*Fuselage shells, wing skins*
Leeds Metal Spinner	Leeds	*Fuel and oil tanks*
Lewin Engineering	Epsom	*Metal fittings*
Lomas, G.	Walton-on-Thames	*Metal fittings*
Marshalls Flying School	Cambridge	*Tail components*
Marston Excelsior	Wolverhampton	*Oil coolers, gun fairings, chutes*
Martin Baker	Denham	*Crew seats*
Martyn	Cheltenham	*Metal fittings*
Massan Seelby	Tulse Hill, London	*Tail units*
Metalair	Wokingham	*Engine mounts*
Metal Components	Portslade	*Metal fittings*
Medspin Engineering	South London	*Machined parts*
Midland Aero	Bloxwich	*Machined parts*
M.L.H. Engineering	Cranford	*Metal fittings*
Mollart Engineering	Kingston	*Universal joints, fittings*
Moulded Components	Croydon	*Canopy mouldings*
Morris Radiator	Oxford	*Radiators, oil coolers*
Moseley Bros.	Birmingham	*Sheet metal fittings, ammunition boxes*
H. J. Mulliners	Chiswick	*Tail planes, fins, wing flaps*
Nico Ltd.	Letchworth	*Pressings*
Novo Bax	Watford	*Machined parts*
Opperman Gears	Tottenham	*Differential control units*
W. Ottway	Ealing	*Control details*
Rubery Owen	Darleston	*Pilot's control column assemblies*

Makers of the Mosquito

Page	Bexhill	*Machined parts*
Paramount Sheet Metal	Kingston	*Sheet metal panels, doors*
Parker Knoll	High Wycombe	*Wing leading edge, spar assemblies*
Parkinson	Birmingham	*Engine bulkheads, gun fairings*
Parkinson & Polson	East Grinstead	*Metal fittings*
G. Parnall	Bristol	*Ailerons and flaps*
A. R. Parsons	London Colney	*Metal fittings*
L. F. Peatty	Tolworth	*Engine cowlings*
Perfecta Motors	Birmingham	*Canopies, windscreens*
G. D. Peters	Slough	*Machined components*
Piece Parts & Assemblies	Luton	*Aileron control gear*
Plant & Machinery	Kensington	*Aerofoil tab screw jacks*
Pobjoys	Chatham	*Machined components*
Pollards Shopfitters	Clerkenwell	*Wing wood details*
W. Potter	Paddington	*Engine mounts*
Pullars	Bridge of Allan, Scotland	*Engine cowlings*
Pye Radio	Cambridge	*Electrical assemblies*
Ransome, Sims & Jeffries	Ipswich	*Control fittings*
Reliance Precision Tools	Staines	*Machined parts*
Royal National Lifeboat Institution	Elstree	*Tail and rudder fittings*
W. A. Rollason	Croydon	*Tail units, oil tanks*
Royston Engineering	Kingston	*Metal fittings*
Rumbold	Kilburn	*Pilot seats*
Frederick Sage	Holborn, London	*Fuselage details*
Salisbury Sheet Metal	Salisbury	*Oil tanks*
Salter	West Bromwich	*Tail undercarriage units*
Sankey	Tipton	*Pressings*
Saunders-Roe	Cowes, Eastleigh	*Undercarriage units*
R. Sanders	Stevenage	*Metal fittings*
Searle Radiators	South London	*Radiators*
Serke Radiators	Birmingham	*Radiators*
Simpsons	Luton	*Electrical details*
Singer Motors	Birmingham and Coventry	*Undercarriage units, engine mounts*
Smiths Jacking System	Edgware	*Flap jacks*
Souther Aircraft Components	Hemel Hempstead	*Metal fittings*
South East Engineering	Romford	*Machined parts*
Strachans	Acton	*Undercarriage door panels*
Sorbo	Woking	*Seat squabs*
Southborough Sheet Metal	Tolworth	*Sheet metal panels*

Makers of the Mosquito

Stanley Smith	Isleworth	*Pre-formed plywood panels*
Starkie Gardner	Putney	*Fitting sub-assemblies*
Stevens & Bulliford	Birmingham	*Control fittings*
Stovis Engineering	S.E. London	*Machined parts*
Straight Light Reflectors	Holloway	*Metal fittings*
Styles & Mealing	High Wycombe	*Fuselage shells*
Such	Cheltenham	*Fittings*
Technical Platings	Teddington	*Plating*
Teleflex Products	Chadwell Heath	*Control assemblies*
Temple Piano	London	*Pre-formed plywood*
Thurgoods Coachworks	Ware	*Fuselage components, jettison tanks*
Thomas	Hanwell	*Metal fittings*
Tipsy Aircraft	Slough	*Metal fittings*
Triplex	Willesden	*Acrillic mouldings*
Ultra Electric	N. Devon and Kings Langley	*Ailerons, assemblies*
Unicam Instruments	Cambridge	*Machined parts*
Universal Engineering	Croydon	*Machined parts*
Universal Tools	Mitcham	*Machined parts*
Vanden Plas	Kingsbury	*Wings, spars, skins*
Vauxhall Motors	Luton	*Machined landing legs*
Vincent HRD	Stevenage	*Manifold Pressings*
Vokes Filters	Guildford	*Sheet metal assemblies*
Walters Electric	Kensal Rise, London	*Electrical harness and assemblies*
Waring & Gillow	Lancaster	*Wing spars, flaps, leading edges*
Watford Electric	Watford	*Electrical assemblies*
Eustace Watkins	Chelsea	*Tail planes, fins, flaps*
Watson Baker	Welwyn Garden City	*Oil filters, electrical assemblies*
Weatherley Oil Gear	Biggleswade	*Hydraulic pipes and fittings*
Wellworthy	Ringwood	*Control machined parts*
Westbourne Engineering	Daventry	*Undercarriage components, control columns*
West Nor	Croydon	*Sheet metal work*
Weyburn Engineering	Elstead	*Gun mounting and chutes*
Weyer	Islington	*Machined components*
J. S. White	Cowes	*Engine mountings*
Whittingham & Mitchell	Putney	*Sheet metal work*
D. Wickham	Ware	*Control fittings*
Wight Allom	N. London	*Fuselage and wing components*

Makers of the Mosquito

Willenhall Motor Radiators	Willenhall	*Engine manifold pressings*
Willis Hole Aviation	Croydon	*Metal fittings*
Wimbush	Belgravia, London	*Machined components*
Williams	Islington	*Fuel tanks, coolers*
Wingham Bennett	West Moseley	*Electrical assemblies*
Wolf & Sabine	Mitcham	*Metal fittings*
Wrighton Aircraft	Walthamstow	*Fuselages and wings*
Wright & Weare	Tottenham	*Machined components*
James Young	Bromley	*Machined parts*
Young & Wilson	Hatfield	*Metal fittings*

APPENDIX 20

Royal Air Force Mosquito Squadrons and Units

BASES

Listed are unit bases and dates of occupation, formation or equipping. Disbanded is indicated by (D), re-equipment (approx.) by (R). Dates of movements generally denote those of Main Party moves. Bracketed after the unit's title are its identity letters.

1. Squadrons

No. 4 (UP post-war)
Sawbridgeworth and Hunsdon	1.44 Aston Down	3.1.44
Gatwick	4.4.44 Reformed Vokel from 605 Sqn.	31.8.44
Gilze Rijen	13.9.45 Gutersloh	8.11.45
Wahn	13.11.47 Celle	19.9.49
Wunsdorf	10.7.50–7.50 (R)	

No. 8 (nil)
Khormaksar (Aden)   1.9.46–5.47 (R)

No. 11 (OM)
Wahn, Germany (107 renumbered)   4.10.48 Celle, Germany   9.49–9.50 (R)

No. 13 (nil)
Ein Shemer (680 renumbered)	1.9.46 Kabrit	11.12.46
Fayid	5.2.47 Kabrit	21.2.51–52 (R)

No. 14 (CX)
Banff (143 renumbered)	1.6.45 Disbanded	1.4.46
Wahn (128 renumbered) reforming	1.4.46 Celle	9.49
Fassberg	11.50–2.51 (R)	

No. 16 (EG)
Celle   20.9.45–31.3.46 (D)

No. 21 (YH)
Sculthorpe	10.43 Hunsdon	31.12.43
Gravesend	23.4.44 Thorney Island	23.6.44

Mosquito Squadrons and Units

Rosières 6.2.45 Melsbroek 18.4.45
Gutersloh 3.11.45 Munster/Handorf 27.6.46
Gutersloh 9.8.46–7.11.47 (D)

No. 22
Seletar (89 Sqn. renumbered) 1.5.46–15.8.46 (D)

No. 23 (YP)
Ford 7.42 Manston 6.8.42
Bradwell Bay 13.9.42 In transit to Malta 7.12.42
Luqa, Malta 28.12.42 Detachments:
    Pomigliano 27.10.43
    Alghero 7.12.43–5.44
At Sea 8.5.44–1.6.44
Little Snoring 19.6.44 Disbanded 25.9.45
Reformed Wittering 10.10.46 Coltishall 23.1.47
Det. Church Fenton 19.11.49 Coltishall 22.9.50
Horsham St. Faith 15.1.52 Coltishall 4.7.52 then (R)

No. 25 (ZK)
Church Fenton 17.5.42 Acklington 19.12.43
Coltishall 5.2.44 Castle Camps 27.10.44
Boxted 11.1.46 West Malling 5.9.46–9.51 (R)

No. 27
Agartala 5.43–3.44

No. 29 (RO)
West Malling 10.42 Bradwell Bay 13.5.43
Ford 4.9.43 Drem 1.3.44
West Malling 1.5.44 Hunsdon 19.6.44
Colerne 22/26.2.45 Manston 11.4.45
West Malling 1.1.46 Tangmere 12.50–1.51 (R)

No. 36
Thorney Island 10.46–10.47

No. 39 (nil)
Fayid 9.49 Kabrit 21.2.51–3.53 (R)
Khartoum 25.9.45–5.10.46

No. 45 (OB)
Yelahanka, India 12.2.44 Dalbhumgarth, India 29.5.44
Ranchi, India 8.44 Kumbhirgram, India 9.44
Joari, India 26.4.45 Cholavaram, India 5.45
Madras, St. Thomas' Mount, India 10.45–5.46 (R)

No. 47 (KU?)
Kumbhirgram 18.2.45 Kinmagan, Burma 28.4.45
Hmawbi, Burma 15.8.45
Butterworth, Malaya 1.46–21.3.46 (D) and detachments

441

Mosquito Squadrons and Units

No. 55 (nil)
Hassani, Greece   7.46–1.11.46 (D)

No. 58 (nil) (OT post-war)
Benson   8.47 Wyton   3.53–1954 (R)

No. 68 (WM)
Castle Camps   7.44 Coltishall   28.10.44
Wittering   8.2.45 Coltishall   27.2.45
Church Fenton   16.3.45–20.4.45 (D)

No. 69 (WI)
Cambrai/Epinoy,   8.8.45–28.3.46 Wahn, Germany   1.4.46–6.11.47 (D)
  France (613 renumbered)   (180 renumbered)

No. 81 (nil)
(684 renumbered)   1.9.46 Seletar   9.46
Changi   10.47 Tengah   2.48–12.55 (R)

No. 82 (UX)
Kolar, India   4.7.44 Ranchi   5.10.44–12.44
Chharra, India   1.45 Kumbhirgram   1.45–4.45
Cholavaram, India   26.5.45–10.45
Madras, St.   12.10.45–15.3.46 (D)
  Thomas Mount

No. 84 (PY)
Yelahanka, India   2.45 Chharra, India   23.4.45
Madras, St. Thomas   8.6.45 Guindy   9.45
  Mount
Seletar, Singapore   9.45 Det. to Soerabaja   9.11.45
Batavia/Kemajoran   1.46 Kuala Lumpur, Malaya   5.46
Seletar, Singapore   9.46–12.46 (R)

No. 85 (VY)
Hunsdon   8.42 West Malling   13.5.43
Swannington   1.5.44 West Malling   14.7.44
Swannington   29.8.44 Castle Camps   27.6.45
Tangmere   9.10.45 West Malling   16.4.47–10.50 (R)

No. 89
Baigachi   21.5.45 Seletar   9.45–30.4.46 (D)

No. 96 (ZJ)
Drem   11.43 West Malling   8.11.43
Ford   1.8.44 Odiham   24.9.44–12.12.44 (D)

No. 98 (VO)
Brussels/Melsbroek,   18.9.45 Wahn, Germany   15.3.46
  Belgium
Celle, Germany   17.9.49–2.50 (R)

No. 105 (GB)
Swanton Morley   11.41 Horsham, St. Faith   9.12.41
Marham   29.9.42 Bourn   23.3.44
Upwood   6.45–1.2.46 (D)

Mosquito Squadrons and Units

No. 107 (OM)
Lasham	2.44	Hartford Bridge	23.10.44
Epinoy/Cambrai	19.11.44	Melsbroek/Brussels	19.7.45
Gutersloh	45	Wahn	1.12.47
(renumbered 11 Sqn.)	4.10.48		

No. 108
Luqa	2.44	Hal Far	1.7.44
Detachments: Alghero	24.7.44 (R)		
Catania			

No. 109 (HS)
Wyton	8.42	Marham	5.7.43
Little Staughton	2.4.44	Disbanded	30.9.45
Woodhall Spa (627 Sqn. renumbered)	1.10.45		
Wickenby	10.45	Hemswell	27.11.45
Coningsby	4.11.46	Hemswell	1.4.50
	7.52 (R)		

No. 110 (nil)
Yelahanka, India	11.44	Joari, India	15.3.45
Kinmagan, Burma	25.5.45	Hmawbi, Burma	16.8.45
Seletar, Singapore			
(det. Kemajoran)	10.45		
(det. Seletar)	1.46		
	15.4.46 (D)		

No. 114 (RT)
Khormaksar, Aden	1.11.45
(renumbered 8 Sqn.)	1.9.46 (R)

No. 125 (VA)
Valley	1.44	Hurn	25.2.44
Middle Wallop	1.8.44	Coltishall	18.10.44
Church Fenton	24.4.45		11.45 (D)
Disbanded	20.11.45		

No. 128 (M5)
Wyton (reformed)	15.9.44	Warboys	22.6.45
Gilze Rijen, Holland	20.9.45		
(absorbed into 226 Sqn.)			
Brussels/Melsbroek	8.10.45	Wahn (re-established)	3.46
(renumbered 14 Sqn.)	1.4.46		

No. 139 (XD)
Horsham	6.42	Marham	29.9.42
Wyton	4.7.43	Upwood	1.2.44
Hemswell	4.2.46	Coningsby	4.11.46
Hemswell	1.4.50		6.53 (R)

No. 140 (nil)
Hartford Bridge	11.43	A.12/Balleroy (Normandy)	3.9.44
Amiens/Glissy	9.9.44	Melsbroek	26.9.44

443

Mosquito Squadrons and Units

Eindhoven	15.4.45	Fersfield	9.7.45
Acklington	9.7.45	Fersfield	25.9.45
	10.11.45 (D)		

No. 141 (TW)

Wittering	10.43	West Raynham	4.12.43
Little Snoring	3.7.45	Disbanded	7.9.45
Reformed Wittering	17.6.46	Coltishall	14.2.47
Church Fenton	23.11.49	Coltishall	22.9.50–3.52 (R)

No. 142 (4H)
Re-formed, Gransden Lodge   25.10.44–28.9.45 (D)

No. 143 (NE)

Banff	11.44	(became 14 Sqn.)	1.6.45 (D)

No. 151 (DZ)

Wittering	4.42	Colerne	30.4.43
Middle Wallop	16.8.43	Colerne	17.11.43
Predannack	1.5.44	Castle Camps	8.10.44
Hunsdon	19.11.44	Bradwell Bay	1.3.45
Predannack	17.5.45	Exeter	6.46
Weston Zoyland	9.46–10.10.46 (D)		

No. 157 (RS)

Debden	13.12.41	Castle Camps	1.42
Bradwell Bay	15.3.43	Hunsdon	13.5.43
Predannack	9.11.43	Valley	26.3.44
Swannington	7.5.44	West Malling	21.7.44
Swannington	29.8.44–8.45 (D)		
Disbanded	16.8.45		

No. 162 (CR)

L.G. 91 (Amriya S.)	1.44	Idku	5.44
Disbanded	25.9.44	Bourn (re-formed)	18.12.44
Blackbushe	10.7.45–14.7.46 (D)		

No. 163
Wyton (reformed) 25.1.45–10.8.45 (D)

No. 169 (VI)

Re-formed Ayr	2.10.43	Little Snoring	8.12.43
Great Massingham	4.6.44		10.8.45 (D)

No. 176
Baigachi   7.45–31.5.46 (D)

No. 180 (EV)

Melsbroek	9.45	Wahn	1.46
(renumbered 69 Sqn.)	1.4.46		

No. 192 (DT)

Gransden Lodge	11.42	Feltwell	5.4.43
Foulsham	25.11.43–22.8.45 (D)		

Mosquito Squadrons and Units

No. 199 (nil)
Hemswell   1.52–3.53 (R)

No. 211
Yelahanka   6.45 St. Thomas Mount   13.7.45
Daun Maung,   9.10.45–22.2.46 (D)
  Siam

No. 219 (FK)
Woodvale   27.2.44 Honiley   15.3.44
Colerne   26.3.44 Bradwell Bay   1.4.44
Hunsdon   29.9.44 Amiens/Glissy   10.10.44
Gilze Rijen   3.45 Twente   7.45
Wittering   8.45 Acklington   11.45
Church Fenton   1.9.46 (D)
Kabrit (re-formed)   4.51–3.53 (R)

No. 235 (LA)
Portreath   6.44 Banff   8.9.44–10.7.45 (D)

No. 239 (HB)
Ayr   12.43 West Raynham 10.12.43–10.7.45 (D)

No. 248 (DM)
Predannack   10.43 Portreath   17.2.44
Banff   12.9.44 Chivenor   19.7.45
Ballykelly   8.12.45 Thorney Island   31.5.46
                                                             30.9.46 (D)

No. 249 (GN)
Eastleigh/Nairobi   2.46 Habbaniya   6.46–12.46 (R)

No. 254 (QM)
North Coates   3.45–5.45 (R)

No. 255 (YD)
Rosignano   9.2.45 Detachments:
                                       Istres   3.45
                                       Falconara   7.3.45
Hal Far, Malta   4.9.45 Gianaclis   1.2.46–31.3.46 (D)

No. 256 (JT)
Ford   4.43 Detachment: Luqa   4.7.43 *et seq.*
Woodvale   9.43 Luqa (re-united)   13.10.43
Detachments:
  Catania   17.12.43–3.44
  Alghero   6.2.44–17.2.44
  Pomigliano   3.44
  La Senia   7.4.44
  Alghero   1.7.44
  Foggia   22.9.44
Forli (re-united)   26.2.45 Aviano   4.6.45
Lecce   1.9.45 El Ballah   13.9.45
Deversoir   16.12.45 Nicosia   13.6.46
                                12.9.46 (D)

445

Mosquito Squadrons and Units

No. 264 (PS
Colerne   5.42   Detachments:
                                                  Trebelzue   1.43–4.43
                                                  Bradwell   1.43–4.43
Predannack   1.5.43   Fairwood   7–12.8.43
Predannack   12.8.43   Coleby Grange   7.11.43
Church Fenton   18.12.43   Hartford Bridge   7.5.44
Hunsdon   26.7.44   A-8/Picauville   11.8.44
B–17/Carpiquet   5.9.44   Predannack   25.9.44
Odiham   30.11.44   B–51/Lille/Vendeville 9.1.45
B–77/Gilze Rijen   26.4.45   B–108/Rheine   7.5.45
B–77/Gilze Rijen   16.5.45   B–106/Twente   7.6.45–6.45 (D)
Church Fenton (re-formed)   20.11.45   Linton-on-Ouse   22.7.46
Wittering   20.4.47   Coltishall   13.1.48
Linton-on-Ouse   24.8.51–2.52 (R)

No. 305 (SM)
Lasham   12.43   Hartford Bridge   23.10.44
Lasham   25.10.44   Hartford Bridge   30.10.44
Epinoy   20.11.44   Volkel   30.7.45
B–77/Gilze Rijen   7.9.45   Melsbroek   11.45
Wahn   3.46–10.46

No. 307 (EW)
Exeter   12.42   Fairwood Common   15.4.43
Predannack   7.8.43   Drem   9.11.43
Coleby Grange   2.3.44   Church Fenton   6.5.44
Castle Camps   27.1.45   Coltishall   1.6.45
Horsham, St. Faith   24.8.45–2.1.47 (D)

No. 333 (ex. 1477 Flight: KK) ('B' Flight only used Mosquitoes)
Leuchars (formed)   10.5.43
Banff   30.9.44   'B' Flight   30.5.45–21.11.45 (D)
                                        became 334 Sqn.

No. 404 (EO)
Banff   4.4.45–25.5.45 (D)

No. 406 (HU)
Winkleigh   4.44   Colerne   9.44
Manston   1.12.44   Predannack   14.6.45–31.8.45 (D)

No. 409 (KP)
Acklington   3.44   Hunsdon   30.4.44
Detachment:
  West Malling   16.5.44–19.6.44
B–17/Carpiquet   24.8.44   St Andre   12.9.44
Le Culot   9.44   Lille/Vendeville   12.10.44
Rheine   21.3.45   Gilze Rijen   13.5.45
Twente   3.6.45–1.7.45 (D)

No. 410 (RA)
Acklington   12.42   Coleby Grange   22.2.43
West Malling   23.10.43   Hunsdon   9.11.43

Mosquito Squadrons and Units

Castle Camps 29.12.43 Hunsdon 28.4.44
Zeals 18.6.44 Colerne 29.7.44
Hunsdon 9.9.44 Amiens/Glissy 22.9.44
Lille/Vendeville 3.11.44 Amiens/Glissy 8.4.45
Gilze Rijen 4.45–9.6.45 (D)

No. 418 (TH)
Bradwell Bay 2.43 Ford 14.3.43
Holmsley South 4.44 Hurn 14.7.44
Middle Wallop 29.7.44 Hunsdon 28.8.44
Hartford Bridge 21.11.44 Coxyde 15.3.45
Volkel 25.4.45–7.9.45 (D)

No. 456 (RX)
Valley 12.42 Middle Wallop 28.3.43
Ford 29.2.44 Church Fenton 31.12.44
Bradwell Bay 16.3.45–31.5.45 (D)

No. 464 (SB)
Sculthorpe 9.43 Hunsdon 31.12.43
Swanton Morley 25.3.44 Hunsdon 9.4.44
Gravesend 17.4.44 Thorney Island 18.6.44
Rosières 7.2.45 Melsbroek 18.4.45
9.45 (D)

No. 487 (EG)
Sculthorpe 10.43 Hunsdon 31.12.43
Swanton Morley 26.3.44 Gravesend 18.4.44
Thorney Island 18.6.44 Rosières 5.2.45
Melsbroek 18.4.45 Cambrai 7.45
Became 16 Sqn. 20.9.45 (D)

No. 488 (ME)
Heathfield 8.43 Bradwell Bay 3.9.43
Colerne 3.5.44 Zeals 12.5.44
Colerne 29.7.44 Hunsdon 9.10.44
Amiens/Glissy 15.11.44 Gilze Rijen 4.4.45
26.4.45 (D)

No. 489 (P6)
Dallachy 5.45–1.8.45 (D)

No. 500 (RAA)
West Malling 2.47–10.48 (R)

No. 502 (RAC)
Aldergrove 8.47–6.49 (R)

No. 504 (RAD)
Hucknall 3.48–7.48 (R)

No. 515 (3P)
Little Snoring 3.44–10.6.45 (D)

No. 521 (nil)
Bircham Newton 8.42.–31.3.43
  (became 1409 Flt)

447

Mosquito Squadrons and Units

No. 540 (nil) (DH post-war)
Benson 'B' flight formed 19.10.42
Leuchars 'A' flight formed 19.10.42
Benson 'A' flight arrived 1.3.44 Coulommiers 29.3.45
Trondheim/Mount Farm 1.10.45 Benson 1.1.46
Wyton 26.3.53–7.53 (R)

No. 544 (nil)
Benson 3.43 'B' Flight detached Gibraltar 3.43
'B' flight re-formed Benson 17.10.43 Detachment Leuchars 3.5.44–45
13.10.45 (D)

No. 571 (8K)
Downham Market (formed) 7.4.44 Detachment: Graveley 12.4–24.4.44
Oakington 24.4.44 Warboys 20.7.45–28.9.45 (D)

No. 578 (?)
Burn 4.45–4.45 (D)

No. 600 (BQ)
Cesenatico 1.45 Campofordino 24.4.45
Aviano 24.7.45–21.8.45 (D)

No. 604 (NG)
Scorton 4.44 Church Fenton 22.6.44
Hurn 3.5.44 Zeals 25.7.44
Colerne 1.8.44 A-8/Picauville 6.8.44
B–17/Carpiquet 9.9.44 Predannack 24.9.44
Odiham 5.12.44 Lille/Vendeville 1.1.45–18.4.45 (D)

No. 605 (UP/RAL)
Ford 2.43 Castle Camps 15.2.43
Bradwell Bay 6.10.43 Manston 7.4.44
Hartford Bridge 21.11.44 Coxyde 15.3.45
Volkel 25.4.45
 became 4 Sqn. 31.8.45
Honiley (re-formed) 5.47–1.49 (R)

No. 608 (6T) (RAO post-war)
Downham Market 1.8.44–28.8.45 (D) Thornaby 8.47–10.48 (R)
 (re-formed)

No. 609 (RAP)
Church Fenton 4.47–9.47 (R) Yeadon 9.47–3.49 (R)

No. 613 (SY)
Lasham 12.43 Swanton Morley 11.4.44
Lasham 24.4.44 Hartford Bridge 23.10.44
Epinoy 20.11.44 Fersfield 28.7.45
Epinoy 4.8.45
 became 69 Sqn. 8.8.45 (D)

No. 616 (RAW)
Finningley 9.47–5.49 (R)

448

Mosquito Squadrons and Units

No. 617 (?)
Woodhall Spa   4.44–3.45 (R)

No. 618 (nil)
Skitten (formed)	1.4.43	Detachments:	
		Turnberry	7.5.43–1.7.43
		Predannack	4.10.43–11.43
Wick	9.7.44	Beccles	21.8.44
Detachment: Turnberry	2.10.43–13.10.44		
At sea	31.10.44–23.12.44		
Australia based	12.44–2.45	Narromine	2.45–25.6.45 (D)

No. 627 (AZ)
Oakington (formed)	12.11.43	Woodhall Spa	15.4.44–30.9.45 (D)

No. 680 (nil)
H.Q. Matariah	2.44	Detachments: Tocra, San Severo
H.Q. Deversoir	20.2.45	Detachments: Habbaniya, Aqir, Teheran, Athens
H.Q. Ein Shemer (became 13 Sqn.)	7.46–1.9.46	

No. 681 (nil)
Dum Dum   8.43–12.43 (R)

No. 684 (nil)
Dum Dum	10.43	Comilla	9.12.43
Dum Dum	31.1.44	Alipore	6.5.44
Detachment: China Bay		Saigon	11.45
Bangkok (became 81 Sqn.)	1.46–1.9.46		

No. 692 (P3)
Graveley (formed)	1.1.44	Gransden	4.6.45–20.9.45 (D)

2. Other Units

Many units other than R.A.F. squadrons used Mosquitoes. A list of the principal users amongst these follows.

Armament Practice School, Acklington, 1947–50.
Air Torpedo Development Unit, Gosport, 1944–50.
Bomber Development Unit, Feltwell, 1943–45.
Bomber Support Development Unit Foulsham 4.44, Swanton Morley 12.44. Disbanded 1945.
Bombing Trials Unit, West Freugh, 1945–51.
Civilian Anti-Aircraft Co-operation Units.
Central Bomber Establishment, 1946–49 (code letters DF).
Central Flying School/Empire Central Flying School.
Central Gunnery School, 1944–51, Leconfield.
Empire Test Pilots School, Boscombe Down, Cranfield, Farnborough.
Empire Air Armament School, Manby.
Photographic Development Unit, Benson.

Mosquito Squadrons and Units

Special Installation Unit, Defford.
Signals Flying Unit, Honiley.
204 Advanced Flying School. Cottesmore, Driffield, Brize Norton, Swinderby, Bassingbourn. Disbanded 1953.
Fighter Interception Unit. Ford 1942, Wittering 3.4.44, Ford 23.8.44, became F.I.D.U. 1.10.44.
Fighter Interception Development Squadron/Night Fighter Development Wing/Fighter Experimental Flight/Ranger Flight/Central Fighter Establishment. Ford 10.44, Tangmere 2.45, Great Massingham 10.45, West Raynham 11.46–52.
No. 228 Operational Conversion Unit, Leeming 1.5.47–54.
No. 229 Operational Conversion Unit, Chivenor.
No. 231 Operational Conversion Unit, Bassingbourn, 1953.
No. 237 Operational Conversion Unit, Leuchars, Benson.
No. 8 Operational Training Unit. Fraserburgh 11.42, Dyce 11.2.43, Haverford West 1.45, Benson 19.10.46, became 237 O.C.U. 31.7.47.
No. 13 Operational Training Unit. Bicester/Finmere 1.44, Harwell/Finmere 13.10.44, Finmere 1.3.45, Middleton St. George 22.7.45. Merged with 54 O.T.U. 1.5.47.
No. 16 Operational Training Unit. Upper Heyford/Barford 31.12.44. Cottesmore 1946–3.48; became 204 A.F.S.
No. 51 Operational Training Unit. Cranfield 7.44–14.6.45.
No. 54 Operational Training Unit. Charter Hall 8.44, East Moor 1.11.45. Merged with 13 O.T.U. 1.5.47.
No. 60 Operational Training Unit. High Ercall 17.5.43, Merged with 13 O.T.U. 11.3.45.
No. 132 Operational Training Unit. East Fortune 15.5.44, Haverford West 6.2.45, East Fortune 6.45–47.
Pathfinder Navigational Training Unit. Upwood and Warboys 6.43, Warboys 5.3.44–45.
No. 1 Photographic Reconnaissance Unit. Benson 13.7.41. Detachments: Wick 9.41–10.42, Leuchars 9.41–10.42, Gibraltar, Middle East; North Russia 13.8.42–23.10.42. Disbanded 19.10.42.
No. 1300 Met. Flt. (briefly existed 1946–47 as 18 Sqn.). No. 1317 Met. Flt.
No. 1401 Flight. Bircham Newton 1942, disbanded into 521 Sqn. 9.42.
No. 1409 Flight. Formed Oakington 1.4.43, Wyton 8.1.44, Upwood 5.7.45, Lyneham 10.10.45–46 (D).
No. 1655 Mosquito Training Unit. (Formed from the Mosquito Conversion Unit 18.10.42; Horsham M.C.U. formed 30.8.42, Marham 29.9.42) Marham—disbanded 1.5.43. Reformed Finmere 1.6.43, Marham 1.7.43, Warboys 7.3.44. Became 16 O.T.U. 31.12.44.
No. 1672 Mosquito Conversion Unit. Yellahanka 1.2.44, Kolar 3.6.44, Yelahanka 29.10.44, disbanded 31.8.45.
No. 1692 Bomber Support Training Flight. Gt. Massingham 17.6.44, disbanded 16.6.45. Unit code 4X.

3. Royal Navy Squadrons

The Royal Navy used Mks. II, III, VI, XIII, P.R. XVI, XXV, 33, P.R.34, 37, 39.

Mosquito Squadrons and Units

No. 728 Squadron, Hal Far 1948–50.
No. 762 Squadron, Ford 2.46, disbanded 11.49.
No. 771 Squadron, Gosport 11.45, Lee-on-Solent 1947–50.
No. 772 Squadron, Arbroath, 1946–47, used P.R. XVIs.
No. 790 Squadron, Dale 10.45, Culdrose 1.48. Disbanded 11.49.
No. 811 Squadron reformed with 15 Mk. VIs at Ford 9.45, reduced to 12 aircraft 11.45, re-armed with T.R.33s 8.46, reduced to 6, and moved to Brawdy 12.46, disbanded 7.47.

4. Foreign Air Forces using Mosquitoes

Belgian Air Force: Mark VI, XXX.
Chinese Nationalist Air Force: Mk. 26.
Czechoslovakian Air Force: Mk. VI.
Dominican Republic: Mk. VI.
French Air Force (L'Armée de l'Air): Mk. VI, P.R.XVI
Israeli Air Force: Mk. VI, P.R. XVI.
Norwegian Air Force: Mk. III, VI.
Swedish Air Force: Mk. XIX.
Turkish Air Force: Mk. VI.
Yugoslav Air Force: Mk. VI, 38.

APPENDIX 21

Brief Notes on the Fighter Interception Unit and its Equipment

A principal user of Mosquitoes for trials and operational purposes was the F.I.U. Its first Mosquito, a Mk. II, arrived from 157 Squadron for handling trials on 24 March, 1942. During August tests with A.I. Mk. V took place, using Mosquito II DD668. S.C.R. 720 radar arrived in Britain towards the end of November, 1942, and was installed in a Wellington VIII of T.F.U. for preliminary trials. Like A.I. VIII it was 10 cm. radar. Installation of S.C.R. 720 was carried out at Hatfield under Pree of Western Electric, then the aircraft DZ659 was sent to F.I.U. on 1 April, 1943, for operational trials. Thirty or so flights were made, ten at night. Brief high-altitude operational trials were also being undertaken at this time using Mk. XVs.

A few defensive experimental night patrols using DD715 and HK166 equipped with A.I. Mk. VIII took place in August, but it was the commencement of offensive patrols that were a greater milestone. These, code named *Mahmoud*, began with Mk. IIs on 25 August, when Wg. Cdr. Chisholm and Plt. Off. Bamford flew in HJ702 to Maastricht and Paris. On 7 September HK195 made the first S.C.R. 720 operational sortie under Durrington G.C.I. station, but in future F.I.U. operated only in conjunction with Bomber Command.

DZ659 arrived for further trials on 26 November and on 23 December HK369 flew in from Defford, a Mk. XVII with A.I. Mk. VIII for comparison tests with the Mosquito XII and XIII. Further important trials, this time with production A.I. Mk. X were undertaken using HK236 an aircraft of the 'crash programme series'. DZ659 differed in that it was a non-standard prototype fitted now with S.C.R. 729 radar, and was christened 'Eleanora'.

When F.I.U. moved to Wittering, 3 April, 1944, it took five Mosquitoes including HJ702:ZQ-D, HJ705:ZQ-P and DZ301:ZQ-F, a Typhoon with A.I. Mk. VI, a Fulmar and several R.A.F. Fireflies. Trial work currently centred on backward-looking A.I. for 100 Group. HK478 had this installed over the rear fuselage main hatch. A Mk. IX MM230 was being used to compare exhaust flame damping in May, 1944, and in July, 1944, NT181 was on loan from B.S.D.U. for *Monica IIIe* tests in conjunction with radar ranging against *Divers*.

F.I.U. was re-established on 1 October, 1944, as part of the Night Fighter Development Wing which became an effective unit on 16 October, 1944, and comprised the Night Fighter Development Squadron, Night Fighter Training Squadron, Fighter Experimental Flight (embracing Ranger Flight) and the Fighter Interception Development Squadron, all a branch of the Central

Brief Notes on Fighter Interception Unit

Fighter Establishment. F.I.D.S. was at once engaged upon intercepting He 111s releasing V-1s, *Flowers*, intruders and by December was using Mosquito XIXs and 30s and Beaufighter VIs.

During twelve days detachment to Swannington, Flt. Lt. Hedgecoe destroyed a Ju 88 and three Bf 110s and damaged three more. The Ranger Flight began operations on 6 December when, using Coltishall as a forward base, it despatched two Mk. VIs to north-east Holland. On a *Day Ranger* on 26 January, 1945, Flt. Lt. E. L. Williams and Flt. Lt. P. S. Compton in Mk. VIs destroyed two Bf 109s. Another successful *Day Ranger* on 14 February, resulted in the destruction of a Ju 88/Fw 190 Mistel and a Ju 88, and two other aircraft on the ground. Similar *Rangers* continued and by the end of March, Ranger Flight claimed 18 enemy aircraft destroyed, 2 probables and 28 damaged. The highlight came on 23 February, when two Mk. VIs destroyed a Ju 52 and damaged thirteen other machines including a Do 24, Me 410, He 111s and Bf 110s.

F.I.D.S. also operated during 1945, at night and assisting 100 Group using Mosquito 30s. Various versions of the Mosquito served F.I.U., and for varying periods, and others used by the N.F.D.W. are also listed here:

Mk. II DZ235 (9.9.43–14.9.43), DZ299 (23.7.43–6.9.43), DZ301 (15.9.43–24.5.44), DZ304 (5.11.43—F.T.R. 26.11.43), DZ680 (1.10.43—F.T.R. 23.10.43), DZ725 (7.8.43–11.9.43), HJ702 (23.7.43–31.7.44), HJ705 (8.10.43–24.5.44).

Mk. VI HR176 (12.7.44–15.4.47), HR250 (22.2.45–14.8.47), HX809 (25.1.44–20.2.44), LR379 (1.11.44–22.3.46), PZ250 (22.3.45–25.10.45), RF875 (10.5.45–3.4.47), RF878 (10.5.45–28.8.47).

Mk. IX MM230 (27.5.44–30.8.45).

Mk. XII HK130 (17.6.43–3.7.43), HK166 (31.5.43–30.11.44), HK195 (Special mods. to Mk. XVII) (11.8.43–17.5.43), HK197 (25.4.44–26.3.45).

Mk. XIII HK464 (21.1.44–6.3.44), HK478 (23.3.44–12.2.45), MM550 (24.5.44–17.2.45), MM564 (24.5.44–26.8.44).

Mk. XV See Chapter 5 footnotes.

Mk. XVII DZ659 (1.4.43–21.4.43; 7.5.43–29.5.45), HK236 (30.11.43–8.3.45), HK311 (4.1.45–4.5.46), HK324 (12.7.45–25.10.46), HK353 (28.6.45–29.1.46), HK369 (23.12.43–2.12.45).

Mk. XIX MM682 (9.44–24.1.46).

Mk. 30 MM695/G (30.3.45–28.7.46), MV554 (17.1.46–10.2.47), NT261 (1.46–10.5.46), NT262 (3.5.45–28.4.47), NT266 (11.12.44–17.5.45), NT268 (17.1.46–3.10.46), NT301 (11.12.45–8.4.46), NT310 (18.12.44–2.5.47), NT442 (2.3.45–3.6.46), NT471 (27.2.45–24.9.47). Other Mk. 30s and N.F. 36s served with C.F.E. post-war, the Mk. 36s until 1953.

APPENDIX 22

Enemy Aircraft Destroyed in Air-to-Air Combat by Mosquito Fighters during Operations over Britain and the Continent

Checking claims to the destruction of enemy aircraft will always be, even with considerable documentation available from many quarters, a difficult task. Where the wreckage of the aircraft involved fell on land it is usually easier to discover the enemy aircraft involved, but when the claim was at sea it is difficult—and many successful Mosquito interceptions were made well out to sea. A further complication arises when, on a dark night, precise identification of the enemy type was uncertain. Sometimes a damaged aircraft was last recorded burning and falling towards the water, perhaps into cloud, only to make its getaway—unknown even to the watching radar stations.

Another complication involves the effectiveness of anti-aircraft fire which, with similar vagaries, complicates the unravelling of the enemy loss records maintained by the Luftwaffe Quartermaster General. One other major complication arises, for some German aircraft limped home only to crash on their territory, or be declared 100 per cent damaged and subsequently written-off.

During the course of preparing the following table it became apparent that claims made could usually be substantiated, and were not excessive. The battle against the night-bomber bore little resemblance to the campaign in daylight, where the milling fighters often made duplicated claims. From the start of 1944 Luftwaffe loss records in detail are not available. After D-Day the claims listed are as they appear in the Squadron O.R.Bs, and can be accepted in good faith as an accurate record, which they invariably were in the earlier phases of the night battle.

Omitted are the claims by Mosquitoes on *Insteps, Day and Night Intruder* and *Ranger* flights, by aircraft of 100 Group, Coastal Command, and those relating to interceptions very far out to sea which were not of a defensive nature.

i) Enemy Aircraft Claimed as Destroyed by Mosquito Fighters Operating in the Defence of Britain during 1942

Date	Sqn.	Acft.	E/A Type	Notes
29.5.42	151	DD608	Do 217 E4	Possibly destroyed over N. Sea; of KG 2
30.5.42	157	W4099	Do 217 E4	Almost certainly destroyed, S. of Dover
24/25.6.42	151	W4097	Do 217 E4	In N. Sea. Dornier of II/KG 40
24/25.6.42	151	W4097	Do 217 E4	In the Wash; of I/KG 2
25/26.6.42	151	DD616	He 111 H6	Almost certainly destroyed in N. Sea
26.6.42	151	DD609	Do 217 E	N. Sea; no confirmation in enemy records
21.7.42	151	W4090	Do 217 E4	N. Sea; probably U5+IH of 1/KG 2
27/28.7.42	151	DD629	Do 217 E4	N. Sea; of I/KG 2

Enemy Aircraft Destroyed

Date	Sqn.	Acft.	E/A Type	Notes
27/28.7.42	151	DD608	Do 217 E4	N. Sea; of I/KG 2
27/28.7.42	157	W4099	—?—	Possibly an He 111 of IV/KG 4; N. Sea
29/30.7.42	151	DD606	Do 217 E4	N. Sea
29/30.7.42	151	DD669	Do 217 E4	N. Sea
30/31.7.42	264	DD639	Ju 88 A4	Shot down near Malvern
22/23.8.42	157	DD612	Do 217 E4	Shot down 20 miles from Castle Camps
7.9.42	157	W4084	Do 217	Probably destroyed at sea; unconfirmed
8.9.42	151	DD669	Do 217 E4	Shot down at Orwell, Cambridgeshire
17.9.42	151	DD610	Do 217 E4	Shot down off East Coast; 4265:U5+UR of III/KG 2
30.9.42	157	DD607	Ju 88 A4	Shot down off Holland, first successful day combat
19.10.42	157	W4094	Ju 88 A4	Destroyed near Southwold
26.10.42	157	DD716	Ju 88 D1	Off Beachy Head in daylight

Unless otherwise stated the aircraft listed above have all been confirmed as lost by Luftwaffe records. Probables have been included in cases where research suggests these to have very likely been successful combats.

ii) Enemy Aircraft Claimed as Destroyed by Mosquito Fighters Operating in the Defence of Britain between 1 January, 1943 and 20 January, 1944

Date	Sqn.	Acft.	E/A Type	Notes
15/16.1.43	151	DD609	Do 217 E4	Dornier of III/KG 2
17.1.43	85	VY-V	Ju 88 A14	Junkers 88 of I/KG 6
22.1.43	410	HJ929	Do 217 E	Claimed as a probable; unconfirmed still
18.3.43	157	W4099	Ju 88 A14	Off Harwich
18/19.3.43	410	HJ936	Do 217 E4	In the Wash, shared with A.A. gunners
28.3.43	157	W4079	Do 217 E4	Shared with 68 Sqn., off Southwold.
14/15.4.43	85	VY-F	Do 217 E4	First to fall to Mk. XII
14/15.4.43	85	VY-L	Do 217 E4	Of II/KG 40
14/15.4.43	157	DD730	Do 217 E4	Of II/KG 40
24.4.43	85	VY-?	Ju 88	Off Lewes; unconfirmed still
13/14.5.43	157	DZ243	Do 217 E4	Of II/KG 40
16/17.5.43	85	VY-A	Fw 190 A	Over Dover ⎫ N.B. Luftwaffe records show that I/SKG 10 lost two 190s without trace 'A4 5838 and 'A5 840043, and two more crashed in France seriously damaged
16/17.5.43	85	VY-G	Fw 190 A	Off Hastings
16/17.5.43	85	VY-L	Fw 190 A	In Channel
16/17.5.43	85	VY-L	Fw 190 A	At Ashford
16/17.5.43	85	VY-D	Fw 190 A	Off Gravesend ⎭
18/19.5.43	85	VY-Z	Fw 190 A5	0856 of I/SKG 10
21.5.43	85	VY-V	Fw 190 A	In sea, 25 miles N.W. of Hardelot
29.5.43	85	VY-S	Ju 88 S1	Fell at Isfield, near Lewes
5.6.43	29	—?—	Ju 88 A14	Ju 88 claimed off Ostend (144327:3E+JT of III/KG 6 ?) Unconfirmed
11.6.43	256	—?—	Do 217	Claimed south of Ford
13/14.6.43	85	VY-C	Fw 190 A5	Came down at Wrotham; 047:CO+LT of I/SKG 10
22.6.43	85	VY-E	Fw 190	Shot down at Strood
3.7.43	F.I.U.	HK166	Ju 88 D1	Shot down in Channel; of 3(F)122
9.7.43	85	VY-Z	Fw 190	Claimed destroyed near Detling
13.7.43	85	VY-T	Me 410 A1	Shot down off Felixstowe; of V/KG 2
13.7.43	410	HJ944	Do 217	Shot down in the Humber; Do 217 K1 or Do 217 M1. Possibly the first of this type in either case claimed over or near U.K. First Do 217 K1 recorded as lost 9.7.43. On 13.7.43 Luftwaffe records show Do 217 M1 56153: U5+EL and Do 217 K1 4478: U5+EM lost, former of I/KG 2, latter of II/KG 2

455

Enemy Aircraft Destroyed

Date	Sqn.	Acft.	E/A Type	Notes
15.7.43	85	VY-G	Me 410 A1	U5+CJ:237 of V/KG 2; off Dunkirk
18.7.43	256	—?—	Fw 190	Claimed shot down in Channel; unconfirmed
26.7.43	85	VY-A	Ju 88	Claimed east of Ramsgate; more likely to have been Do 217 M1 6045:U5+GK of I/KG 2
29/30.7.43	256	—?—	Me 410 A1	Shot down 20 miles S. of Beachy Head. 11+11 of V/KG 2
?.7.43	29	—?—	Ju 88	Claimed on date unknown as shot down N.E. of Foreness. Unconfirmed
15.8.43	410	—?—	Do 217	Off Beachy Head
15.8.43	256	—?—	Do 217?	Claimed as Me 410
15.8.43	256	—?—	Do 217	South of Ford
15.8.43	256	—?—	Do 217	In Channel
15.8.43	256	—?—	Do 217	S. of Selsey Bill

Enemy records show: Lost: Do 217 M1s 40702:U5+EH of II/KG 2, 722852:U5+ET of 9/KG 2, 56160:U5+GT of 9/KG 2, 722753:U5+LR (written off after return) and Do 217 E4 5585:U5+AN of 11/KG 2

Date	Sqn.	Acft.	E/A Type	Notes
23.8.43	85	VY-V	Me 410 A1	Shot down near Shotley
23.8.43	85	VY-R	Fw 190	Claimed near Dunkirk; unconfirmed
23.8.43	29	HK197	Me 410 A1	Shot down east of Manston
23.8.43	29	HK175	Me 410 A1	Shot down near Dunkirk
23.8.43	29	HK164	Me 410 A1	Shot down north of Knocke; 214 of 15/KG 2

Of V/KG 2; 10120:U5+HF and 228:U5+EE

Date	Sqn.	Acft.	E/A Type	Notes
23/24.8.43	85	VY-G	Me 410 A1	Nr. 274:U5+DG of 16/KG 2
6.9.43	85	VY-K	Fw 190 A5	Came down in France; of I/SKG 10
6.9.43	85	VY-V	Fw 190 A5	Came down in France; of I/SKG 10
8.9.43	85	VY-R	Fw 190 A5	One came down off Aldeburgh, two off the Foreland, all of I/SKG 10:— 009 (Red 12), 1438 (Yellow 2) and 1458 (Yellow 2)
8.9.43	85	VY-L	Fw 190 A5	
8.9.43	85	VY-L	Fw 190 A5	
15.9.43	85	VY-T	Ju 88 A14	
15.9.43	85	—?—	Ju 88 A14	Damaged over Tenterden
15.9.43	488	HK204	Do 217 M1	East of Foreness
15.9.43	488	HK203	He 111	Claimed north of Bradwell

Enemy records show:— Aircraft lost were Me 410 A1 U5+AF:10171 of 15/KG 2, Do 217 M1 56162:U5+ET of 3/KG 2, and two Ju 88 A14s of II/KG 6 144512:3E+GM and 550227:3E+EP. The loss of a He 111 is most unlikely

Date	Sqn.	Acft.	E/A Type	Notes
16.9.43	29	HK189	Me 410	Claimed over the Channel
22.9.43	85	—?—	Me 410	Claimed as Me 410 off Orfordness, more likely a Do 217 M1 of III/KG 2.
6.10.43	488	—?—	Do 217 M1	Damaged over Canterbury, probably shot down
7.10.43	157	DZ260	Me 410	Damaged off Shoeburyness, probable
7.10.43	85	VY-E	Me 410	Shot down off Hastings (see footnote to page 166)
7.10.43	85	—?—	Me 410	Damaged over Gravesend
7.10.43	85	VY-A	Me 410	Shot down near Dungeness.
8.10.43	85	—?—	Ju 88 S1	South of Bradwell; of III/KG 6
8.10.43	85	—?—	Ju 188 E1	At sea; of Erprob. St./KG 6
12.10.43	151	—?—	Me 410?	Shot down N.E. of Cromer; Luftwaffe recorded loss of an aircraft of III/KG 6 type unknown
15.10.43	85	VY-E	Ju 188 E1	Wreck fell at Woodbridge
15.10.43	85	VY-K	Ju 188 E1	Wreck fell at Birchington

Enemy Aircraft Destroyed

Date	Sqn.	Acft.	E/A Type	Notes
15.10.43	85	—?—	Ju 188 E1	Claimed as an Me 410, this was possibly the third Ju 188 recorded as lost this night, 260173:3E+RH of I/KG 6. Shot down south of Clacton.
17/18.10.43	85	VY-V	Me 410 A1	Wreck fell at Hornchurch
20.10.43	29	—?—	Fw 190 A	Destroyed south of Beachy Head, one of two lost by I/SKG 10
20.10.43	29	—?—	Me 410 ?	Over Channel; possibly second Fw 190?
22.10.43	29	—?—	Fw 190	Claimed east of Beachy Head; unconfirmed.
30.10.43	85	VY-G	Ju 88 S1	Shot down S.E. of Rye
30.10.43	85	VY-J	Ju 88 S1	Shot down 20 miles S. of Shoreham. (Luftwaffe records show loss of 3E+KS:140485 and 140585:3E+AS of III/KG 6)
1.11.43	29	—?—	Ju 188 E1	Wreck fell near Andover; of KG 6
2.11.43	85	VY-K	Fw 190	South of Canvey Island; unconfirmed
5/6.11.43	410	HJ917	Ju 188 ?	Destroyed 15 miles S. of Dungeness; possibly Ju 88 S1 140583:3E+GS of III/KG 6. No Ju 188 loss recorded in German records
6.11.43	85	VY-W	Fw 190 A	Destroyed 2–3 miles south of Hastings, one of two lost by I/SKG 10
8.11.43	29	HK163	Me 410 A1	Destroyed nr. Beachy Head ⎫ Luftwaffe records show loss of 3 Me 410:— 10262:U5+JF of 14/KG 2, 10311:U5+HE of 14/KG 2, 10244:U5+BF of 15/KG 2
8.11.43	85	VY-E	Me 410 A1	Fell nr. Eastbourne
8.11.43	488	HK367	Me 410 A1	Off Clacton ⎭
20.11.43	29	HK403	Fw 190	Claimed near Horsham
20.11.43	151	HK177	Me 410	Claimed off Esparto Point; unconfirmed
25.11.43	488	HK228	Me 410	Claimed near Calais; unconfirmed
9.12.43	307	EW-R	Ju 88 D1	1310:8H+AH of 1(F)33 on met. flt.
10.12.43	410	DZ292	3 Do 217 M1	Three Do 217 M1 of I/KG 2 (56157:U5+BB, 56059:U5+IK and 722747:U5+CK) were lost, also a Do 217 K1 4476:U5+AS of III/KG 2
10.12.43	410	HJ944	Do 217	Damaged near Chelmsford
20.12.43	488	HK457	Me 410 A1	420085:U5+HE of 14/KG 2; near Rye
2.1.44	488	HK461	Me 410	Straits of Dover
2/3.1.44	85	HK374	Me 410	Off Le Touquet

iii) **Enemy Aircraft Claimed as Destroyed by Mosquito Fighters Operating in the Defence of Britain between 21 January, 1944 and 5 June 1944**

Date	Sqn.	Acft.	E/A Type	Notes
21/22.1.44	29	HK197	Fw 190	Engaged south of Beachy Head
Main	29	HK168	Ju 88	Probable; south of Beachy Head
Target:	85	VY-N	Ju 88	Off Rye
London	96	HK414	Ju 88	Near Tonbridge
	96	HK425	Ju 88	Near Paddock Wood
	96	HK372	Ju 88	Claimed two Ju 88s, one over Sussex and one south of Bexhill
	151	HK193	He 177	Near Hindhead
	488	HK380	Do 217	At sea, off Dungeness
	488	HK380	Ju 88	Fell near Sellindge, Kent
28/29.1.44	410	HK432	Ju 88	HK372 of 96 Sqn. also fired at this aircraft
London	96	HK397	Ju 88	Near Biddenden
3/4.2.44	85	VY-P	Do 217	Over Kent
London	410	HK463	Do 217	At sea, off Orfordness
	488	HK367	Do 217	At sea, off Foreness

Enemy Aircraft Destroyed

Date	Sqn.	Acft.	E/A Type	Notes
12/13.2.44	29	413	Me 410	Claimed off Fécamp
13/14.2.44	96	HK426	Ju 188	Near Whitstable
London	410	HK466	Ju 88 S	Near Romford
	410	HK429	Ju 188	Thames Estuary
	488	MM476	Ju 188	At sea
19/20.2.44	96	HK396	Me 410	South of Dungeness
20/21.2.44	25	HK285	Ju 188	Off East Anglia
London	25	HK255	Do 217	Off East Anglia
22/23.2.44	25	HK283	Do 217	Over Norfolk
London	85	VY-Y	Me 410	Over the Channel
	85	—?—	Me 410	Over the Channel
	410	HK521	Ju 88	Earls Colne
	410	HK521	Ju 188	Fell near Rochford
23/24.2.44	25	—?—	He 177	Fell near Yoxford, Norfolk
London	25	—?—	Do 217	Off East Anglia
	25	HK293	Ju 188	Off Yarmouth
	96	HK370	Me 410	At sea
	85	VY-O	Ju 188	Probable; near the Needles, I.O.W.
24/25.2.44	29	HK413	He 177	Engaged over Kent
London	29	HK515	Ju 88	At sea
	29	HK515	Ju 188	At sea
	29	HK422	Ju 88	At sea
	29	HK168	Do 217	Half credited to A.A. fire
	29	HK168	Me 410	Claimed as a probable
	96	HK405	Ju 88	Claimed as a probable
	96	HK415	Me 410	At sea
	488	HK228	Do 217	Near Brighton
	488	HK228	He 177	Fell near Tonbridge, Kent
	F.I.U.	—?—	He 177	At sea; probable.
25/26.2.44	25	HK293	Ju 188	Off Yarmouth
27/28.2.44	456	HK286	Ju 88	Wreck fell at Beer, near Portland
	456	HK286	Ju 88	Off the S.W. coast
	456	HK323	Ju 88	In sea, off French N.W. coast
29/30.2.44	85	VY-S	He 177	In the English Channel
	96	HK469	Fw 190	In the sea, off Dieppe
1/2.3.44	96	HK499	Me 410	Chased into France, destroyed there
London	151	HK377	Ju 188	Claimed two Ju 188s
	151	MM448	Ju 188	At sea
	151	HK232	Ju 88	At sea
14/15.3.44	96	HK406	Ju 88	Off South Coast
London	96	HK406	Ju 188	Off South Coast; claimed as probable
	410	HK466	Ju 88	Off the East Coast
	410	HK521	Ju 188	Off the East Coast
	410	HK432	Ju 88	Off the East Coast; claimed as probable
	488	MM476	Ju 188	Wreckage fell at Great Leighs, Essex
19/20.3.44	25	HK255	Ju 188	Claimed three Ju 188s off East Anglia
Hull	25	HK278	Do 217	At sea, heading for Hull
	25	HK278	He 177	At sea, off East Coast
	307	HK119	He 177	At sea, near Humber; unconfirmed
	264	—?—	Do 217	At sea
	264	—?—	—?—	At sea; unconfirmed claim
	264	—?—	Ju 88	At sea

Enemy Aircraft Destroyed

Date	Sqn.	Acft.	E/A Type	Notes
21/22.3.44	25	HK322	Ju 188	At sea, heading for London
London	25	HK322	Ju 188	At sea, heading for London
	410	HK456	Ju 88	At sea
	488	HK365	Ju 88	At sea
	488	HK365	Ju 88	Near Herne Bay
	488	HK380	Ju 88	Near Earls Colne
	488	MM476	Ju 88	Near Cavendish
	488	MM476	Ju 188	Shot down at Rochford, Essex
	456	HK359	Fw 190	Off the South Coast
	456	HK297	Ju 88	Destroyed off South Coast, confirmed
23.3.44	85	VY-R	Fw 190	At sea
	96	MM451	Fw 190	Off Hastings
24/25.3.44	25	HK293	Ju 188	Destroyed 45 miles off Yarmouth
London	85	VY-O	Ju 188	At sea
	85	VY-M	Ju 88	At sea
	410	HK466	Me 410	Listed as probable, at sea
	456	HK286	Ju 88	Fell near Walberton, Sussex
27/28.3.44	96	HK425	Fw 190	At sea
Bristol	219	HK260	Ju 88	Fell near Ilminster
13/14.4.44	96	MM497	Ju 88	Destroyed near Le Touquet
London	96	HK497	Me 410	At sea
	96	HK415	Me 410	At sea
18/19.4.44	85	VY-B	Ju 88	At sea
London	85	VY-R	Ju 188	At sea
	96	MM499	Me 410	Wreck fell at Brighton
	96	MM495	Ju 88	Wreck fell near Margate
	488	MM551	Ju 88	At sea
	488	MM813	Ju 88	At sea
	25	HK237	Me 410	At sea
	410	MM456	He 177	Wreckage fell near Saffron Walden
	456	—?—	Me 410	Claimed near Horsham
20.4.44	25	HK354	Ju 188	At sea
Hull	264	HK480	He 177	Claimed 40 miles E.N.E. of Spurn Head
23/24.4.44	125	HK355	—?—	Claimed off S.W. Coast
Bristol	125	HK299	—?—	Claimed off S.W. Coast
	125	HK301	—?—	Claimed off S.W. Coast
	456	HK317	Ju 88	In sea, near Swanage
25/26.4.44	85	VY-B	Me 410	At sea
Portsmouth	125	HK299	Do 217	At sea
26/27.4.44	456	HK286	Ju 88	At sea; unconfirmed
Portsmouth	456	HK297	Ju 88	At sea
	456	HK264	Ju 88	At sea
	125	HK346	Ju 188	Off St. Catherine's Point
28.4.44	125	HK346	Ju 88	Claimed off Cherbourg
Plymouth	456	HK286	Do 217	Classified as probable
29/30.4.44	406	'O'	2 Do 217	At sea ⎫ Possibly these 217s were amongst
Plymouth	456	—?—	Do 217	At sea ⎬ those using PC 1400 FX radio-controlled bombs against naval ⎭ ships at Plymouth
14/15.5.44	125	HK325	Ju 88	Off Cherbourg
Bristol	125	HK318	Me 410	North of Portland Bill
	406	'D'	Ju 88	At sea

Enemy Aircraft Destroyed

Date	Sqn.	Acft.	E/A Type	Notes
14/15.5.44 Bristol	456	HK246	Ju 188	Wreckage fell on Salisbury Plain
	456	HK297	Ju 88	Wreckage fell at Medstead
	488	MM551	Ju 188	At sea
	488	HK381	Ju 88	At sea
	604	HK527	Do 217	At sea
	604	MM526	Ju 188	At sea
22.5.44 Portsmouth	125	HK316	Ju 188	Fell near Southampton
	125	HK252	Ju 88	Fell near Southampton
	456	HK286	Ju 88	Fell near Southampton
	456	HK353	Ju 88	Fell near Southampton
27.5.44 Weymouth	456	HK346	Me 410	Near Cherbourg
28.5.44	410	HK462	Ju 88	Chased into France, fell nr. Lille
29.5.44	25	HK257	Me 410	50 miles east of Cromer
	456	HK286	Me 410	Two engaged over Southern England; no confirmation of claimed success

Total enemy sorties despatched on nights of major operations

January 21st (447), 29th (285)
Total lost: 57

February 3rd (240), 13th (230), 18th (200), 20th (200), 22nd (185), 23rd (161), 24th (170)
Total lost: 72

March 1st (165), 14th (187), 19th (131), 21st (144), 24th (143), 27th (139)
Total lost: 75

April 18th (125), 20th (130), 23rd (117), 25th (193), 26th (78), 27th (60), 28th (58), 29th (101)
Total lost: 75

May 14th (91), 15th (106), 22nd (104), 27th (28), 28th (?), 29th (?)
Total lost: About 50

Defensive Mosquitoes claimed 129 enemy aircraft. Others fell to intruders and Mosquitoes of 100 Group, Beaufighters, anti-aircraft defences and technical defects.

iv) **Enemy Aircraft Claimed by Mosquito Fighters of A.D.G.B. and 2nd T.A.F. Squadrons at Night during Defensive Operations mainly over France, Belgium, Holland and Germany between 6 June, 1944 and 7 May, 1945**

Date	Sqn.	Acft.	E/A Type	Notes
5/6.6.44	409	—?—	Ju 188	Probable, over East Coast of Britain
6/7.6.44	456	HK303	He 177	Off Normandy
7.6.44	604	MM500	Me 410	N.E. of Laval
7.6.44	219	HK319	Me 410	Over France
7.6.44	219	HK248	Ju 188	
7.6.44	456	HK290	He 177	Off Normandy; carried glider bombs
7.6.44	456	HK290	He 177	Off Cherbourg
7/8.6.44	406	'D'	Do 217	Off Normandy
8.6.44	456	HK323	He 177	Off Normandy
8.6.44	456	HK323	He 177	Off Normandy
8.6.44	456	HK302	He 177	Off Normandy
8.6.44	25	HK354	Me 410	Off Southwold
9/10.6.44	456	HK353	He 177	Off Cape Levy; carried Hs 293s
9/10.6.44	456	HK353	Do 217	Off Cap de la Hague, fell on shore
9/10.6.44	409	MM460	Ju 188	Destroyed 40 miles S.E. of Le Havre, first E/A known to have been definitely shot down over land since D-Day

Enemy Aircraft Destroyed

Date	Sqn.	Acft.	E/A Type	Notes
10/11.6.44	264	—?—	Ju 88	Probably destroyed a Fw 190 too
10/11.6.44	409	MM453	Ju 188	
10/11.6.44	456	HK249	He 177	At sea
10/11.6.44	409	MM547	Ju 188	
10/11.6.44	409	MM547	Ju 188	
11.6.44	409	MM555	Ju 188	Over France
11.6.44	409	MM523	Do 217 E	Over France
12/13.6.44	264	HK512	Ju 188	
12/13.6.44	456	HK286	Ju 88	
12/13.6.44	264	HK475	Ju 188	
13.6.44	264	HK502	He 177	Over the Channel
13.6.44	264	MM549	Ju 188	West of Cherbourg
13.6.44	410	HK366	2 Do 217	Over the Beachhead
13.6.44	410	HK459	He 177	Over Channel
13.6.44	410	HK466	Ju 88	10 miles off Le Havre
13.6.44	409	MM560	He 177	10 miles east of Le Havre
14.6.44	—?—	—?—	Me 410	Claimed during sortie over U.K.
14.6.44	456	HK282	He 177	Minelayer; off Fecamp
14.6.44	410	MM499	—?—	Unidentified claim
14.6.44	410	MM570	—?—	Unidentified claim
14/15.6.44	410	HK476	Mistel	Over the Channel
14/15.6.44	456	HK356	Ju 88	At sea
14/15.6.44	264	HK502	Mistel	Off Normandy
15.6.44	219	HK315	Ju 88	
16.6.44	410	—?—	Ju 88	S.E. of Valonges
16.6.44	219	HK344	Me 410	
17/18.6.44	410	HK466	Ju 188	6 miles off Le Havre
18.6.44	410	MM499	Ju 188	Near Caen
18/19.6.44	410	HK470	Ju 88	Over Beachhead
18/19.6.44	410	MM571	Ju 88	Over Beachhead
20.6.44	409	MM573	Ju 188	Claimed as probable
22.6.44	604	MM552	Ju 188	
22/23.6.44	125	HK238	3 Ju 88	At sea
22/23.6.44	125	HK262	2 Ju 88	At sea
23.6.44	25	HK257	Ju 188	
23.6.44	604	MM527	Ju 188	
23.6.44	409	MM554	Ju 188	
24.6.44	125	HK262	Ju 88	Near Isle de San Marcour
24.6.44	125	HK310	Ju 88	West of Le Havre
24.6.44	410	HK500	Ju 188	15 miles N.W. of Beachhead
25.6.44	125	HK287	Ju 88	Le Havre area
25.6.44	409	MM518	Ju 188	Also Do 217; unconfirmed
28.6.44	409	MM589	Ju 188	
2.7.44	604	MM526	Ju 188	15 miles W. of Le Havre
2.7.44	604	MM465	Ju 88	10 miles N. of Ouis
3.7.44	604	MM517	Ju 88	15 miles N. of Le Havre
3.7.44	410	MM447	Ju 88	At Pointe et Raz de la Percee
3.7.44	410	MM570	Ju 188	Also Me 410. MM570 F.T.R.
4.7.44	125	HK325	Do 217	Le Havre; possibly a Ju 188
4/5.7.44	456	HK356	He 177	At sea; carried PC 1400 FX bombs
4/5.7.44	456	HK249	He 177	North of Cherbourg; carried PC 1400 FX bombs
4/5.7.44	456	HK282	He 177	In Channel
4/5.7.44	456	HK282	Do 217	At sea
5.7.44	456	HK312	He 177	In Channel; carried PC 1400 FX bombs
5.7.44	604	MM552	Me 410	15 miles S.W. of Caen
7.7.44	409	MM504	Ju 88?	
8.7.44	604	MM465	Ju 88	In Channel; also Do 217 claimed
11.7.44	219	—?—	Ju 88	

Enemy Aircraft Destroyed

Date	Sqn.	Acft.	E/A Type	Notes
12.7.44	264	HK481	Ju 88	Into Seine Estuary
18.7.44	409	MM512	Do 217	
18.7.44	409	MM589	Ju 88	
19.7.44	219	—?—	Ju 188	Over Beachhead
21/22.7.44	406	MM731	Do 217	
24.7.44	409	MM504	Ju 88	
25.7.44	409	MM587	Ju 88	
26.7.44	409	HK462	Ju 88	Over Caen. Mosquito F.T.R.
28.7.44	604	MM500	Ju 88	Near Lisieux
28.7.44	488	MM513	2 Ju 88	
28.7.44	488	MM439	Ju 188	
29.7.44	488	MM466	Do 217	Also claimed two Ju 88s
30.7.44	409	MM589	Ju 88	
30.7.44	410	MM501	Ju 88	
30.7.44	604	MM621	Ju 88	30 miles south of Cherbourg
31.7.44	219	—?—	Ju 188	
1.8.44	488	MM498	Ju 88	10 miles east of St. Lo
1.8.44	604	—?—	Ju 88?	S.E. of Caen
2.8.44	488	HK532	Do 217	4 miles east of Avranches
2.8.44	488	MM439	Ju 188	8 miles south of Avranches
3.8.44	219	—?—	2 Ju 188	
3.8.44	604	MM552	Do 217	8 miles south of Granville
3.8.44	488	MM466	Ju 88	Also claimed another Ju 88
3.8.44	488	HK504	Ju 88	E.N.E. of Vire
3.8.44	488	MM513	Do 217	N.W. of Rennes
3.8.44	488	MM502	Do 217	W. of Angers
3.8.44	488	HK420	Ju 88	Two other aircraft claimed, too
3.8.44	409	MM554	Ju 188	
3.8.44	409	MM508	Ju 188	
4.8.44	409	MM512	Ju 188	
4/5.8.44	604	MM514	Ju 88	Near Rennes
4/5.8.44	410	MM449	Hs 126	
4/5.8.44	604	MM514	Ju 188	Near Rennes
6.8.44	604	MM500	Ju 88	
6.8.44	604	MM465	Ju 88	Off Avranches
6.8.44	604	MM449	2 Do 217	Also claimed a Bf 110
7.8.44	409	MM555	Ju 88	
7/8.8.44	604	MM429	2 Do 217	Near Rennes
7/8.8.44	604	HK525	Ju 188	East of Nantes
7/8.8.44	604	MM517	Ju 188	East of Falaise
7/8.8.44	604	MM517	Ju 88	Near Condé
8.8.44	604	MM528	Do 217	
8.8.44	219	—?—	2 Ju 188	
9.8.44	219	—?—	Fw 190	
10.8.44	219	—?—	Ju 88	Also claimed two Fw 190s
10.8.44	409	MM504	Fw 190	
10.8.44	409	MM523	Do 217	
10.8.44	409	MM508	Ju 88	Claimed as a probable
11.8.44	219	—?—	Ju 88	
11.8.44	409	MM619	Fw 190	
14.8.44	409	MM491	Ju 88	
15.8.44	488	MM466	Ju 88	S. of Caen
15.8.44	219	—?—	Ju 88	Also claimed a Ju 188
16.8.44	409	MM590	Ju 188	
16.8.44	488	HK377	Ju 88	S.E. of Caen
18.8.44	409	MM560	Ju 88	Also claimed a Ju 188
18.8.44	488	MM622	Do 217	
19.8.44	409	MM589	Do 217	
19.8.44	410	MM744	2 Ju 88	Possibly first to fall to Mk. XXX

Enemy Aircraft Destroyed

Date	Sqn.	Acft.	E/A Type	Notes
20.8.44	488	MM439	Ju 188	15 miles E. of Caen
27.8.44	25	HK304	Bf 109	Over Northern France
1.9.44	488	MM566	Ju 188	10 miles W. of Le Havre
25.9.44	409	MM589	He 111	V-1 launcher, over North Sea
25.9.44	25	—?—	He 111	V-1 launcher, over North Sea
27.9.44	25	—?—	Ju 188	
28.9.44	219	—?—	Ju 87	Over Low Countries
29.9.44	25	HK357	2 He 111	V-1 launchers, over North Sea
2.10.44	219	—?—	3 Ju 87	In the Nijmegen area
5.10.44	25	HK239	He 111	V-1 launcher, over North Sea
5.10.44	409	—?—	Ju 88	
5.10.44	409	—?—	Bf 110	
5.10.44	410	—?—	Ju 88	
6.10.44	456	HK317	Ju 188	20 miles N.W. of Nijmegen
6.10.44	219	—?—	Ju 87	Also claimed a Bf 110
14.10.44	125	HK245	He 219	Claimed near Duisburg
25.10.44	125	HK310	He 111	V-1 launcher, over North Sea
30.10.44	125	HK240	He 111	V-1 launcher, over North Sea
4.11.44	488	MM820	Bf 110	
5.11.44	68	TA389	He 111	V-1 launcher, at sea
10.11.44	125	—?—	He 111	V-1 launcher, at sea
18.11.44	219	MM813	Ju 87	
19.11.44	456	—?—	He 111	V-1 launcher, claimed 75 miles E. of Lowestoft
25.11.44	409	HK425	Ju 88	
25.11.44	456	HK290	He 111	V-1 launcher, claimed 10 miles off Texel
29.11.44	409	MM622	2 Ju 88	
4.12.44	219	MM790	Bf 110	
18.12.44	409	MM569	Bf 110	
18.12.44	409	MM456	Ju 88	
18.12.44	409	HK415	Ju 88	
22/23.12.44	219	MM792	Ju 88	
23/24.12.44	219	MM702	Ju 88	
23/24.12.44	488	NT263	Ju 188	
23/24.12.44	488	MM822	2 Ju 88	
23/24.12.44	488	MT570	Me 410	
23/24.12.44	219	MM706	Ju 88	
23/24.12.44	219	NT297	Ju 88	
23/24.12.44	125	HK247	He 111	V-1 launcher, at sea
23/24.12.44	409	MM461	Ju 188	
24.12.44	604	MM462	He 219	
24.12.44	219	MM698	2 Ju 188	
24.12.44	219	MM790	Bf 110	
24.12.44	—?—	—?—	2 Ju 87	
25.12.44	219	MM706	Bf 110	
26.12.44	219	MM792	Ju 87	
27.12.44	409	MM466	2 Ju 88G	
30.12.44	409	MM560	Ju 88G	
?.12.44	68	TA389	He 111	V-1 launcher, at sea
31.12.44	604	MM569	2 Ju 87	
1/2.1.45	219	MM790	Bf 110	Near Munchen Gladbach
1/2.1.45	604	HK529	He 219	
1/2.1.45	604	HK526	3 Ju 88	
4.1.45	604	MM563	Ju 88	
6.1.45	68	HK296	He 111	V-1 launcher; Mosquito F.T.R.
6.1.45	488	—?—	Bf 110	Shot down in Holland
?.1.45	219	MM792	Ju 87	
13.1.45	604	MM459	Ju 188	Rotterdam area
17.1.45	219	MM696	Ju 88	East of Aachen

Enemy Aircraft Destroyed

Date	Sqn.	Acft.	E/A Type	Notes
?.1.45	604	MM461	2 Ju 87	Over Ruhr
22.1.45	219	MM703	2 Ju 88	
23.1.45	409	MM466	Ju 88	Over the mouth of the Scheldt
23.1.45	409	MM456	Ju 188	East of Brussels
?.1.45	604	MM517	Ju 88	Dutch/German frontier
31.1/1.2.45	410	—?—	Ju 188	Four claimed by unrecorded aircraft of the squadron during the night; also claimed a Ju 87, two V-1s and damaged a Ju 87 and Ju 88
21.2.45	488	NT263	Ju 88	Fell nr. Nijmegen
24.2.45	219	MM792	Ju 87	
3.3.45	68	NT368	Ju 188	Intruder; claimed at sea
3.3.45	68	NT381	Ju 188	Intruder over U.K.; at sea
3/4.3.45	125	NT415	Ju 188	Intruder over U.K.; at sea
20.3.45	125	NT450	Ju 188	
21.3.45	409	MM466	Bf 110	
22.3.44	—?—	—?—	Bf 110	Nr. Dhunn
24.3.45	264	HK 403	Ju 88	West of Nijmegen
24.3.45	264	HK409	Ju 88	Near Düsseldorf
25.3.45	409	MM513	Ju 88	
26.3.45	604	MM497	Ju 88	
26/27.3.45	488	NT263	He 111	Also claimed Bf 110
26/27.3.45	488	NT314	Ju 88	
30.3.45	264	—?—	Fw 190	S.E. of Münster
20.4.45	264	MM521	Ju 88	20 miles West of Berlin
21.4.45	488	—?—	Ju 52	
23/24.4.45	488	NT512	Ju 52	
23/24.4.45	488	NT327	Ju 88	
23/24.4.45	409	HK506	2 Ju 52	
23/24.4.45	409	HK429	2 Ju 87	Also claimed a Fw 190
23/24.4.45	409	MM588	Ju 52	
23/24.4.45	409	MM459	Fw 190	
24/25.4.45	409	MM517	Ju 290	
25.4.45	264	HK466	Fw 190	West of Berlin
25.4.45	488	NT527	Fw 189	

v) Enemy Aircraft Claimed in Air-to-Air Combat by Mosquito Fighters of 100 Group, Bomber Command, during Offensive Operations between November, 1943, and May, 1945. Discrepancies between these claims and the tabulated material between pages 300 and 301 is probably due to up rating of enemy aircraft claimed, at the time of combat, as only damaged but later known to have been destroyed

Date	Sqn.	Acft.	E/A Type	Notes
28.1.44	141	HJ941	Bf 109	Berlin area
28.1.44	239	HJ644	Bf 110	Berlin area
30.1.44	169	HJ711	Bf 110	Berlin
5.2.44	169	HJ707	Bf 110	
15.2.44	141	DZ726:Z	He 177	
20.2.44	239	DZ270	Bf 110	Stuttgart area
25.2.44	169	DZ254	Bf 110	Stuttgart area
18.3.44	141	HJ710:T	Ju 88	Frankfurt
18.3.44	141	DZ761:C	Ju 88	Frankfurt
24.3.44	141	DD717:M	Fw 190	Berlin area
31.3.44	239	DZ661	Ju 88	Over Ruhr
11.4.44	239	DZ262	Do 217	
18.4.44	169	DD799	Bf 110	Compiegne area
20.4.44	141	DD732	Do 217	
22.4.44	169	W4076	Bf 110	S.E. Ruhr

Enemy Aircraft Destroyed

Date	Sqn.	Acft.	E/A Type	Notes
23.4.44	141	HJ712:R	Fw 190	Baltic patrol
23.4.44	169	W4085	—?—	S.E. of Cologne
27.4.44	169	W4076	Bf 110	
27.4.44	239	W4078	Bf 110	
28.4.44	239	DZ309	Ju 88	
28.4.44	239	DD622	Bf 110	
10/11.5.44	239	W4078	Bf 110	Courtrai
12/13.5.44	239	W4078	Bf 110	
12/13.5.44	239	W4092	Ju 88	
15.5.44	169	DZ478	2 Ju 88, Bf 110	Kiel area
22/23.5.44	239	DZ309	Bf 110	
24/25.5.44	239	DZ265	Bf 110	
24/25.5.44	239	DZ297	Ju 88	
27.5.44	141	HJ941	Bf 109	Aachen area
27/28.5.44	239	DD622	Bf 110	Leeuwarden
30.5.44	239	DZ297	Bf 110	Trappes
1.6.44	239	DZ265	Bf 110	
4/5.6.44	239	DD789	Ju 88	
5/6.6.44	239	DZ256	Bf 110	North of Aachen
9.6.44	169	DZ241	Do 217	
11/12.6.44	239	DZ256	Bf 110	Near Paris
12/13.6.44	157	MM630:E	Ju 188	Compiegne
13.6.44	169	DZ254	Ju 88	Near Paris
14.6.44	141	DZ240:H	—?—	
14/15.6.44	157	MM671:C	Ju 88	Juvincourt
14/15.6.44	85	VY-J	Ju 88	
15/16.6.44	157	MM671:C	Ju 88	Creil
16/17.6.44	85	VY-C	Bf 110	
17.6.44	169	W4076	Ju 88	Paris area
17.6.44	239	W4092	Ju 88	Near Eindhoven
21.6.44	515	203	Bf 110	On day ranger
21/22.6.44	239	DZ290	He 177	
24/25.6.44	239	DD759	Ju 88	Near Paris
27.6.44	141	DZ240:H	Ju 88	
27.6.44	141	HJ911:A	Ju 88	
28.6.44	239	DD759	Me 410	Near Paris
28.6.44	239	DD749	Ju 88	Near Brussels
28.6.44	169	NT150	Bf 110	
30.6.44	515	PZ203	He 111, Ju 34	
30.6.44	515	PZ188	He 111	
1.7.44	239	DZ265	Ju 88	
4/5.7.44	239	DZ298	Bf 110	Near Paris
4.7.44	169	NT121	Bf 110	
4.7.44	141	DD725:G	Me 410	Near Orleans
4/5.7.44	515	PZ163	Ju 88	Coulommiers
5.7.44	169	NT121	Ju 88	
7.7.44	141	HJ911:A	Bf 110	
7/8.7.44	239	DD789	Fw 190	
7/8.7.44	239	W4097	2 Bf 110	
7/8.7.44	239	DZ298	Bf 110	Near Charleroi
10.7.44	515	188	Ju 88, Ju 188	On a day ranger
14.7.44	169	NT112	Bf 110	
14.7.44	515	420	Ju 34	On a day ranger
18.7.44	141	HJ659:B	Bf 110	
20.7.44	141	DZ267:V	Bf 110	
20/21.7.44	169	NT113	Bf 110	
21.7.44	169	NT121	Bf 110	Courtrai
23.7.44	141	HJ710:T	Ju 88	
23.7.44	169	NS997	Bf 110	Kiel area

Enemy Aircraft Destroyed

Date	Sqn.	Acft.	E/A Type	Notes
24/25.7.44	239	DZ661	Bf 110	
28.7.44	141	HJ712:R	2 Ju 88	
28.7.44	141	HJ741:Y	Ju 88	Near Stuttgart
7/8.8.44	239	W4092	Bf 110	
8/9.8.44	239	DZ298	Bf 109	
8/9.8.44	239	DZ256	Fw 190	
9.8.44	169	NT156	Fw 190	East of Abbeville
10.8.44	169	NT176	Bf 109	
16.8.44	141	HR213:G	Ju 88	
26.8.44	169	NT146	Ju 88	Bremen area
29/30.8.44	239	W4097	Ju 88	Stettin area
11/12.9.44	141	HR180:B	Bf 110	S.W. of Mannheim
12.9.44	85	VY-A	Bf 109	
13.9.44	85	VY-D	Bf 110	
17.9.44	85	VY-J	2 Bf 110	
27.9.44	85	VY-J	Ju 88	
28.9.44	85	VY-Y	Ju 88	
6.10.44	141	NT234:W	Ju 88	
8.10.44	515	181	Bf 109	On a day ranger
14.10.44	125	HK245	He 219	Near Duisburg
14/15.10.44	239	PZ245	Fw 190	
15.10.44	85	VY-D	Bf 110	
19.10.44	141	NT250:Y	Ju 88	
19.10.44	141	PZ175:H	Ju 88	
19.10.44	85	VY-Y	Ju 88	
19.10.44	141	NT250:Y	Ju 88	Near Karlsruhe
19/20.10.44	239	PZ275	Bf 110	
29.10.44	515	PZ344	Fw 190, Ju 34	On a day ranger
29.10.44	515	PZ217	Bf 110	
1.11.44	85	VY-R	Ju 88	
1.11.44	85	VY-N	Ju 88, Ju 188, Bf 110	
1.11.44	85	VY-Q	—?—	
4.11.44	157	TA401:D	Bf 110	
6.11.44	85	VY-N	Ju 188	
6.11.44	85	VY-Y	Bf 110	
6.11.44	85	VY-A	Ju 88	
6.11.44	157	TA391:N	Ju 188	
6.11.44	157	TA404:M	Bf 110	
10.11.44	157	TA402:F	Ju 88	
11.11.44	85	VY-R	Fw 190	
21.11.44	157	MM630:E	Ju 88	
2.12.44	85	VY-A	Bf 110	
2.12.44	157	MM671:C	Ju 88	
4.12.44	157	MM671:C	Ju 88	
4.12.44	85	VY-C	2 Bf 110	
4.12.44	85	VY-B	Bf 110	
4.12.44	85	VY-D	Ju 88	
6.12.44	85	VY-O	Bf 110	
6.12.44	157	MM671:C	Bf 110, Ju 88	
12.12.44	85	VY-A	Ju 88	
12.12.44	85	VY-O	2 Bf 110	
12.12.44	85	VY-Z	—?—	
17.12.44	157	MM627:H	Bf 110	
17.12.44	157	MM653:L	Bf 110	
18.12.44	157	MM640:I	He 219	
21.12.44	157	TA401:D	Ju 88	
22.12.44	85	VY-B	2 Ju 88, Bf 110	
22.12.44	85	VY-P	Bf 110	
22.12.44	157	TA404:M	Ju 88	

Enemy Aircraft Destroyed

Date	Sqn.	Acft.	E/A Type	Notes
23.12.44	157	TA404:M	Ju 88	N.E. of Koblenz
23.12.44	85	VY-N	Bf 110	
24.12.44	157	MM671:C	Ju 88	
24.12.44	157	TA404:M	2 Bf 110	
24.12.44	157	MM676:W	Bf 110	
24.12.44	85	VY-A	Bf 110	
31.12.44	515	RS518	Ju 88	
1.1.44	85	VY-E	Ju 188	
1.1.45	85	VY-R	Ju 88, Ju 188	
1.1.45	23	RS507	Ju 88	Ahlhorn, First *ASH* success
1.1.45	157	TA393:C	Ju 88	
1.1.45	406	MM732	Ju 188	
1.1.45	406	NT283	Bf 110	
2.1.45	85	VY-X	Ju 88	
2.1.45	85	VY-N	Ju 188	
5.1.45	85	VY-B	Bf 110	
5.1.45	157	TA394:A	He 219	
14.1.45	141	HR294:T	—?—	
14/15.1.45	157	MM653:L	Bf 110	
16.1.45	85	VY-R	He 219	
16.1.45	141	HR200:E	Bf 110	
16.1.45	141	HR213:G	Bf 110	
16/17.1.45	157	TA446:Q	Ju 188	
1/2.2.45	85	MV557:Y	Bf 110	
1/2.2.45	239	NT309	Bf 110	
2.2.45	85	MT548:Z	Ju 88	Wiesbaden area
7/8.2.45	239	NT361	Bf 110	Hamm area
7/8.2.45	239	NT330	Bf 110	Goch area
14.2.45	85	MT532:S	Ju 88	
5/6.3.45	239	NT361	Ju 88	Chemnitz area
7/8.3.45	23	PZ288	Fw 190	Stendaal
8.3.45	85	MT555	Ju 188	Hamburg area
15.3.45	85	NT309	Ju 88	Musburg area
17.3.45	239	NT330	Ju 188	Nuremburg area
18.3.45	157	NT364:K	Ju 88	
18/19.3.45	85	MT548:Z	Bf 110	
19.3.45	239	NT271	He 219	
20/21.3.45	85	NT324:T	Bf 110, He 219	W. of Magdeburg
4/5.4.45	85	455:C	Ju 188	
7.4.45	85	VY-Q	Fw 190	Near Lutzendorf
8/9.4.45	85	NT494	Ju 88	
9/10.4.45	515	RS575	Ju 188	
10/11.4.45	239	NT331	He 111	Kiel area
13/14.4.45	85	NT334:S	He 219	Prenzlau area
14/15.4.45	406	NT283	Ju 88	Schleisheim area
15/16.4.45	515	PZ398	Ju 52	
17.4.45	85	MV557	Fw 190	
23.4.45	85	VY-Z	Ju 88	Wittstock
23/24.4.45	406	MM727	Ju 88	
24/25.4.45	515	RS575	Do 217	

Index

Where photograph numbers are indexed, see the photograph number in the list of full photograph captions, pages 375–90.

Aachen, 304
Aalesund, 133, 277, 278
Aarhus, 254, 311, 386, 414
A.B. Aerotransport, Sweden, 329-33
Abbeville, 132, 137, 138, 358
Abancourt, 245, 248
Acheres, 358
Ackroyd, Sqn. Ldr., 361
Acott, Flt. Lt. W. R., 127, 130
Acton, Mosquito KB328, 102
Adams, D. R., of de Havilland, 35
Adcock, Flt. Lt. D., 144
Addison, Air Commodore E. B., 287
Aerial torpedo, 399, 409, 410
Aeroembolism, or 'bends', 186, 309
Aeroplane and Armament Experimental Establishment, Boscombe Down, 44-47, 49-52, 56, 58, 60, 64, 66, 67, 102, 117, 148, 149, 187, 235, 236, 271, 274, 276, 307, 377, 396
Aerotransport, A.B., 329-33
Africa, North, 133, 219, 221, 227, 232, 332
A.I.D., 86
Ailerons, convex trailing edge, 57
Ailerons, dropped, 57
A.I. Mk. IV Radar, 287, 288, 291, 294 (for A.I., and all types of radar, *see* Radar and Radio)
A.I. Mk. V Radar, 148, 160, 452
A.I. Mk. VI Radar, 452
A.I. Mk VI II Radar, 59, 60, 66, 89, 157, 170, 291, 351, 452
A.I. Mk. X Radar, 59, 60, 170, 177, 291, 292, 294, 300
A.I./SCR 720 (British A.I. Mk. X) Radar, 59, 60, 170, 177, 291, 292, 294, 300, 452
A.I./SCR 729 (Eleanora) Radar, 452
A.I. Mk. XV 3 cm. Radar (*ASH*), 246, 294, 298, 299
A.I. Radar (and, for all types, *see* Radar and Radio), 154, 166, 167, 171, 338, 452
Aircraft Inspection Directorate, 86
Aircraft Supply Council, 86
Air Defence of Great Britain ('Home Guard'), 168

Airfields in *Britain*:
Abingdon, 140
Acklington, 155
Alconbury, 134
Angle, 74
Aston Down, 136
Ayr, 289
Banff, 276
Bardney, 142
Beccles, 77
Benson, 51, 52, 117, 118, 122, 125, 127, 130, 131, 133, 134, 138, 140, 142, 144, 146, 218, 377, 383
Biggin Hill, 177, 359
Bircham Newton, 214, 386
Blackbushe, 258
Boscombe Down, 44-47, 49-52, 56, 58, 60, 64, 66, 67, 102, 117, 148, 149, 187, 235, 236, 271, 274, 276, 307, 377, 396
Bourn, 318, 326, 370, 371, 388
Bradwell Bay, 162, 218, 238, 248, 341
Bramcote, 335
Burtonwood, 134
Cambridge (M.F.S.), 59, 77, 103, 167, 246, 296
Carnaby, 184
Castle Camps, 56, 148-50, 165, 171, 182, 236, 298
Chester, 95
Church Fenton, 158, 387
Coleby Grange, 167, 345
Colerne, 151, 156, 174, 289
Coltishall, 165, 173, 174, 182, 184, 205, 288, 290, 453
Coningsby, 365
Crail, 77
Cranfield, 340
Cranwell, 67
Croydon, 335
Dale, 282
Dallachy, 276, 293
Debden, 63, 148, 150, 151
Defford (T.F.U.), 59, 67, 452
Downham Market, 309, 312
Drem, 168
Dunsfold, 251

Index

Airfields in *Britain—cont.*
  Duxford, 187, 190, 377
  Errol, 381
  Exeter, 60, 155, 236, 271
  Farnborough (R.A.E.), 33, 45, 46, 48, 49, 67, 151, 158, 168, 187
  Feltwell, 288
  Fersfield, 256
  Ford, 60, 67, 162, 165, 227, 238, 282, 288, 289, 296, 338, 339, 346-8
  Foulsham, 288, 294
  Fraserburgh, 127
  Friston, 239, 290
  Gatwick, 136
  Gransden Lodge, 315
  Graveley, 305, 309, 318
  Gravesend, 176, 251, 389
  Halesworth, 282
  Halton, 67
  Hanworth, 50
  Harrington, 142
  Hartford Bridge, 48, 135, 237, 253
  Hatfield, 30, 38, 50, 53, 56, 58, 60, 62, 66, 81-3, 85, 86, 88, 89, 91, 93, 97-102, 111, 121, 157, 162, 179, 186, 260, 262, 263, 276, 306, 340, 342, 345
  Henlow, 102, 104, 105
  Heston, 74, 157
  High Ercall, 340
  Horsham St. Faith, 187-9, 192, 194
  Hucknall, 56
  Hullavington, 158
  Hunsdon, 67, 161, 174, 239, 241, 242, 340, 351, 384
  Langford Lodge, 134
  Langley, 43, 54, 96
  Lasham, 238, 239, 249, 251, 253, 350, 385
  Leavesden, 59, 81-3, 86, 88, 91, 95
  Lee-on-Solent, 282
  Leuchars, 119, 120, 126-8, 130-2, 139, 328-37, 387
  Linton-on-Ouse, 298
  Little Snoring, 289, 290, 293, 387
  Little Staughton, 370, 371, 388
  Lossiemouth, 367
  Lympne, 170, 192
  Lyneham, 332
  Manston, 74, 134, 176, 182, 237, 298, 303, 339, 344, 383
  Marham, 74, 195, 198, 199, 355, 357, 384, 388
  Martlesham, 201
  Metfield, 184
  Methwold, 236
  Middle Wallop, 170
  Mildenhall, 29
  North Coates, 254, 280
  Northolt, 64, 66, 67, 144
  North Weald, 136
  Oakington, 192, 216-17, 308, 309, 366
  Odiham, 249
  Oulton, 236
  Ouston, 289
  Panshanger, 58
  Poole, Dorset, 98
  Portreath, 187, 219, 272, 274
  Portsmouth, 81
  Predannack, 236, 238, 271, 288
  Prestwick, 97, 99, 102-4
  Renfrew, 77, 390
  Rougham, 318
  Sawbridgeworth, 136
  Scampton, 184
  Sculthorpe, 236-9
  Shawbury, 75, 218
  Shipdham, 163, 198
  St. Athan, 218
  St. Eval, 122, 187
  St. Mawgan, 103
  Stradishall, 201, 354
  Sumburgh, 74, 127
  Swannington, 291, 292, 295, 453
  Swanton Morley, 186, 244
  Tangmere, 290
  Thorney Island, 251, 255
  Trebalzue, 341
  Turnberry, 74, 75, 77
  Turnhouse, 67
  Twinwoods, 340
  Upwood, 308, 314, 365, 389
  Valley, 156
  Warboys, 389
  Watton, 103, 118, 142
  West Malling, 63, 129, 161, 167, 173, 177, 239, 290, 292
  West Raynham, 61, 289, 290, 350
  Weybridge, 77
  White Waltham, 85
  Wick, 75, 77, 118, 119, 156, 333
  Winkleigh, 293
  Witney, 82
  Wittering, 148, 157, 287, 364, 452
  Woodbridge, 293, 312, 314, 319, 365
  Woodhall Spa, 308, 366
  Woodhaven, 130
  Wyton, 102, 194, 216, 314, 317, 354, 355, 387
  (*See also* Appendix 21)
Airfields in *Europe:*
  Alborg, 343
  Achmer, 344, 345
  Ahlhorn, 298
  Alghero, Sardinia, 229, 231, 348
  Ansbach, 347
  Ardorf, 183
  Avord, 339, 343
  B17 (Caen/Carpiquet), 183
  Bad Doberan, 349
  Barkaby, near Stockholm, 328-37
  Barth, 349
  Beaumont, 245
  Beauvais, 134, 339, 383
  Belp/Berne, 126

470

Index

Airfields in *Europe—cont.*
Bo Rizzo, 133
Bourges, 343, 346
Bretigny, 290, 342
Bromma, Stockholm, 328-37
Budojovice, 146, 351
Cambrai, 268
Castel Vetrano, 222
Catania, 131, 221, 223, 231
Caudebac, 251
Celle, 258
Chartres, 134, 162, 245, 339
Châteaudun, 134, 138, 139, 170, 172, 248, 339, 350
Chièvres, 346, 347
Comisso, 223
Compiegne, 292
Concarneau, 343
Conches, 245
Conde, 414
Courtrai, 291
Creil, 383, 414
Dedelsdorf, 345
Deelen, 287, 291, 339, 383
Dijon, 289, 296
Dole/Tavaux, 347
Dubendorf, 352
Eferding, 351
Eggebeck, 321
Eindhoven, 170, 257, 287, 291
Epinoy, 253, 255
Erding, 351
Etampes, 339
Evreux, 339, 345, 350
Fermo, 142
Flensburg, 300
Florennes, 248, 290, 292
Foggia, 131, 226, 231, 232
Forus, 146
Fürth, 258
Gardelegen, 348
Gilze Rijen, 287, 291, 347
Griefswald, 349
Grove, 349, 414
Gutersloh, 258
Hagenau, 348
Handorf, 347
Hohn, 300
Holzkirchen, 352
Hopsten, 344
Husum, 321
Ingoldstadt, 126, 321
Jagel, 351, 352
Jasmunder Bay, 352
Jesi, 351
Juvincourt, 292, 347, 413, 414
Kastrup, 348
Kjevik, 146, 280
Lake Biscarosse, 344, 346, 347
Lake Bracciano, 226
La Marsa, 133, 140
Laon, 39, 170
Larissa, 228
Le Bourget, 125
Le Culot, 183, 352, 413
Leeuwarden, 351
Lille, 145, 164
Lister, 146
Lubeck/Blankensee, 184, 300
Luxeuil, 348, 349
Marx, 170
Melsbroek, 172, 257, 347
Melun, 290, 342
Metz, 348
Mont-de-Marson, 350
Montdidier, 138, 245, 339
Morlaix, 274
Munster/Handorf, 166
Nemecky Brod, 351
Neubiberg, 300, 352
Ober-Faffenhofen, 130
Orly, 388
Palermo, 226
Pau-Pont Long, 348
Perpignan, 229
Politz, 131
Pomigliano, 228, 229, 231
Putlitz, 349, 350
Quackbruck, 295
Rechlin, 131, 353
Rheine, 143, 174
Ribnitz, 349
Rosiere-en-Sauterre, 255
Rosignano, 232
San Severo, 144, 230, 231, 385
Sarenas, 335
Schiphol, 162
Schlessheim, 321
Schleswig, 300, 351, 352
Soesterberg, 291, 292
Sola, 146
St. Dizier, 300, 343, 351
St. Trond, 248
St. Yan, 347-9
Steenwijk, 246, 294
Tarnewitz, 146
Rorslanda, Göteburg, 332, 333
Toul, 349
Valkenburg, 370
Venlo, 343
Volkel, 360
Waalhaven, 370
Wahn, 258
Westerland, 300
Wittstock, 321, 353
Woensdrecht, 197
Wunsdorf, 258
Yagodnik (Russia), 142
Ypenburg, 371
Zempin, 132
Airfields in the *Far East:*
Alipore, 262, 265

Index

Airfields in the *Far East—cont.*
  Allahabad, 261
  Baigachi, 267
  Bangkok, 267
  China Bay, 265, 267
  Chittagong, 261
  Cocos Islands, 266
  Comilla, 261
  Dum Dum, 260-2
  Joari, 264
  Kallang, 266
  Karachi, 262
  Kemajoran, 268
  Kolar, 262, 264
  Kumbhirgram, 262
  Maiktila, 263, 264
  Mauripur, 262
  Narromine, 78
  Ranchi, 261, 262
  Saigon, 267, 268
  Seletar, 266
  Yelahanka, 262-4
Airfields in the *Middle East:*
  Ariana, 226
  Gerbini, 228
  Kasfareet, 225
  Krendi, 227
  LG91, 229
  LG219, 228
  Luqa, 220, 225, 227, 228, 231
  Rabat/Sale, 268
  Rasel Ma, 227
  Signella, 228
  Tocra, 228
Air Member for Development, 33
Air-Sea Rescue, 278
Airspeed, 68, 77, 81, 91, 93, 99
Airspeed Oxford trainer, 32, 33, 38, 82, 84, 85, 87, 88, 264
Air Staff, 34, 273
Air Training Corps, 189
Air Transport Auxiliary, Women Ferry Pilots Pool, 86
Aitken, Digger, 219
Akyab, 260, 267
Alan Muntz, 157
Albatross, de Havilland D.H.91, airliner, 29-32, 35, 53
Albemarle, Armstrong-Whitworth bomber, 31, 85, 330
Albert, 241
Aldenham motor-bus depot, 85
Alderney, 125
Alderton, Sqn. Ldr. A. M. L., 238
Alencon, 251
Alessandria, 251
Algiers, 227
Alkmaar, 201
Allan, Sqn. Ldr. J. W., 227
Allardyce, Jock, 86
Allen, Plt. Off., 174

Alliance Works, Western Avenue, Acton, 85
Allied Expeditionary Air Force, Group 85, 168, 249
Allied Expeditionary Force, 185
Allied Intelligence, 251
Allington, Wg. Cdr., 344
Alten Fiord, 74
Ambarawa, 268
Ambassador, British, to Sweden, 329
Amersfoort, 146
Amiens, 125, 136, 138, 241, 243, 251, 358
Amsterdam, 146, 197
Ancona, 229
Andaman Islands, 261, 265, 267
Anderson, Sqn. Ldr. C. A. S., 172
Andrews, Flt. Lt. C. G., 267
Andrews, Sqn. Ldr. Clif, 266
Annecy, 367
Anson, Avro Navigational trainer, 82, 87, 96, 262, 312
Antheor, 131
Anti-Diver Committee, 177
Antwerp, 237
Anzio, 231
A.R. 5513 (*Oboe*), *see* Radar and Radio, for this and all types
Araxos, 228
Arbon, Sqn. Ldr. P. W., 162
Ardennes, 254, 255
Ardley, Plt. Off. G. H., 138
Argus, H.M.S., 219, 220
Arino, 233
Armitage, Flt. Sgt., 188
Armour plating, 271
Armstrong, Maj. H. L., D.G.A.P., 190
Armstrong Whitworth Albemarle bomber, 31, 85, 330
Armstrong Whitworth Whitley bomber, 31, 38, 330
Army, 250
Army, 14th, 263
Arnhem, 253, 294, 386
Arnold, General, 96
Arnsberg Viaduct, 362, 370
ASH Radar (A.I. Mk. XV, 3 cm.), 246, 294, 298, 299 (and, for all types, *see* Radar and Radio)
Ashfield, Sqn. Ldr., 149-52
Ashford, 161
Ashley Walk bombing range, 74
Askevold Fiord, 278
Asten, 214
Aston, Sqn. Ldr. B. G., 133
Atcherley, David, 236
Atfero, 110
Ath, 136
Athens, 231, 232
Athlone, Earl of, 98
Atlantic, 217
Aube-sur-Rile, 238
Auburn, 78

Index

Augusta, 231
Augsburg, 125, 130, 230
Auloyne, 200, 388
Austin, Flg. Off. A. E., 362
Australia, 24, 25, 82, 91, 97, 98, 109-14, 194
Australia, de Havilland, 99, 109-14
Australian Air Board, 109-14
Australian Navy, Royal, 78
Austria, 231, 338, 351
Avenger bomber, 77
Avranches, 137, 413
Avro Anson navigational trainer, 82, 87, 96, 262, 312
Avro Lancaster bomber, 100, 105, 138, 293, 295, 332, 355, 356, 359-62, 368, 370, 371
Avro Manchester bomber, 38, 46
Avro Vulcan bomber, 25
Avro York transport, 332, 334, 336, 344

B-17 Boeing bomber, 139, 224
B-24 Consolidated Liberator bomber, 99, 105, 110, 139, 217, 274, 332-4, 336
B-25 Mitchell bomber, 77
B-26 Marauder bomber, 77
Babe Barkell, 220
Babington, Flg. Off., 149, 150
Babington-Smith, Flt. Off., 132
Baby Blitz, 173, 174, 347
Badley, Flg. Off. D., 229
Bad Oldeslow, 257
Baedeker raids, 149, 155
B.A.F.O., British Air Forces of Occupation, 258
Bagguley, Flt. Lt., 189, 199, 200
Baird, Flg. Off. S. I., 132
Baker, of de Havilland, 249
Baker, Flg. Off. John, 415; photograph, 57
Bâle, 369
Balkans, 230, 231
Ball, Wg. Cdr. H. W., 143
Ball-bearings, Swedish output of, 328
Balloons, barrage, 206
Balloon cutters, 57
Baltic, 131, 338, 349, 350, 353
Baltimore, Glenn Martin bomber, 98, 224
Bamford, Plt. Off., 452
Bangkok, 261, 262, 265
Banks, of de Havilland drawing office, 249
Bankstown, 109-14
Bannister, W. Off., 265
Bannock, Sqn. Ldr. Russell, of de Havilland, Canada, later Sales director, 178; photograph, 56
Barbe, St., F. E. N., 82, 87, 91
Barber, Flg. Off. C. J., 227
Barges, 256, 344, 352
Barkell, Babe, 220
Barnes, Sqn. Ldr., 274
Barnes Wallis, Dr., 74-78
Barracuda bomber, 77

Barry, Flt. Lt. J. E., 172
Barton, Lt. W., 300
Bastogne, 255
Batenburg, Flg. Off., 137
Bateson, Grp. Capt. R. N., 256
Bateson, Wg. Cdr., 244
Bateson, Sqn. Ldr. R. N., 238
Battle, Fairey bomber, 46
Battle of Britain, 83
Battye, R/O G. A., of B.O.A.C., 337
Bavaria, 125, 230, 231, 350
Bayley, Flg. Off. K. H., 125, 128
Beachy Head, 139
Beaufighter, 46, 52, 56, 155-7, 163, 222, 228, 232, 236, 261, 264, 271, 273, 274, 276-8, 280, 287, 330, 341
Beaufort bomber, 74
Beaumont, R/O N. H., of B.O.A.C., 332, 335, 337
Beauvoir, 368
Beaverbrook, Lord, 35, 38, 43, 86
Bedford, 57
Beer, Devonshire, 413
Beilen, 295
Belgian Air Force, 60, 180
Belgium, 63, 176, 180, 214, 237, 253, 257, 258, 341
Belgrade, 25
Bell, Flt. Lt., 315
Bell Punch, 85
Bell, Ralph P., Canadian Director of Aircraft Production, 97-99; photograph 54
Bellencombe, 136
Belleville-en-Caux, 248
Bends (embolism), 186, 309
Bengal, 261
Benn, Flt. Lt. M. J., 243
Bennett, Grp. Capt. D. C. T. (later Air Vice-Marshal), 301, 309, 354
Benson, Flt. Lt. J. G., 160, 292
Berchtesgaden, 370
Bergen, 118, 119, 293, 295, 329, 331, 334, 369
Bergen-Am-Zee, 382
Berggren, Sqn. Ldr., 200, 204
Berlin, 23, 29, 32, 102, 103, 104, 128, 129, 132, 135, 136, 140, 185, 192, 198, 199, 255, 257, 287, 289, 290, 295, 299, 301-6, 309, 310, 312, 314-22, 329, 362-5, 384, 387-9, 414, 417
Berlin, Largest Mosquito raid on, 318, 320
Bermuda, 103
Berne, 126
Besançon, 382
Betty Grable, 331
Bevington, Flg. Off. R. C., 102
Beziers, 229
Bf 109 and 110 aircraft, *see* Messerschmitt Bf/Me 109, 110
Bielefeld, 143, 361
Biggest bombing raid, 362

Index

Billancourt (Renault Works), 120, 121, 125, 382
Birchington, 166
Birkin, Wg. Cdr. J. M., 309
Biscay, 271–3, 275, 288, 338, 341, 343
Bishop, Flg. Off. J. G., 137
Bishop, Sqn. Ldr., 359
Bishop, R. E., of de Havilland, chief designer, 34, 35, 43, 55, 68, 270, 306; photograph 14
Bissby, Sqn. Ldr. P., 137
Blackburn, R/O R. J. G., of B.O.A.C., 337
Blackburn Botha, 312
Blair, Wg. Cdr. K. H., 238
Blenheim, Bristol bomber, 31, 38, 56, 262
Blenheim, Bristol fighter, 54, 194
Blessing, Flt. Lt. W. C. S., 200, 201
Blessing, Sqn. Ldr., 205, 357
Blizna, 230
Blohm und Voss Bv 138 flying-boat, 280, 343, 344, 346, 352
Blohm und Voss Bv 222 flying-boat, 344, 346
Bloomfield, Flg. Off., 136
Blyth, Flt. Lt. R. L. C., 130, 131
B.O.A.C. Mosquito service, 25, 191; photographs 117, 118
Boal, Flg. Off., 347, 348
Bocholt, 289
Bochum, 356
Bochumer Verein, 357
Bodien, Flt. Lt., 154
Bodington, Flt. Lt., 239
Bodington, Flt. Lt. W. J., 248
Bodo, 133
Boeing Flying Fortress bomber, 23, 105, 135, 217, 224
Boeing B-17, 139, 224
Boeing, Vancouver, 98
Bofors gun, 207
Bohlen, 299, 318
Bois Carré, 132
Bologna, 229
Bolton, 82, 85
Bolton, Sqn. Ldr., 312
Bolton, Wg. Cdr. J. D., 364
Bolzano, 226
Bomber Command, 34, 121, 132, 139, 156, 186, 196, 197, 204, 215, 216, 287, 290, 293, 300, 301, 306, 346, 350, 354, 360, 370
Bomber Command aircraft types, effort and effectiveness, 324
Bomber Command, Commander-in-Chief, 197, 306
Bombs, radio controlled, 175, 179
Bombsight, Mk. XIV, 188
Bonaventure, 74
Bone, Plt. Off. F. C., 290
Bonn, 216, 295, 318, 414
Bonnett, Flg. Off. A. L., 271, 272

Bonneuil Matours, 251
Bolls, W. Off., 192
Booth, Flt. Lt., 289
Booth, Sgt., 192
Booth, Flt. Lt. K. R., 231
Booth, Sqn. Ldr. J. S., 414
Boothman, Air Cdre. John N., 133
Boozer radar, 134 (and, for all types, *see* Radar and Radio)
Bordeaux, 118, 129, 293, 343
Boston bomber, 197, 232
Botha aircraft, 312
Bottrop, 362
Boulogne, 63, 360, 388
Boulton Paul, Defiant, 151, 290
Bourg Leopold, 367
Bourke, W. Off., 174
Bovril, 221
Bowen, Wg. Cdr., 132
Bower, Grp. Capt. L. W. C., 252
Bowman, Plt. Off., 149
Boyle, Flg. Off., 343
Brabazon, Lord (formerly Colonel Moore-Brabazon), 86
Brad, 232
Bradshaw-Jones, Sqn. Ldr., 161
Braham, Wg. Cdr. J. R. D., 350
Braithwaite, Flt. Lt., 214, 215
Bradenburg, 185
Brant, A. J., 81, 191, 379 (caption to photograph 39)
Brazil, 97
Bratislava, 231
Brauweiler power station, 195, 357
Breakey, Air Cdre., 52
Bremen, 128, 138, 194, 198, 255–7, 288, 317, 343
Brenner Pass, 133
Breslau, 133, 231
Brest, 118, 137, 272, 360, 366, 368
Bridges, 262
Brigand aircraft, 269
Briggs, Flt. Lt. Maurice, 415; photograph 57
Briggs Motor Bodies, photograph 44
Brighton, 174
Bristol, 62
Bristol Beaufighter, 46, 52, 56, 155–7, 163, 222, 228, 232, 236, 261, 264, 271, 273, 274, 276–8, 280, 287, 330, 341
Bristol Beaufighter II (Merlin), *see* Appendix 20
Bristol Beaufort bomber, 74
Bristol Blenheim bomber, 31, 38, 56, 262
Bristol Blenheim fighter, 54, 194
Bristol engines, *see* Hercules, Perseus, etc.
Bristow, Alec, 192, 196
Bristow, Flg. Off., 192
British Overseas Airways Corporation, 25, 191, 328–37
Brittany, 201
Britten, Flg. Off. R. E., 413

474

Index

British Ambassador to Sweden, 329
Broadley, Flt. Lt. J. A., 241
Brochocki, Plt. Off. J., 173
Brockbank, Plt. Off. J. R., 172
Brodie, John L. P., 81
Bromley, Wg. Cdr. N. B. R., 291
Brooking, W. Off., 182
Broom, W. Off., 188
Broome, W. Off., 195
Brown, Flg. Off., 200
Brown, Flg. Off. E. A., 346
Brown, Flg. Off. H. J., 199
Brown, Flg. Off. J. H., 204
Browne, Flg. Off., 146
Bruce, Flg. Off. Bob, 178
Bruneval, 358
Brunsbuttelkoog, 389
Brunswick, 314
Brussells, 32, 293, 299, 361
Brux, 129, 231, 369
Bryan, P. F., of de Havilland, chief draughtsman, photograph 14
Buchanan, J. S., Deputy Director General of Production, M.A.P., 84
Bucharest, 131
Budapest, 131, 226, 231
Buffalo aircraft, 46
Buffalo, N.Y., 98
Bufton, Sqn. Ldr. H. E., 354, 355
Bugge, Peter (later de Havilland Chief Development Test Pilot), 328
Bull, Flg. Off. E. G., 77
Bull, Flg. Off. R. M., 342
'Bull-nosed' Mosquito, 59
Bulpitt, Flg. Off. A. N., 200
Bunting, Flt. Lt., 154, 163, 171, 173
Bunting, Flt. Lt. E. N., 67
Bunzlau, 230
Burbridge, Flg. Off., 172
Burge, Flt. Sgt., 149
Burgess, Charles R., 82
Burhill, 75
Burke, Plt. Off. D. T., 359
Burley, W. Off. F. A., 267
Burma, 113, 219, 261, 262, 265
Burma-Siam Railway, 261, 266
Burmeister Wain Diesel Works at Copenhagen, 198
Burnett, Flg. Off., 162
Burnett, R/O J., of B.O.A.C., 337
Burrell, Pepe, 98, 103; photograph 54
Burt, Sqn. Ldr. J. B., 361
Burton-Gyles, Wg. Cdr., 228
Butler, Mrs. Alan, 86, 381 (caption to photograph 61)
Butler, Alan S., 98; photographs 5, 6, 12, 15, 16
Butt, Flt. Lt. L. T., 144
Bv aircraft, 418 (Appendix 12)
Bv 138 flying-boats, *see* Blohm und Voss
Bryne, John, 109, 110

Cabourg, 135
Caen, 131, 137, 139, 246, 249, 250, 339, 350, 358, 359, 388, 413, 414
Cagny, 388
Caicos, H.M.S., 182
Caine, Flt. Lt. J. T., 347-9
Cairnbawn, Loch, 74
Cairo, 144
Calais, 129, 288, 360, 388
Calcutta, 262
Camberley, 62
Cambodia, 267
Cambrai, 138
Cambridge, 59, 77, 103, 246
Cambridgeshire, 165
Camera, photograph 106
Camera, 5-inch, 256
Camera F24, 34, 58, 119
Camera F52, 125
Camera Fairchild, 10-inch, 138
Camera K8, 58
Camera K19, 137, 139
Camera, oblique, 135
Campbell, Flt. Lt., 257
Campbell, Flt. Lt. F. A., 183
Canada, 24, 56, 86, 87, 96-108, 353
Canada, de Havilland, 81, 86, 88, 91, 96-108, 307, 363
Canadian Department of Munitions and Supply, 43, 96-108
Canadian Power Boat, 98
Canal Zone, 234
Cannes, 131
Cannon, 40 mm., 57
Cant 1007 floatplane, 227
Canterbury, 155
Cap de la Hague, 176
Cap de la Heve, 251
Cape levy, 176
Caproni Ca 506 floatplane, 226
Card (*Highball* Development), 431
Carlson, G. K., 86
Carpenter, Flt. Lt., 361
Carpetbagger Operation, 142
Carroll, Capt. A. M., of B.O.A.C., 337
Carter, of de Havilland, 93, 155
Carter, Flt. Sgt., 192
Casein glue, 218
Casey, Flg. Off., 361
Cassells, Flg. Off., 364
Catalina flying-boat, 130
Caumont, 368
Cavarzere, 233
Cave, Flg. Off. E. H., 155
Cave-Brown, 224
Cavity Windscreen, 187
Ceylon, 25
Chandler, Flt. Sgt. W. S. G., 228
Channer, Sqn. Ldr., 188
Channer, Sqn. Ldr. R. J., 358
Chaplin, Flg. Off., 251

Index

Chapman, Sqn. Ldr., 251
Charing Cross, 170
Charleroi, 214, 293
Château Maulny, 251
Chatellerault, 250
Cheltenham, 62
Chelmsford, 160, 165, 168
Chemnitz, 299, 318, 364
Cherbourg, 136, 137, 175, 238, 350, 368, 414
Cheshire, Grp. Capt. G. L., 365, 366
Chester, 95
Chiang Kai Shek, General, 105
Chimes, Sgt. D. I., 154
Chimney pot, 190
China, 105, 106, 265
Chisholm, Wg. Cdr., 452
Chiswick, 85
Chivers, 63
Chrysler, Flt. Sgt. C. K., 356
Chubb, Flt. Sgt., 128
'Chuff bombs' (V-1s), 177
Churcher, Sqn. Ldr. R. G., 369
Churchill, Winston (later Sir Winston Churchill), 25, 49, 88, 144, 217, 224, 225; photograph 17
Clacton, 154, 155, 164
Clark, Sir Kenneth, 331
Clarion Operation, 255 (and, for other operations *see* Operation)
Clarkson, R. M., of de Havilland, assistant chief engineer, later chief engineer, 33, 35, 43, 44, 46, 48, 49, 56, 63, 330; photograph 14
Clause, Flt. Lt., 279
Clayton-Graham, Flt. Lt. W. F., 278
Clear, Flg. Off., 201
Clerk, 349
Clerke, Sqn. Ldr. Rupert F. H., 117, 118, 154, 218
Cleve, 238, 253, 361, 388
Cleveland, Flt. Lt. A. D., 347, 349, 350
Cloppenburg, 295
Clunes, Flt. Sgt., 224
Clutterbuck, Flt. Lt. T. M., 77, 130
Clyde, 126
Cobley, Flt. Lt. P. C., 244
Cockpit layouts, photographs 40, 41, 42
Coesfeld, 238
Cohen, Flg. Off. R., 244
Coke-Kerr, Flt. Lt., 365
Colchester, 136
Cologne, 188, 189, 253, 288, 302-6, 309, 315, 355, 357, 363, 364, 388, 414
Comet, de Havilland D.H.88, racer, 29, 30, 32, 35
Commercial Aviation, 189
Compeign, 136
Concarneau, 274
Conches, 358
Condor, Focke Wulf aircraft, 54, 168, 343, 346, 347

Consolidated B-24, 139
Consolidated B-24 Liberator bomber, 99, 105, 110, 139, 217, 274, 332-4, 336
Constance, Lake, 127
Constant-speed unit, 135
Contracts, Director of, M.A.P., 84
Contrails, 133
Controller-General, Ministry of Aircraft Production, 86
Controller of Research and Development, 68, 86, 271
Cooke, Sqn. Ldr. L. A., 153
Cooper, Wg. Cdr., photograph 57
'Cope' (successful *Oboe* sortie), 356 (footnote), 362
Copeland, Capt. J. R. G., of B.O.A.C., 337
Copenhagen, 119, 135, 198, 256
Copparo, 233
Cornes, Plt. Off., 338
'Corpet', 75
Corre, Flt. Lt., 176
Costello-Bowen, Plt. Off., 188, 195
Coswig, 138
Cougars, *see* 410 Squadron
Coustances, 251
Coutances, 359
Coventry, 81
Cox, Flt. Lt., 172
Crampton, Flg. Off., 201
Craven, Sir Charles, Controller General, M.A.P., 86-88, 96
Crete, 228, 231, 233
Crew, Sqn. Ldr., 161, 162
Crew, Wg. Cdr. E. D., 174, 177, 178
Cripps, Sir Stafford, Minister of Aircraft Production, 91; photograph 17
Crisbecq, 358
Crisham, Wg. Cdr. Paddy, 338
Croft, Plt. Off. Tony, 342
Cromer, 163, 168
Crookes, Flg. Off. A. N., 413
Crosbie, Wg. Cdr. D. S. K., of de Havilland, 262
Cross, Flt. Lt. A. M., 144
Cross, Brian, of de Havilland, 48, 57
Crossbow Operation, 132, 133, 136 (and, for other operations, *see* Operation)
Crossley, Mrs. Winifred, 86
Crow, Flt. Lt. G. C., 316
Crowe, J. K., of de Havilland, 35; photograph 14
Cruickshank, of de Havilland Propellers, 155
Cumming, Flt. Sgt. G., 302
Cummins, Sgt., 200, 204
Cundall's assembly shop, photograph 43
Cundall, Wg. Cdr. H. J., 358
Cunliffe-Lister, Flt. Lt. P., 216
Cunningham, Wg. Cdr. John, 67, 157, 160-3, 168
Cunningham, Flt. Lt. L., 298
Curlew, 195

Index

Curry, Wg. Cdr. G. W., 369
Curtiss-Wright, 96
Cussens, Sqn. Ldr. A. S., 238
Custance, Flt. Sgt. M., 129
Cutters, balloon, 57
Cuxhaven, 119, 145
Cybulski, Flt. Lt. M. A., 345
Cyclades, 228
Czech, 339
Czechoslovakia, 62, 231, 338, 350, 351

DB-7 Douglas Havoc night-fighter, 46
Dagger engine, 32, 33
Daily Telegraph, 319
Dakar, 103
Dakota, Douglas transport, 329, 331-3, 336
D'Albiac, Air Vice Marshal, 236, 301
Dalcom, Flt. Lt. D. P., 371
Dale, Wg. Cdr. I. G., 243
Dalglaish, R/O I. A., of B.O.A.C., 337
Dancer & Hearne Ltd., High Wycombe, 85; photograph 28
Danzig, 119, 190
Daphne, Major P. P., 232
Darling, Flt. Lt., 148, 152
Darling, Sqn. Ldr. D. F. W., 198
Darmstadt, 129, 143, 318
Darrall, Plt. Off. M. L. S., 243
Darwen, Sqn. Ldr., 187
Datchworth, 184
David, Grp. Capt., photograph 17
Davies, Flg. Off., 292
Davies, Flt. Lt. L. R., 174
Davies, O. G., 224
Davis, 215
Dawlish, Sqn. Ldr., 316
Dawson, Sgt., 226
Dawson, Plt. Off. R. G., 256
Day, Flt. Sgt. Frank, 347, 349, 350
Day Intruder Operation, 350 (and, for other operations, *see* Operation)
Daylight bombing operations, Table of, 209
Day Ranger Operation, 158, 229, 232, 245, 248, 341-3, 346-8, 351-3 (and, for other operations, *see* Operation)
DB-7 Douglas fighter, 46
D-Day, 24, 95, 117, 137, 138, 140, 174, 175, 183, 217, 235, 250, 252, 257, 273, 311, 312, 364
Dead reckoning, 306, 311, 329, 356
Deakin, Flg. Off., 160
Dean, Flg. Off., 204
Defiant, Boulton Paul fighter, 151, 290
de Havilland, Australia, 81, 89, 101, 109-14
de Havilland, Canada, 81, 86, 88, 91, 96-108, 109, 126
de Havilland Forge, 82
de Havilland, D.H.1, fighter, 29
de Havilland, D.H.82, Tiger Moth trainer, also Queen Bee pilotless target, 32, 38, 82, 84, 88, 96

de Havilland, D.H.84, Dragon, navigational trainer, 82
de Havilland, D.H.88, Comet, racer, 29, 30, 32, 35
de Havilland, D.H.89, Dominie, navigational trainer, 82, 84, 88
de Havilland, D.H.91 Albatross airliner, 29-32, 35, 53
de Havilland, D.H.93, Don trainer, 33
de Havilland, D.H.94, Moth Minor light aeroplane, 29, 32, 38, 71
de Havilland, D.H.95, Flamingo airliner, 31, 32, 43
de Havilland, D.H.99 (twin Sabre) bomber project (renamed D.H.101), 69-71
de Havilland, D.H.100, Vampire, jet-fighter, 71, 72
de Havilland, D.H.101 (twin Sabre) bomber project (provisionally called D.H.99), 71
de Havilland, D.H.102, Enlarged Mosquito, twin Merlin project, 63, 71, 72
de Havilland, D.H.103, Hornet, single seater fighter, 72, 73
de Havilland, D.H.104 Dove, light transport, 113
de Havilland, Capt. Geoffrey, later Sir Geoffrey, 17, 30, 43, 44, 68, 196; photographs 15, 16, 17
de Havilland, Mrs. Geoffrey, later Lady de Havilland, photograph 16
de Havilland, Geoffrey R., 39, 46-48, 50, 55-58, 98, 186; photographs 12, 16, 71
de Havilland, Major Hereward, 81, 101, 109, 112, 117, 129, 149, 156, 160, 187, 193, 197, 237, 301, 332, 340, 356, 357; photograph 5
de Havilland, John, 46, 48, 93, 111; photograph 18
de Havilland name, elimination of, 25
De Jace, Flt. Lt., 214
Delft, 244
Denain, 137, 417
Denby, Flg. Off. D. S., 233
Den Helder, 138
Denmark, 119, 129, 137, 143, 144, 279, 281, 293, 328-37, 338, 345, 350
Dennis, Wg. Cdr., 253
Denny, Flg. Off., 304
Denton, Sqn. Ldr., 254
Denton, Wg. Cdr., 256
Depth charge, Mk. XI, 274
Deputy Director General of Aircraft Production, 84
Deputy Director, Research and Development, Aircraft, 270
Dessau, 318, 365
Destroyer, Jaguar-class, 273
Deutsche Lufthansa, 329
Deventer, 128, 145
D/F, 329 (and, for all types, *see* Radar and Radio)

477

Index

Dickinson, Flg. Off. T., 364
Dieppe, 132, 136, 239, 242, 388
Dillon, Flg. Off. P. F., 308, 309, 312
Dinsdale, Flg. Off. W. G., 171, 176
Director of Aeronautical Inspection, Australia, 111, 112
Director of Aircraft Production, Canada, 97, 98
Director of Contracts, 84
Director General of Aircraft Production, 190, 191
Director of Technical Development, 45, 63, 306
Director of Sub-contracts, 84
Directorate of Aircraft Production, Britain, 83, 91
Diver, see V-1 flying-bomb
Do 24, 215, 217 aircraft, *see* Dornier
Dobie, Flt. Lt. I. A., 178
Dodd, Flt. Lt. F. L., 139, 142, 143
Dodecanese, 231
Dominican Republic, 419
Dominie, de Havilland D.H.89, navigational trainer, 82, 84, 88
Domleger, 388
Don, Flt. Lt., 313
Don, Sqn. Ldr. R. C., 315
Don, de Havilland D.H.93 trainer, 33
Doren, 216
Dornier, 153; *also see* App. 12
Dornier Do 24 aircraft, 130, 352
Dornier Do 215 aircraft, 62
Dornier Do 217 bomber, 62, 149-55, 157, 158, 160-2, 165, 168-76, 179, 229, 275, 339
Dortmund, 214, 303, 306, 316, 356, 362, 363
Dortmund Emms Canal, 345, 360, 369
Dorum, 189
Douai, 136
Doube, Flg. Off. G. M. L., 253
Doughty, Flt. Sgt., 272
Douglas Dakota transport, 329, 331-3, 336
Douglas DB-7 Havoc night-fighter, 46, 55
Douglas, Air Chief Marshal Sholto (later Lord Douglas), 273, 287
Douglas-Hamilton, Wg. Cdr. Malcolm, 127, 129, 135, 228
Dove, de Havilland D.H.104, light transport, 113
Dover, 63
Downs, Flg. Off., 196
Downsview, Ontario, 87, 97-108
Dracula Operation, 265 (and, for other operations, *see* Operation)
Dragon, de Havilland D.H.84 navigational trainer, 82
Dresden, 125, 133, 232, 299, 318, 369
Drummond, Wg. Cdr. A. H., 232
Drury, G. W., of de Havilland, photograph 14

Duisburg, 302, 303, 306, 309, 315, 318, 355, 356, 360, 361, 389
Duncan Sandys, 177
Dunkirk, 35, 38, 96, 384
Dunne, Plt. Off. O'Neille, 155
Dupee, Flg. Off., 260
Duport Street, Toronto, 98
Duppel radar device, 166, 170, 171, 232 (and, for all types, *see* Radar and Radio)
Dusseldorf, 288, 302, 303, 305, 306, 308, 309, 313, 355-7, 364
Dutch, 267
Dutch Central Population Registry den Haag, 244
Dykes, Capt. A., of B.O.A.C., 337

E.6 Jet fighter, 71
Eagle's Nest, 370
Early Window Radar device, 305, 309, 317, 367 (and, for all types, *see* Radar and Radio)
East Anglia, 134, 149, 150, 152, 164, 167, 182, 310
East Indies, 25
Edgware, 82
Edwards, Flg. Off. G. W., 362
Edwards, Wg. Cdr. Hughie, 191, 193-5, 197, 198
Effectiveness of Bomber Command aircraft types, 324
E.F.T.S., 83, 85, 157
Egero, 276
Egletons, 235, 252
Egypt, 52, 218
Eiffel Tower, 124
Eindhoven, 197, 201, 339, 367, 383
Eisenhower, General Dwight D. (later President of the United States of America), 224
El Agheila, 224
Eleanora radar, SCR 729, 452
Elliott Brothers, 111
Elliott, Wg. Cdr. R. P., 367
Elliott, Flt. Sgt. S. H. J., 176
Embolism, or 'bends', 186, 309
Embry, Grp. Capt. Basil, 66, 148, 150, 151, 236, 237, 240, 251
Emden, 128, 143, 215, 257, 357
Emden (cruiser), 369
Emery, Flt. Lt. C., 263
England-to-New Zealand Air Race, 113
Epinal, 214
Erfurt, 361
Errington, George, 99
Esbjerg, 119, 120, 132
Esparto Point, 346
Essen, 196, 295, 302, 313, 315, 342, 343, 355-7, 360, 362, 388, 389
Essex, 162, 163, 165
Europe, Southern, 230, 231
Evans, Flt. Lt., 178

Index

Evans, Flt. Lt. C. J., 350
Evans, Sgt. H. W., 128
Everest, Himalayan peak, 266; photograph 103
Evian, 238
Ewing, Flg. Off. P., 263
Exhaust propulsion, 31

F24 Camera, 34, 58
F52 Camera, 125
Fairey Aviation, 246
Fairey Battle bomber, 46
Fairey Firefly fighter, 452
Fairey Fulmar aircraft, 452
Falaise, 238, 255, 414
Falkenburg, 215
Falun, 331
F.A.N.Y., 229
Farben Chemical Works at Leverkusen, 305, 310, 357; at Mannheim, 310
Far East, 260-9
Farnborough, Royal Aircraft Establishment, 33, 45, 46, 48, 49, 67, 151, 158, 168, 187
Farren, W. S., 44
Farrer, Mrs., 86
Fascists, Spanish, 305
Fast Night Striking Force, *see* Light Night Striking Force
Fawcett, A.W., of de Havilland, 376 (caption to photograph 14)
Fawke, Flt. Lt. G. A., 365, 366
Fecamp, 248
Federal Aircraft, 102
Felixstowe, 163
Fencer, H.M.S., 77
Fenwick, Flt. Lt. C. R., 261
Ferguson, Earl, 103
FIDO flare path, 175, 309
Fielden, Flg. Off. J. F., 133
Fielden, Sgt., 126
Fielding, Plt. Off., 153
Fieseler Fi 156 Storch, 184, 348
Fifth Panzer Army, 255
Fillingham, P. W. F., 46, 99, 101, 111; photograph 50
Film 'Mosquito' released by Metro-Goldwyn-Mayer, 91
Film Production Unit, R.A.F., 242, 254, 256, 369
Finlayson, Flg. Off. C. G., 348
Finlayson, Flg. Off. J., 347, 349, 350
Firebash Operation, 300 (and, for other operations, *see* Operation)
Firedog Operation, 268 (and for other operations, *see* Operation)
Fisher, 99
Fisher, Plt. Off., 153
Fishermen's Bend, 78
Fitch, R. W., Ministry resident technical office, photograph 12

Fiume, 126, 127
Flak ships, 331, 334
Flame damping shroud, 56, 57, 127, 179, 188, 193, 333
Flamingo, de Havilland D.H.95 airliner, 31, 32, 43
Flare path (FIDO), 175, 309
Flares, yellow, drip, 304
Flekkefiord, 277
Flensburg, 189, 190, 191, 215
Fletcher, Sgt. R. C., 198; photograph 87
Flight refuelling for Mosquito, 262
Florence, 229, 232
Flowers Operation, 238, 245, 246, 288, 348, 350 (and, for other operations, *see* Operation)
Flushing, 360, 369
Flying-bombs V-1, *see* V-1
Flying Fortress Boeing bomber, 23, 105, 135, 217, 224
Focke-Wulf aircraft, 418, Appendix 12
Focke-Wulf Fw 189, 185
Focke-Wulf Fw 190, 52, 128, 129, 134, 138, 158, 161-3, 165, 168, 169, 171, 176, 184, 185, 186, 189, 191-3, 196, 197, 200, 204, 229, 233, 242, 277, 278, 303, 330, 332, 345, 348-50, 361, 453
Focke-Wulf Condor Fw 200, 54, 168, 343, 346, 347
Foix-Grenoble, 133
Food Minister, 54
Fôret de Helles, 244
Fôret de Moulliere, 251
Fôret de Nieppe, 144, 359
Fôrge, de Havilland, 82
Formaldehyde glue, 218, 263
Forrow, Grp. Capt. H. E., 43
Fortress Europe, 168
Fort William, Ontario, 104
Fourteenth Army, 263
Four-thousand-pound bomb, 23, 24, 49, 72, 103, 104, 306-27, 357, 363; photographs 113, 114
Fowler, Plt. Off. D. R., 243
Fownes, Flt. Lt. L. E., 413
Fox, Flg. Off. E. S. P., 171
Fox, Flt. Lt., 175
Fox, Plt. Off. J. D., 233
Foxley-Norris, Wg. Cdr., 280, 281
France, 35, 39, 62, 130-2, 135, 136, 138, 146, 173, 175, 176, 179, 196, 219, 235, 238, 249, 250, 252, 253, 268, 290, 338, 340, 343, 346, 349-51
Frankfurt, 128, 288, 298, 303, 305, 314, 317, 318, 363
Franklin, Flg. Off., 348
Frape, R/O F., of B.O.A.C., 330, 337
Fraserburgh, 127
Freeman, Sir Wilfrid, 30, 33, 34, 35, 39, 43, 68, 72, 101, 117
Freiburg, 289

Index

Freisner, Sgt., 343
French Air Force Groupe de Chasse 1/6 Corse, 268
French Air Force 10/Groupe de Chasse 1/3 'CORSE', 268
Freteval, 138
Friedrichshafen, 140, 310, 414
Friedrichstadt, 257
Frisians, 199
Frugal Operation, 144 (and, for other operations, *see* Operation)
FuG 200 radio, 295
FuG Radar, 295 (and, for all types, *see* Radar and Radio)
Fulmer, Fairey aircraft, 452
Fulton, Flg. Off. S. J., 155
Furnes, 199
Fw 189, 190, 200 aircraft, *see* Focke-Wulf
Fx 1400 PC radio-controlled bombs, 175, 179

Gabes, 222
Gaffeny, R/O J. C., of B.O.A.C., 334, 336, 337
Gallienne, Sqn. Ldr., 364
Galloway, Flt. Lt., 313
Galloway, Flt. Lt. T. H., 316
Gander, 103
Gardiner, G. C. I., of de Havilland, designer, later chief designer (propellers) and still later chief executive (rockets, missiles), 43; photograph 14
Garratt, Philip C., 87, 96–108
Gatonski, 155
GCI exercises, 230
Gdynia, 119, 120, 125, 138, 382
Geary, Flt. Lt., 171
Gebelstadt, 142
Gedong, 266
Gee Radar, 133, 139, 233, 238, 255, 272, 287, 295, 302, 304, 306, 311, 356 (and, for all types, *see* Radar and Radio)
Gelsenkirchen/Scholven, 311, 314, 317, 357, 358
General Aircraft Limited, 50, 326
General Motors-Holdens, Australia, 82, 111
General Motors, Oshawa, Ontario, 98
Gennevilliers, 367
Genoa, 25, 117, 127, 229
George V Dock, Glasgow, 77
George VI, H.M. King, 118, 217, 344; photograph 15
German High Command, 189
Germany, 138, 143, 176, 253, 338, 341, 351, 352, 357, 358, 361, 362
Gestapo, 192, 235, 251, 254–7, 305, 369
Ghent, 166
Ghislain, 245
G-H Radar, 248, 250, 305, 306, 363 (and, for all types, *see* Radar and Radio)

Gibb, Flt. Lt. W. F., 341
Gibbins, George V., 46, 56, 93
Gibbons, Flt. Lt., 347
Gibraltar, 122, 126, 127, 131, 133, 227, 228
Gibson, Wg. Cdr. G. P., 369
Giessen, 295, 299, 414
Gieswald, 132
Gill, Maurice, 104
Gill, Sqn. Ldr., 174
Gipsy Engines, 82
Gironde Estuary, 196, 274, 275, 345
Gladbach, 216
Gladbeck, 362
Glasgow, 77
Glassco, Grant, 100
Gleiwitz, 129
Glen Esk, 332, 335
Glenn Martin Baltimore bomber, 98, 224
Glenn Martin Maryland bomber, 46, 224
Glider bombs, 275
Glider train, 347, 348
Gloster E28/40 (experimental aircraft), 49
Gloster F.9/37 fighter, 43
G.M.2 day reflector sight, 223
Gneisenau, 119
Goblin engine (H1), 68, 72
Goch, 361
Goebbels, 191, 198
Goering, Hermann, 198
Goldfinch, H.M.S., 234
Gomme, E., Ltd., High Wycombe, 85
Goodman, Wg. Cdr., 296, 349, 367
Goodman, Wg. Cdr. G. H., 173
Goodwin, Flg. Off., 308
Goodwin, Flt. Lt. C. R., 264
Goodwood Operation, 359 (and, for other operations, *see* Operation)
Goose Bay, 101, 104
Gordon, Flt. Lt. J., 198; photograph 87
Gosnell, Wg. Cdr. R. J., 315
Gossen, 277
Goteburg (Torslanda), 332, 334, 336
Gotha, 362
Gotha 242 glider, 347, 348
Gottingen, 311
Gouda, 244
Gough, Flg. Off., 173
Gower, Miss Pauline, 86; photograph 61
Grable, Betty, 331
Gracie, Wg. Cdr. E. J., 289
Graf Zeppelin aircraft carrier, 125; photograph 67
Graham, Flt. Lt. A. C., 136
Graham-Little, Flg. Off., 149
Graves, Flg. Off., 196
Gravesend, 161, 176, 251
Great Leighs, Essex, 173
Greece, 230
Green, Sqn. Ldr., 160, 161, 173, 174
Green, Sqn. Ldr. F. A., 356
Green, Plt. Off. N. W. F., 214

Index

Greenland, 101, 102
Greenleaf, Sqn. Ldr. E. J., 314
Grey, Flg. Off. W. G., 179
Grieve, Flt. Sgt., 152
Griffiths, Sqn. Ldr., 300
Griffiths, Flt. Lt. N. J., 413
Griffiths, Flt. Sgt. T., 229
Griffon engine, 32
Grimsby, 150
Grimstone, Flt. Sgt., 161
Grimstrup, 198
Grinham, Ernest G., 86, 88, 89, 93, 378 (caption to photograph 34)
Groix, 238
Grover, Flt. Lt. G. E., 143
Gun, 3.7 in. anti-aircraft, 270
Gun, German, 7.9 mm., 163
Gun, German, 13 mm. MG 131, 163
Gun, German, 88 mm. anti-aircraft, 303
Gun, Six-pounder (57 mm.), 24, 60, 270, 276
Gunnis, Sqn. Ldr. A. H., 280
Guthrie, Plt. Off., 155
Gunzburg, 230

H Radar (*Oboe* in reverse) explanation, 363 (and, for all types, *see* Radar and Radio)
H1 (Goblin Engine), 68, 72
H2S Radar, 50, 134, 309, 312, 317, 319, 354, 363, 364 (and, for all types, *see* Radar and Radio)
H2X, American form of *H2S* Radar ('Mickey'), 134 (and, for all types, *see* Radar and Radio)
Hagen, 316
Hague, 204, 208, 244, 246, 289, 370, 371, 386
Hailfingen, 351
Haine St. Pierre, 245
Hale, Mr. and Mrs. Bertram, photograph 48
Halford, Frank, of de Havilland, 49, 68
Halifax, Handley Page, bomber, 38, 43, 46, 105, 288, 295, 359, 360
Halifax, Nova Scotia, 98
Hall, Flt. Lt. J. A. S., 170, 174
Hall, Flg. Off. K., 256
Hall, Flt. Lt. P. F., 172, 413
Halmaheras, 113
Halske, Berlin, 317
Hamborn, 316, 357, 363
Hamborn/August Thyssen Foundry, 357
Hamburg, 102, 145, 192, 215, 255, 256, 289, 298, 300, 302, 303, 313, 316, 319, 357, 363
Hamilton, F. J., of de Havilland, senior designer, photograph 14
Hamilton propellers, 57
Hamm, 143, 214
Hammond, Sqn. Ldr. G. W., 233
Hampden, Handley Page, bomber, 31

Hampshire, Wg. Cdr. K. M., 413
Hampson, Flt. Lt., 138
Hamre, Capt. Martin, Royal Norwegian Air Force, and B.O.A.C., 332, 335, 337
Hammstede, 339
Hamstede, 154
Hanafin, Flt. Lt. B. D., 243, 369
Habau, 316
Handley, 201
Handley Page Halifax bomber, 38, 43, 46, 105, 288, 295, 359, 360
Handley Page Hampden bomber, 31
Handley Page slots, 39
Handley Page Victor bomber, 25
Hanover, 102, 145, 199, 215, 288, 299, 303, 313, 315–17, 363
Harbours, 24
Hare, R. M., of de Havilland, senior stressman, later chief stressman, 35; photograph 14
Harler, Flg. Off. F. J., 265
Harper, R. H. of de Havilland, became chief structural engineer, photograph 14
Harper-Rees, 349
Harrington, 142
Harrington, Lt., 173
Harris, Air Marshal (later Sir Arthur Harris), 204, 287, 306, 356
Harrison, 349
Harrison, Sqn. Ldr., 173
Harrogate, 84
Harz Mountain 'National Redoubt', 370
Hassell, Flg. Off., 308
Haster, Flt. Lt. V. A., 253
Hatfield, *see* under Airfields in Britain
Haug, R/O Serre, Royal Norwegian Air Force, and B.O.A.C., 332, 335, 337
Haverfordwest, 75
Havering-atte-Bower, 171
Havoc, Douglas DB-7 night-fighter, 46, 55
Havoc/Boston, 232, 417
Havstund, 332
Hawker Hurricane fighter, 38, 43, 46, 54, 56, 82, 85
Hawker Tempest fighter, 49, 177
Hawker Tornado fighter, 43, 46
Hawker Typhoon fighter, 46, 156, 160, 161, 238, 242, 243, 452
Hay, Flg. Off. F. E., 199, 363
Haycock Operation, 144 (and, for other operations, *see* Operation)
Hayes, Flg. Off. R. G., 198; photograph 87
Hays, Flg. Off. R. M., 145
He 111, 111z, 115, 177 aircraft, *see* Heinkel
Hearle, F. T., 56, 72, 81, 86, 89; photographs 6, 12
Hearn, Martin, 81, 150, 339
Heat and exposure, 156, 260
Heath, Sgt., 150
Hedgcoe, Flt. Lt., 172, 453
Heide, 257

481

Index

Heilbronn, 143, 299
Heinkel aircraft, 418 (Appendix 12)
Heinkel He 111 bomber, 62, 150, 152, 175, 182, 183, 227, 231, 296, 299, 339, 343, 347–9, 351
Heinkel He 11z, Bi-Heinkel, 347, 349
Heinkel He 115, 349
Heinkel He 177, 63, 158, 168–70, 172–6, 179, 275, 290, 343, 346, 347, 348, 350, 353
Heligoland, 118, 198, 310, 382, 414
Helmond, 216
Hemel Hempstead, 136
Hemmingstadt, 382
Hendon, 85
Hengelo, 195, 198, 204, 237, 244
Hennesey, Patrick, 38, 84
Henry, A. T., of Rolls-Royce, 34
Henschel 293, glider bombs, 175, 176
Hercules Engine, 30, 43
Herrod-Hempsall, M., of de Havilland, senior aerodynamicist, 35; photograph 12
Hertfordbury, 39
Hesepe, 342
Hesketh, Grp. Capt., 193
Hespe, 184
Hester, Flt. Lt. V. A., 244
Heston Aircraft, 168
H/F Radio, 329 (and, for all types, *see* Radar and Radio)
Hibbert, Plt. Off. W., 63
Highball, 24, 74–78, 366; photograph 111
High-tension cables, 204, 205
High Wycombe, 85
Hill, Flt. Lt., 316
Hill, Flt. Sgt. E., 139, 143
Hill, Air Marshal Roderic, 168
Hillock, Wg. Cdr., 155
Hirson, 245
Histon, 63
Hitler, 29, 149
Hitler's birthday, 417
Hitler's Chalet, 370
Hives, E. W. (later Lord Hives), 101
Hoare, Wg. Cdr., 346, 349
Hoare, Sqn. Ldr. S., 338–40, 345
Hobson, Flt. Lt. H. G., 257
Hodgkinson, Flt. Lt., 224
Hoffman, ball-bearing works, 160
Hogan, Flt. Lt. E. E., 243
'Hogsnorton', 58
Holdaway, Flt. Lt. E. A., 363
Holland, 125, 130, 154, 166, 176, 182, 219, 237, 253, 258, 302, 339, 341–4, 355, 360, 370
Holland, Flg. Off., 135
Homberg, 312, 358, 360, 388
'Home Guard' (Air Defence of Great Britain), 168
Hood, Sqn. Ldr. M. D. S., 129
Hoopers, coachbuilders, photograph 47

Hornet, de Havilland D.H.103, single-seater fighter, 72, 73
Horsfall, Sgt., 118
Houlston, Flt. Lt., 188, 189
Hosking, Plt. Off. R. A., 129, 130, 132
Houffalize, 235, 255
Houlder, Capt. C. B., of B.O.A.C., 330, 337
Houston, Flt. Lt., 178, 187
Houston, W., 100
Howard, Sqn. Ldr. D., 175
Howard, Flg. Off. Lee, 417
Howe, The Hon. C. D., Canadian Minister of Munitions and Supply, 43, 96
Howitt, Flt. Lt., 161, 163
Hs 293 glider bombs, 176
Hudson, Lockheed bomber and transport, 46, 312, 329, 334, 336
Hughes, Flt. Lt., 189, 190
Hughes, Sqn. Ldr. D. L., 299
Hughes, Sqn. Ldr. G. E., 128, 133
Hughes, Wg. Cdr., 228
Hugo, Flg. Off. P., 129, 133
Huletsky, Pete, 347, 348
Hull, 173
Humblestone, Flg. Off. S., 350
Humphrey, Flg. Off. W. E. G., 357
Humphreys, Plt. Off., 343
Hunt, Capt. V. A. M., of B.O.A.C., 337
Hunter, Flg. Off., 239
Hunter, W. D., 97–108
Hunter Field, U.S.A., 103
Huntingdonshire, 165
Huppert, Flg. Off. S. B., 174
Huppert, Flt. Lt., 179
Hurricane, Hawker, fighter, 38, 43, 46, 54, 56, 82, 85
Hutchinson, Wg. Cdr., 77
Hutchinson, R., of de Havilland, 35; photograph 14
Hyland, A. E., 110

Iceland, 102, 219
I.G. Farben Chemical Works at Leverkusen, 305, 310, 357; and at Mannheim, 310
Ijlist, 342
Ijmuiden, 128, 144, 366, 368, 382
Ile de Groix, 274
Implacable, H.M.S., 77
Incendiaries in 4,000-lb. bomb casing, 365
India, 260, 261, 267
Indian Ocean, 113
Indo-China, 267, 268
Indonesia, 267
Innsbruck, 133
Inskeep, Flt. Sgt., 138
Instep operation, 168, 173, 341, 346–9 (and, for other operations, *see* Operation)
Intelligence, 251, 332
Interview Isle, 261

482

Index

Intruder operation, 346-8, 351-3 (and, for other operations, *see* Operation)
Invermairk, Glen Esk, 335
Ipswich, 163
Iredale, Wg. Cdr., 243, 244, 250
Ireland, 54
Irish Sea, 216
Iron Curtain, 268
Irvine, Flt. Lt., 265, 266
Isfield, nr. Lewes, 162
Isle Brewer, 413
Israel, 419
Italian Home Guard, 224
Italy, 131, 218, 351
Itzeloe, 257

Jacks, Eddy, 105
Jackson, C. H., of B.O.A.C., Chief Project Engineer, 330
Jackson, Sgt., 192
Jackson, C. C., of de Havilland, 376 (caption to photograph 14)
Jackson-Smith, Sqn. Ldr. Norman, 277
James, Flt Lt. A. W. D., 315
Jameson, Flt. Lt. G. E., 179, 413
Japan, 97, 112, 261, 267-9
Jasper, Martin, 349
Java, 267, 268
Jaworski, Flt. Lt., 168
Jena, 130, 205, 359, 362, 417
Jenkins, Flt. Sgt., 227
Jennings, Flt. Sgt. S., 243
Jericho operation, 241 (and, for other operations, *see* Operation)
Jet fighters, German, 142, 143, 184, 313, 315
John Cockerill Armament Works, Liège, 200
Johnson, Plt. Off., 188
Johnson, Flt. Lt. J. S., 347
Jones, Allan Murray, 109-14
Jones, Grp. Capt. C. Williamson, 186
Jordan, Wg. Cdr., 186
Jostle jamming of V.H.F. signals, 295
Jugoslavia, 61
Junkers aircraft, 418 (Appendix 12)
Junkers Ju 34 trainer, 346, 347, 349, 351
Junkers Ju 52, 184, 222, 231, 329, 348
Junkers Ju 86, 62-67, 347, 349, 352
Junkers Ju 87, 183, 184, 232, 233, 349, 352
Junkers Ju 88 bomber, 39, 62, 150-5, 157, 160, 162, 165, 168-76, 179, 180, 183, 201, 222, 226, 227, 231, 238, 246, 272, 273, 275, 280, 291-3, 296, 341, 342, 343, 347, 349, 351-3, 453
Junkers Ju 88D reconnaissance aircraft, 62, 155
Junkers Ju 88 fighter, 184, 343, 344, 345, 346
Junkers Ju 188 bomber, 165, 166, 169, 171-6, 179, 183, 184, 201, 229, 232, 274, 278, 280, 299, 346, 350
Junkers Ju 290, 184, 346, 347

Jupiter operation, 126 (and, for other operations, *see* Operation)
Jutland, 254
Juvincourt, 248
Juvisy, 365

K8 Camera, 58
K19 Camera, 137, 139
Kahle, 214
Kaiserslautern, 369
Kamen, 388, 414
Kampong, 268
Karachi, de Havilland Branch, 262
Karlsruhe, 143, 295, 343
Karnoy, 130, 277
Kassel, 145, 204, 216, 288, 303, 363
Kattegat, 278-80, 328-37, 348
Kelly, Flg. Off. P., 365
Kelsey Wheels of Windsor, Ontario, 97
Kembs Barrage, nr. Bale, 369
Kemp, W. Off. H. K., 170
Kempe, Wg. Cdr. J., 232
Kennard, Plt. Off., 188
Kennedy, W. Off. W., 134
Kent, 165
Kent, Flg. Off. M. G., 413
Keohane, Flt. Lt. J. R., 280
Kerr, Hamish, Wg. Cdr., 151, 156
Kerr, Wg. Cdr., 66
Kiel, 118, 119, 255, 281, 302, 308, 319, 321, 364
Kiel Canal, 24, 29, 299, 310, 314
Killingworth, Flt. Lt., 257
Kimber, Wg. Cdr., 298
King, D., of de Havilland, photograph 14
King, E. H., of de Havilland, photograph 14
King, Rex, of de Havilland, 60; photographs 12, 14
King's Lynn, 136
Kipp, Bob, 347, 349
Kipp, R. A., 348
Kirchbarkau, 255
Kjeller, 367
Klagenfurt, 226
Kleboe, Wg. Cdr. P. A., 256
Knaben, 128, 200
Knapsack power station, 357
Knickebein, radio beam, 166
Koblenz, 295, 316, 318, 414
Kochem, 316
Koh Si Cangs Island, 262
Koln (cruiser), 369
Konigsburg, 25, 119, 351
Konopasek, Wg. Cdr., 245
Korini, 228
Kraakhellesund, 277
Kraslice, 55
Krawiecki, Plt. Off., 156
Krefeld, 356, 415
Kristiansund, 118

Index

Krupp Works at Rheinhausen, 357
Kuala Lumpur, 266
Kuttelwascher, Flt. Lt., 339
Kyle, Crp. Capt., 193

La Chapelle, 358, 367
Ladbrook, Flg. Off. H. H., 345
La Guardia, 104
Lake Constance, 127, 143
Lake Havel, 302
Lake Togel, 302
Lamb, Plt. Off., 343
Lambert, Wg. Cdr. E. F. F., 290, 293
Lamp-black finish, 57
Lanbrechts, Cmr., 130
Lancaster, Avro, bomber, 24, 100, 105, 138, 293, 295, 332, 355, 356, 359–62, 368, 370, 371
Laon, 39, 358
La Pallice, 118, 366, 368, 383
La Peronelle, 358, 368
Lapstrup, 205
La Roche, 238
La Rochelle, 128
Larrard, Directorate of Aircraft Production, M.A.P., 83
Larsen, Riisar, 328
Launching sites, 24, 131
Laval, 245
Laverton, 110
Law, Sqn. Ldr. R. C. E., 244
Lawrence, W. Off. G. H., 183
Laws, Grp. Capt., 52
Leach, Sgt., 127
Leach, Flt. Lt., 130
Leach, Flg. Off. J. E., 342
Leach, Plt. Off., 128
Leatham, Flt. Lt. E. G. C., 130, 143
Leavesden, near Watford, *see* under Airfields in Britain
Leba, 119
Le Creusot, 127
Ledeboer, Mrs. D., of de Havilland, 35
Lee, Wg. Cdr. Gibson, 110 (caption to photograph 49)
Lee Howard, Flg. Off., 417
Leech, Flt. Lt. H. C. L., 146
Lees, Air Vice-Marshal, 193
Leghorn, 127, 229, 233
Le Havre, 138, 175, 176, 339, 360, 366, 388, 414
Le Haye de Puits, 250
Leigh Mallory, Air Marshal, 340
Leipheim, 230
Leipzig, 133
Leipzig, 140, 145, 230, 304, 346, 362, 370
Leirvick, 277
Leland, Flt. Lt. J. H., 299
Le Long, Flg. Off., 350, 352
Le Mans, 200, 388
Lend-Lease Agreement, 98, 139, 167

Lens, 358
Le Ploy, 239
Le Rossignil, Flg. Off. J. A., 227
Le Touquet, 136, 168
Le Treport, 136, 248
Leuna, 300
Leverkusen, 305, 310, 357, 388, 389
Lewis, Ted, 223
Lhotse, Himalayan peak, 266
Liberator aircraft, 99, 105, 110, 217, 274, 332–4, 336
Liège, 199, 200, 385
Lightcap, Jack, photograph 57
Lightning Lockheed P38 fighter, 139, 231
Ligurian Sea, 232
Lille, 136
Limburg, 414
Limoges, 137, 382
Lindesnes, 276
Lingen, 198, 238, 295, 382
Linnell, Air Marshal F. J., 68, 86
Lintott, Flg. Off., 161, 162
Linz, 131
Lisieux, 138, 359, 413
Lisson, Sqn. Ldr., 348
Lister, 130, 278
Lithuania, 119
Little, Flt. Sgt., 125
Little, Flg. Off., 151
Little Hampton, 242
Little Walden/Saffron Walden, 174
Llewellin, Col., Minister of Aircraft Production, 91
Lobeda, 205
Loch Cairnbawn, 74
Loch Ness Monster, 136
Loch Striven, 75, 77
Lockhart, Flt. Lt., 313
Lockhart, Wg. Cdr., 305
Lockhart-Ross, Lt. Archie, 230; photograph 94
Lockhood 14, 329
Lockheed Hudson aircraft, 46, 312, 329, 334, 336
Lockheed P38 Lightning fighter, 139, 231
Lockheed Lodestar, 328, 329, 333, 334, 336
Lockheed Ventura, 236, 329
Locomotives, 199, 201, 352
Lodestar, Lockheed transport, 328, 329, 333, 334, 336
Lodge, W. Off., 146
Lofoten Islands, 139
London, 30, 43, 44
London Aeroplane Club, 83
London Colney, 35
Lindon, Ontario, 105
Long, Cyril G., of de Havilland, chief development and purchasing engineer, 38, 84, 87, 89 (captions to photographs 14, 45)
Long, Don, 98

Index

Longbottom, Sqn. Ldr., 74, 75, 77
Longden, Capt. C., of B.O.A.C., 337
Longfield, Wg. Cdr., 199, 200
Longuemont, 239
Longues, 358
LORAN Radar, 317 (and, for all types, *see* Radar and Radio)
Lorenz Radar, 221 (and, for all types, *see* Radar and Radio)
Lorient, 216, 238, 341, 356, 368, 417
Losses, Mosquito night bombers, 323
Lostock, nr. Bolton, 82, 85
Louvred shrouds, 179
'Lovely Lady', Mosquito of 60 Sqn., S.A.A.F., 233; photograph 93
Lovesey, Cyril, of Rolls-Royce, 34, 51
Low Attack by Wg. Cdr. John Wooldridge (Sampson Low), 208
Low Countries, 132, 280, 290, 293, 340, 346
Lowry, Wg. Cdr., 265, 266
Lowry, Wg. Cdr. W. E. M., 266
Lowther, Flt. Lt. L. W., 136
Lübeck, 198, 302, 313
Ludwigshaven, 146, 364
Luftwaffe:
　Aufkl. Gr. 106, 153
　Aufkl. Gr. 3 (F) 123, 155
　Aufkl. Gr. 1 (F) GR 129, 343, 344
　I/FAG 5, 346
　Fliegerkorps IX, 175
　Fliegerkorps X, 175
　4 (F) Ob d.L., 62
　NJG 4, 184
　NJG 5, 184
　Kampfgeschwader:
　KG 1, 227
　KG 2, 152-4, 160, 163-8, 173
　KG 3, 182
　KG 6, 154, 155, 157, 160, 162, 166, 172, 413
　KG 30, 170
　KG 40, 152, 158, 160, 170, 343-6
　KG 51, 174
　KG 53, 182
　KG 54, 170, 174, 413
　KG 66, 162, 171
　KG 77, 39
　KG 100, 172, 174
　SKG 10, 161, 165
Lukhmanoff, Sgt. Boris, 119, 121, 126, 382 (caption to photograph 63)
Luma, 1st Lt. J. F., 348
Luton, 63, 81
Lutterade, 355
Lyons, 125, 131, 229, 348, 382

M46 American photoflash, 130
Maastrict, 452
MacDonald, Grp. Capt., 189
MacDonald, Sqn. Ldr. D., 346, 347
MacFadyen, Flt. Lt. D., 176

MacFadyen, Sqn. Ldr. D. A., 178
MacFadyen, 349
Mack, Wg. Cdr., 172
Mackay, Flg. Off. H. M., 238
MacMullin, Wg. Cdr., 354
MacRobertson International Air Race, 29
Mae West life jacket, 331
Magdeburg, 143, 298, 313, 317-19, 363
Maguire, Sqn. Ldr., 166
Magwe, 260
Mahmoud operation, 288 (and, for other operations, *see* Operation)
Maidment, Flt. Sgt., J. R., 170
Mailly-le-Camp, 366, 367, 388
Maintenay, 239
Mainz, 317, 318
Maisy, 358
Majer, Flg. Off. J. E., 238
Makin, Plt. Off., 198
Makin, Plt. Off. E. D., photograph 87
Malacca, 266
Malaya, 267-9
Malaya, H.M.S., 77
Malines, 201, 204
Mallender, Flt. Lt. P. F., 369
Mallory, Air Marshal Leigh, 340
Malta, 52, 127, 144, 218-34, 236
Malton, 99
Malvern, 153
Manchester, Avro bomber, 38, 46
Mandalay, 260, 263
Mandrel Radar, 290 (and, for all types, *see* Radar and Radio)
Manna Operation, 371 (and, for other operations, *see* Operation)
Manners, Flt. Lt. J. R., 267
Mannheim, 128, 143, 288, 298, 302, 303, 309, 310, 313, 315, 317, 364, 414
Maquis, 252
Marauder bomber B-26, 77
Marec, 1st Lt., 225
Mareth Line, 224, 225
Markers, red route, 319
Markers, visual (*see also* Flares and Target Indicators), 310, 318; photographs 115, 126
Marseilles, 125, 215, 229
Marsh, Plt. Off., 152
Marshall, Flt. Sgt., 304
Marshall, Plt. Off., 195, 196
Marshall, W. Off. J. W., 183
Marshalling yards, 312, 344
Marshalls Flying School, 50, 59, 60, 77, 103, 246, 335
Marstein, 277
Marstrand tailwheel tyre, 149, 219
Martin Hearn, 81, 150, 339
Martin, Lt. Oliver, 225
Martin Baltimore bomber, 98, 224
Martin Maryland bomber, 46, 224
Martlet fighter, 46

Index

Maryland Gleen Martin bomber, 46, 224
Mascot, 110
Massey Harris, 98
Massey, Sgt. J., 198, 199, 201; photograph 87
Master bomber, 300, 311, 312, 368 (explanation)
Master trainer, 188
Mattingley, Flt. Lt., 219, 220
Maule, Adrian, of de Havilland, 218
Maxwell, Wg. Cdr., 179, 341
Mayen, 255, 316
McClelland, Flt. Lt. W. R. M., 257
McCulloch, Flt. Lt., 260
McEwan, Flg. Off. B. D., 313
McGeehan, Flt. Sgt., F. I. D., photograph 87
McGeehan, Flt. Sgt., P. J., 198, 201
McGoldrick, Flt. Sgt., 138
McCreal, Flt. Lt., 361
McIlvenny, Sgt., 150
McKeard, Flt. Lt., 308
McKenzie, W. Off., 341
McLaren, Sgt., 350
McLeod, Plt. Off., 128
McPhee, Flt. Lt. T., 243
McPhee, Sqn. Ldr., 245
McRitchie, Sqn. Ldr., 242, 243
Me 108, 109, 110, 163, 210, 262, 410 aircraft, *see* Messerschmitt
Meakin, Wg. Cdr., 236, 237
Meakin, Wg. Cdr. H. J. W., 252
Mediterranean, 234
Medstead (Alton), 174
Meiderich/Gessellschaft Teerverwertung, 315, 316
Melbourne, 29, 78
Meldorf, 256
Mellersh, Flt. Lt., 177
Memel, 144
Menkes, Flg. Off., 226
Mergui, 262
Merifield, Flt. Lt., 119, 120, 129, 131, 132, 137; photograph 65
Merville, 358
Mesareing, 265
Messerschmitt aircraft, 418 (Appendix 12)
Messerschmitt factory, 143, 230
Messerschmitt Me 108, 184
Messerschmitt Bf/Me 109 fighter, 62, 99, 120, 126, 128, 129, 139, 145, 151, 176, 215, 277–80, 352, 453
Messerschmitt Bf 110, 153, 180, 183, 184, 237, 289, 291–3, 298, 299, 351, 352
Messerschmitt Me 163, 144, 145
Messerschmitt Me 210, 226
Messerschmitt Me 262, 142–5, 184, 230, 231, 385
Messerschmitt Me 410 bomber, 163–9, 172–6, 179, 183, 253, 293, 345, 347, 350
Messervy, Sqn. Ldr., 192

Messervy, R. F., of Rolls-Royce, 34
Messiter, Grp. Capt., 249
Mesuil, 251
Meteorological Office, 156
Meteorology, 214–17; photograph 119
Metz, 237, 312
Meulan-les-Mureaux, 358
Mezidon, 250
Midget submarines, 280
Midgley, Sgt., 162
Midgulen, 278
Milan, 127, 226, 229
'Milk run', 318
Miller, Plt. Off., 290
Miller, W. Off., 171
Miller, R/O J., of B.O.A.C., 333
Miller, Flt. Lt. R. A., 176
Mills, John, 109, 110, 112
Mimoyecques, 366
Mine Mk A. VIII, 274
Minelaying, 148
Mines, 24, 175
Ministry of Aircraft Production, Britain, 68, 83, 86, 91, 99
Ministry of Economic Warfare, Britain, 333
Ministry of Food, 54
Misfelt, Charles, 97
Mistel combination, 175
Mitchell bomber B-25, 77, 235, 240, 250, 252, 261
Mitchell, Wg. Cdr. L. J., 182
Mitchell, Flg. Off. T. M., 363
Mittelland Canal, 369
Mitsubishi Zero fighter, 261
Modane, 131, 140
'Model Aeroplane Club', 366
Mohawk fighter, 46
Möhne Dam, 74, 366
Molony, Flt. Lt., 161
Molybdenum, 200
Monaghan, Flg. Off. K. L., 243
Monica I Radar, 291, 292, 294, 295, 298, 300 (and, for all types, *see* Radar and Radio)
Monroe bomb, 370
Mons, 136, 245
Mont Blanc, 215
Mont Fleury, 358
Montgomery, General (later Field-Marshal Earl Montgomery), 224, 226
Montigny, 242
Montreal, 97
Moody, Flt. Lt., 152
Moonshine Radar jamming equipment, 290 (and, for all types, *see* Radar and Radio)
Moore, Flt. Lt. V. S., 308, 309, 312
Moore-Brabazon, Col. (later Lord Brabazon), 86
Moran, Sqn. Ldr., 344
Moran, Sqn. Ldr. C. C., 343
Morane aircraft, 352

Index

Morgan, Flg. Off. A. P., 134
Morgan, Flt. Lt. E. W. M., 290
Morib, 266
Morley, Sqn. Ldr., 352
Morris Motors, 84, 88
Morris, Flg. Off. R. C., 198, photograph 87
Mortain, 245
Morton, Earl, 342
'Mosquito' documentary film made by de Havilland, world-released by Metro-Goldwyn-Mayer, 91
Mosquito mock-up, photograph 2
Mosquito Prototype under construction, photographs 1, 3, 4, 5, 6, 7, 9, 10, 11
Mosquito Repair Organization, 85, 103; photograph 38
Mosquito Series II (proposed Mosquito replacement, DH 102), 71
Moth Minor, de Havilland DH94 light aeroplane, 29, 32, 38
Motor boats, 328
Motor torpedo boats, 264
Motteville, 386
Moulgate, 358
Moulmein, 261
Moult, Dr. Eric S., 81
Mountbatten, Earl, 262
Mount Everest, 266, 386
Mount Lhotse, 266
Mount Makalu, 266
Muir, Flg. Off. R. M., 341
Mulheim, 302
Mulhouse, 125
Mulliner, H. J., Chiswick, 85
Munchen Gladbach, 143, 363, 368
Munich, 30, 84, 131, 133, 137, 140, 143, 202, 215, 230–2, 300, 321, 352, 362, 364, 365, 367, 369, 385, 388
Munro, Flt. Sgt. N., 155
Munster/Handorf, 143
Muntz, Alan, 157
Murmansk, 126
Murphy, Wg. Cdr., 293
Murray, Flt. Lt., 245
Murray Jones, Allan, 109–14
Murray, L. C. L., of de Havilland, 35, 38, 83, 86–88, 96, 98, 99, 109, 112, 191; photograph 15
Musgrave, Flt. Lt. J. G., 176
Mustang fighter P 51, 49, 136, 137, 230, 251, 254, 256, 257, 277–81, 289, 350, 366
Myers, of de Havilland, 260, 261

Naerbo, 278
Nairn, Flt. Lt. G. D. T., 316
Name, Elimination of de Havilland, 25
Namsos, 117
Namur, 204
Nancawry, 262
Nancy, 382
Nantes, 201, 204, 236, 342

Napalm, 300
Napier engines, *see* Dagger, Sabre, etc.
Napier engines, design Nos. E108 and E112, 33
Naples, 144, 226, 231
Narromine, 78
Narvik, 125, 133, 139
Nash, Flg. Off. F. J., 369
Natal, 103
'National Redoubt' in the Harz Mountains, 370
Naze, 278
Nazi leaders, 370
Neal, Sgt., 182
Nelles, Sqn. Ldr., 367
Nelson, Plt. Off., 125
Nelson, R. B., 103
Neinburg, 364
Netherlands East Indies, 267
Neufahrwasser, 119
Neustadt, 146
Newbigin, S., 111
New Glasgow, Mosquito KB162, 102; photograph 125
Newman, Sqn. Ldr. C. W. M., 244, 252
Newman, D. R., of de Havilland, senior aerodynamicist, later chief aerodynamicist, photograph 14
Newman, Sqn. Ldr. K. J., 265
Newport, 62
New Zealand, 148, 179
New Zealand Air Race, 113
Nicobars, 262, 265
Nienburg, 146
Night Intruder Operation, 342, 346, 349 (and, for other operations, *see* Operation)
Night Ranger Operation, 232, 239, 249, 295, 341–3 (and for other operations, *see* Operation)
Nijmegen, 183, 253
Nimmo, Flt. Lt. Neil (of D-Dog), 318, 319
Nitelight Operation, 250, 252 (and, for other operations, *see* Operation)
Nixon, R. J., of de Havilland, photograph 14
Nixon, W. E., 81, 82
N_2O installation, 168, 173
Noball Operation, 239, 244–6 (and, for other operations, *see* Operation)
Nomeny, 252
Norfolk, 165
Normandy, 24, 138, 176, 179, 246, 251, 359, 364
North, Wg. Cdr. R. M., 239
North African Landings, 126, 127, 156
Northern Aluminium Company, 82, 97
Northern Lights, 152, 153, 328–37
North Sea, 350, 356
Norway, 24, 117–19, 124, 129, 131, 158, 192, 193, 200, 276, 278, 280, 281, 284–6, 311, 328–37, 338, 346, 351, 419
Norwegian Squadron 333, 130, 270, 279

487

Index

Norwich, 149, 187
Nose searchlight, 57
Nunn, Flt. Lt. S. G., 275
Nuremburg trials, 258
Nürnberg, 128, 129, 143, 299, 302, 313, 316, 318, 363, 414, 417

Oakeshott, Sqn. Ldr., 113
Oboe, see Radio and Radar
Oboe radar, explanation, 354–6 (footnote)
Obrestad, 276
Odense, 128, 257
Ofot Fiord, 127
Oisemont, 359
Olaf, H.R.H. Crown Prince of Norway, 257
Omholdt, R/O, of B.O.A.C., 332
O'Neille Dunne, Plt. Off., 155
Openshaw, Flg. Off., 204
Operation *Carpetbagger*, 142
Operation *Clarion*, 255
Operation *Crossbow*, 132, 133, 136
Operation *Day Intruder*, 350
Operation *Day Ranger*, 158, 229, 232, 238, 245, 248, 296, 341–3, 346–8, 350–3, 453
Operation *Dracula*, 265
Operation *Firebash*, 300
Operation *Firedog*, 268
Operation *Flowers*, 245, 246, 238, 288, 348, 350, 453
Operation *Frugal*, 144
Operation *Goodwood*, 359
Operation *Haycock*, 144
Operation *Instep*, 168, 173, 341, 346–9
Operation *Intruder*, 346–8, 351–3
Operation *Jericho*, 241
Operation *Jupiter*, 126
Operation *Mahmoud*, 288
Operation *Manna*, 371
Operation *Night Intruder*, 342, 346, 349
Operation *Night Ranger*, 232, 239, 249, 295, 341–3
Operation *Nitelight*, 250, 252
Operation *Noball*, 239, 244–6
Operation *Outmatch*, 275
Operation *Outstep*, 346
Operation *PAMPA*, 214–16
Operation *Ploughman*, 304
Operation *Plunder*, 257
Operation *Ranger*, 160, 238, 248, 341, 342, 344, 346, 348, 351, 352, 354 (explanation, 341),
Operation *RHOMBUS*, 215
Operation *Rhubarb*, 248, 263
Operation *Rover*, 276
Operation *Siren Tours*, 304, 363, 364
Operation *Spoof*, 295, 300, 301
Operation *Spook*, 301
Operation *Steinbock*, 168, 169
Operation *TacR*, 268
Operation *Totalize*, 359
Operation *Tractable*, 360

Operation *Zipper*, 267
Oranienburg, 362
Orford, 154
Orfordness, 152
Orleans, 248, 339, 415
Orrock, Wg. Cdr., 278
Orton, Capt. B. W. B., of B.O.A.C., 334
Orwell, 154
Oscar II fighter, 264
Oshawa, Ontario, 98
Oslo, 118, 127, 192, 193, 331, 369
Oslo Fiord, 192, 369
Osnabruck, 202, 309, 311, 364
Ottawa, 98
Ouin, Flg. Off. C. K., 371
Ouistreham, 137, 358
Outmatch Operation, 274 (and, for other operations, *see* Operation)
Outstep Operation, 346 (and, for other operations, *see* Operation)
Ouville-la-Rivier, 245
Over-Flakkee, 244, 313
Overseer, M.A.P., 43
Oxford, 165
Oxford, Airspeed trainer, 32, 33, 82, 84, 85, 87, 88, 264
Oxlade, Sqn. Ldr. A. G., 249
Oxtail, 77

P-47 Thunderbolt fighter, 49, 138
Pacific, 77, 95
Pacific Fleet, Royal Navy, 78
Packard Merlin, 60, 96, 100, 101, 103, 109–11, 113
Paderborn, 200, 201, 204
Padua, 232
Pagewood, 111
Paimbouf, 238
Paimpol, 293
Palembang, 267
Palestine Emergency, 234
Palmer, Geoff, 220
Palmer, Flg. Off. J. H., 256
Palmer, Sqn. Ldr. R. A. M., 361
PAMPA meteorological reconnaissance operation, 214, 215, 216 (and, for other operations, *see* Operation)
Panitz, Wg. Cdr. G., 345
Panzer, 255, 367
Parametta Radio, 311 (and, for all types, *see* Radar and Radio)
Pargeler, Flt. Lt. R. C., 172
Paris, 121, 134, 137, 292, 293, 367, 375, 382, 383, 452
Park, Wg. Cdr., 163, 165
Parker, of B.O.A.C., 330, 337
Parker, Flg. Off. A. E., 263
Parker-Rees, Sqn. Ldr., 172
Parkes, John J., 82
Parry, Flt. Lt., 191, 192, 330
Parry, Sqn. Ldr. D. A. G., 192, 197

Index

Pas de Calais, 132, 291
Pasing, 321
Passenger in Mosquito bomb bay, 331
Pateman, Flt. Sgt., 266
Paterson, Flt. Lt., 224
Pathfinder role, 301
Patient, Plt. Off., 303
Patrus, 228
Patterson, Flt. Lt., 197, 236
Patterson, Sqn. Ldr. C. E. S., 417
Payne, R/O J., of B.O.A.C., 331, 332–7
Pearce, Flg. Off., 175
Pearson, W. Off. A. C., 305
Peenemünde, 128, 131, 132, 135, 217
Pelly, Capt. C. N., of B.O.A.C., 331–3, 337; photograph 117
Peltz, 172
Penang, 266, 267
Pennington, Flt. Lt., 150, 151, 153
Percival Aircraft, 81, 93
Percival Proctor trainer, 58
Peregrine engine, 43
Pereira, Flg. Off., 204
Perfecta Motors, photograph 46
Perfectos Radio, 294, 295 (and, for all types, *see* Radar and Radio)
Peters, A. G., of de Havilland, designer, later general manager, photograph 14
Pforzheim, 388
Philips radio factory, Eindhoven, 197, 201, 367
Philips & Powis fighter, 43
Photoflash, 24, 130, 137–9, 142
Photo-reconnaissance, 24
Pickard, Grp. Capt. P. C., 236, 237, 239, 241, 242
Picknett, Flt. Lt., 260
Pickup, Flg. Off., 143
Pienaar, Capt., 230
Pierce, Flt. Lt., 239
Pike, Wg. Cdr. C. A., 157
Pilos, 228
Pilsen, 125, 129, 350, 370
Piltingsrund, Capt., Norwegian, 333
Pinsent, Flt. Lt. H. R., 137
Piperack Radar jamming device, 299 (and, for all types, *see* Radar and Radio)
Playford, 349
Ploughman Operation, 304 (and, for other operations, *see* Operation)
Plumb, F., of de Havilland, 35, 45, 55, 66, 105; photographs 12, 14
Plunder Operation, 257 (and, for other operations, *see* Operation)
Pointe d'Ailly, 136
Point Pinto, 266
Poissy, 125
Poitiers, 251
Pola, 126, 127, 226
Poland, 119, 129
Poles, 240, 245, 251, 258

Polesella, 233
Politz, 369
Pollard, Wg. Cdr., 245
Pollards, 85
Pommereval, 239
Poole, Dorset, 98
Pont-Château, 236
Pontoise, 138
Pordenone, 233
Porsgrunn, 278, 279
Port Campbell, 261
Port Swettenham, 266
Portal, Admiral, 78
Portal, Air Chief Marshal Sir Charles, 45
Port en Bassin, 132
Porteous, Wg. Cdr. R. C., 252
Portreath, 219
Portsmouth, 81
Potsdam, 298, 362
Potsdam Conference, 145
Povey, Harry, 86, 97–108; photograph 54
Poznan, 131, 382
Prag, Sgt., 214
Prague, 127, 129, 146, 230–2, 353
Pree of Western Electric, 452
Prewzlau, 180
Prinz Eugen, 125
Proctor, Percival trainer, 58
Prome, 260
Propeller constant-speed unit, 135
Propellers, braking, 57
Propellers, Hamilton, 57
Propellers, Hydromatic, 47
Propellers, paddle-blade, 113, 135; photograph 106
Provan, Flg. Off. W. W., 172
Province of Brabant, 368
Prum, 255
Prussia, East, 351
Puisseauville, 239
Puket Sound, 266
Pyinmana, 264
Pyrenees, 131, 382

Quarry, Wizernes, 135
Queen Bee pilotless target DH.82, 32, 82
Quiberon Bay, 293

Rabone, Sqn. Ldr. P. W., 226, 351
Rad, Lt., 163
Radar and Radio:
    A.I., 154, 166, 167, 171, 338, 452
    A.I. Mk. IV, 287, 288, 291, 294
    A.I. Mk. V, 148, 160, 452
    A.I. Mk. VI, 452
    A.I. Mk. VIII, 59, 60, 66, 89, 157, 170, 291, 351, 452
    A.I./SCR 720, British A.I. Mk. X, 59, 60, 170, 177, 291, 292, 294, 300, 452
    A.I./SCR 729 (Eleanora), 452
    A.R. 5513 (Oboe), *see* Oboe

Index

Radar and Radio—*cont.*
  A.I. Mk. XV 3 cm. Radar (*ASH*), 246, 294, 298, 299
  ASH, 246, 294, 298, 299
  BOOZER, 134
  D/F, 329
  Duppel, 166, 170, 171, 232
  Early Window, 305, 309, 317, 367
  FuG, 295
  Gee, 133, 139, 233, 238, 255, 272, 287, 295, 302, 304, 306, 311, 356
  GH., 248, 250, 305, 306, 363
  H/F, 329
  H2S, 50, 134, 309, 312, 317, 319, 354, 363 364
  H2X, 134
  Lorenz, 221
  LORAN, 317
  Mandrel, 290
  Monica, 291, 292, 294, 295, 298
  Moonshine, 290
  Oboe (AR 5513), 23, 50, 244, 248, 289, 301, 305, 306, 309, 311, 316, 317, 319, 321, 322, 326, 354–71; *Oboe* (explanation), 354, 355, 356 (footnote)
  Perfectos, 294, 295
  Parametta, 311
  Piperack, 299
  Rebecca H, 134, 139
  H (explanation, *Oboe* in reverse), 363
  Serrate, 287, 289, 290, 292, 293, 294, 295
  SN2, 294
  VHF, 242, 272, 300, 359
  Window, 166, 302, 304, 305, 312, 314, 317, 318, 364
  Wurzburg, 134, 295
Radio 1143, 57
Radio R3090, 57
Radio Telephony, 161
Rae, Capt. Gilbert, of B.O.A.C., 330, 331, 332, 334, 337
R.A.E. Farnborough, 33, 45, 46, 48, 49, 67, 151, 168, 187, 191, 246
Railway, 250, 262, 344, 358
Railway tunnels, 24, 316
Rajah, H.M.S., 77
Ralston, Sqn. Ldr. J. R. G., 190, 196, 198, 200; photograph, 85
Ramenskoye, 144
Ramsgate, 63
Randall, Flg. Off. M.D., 263
Random Range, 263
Range indicators, photo-cell, 177
Ranger Operation, 160, 238, 248, 341, 342, 344, 346, 348, 351, 352; explanation 341 (and, for other operations, *see* Operation)
Rangoon, 260, 261, 263, 265
Raphael, Wg. Cdr., 153
Rawnsley, J., 160, 163
Rayski, Flt. Lt., 245

R-boat, 281
Reading, 39
Rebecca H Radar, 134, 139 (and, for all types, *see* Radar and Radio)
Reculver, 74, 75
Red Army, 185
Rees, 177
Reeves, Plt. Off., 150
Regensburg, 125, 131
Reid, Sqn. Ldr., 278
Reims, 137
Remscheid, 357
Renault works, 121
Rendall, Capt. A., of B.O.A.C., 337
Rennes, 199, 384
Research and Development, 34, 43, 86, 270, 271
Resistance movement, 243, 244, 258
Reynolds, Wg. Cdr. R. H., 205, 251
Reynolds, Sqn. Ldr. R. W., 198, 199; photograph 87
Reynolds, Wg. Cdr. R. W., 302
Rheinhausen, 357, 389
Rheinmettal Borsig works, Dusseldorf, 357
Rhineland, 117
Rhodes, 231
Rhodesia, 222, 233
RHOMBUS meteorological reconnaissance operation, 215 (and, for other operations, *see* Operation)
Rhubarb operation, 248, 263 (and, for other operations, *see* Operation)
Richards, Sqn. Ldr. S. P., 156
Richter, Karel Richard, 55
Ricketts, Flt. Lt. Victor, 119, 120, 121, 125, 126; photograph 63
Riley, Charles, 221
Rimini, 233
Ring, Wg. Cdr., 125
Ringkjobing, 293, 350
Ritchie, Flt. Lt., 153, 154
River Elbe, 369, 370, 387
River Loire, 348, 350
River Ouse, 310
River Po, 232
River Rhine, 253, 362
River Seine, 121, 252, 350
River Weser, 369, 370, 387
Rivers, Flg. Off. R. N., 155
RM3SM engine, 34
Roanne, 133
Roberts, R/O D. T., of B.O.A.C., 334, 336, 337
Robertson, Flt. Sgt., 166
Robinson, Flg. Off., 162
Robinson, Plt. Off., 191, 330
Rochford, 172
Rochford, Flt. Lt. D. G., 77
Rocket fighters, 143
Rocket projectile, 24, 273, 276, 278, 280 photograph 106

Index

Roe, Sgt., 152
Rogers, Carl, 332
Rogers, Sgt. G. V., 343
Rohmer, Dr., 137
Rolls-Royce, 101, 110, 111, 142
Rolls-Royce engines, *see* under Griffon, Vulture, etc.
Rome, 226, 227, 229, 232
Rome/Romulus USAAF, 103
Rommel, General, 224
Ronszek, Sq. Ldr., 156
Roosevelt, Colonel (son of the President of the U.S.A.), 227
Roosevelt, President, 328
Rose Brothers, Scampton, 194
Rose, Bruce, 110
Rose, Flg. Off. C. F., 214, 272
Rose, Sqn. Ldr., 271
Rose, Sqn. Ldr. C. F., 74, 75
Rossnitz, 318
Rostock, 132, 215
Rothesay, 77
Rotterdam, 195, 371
Roubaix, 136
Rouen, 251
Rover Operations, 276 (and, for other operations, *see* Operation)
Rowe, N. E., 63, 66, 68, 70, 306
Rowlands, Sgt., 190
Royal Aircraft Establishment, Farnborough, *see* Farnborough, Airfields section
Royal Air Force:
  Aeroplane and Armament Experimental Establishment, Boscombe Down, *see* Boscombe Down, Airfields section
  Air Fighting Development Unit, 49, 187, 190, 194
  Bentley Priory, Headquarters, Fighter Command, 219
  Bomber Command, 34, 121, 132, 139, 156, 186, 196, 197, 204, 215, 216, 287, 290, 293, 300, 301, 306, 346, 350, 354, 360, 370
  Bomber Support Development Unit, 291, 294, 452
  Central Fighter Establishment, 61, 452, 453
  Central Gunnery School, 384
  Coastal Command, 127, 130, 270, 281, 291
  Elementary Flying School, No. 1, 157
  Ferry Command, 99, 156
  Fighter Command, 60, 154, 160, 161, 165, 218, 219, 276, 287, 298, 340
  Fighter Interception Development Unit, 55, 67, 182, 288, 294, 452, 453
  Fighter Squadrons, 185
  2nd Tactical Air Force, 136, 235, 236, 249–52, 254, 289, 316, 350, 352
  Film Production Unit, 241–3, 254, 256, 369
  Far East Air Force, 140

  Group 1, 370
  Group 2, 187, 193, 194, 197, 202, 253, 257, 258, 301, 350, 356; Analysis of daylight raids 5.42–5.43, 209, 235, 240, 252
  Group 5, 308, 367, 369, 370
  Group 8, 50, 75, 103, 216, 289, 299, 301, 310, 311, 312, 314, 321, 357, 370; Analysis of sorties, 216
  Group 9, 156
  Group 10, 169
  Group 11, 161
  Group 16, Coastal Command, 127
  Group 19, 272
  Group 43, 81
  Group 84, Support Unit, 413
  Group 85, Allied Expeditionary Air Force, 175, 176, 183–5, 249
  Group 100 (Bomber Support), 60, 233, 259, 287–300, 310, 351, 353, 452, 453
  High Altitude Flight, Fighter Command, 66, 67
  Light Night Striking Force, 301–25, 363, 365
  Night Fighter Development Squadron Wing, 294, 452, 453
  Pathfinder Navigational Training Unit, 389
  P.R. Development Unit, 142
  Royal Aircraft Establishment, 33, 45, 46, 48, 49, 67, 362
  Special Installations Unit, Defford, 67
  Squadrons and Units (for their bases and dates *see* Appendix 21)
  Squadrons of the Royal Air Force (*see* also Appendix 20 and Appendix 21):
    4, 136, 386
    11, 246, 258
    13, 386
    14, 326
    16, 258
    18, 450
    19, 158, 281
    21, 236–9, 241–5, 248, 249, 253–8, 350, 385, 389, 416
    22, 430
    23, 180, 218–24, 226–9, 231, 236, 246, 259, 293, 294, 298–300, 338–40, 383, 393, 420, 423
    25, 107, 155, 157, 170–5, 180, 182, 184, 288, 340–3, 385, 414, 420
    29, 107, 161–3, 168, 170–2, 180, 184, 296, 351, 384, 390, 420
    39, 234
    45, 262–5
    47, 263–5, 268
    58, 386
    64, 107, 256
    65, 107, 281
    68, 151, 160, 182–4
    69, 246, 258

491

Index

Royal Air Force—*cont.*
81, 140, 269
82, 262–5, 268
83, 355
84, 264, 268
85, 59, 67, 107, 153–7, 160–3, 165–8, 170–3, 177, 180, 291, 292, 294–6, 299, 340, 420
89, 267, 430
91, 281
96, 168, 170, 172, 173, 177, 178, 296, 379, 390
98, 250
105, 49, Chapter 13, 301, 302, 306, 308, 321, 324–6, 357–3, 370, 371, 377, 381, 383–5, 388, 415, 420
107, 240, 245, 250, 251, 253–8, 386
108, 227, 228, 259
109, 49, 194, 197, 244, 245, 248, 301, 305, 306, 319, 324–6, 354–63, 370, 371, 379, 388, 389, 415, 420, 429
110, 264, 267, 268
121, 63
124, 62, 63, 134
125, 174, 179, 182, 184, 387
128, 314–16, 319, 320, 324, 325, 387
139, 49, 102–4, Chapter 13, 301–6, 308–12, 315, 317, 319–21, 324, 325, 357, 358, 363–5, 381, 383–5, 387, 389, 415, 420, 429
140, 134, 135, 137–9, 144, 145
141, 125, 158, 180, 246, 287, 289, 293–5, 298–300, 339, 383
142, 315, 318–20, 324, 325
143, 276, 278, 279, 281, 284–6, 386, 387, 422
144, 273, 274, 276, 284, 286
151, 64, 66, 107, 148–50, 152–4, 157, 158, 160, 165, 170, 173, 223, 241, 296, 298, 340–3, 349, 383, 387, 414, 420
152, 186, 187
157, 56, 148–55, 157, 158, 160–3, 177, 291, 292, 294–6, 299, 340–3, 346, 347, 384, 390, 414, 420, 452
162, 104, 229, 230, 317–20, 324, 325, 364–5
163, 104, 317, 320, 324, 325
169, 289–91, 293–6, 298–300, 352, 353, 390
176, 267, 430
180, 250
192, 202, 287, 290, 295, 299, 324, 420
198, 241
206, 274
211, 267
214, 293
219, 174, 177, 179, 183
235, 274–6, 278, 279, 281, 284–6, 386, 416, 422
239, 246, 289, 292–5, 298–300, 384, 414
248, 77, 158, 271–81, 284–6, 386, 422, 425, 428
249, 107, 234
254, 280, 422, 428
255, 232, 233, 430
256, 162, 163, 165, 227, 231–4, 420
264, 64, 107, 151–3, 156, 158, 174–6, 180, 184, 185, 340, 341, 343, 344, 383, 384, 387, 420
268, 258
279, 278
305, 240, 245, 249, 250, 252–4, 256–8
307, 155, 156, 158, 168, 173, 183, 298, 341, 343, 345, 346, 387, 420
310, 62
313, 62
315, 277, 285
332, 63
333, 63, 130
400, 134, 137
401, 62
404, 273, 274, 276, 280, 281, 284–6, 422
406, 175, 179, 180, 298, 351, 353
409, 175–7, 179, 182–4, 384, 414
410, 155, 158, 167, 168, 170–3, 176, 179, 183, 341, 342, 345, 384, 420
418, 176–8, 236, 238, 259, 342–53, 383, 385, 416, 420
455, 277
456, 125, 156, 170, 173–6, 178, 288, 296, 298, 341–3, 379, 413, 420
464, 346–239, 241–3, 245, 248–51, 254–8, 379, 384, 389, 416
487, 236–9, 241–3, 245, 248, 250, 252, 254–8, 385, 416
488, 167, 170, 171, 173, 174, 179, 183, 384, 413
515, 290, 293, 294, 298–300, 351, 352
521, 214, 377, 420
524, 280
532, 157
540, 122–46, 227, 377, 382, 383, 414, 420
541, 131
544, 130–46, 382, 414
571, 309–11, 314–20, 324, 325, 389
600, 232, 233, 430
604, 148, 174, 179, 381, 414
605, 176, 236, 238, 259, 287, 290, 342–6, 348–53, 383
608, 103, 104, 312, 315, 316, 319, 320, 324, 325
609, 180
613, 238, 239, 244, 246, 248–50, 252, 253, 255, 258, 379, 385, 386
614, 104, 234
616, 63, 180
617, 24, 78, 131, 138, 324, 345, 362, 365, 366, 382
618, 74–78, 271, 387, 424
627, 102–4, 202, 305, 306, 308, 310, 324, 325, 366–70, 384, 385, 387, 388
680, 228, 230, 231, 233
681, 260

492